Methodologies and Results in Grapevine Research

Serge Delrot · Hipólito Medrano · Etti Or ·
Luigi Bavaresco · Stella Grando
Editors

Methodologies and Results in Grapevine Research

Springer

Editors
Prof. Serge Delrot
Université de Bordeaux
UMR Ecophysiologie et
Génomique Fonction
nelle de la Vigne
Institut des Sciences
de la Vigne et du Vin
Chemin de Leysotte CS 50008
33882 Villenave d'Ornon
France
serge.delrot@bordeaux.inra.fr

Dr. Etti Or
Department of Horticulture
Agricultural Research
Organisation
The Volcani Center
Department of Fruit Tree Sciences
50250 Bet Dagan
Israel
vhettior@agri.gov.il

Stella Grando
IASMA Research and Innovation
Centre
Fondazione Edmund Mach
Via E. Mach 1
38010 San Michele alll'Adige (TN)
Italy
stella.grando@iasma.it

Dr. Hipólito Medrano
Universitat de les Illes Balears
IMEDEA-Dept. de Biologia
Carretera Valldemossa Km. 7,5
01722 Palma de Mallorca
Spain
hipolito.medrano@uib.es

Prof. Luigi Bavaresco
CRA-Research Centre for Viticulture
Viale XXVIII Aprile, 26
31015 Conegliano (TV)
Italy
luigi.bavaresco@entecra.it

Università Cattolica del Sacro
Cuore
Istituto di Frutti-Viticoltura
Via E. Parmense 84
29100 Piacenza
Italy
luigi.bavaresco@unicatt.it

ISBN 978-90-481-9282-3 e-ISBN 978-90-481-9283-0
DOI 10.1007/978-90-481-9283-0
Springer Dordrecht Heidelberg London New York

Library of Congress Control Number: 2010932512

Printed on acid-free paper

Springer is part of Springer Science+Business Media (www.springer.com)

Preface

Grapevine is the most cultivated fruit tree worldwide, and wine also represents a multicultural heritage which has grown since several milleniums in Europe and later spread to other continents. Viticulture and oenology are a subtle blend of science and art, tradition and innovation. There is a huge diversity in grapevine genomes (rootstocks and scions), climates, soils, viticultural practices and wine-making processes. This results in a tremendous range of wine types all around the world. However, in spite of its historical, cultural and economical value, viticulture must face several challenges.

Global climate change already impacted significantly grapevine physiology and berry composition in the last decades. Harvest occurs sooner and sooner, although grapegrowers tend to wait longer for ripeness. Berry sugar content (and alcohol in the wine) tend to increase whereas phenolic and aromatic ripeness are not always achieved. Acidity tends to decrease with potential effects on wine ageing capacity. Water supply is becoming shorter in many regions, and there is an increasing competition from many human activities for limited water resources. For France, the Intergovernmental Panel on Climate Change forecasts a rise in mean temperature of +2.4°C, a decrease in summer rainfall, a 30% increase in potential evapotranspiration and a doubling of atmospheric CO_2 by the end of this century, and similar trends may be forecast for many other viticultural countries. This may obviously affect the physiology of grapevine, fruit yield and composition, sensitivity to diseases, and ultimately wine quality and typicity. Should the present climatic trend continue in the next decades, we may have to switch to varieties that are better adapted to new climatic conditions, which may affect the traditional typicity characterizing given regions.

National and international regulations are more and more restrictive regarding phytochemical treatments and residues, and the consumer pays more and more attention to environmentally safe practices. In this context, new methods aiming at a strong reduction of phytochemical treatments (but maintaining yield and typicity) must be designed. They include new viticultural practices, precision viticulture, sophisticated alert systems based on epidemiological studies, use of compounds that may stimulate natural plant defence, biological pest control, and new grapevine genotypes obtained by breeding.

The issues of climate change and phytochemical treatments are related in some ways. On one hand, global warming and extended drought may decrease the impact of some fungal diseases that normally spread under humid conditions. On the other hand, it may also affect the natural resistance of the plants.

The sequence of the grapevine genome, released in 2007, was a major breakthrough that boosted grapevine research. It facilitated gene mapping, allowed the design of performing microarrays, and the adaptation of tools mapping the transcriptional data on metabolic pathways. Continuous improvements in high throughput DNA and RNA sequencing, gene expression studies, metabolomic analysis now result in the development of systems biology approaches, which require a major investment in bioinformatic tools. This will undoubtly make it easier to face the practical challenges of adaptation to climate change and pesticide limitation, provided that these new tools are constantly combined with field observations and measurements.

The European Union is ranked number 1 all over the world for viticultural area (4,139,975 ha = 55%), grapes (29,050,923 t = 43%) and wine production (191,015,000 hl = 67%). In the context briefly summarized above, the European Union thus supported COST Action 858 "Viticulture - biotic and abiotic stress - grapevine defence mechanisms and grape development" between 2003 and 2009. COST 858 aimed to create an active network of scientists working in different scientific areas (ecophysiology, agronomy, plant physiology, cellular and molecular biology, phytopathology, genetics, chemistry) related to viticulture. From 9 participating countries in 2003, COST 858 grew up to 17 European countries in 2009, and was also joined by Australia and New Zealand. A similar network (GRCN, Grape Research Coordinated Network) was initiated in 2009 in the USA, showing that both the challenges faced by viticulture, and the possible ways to face them are worldwide issues. My intimate belief is that high tech molecular biology is not very useful without every day field experience, and that field experience should always be open to the rapid progress made in technological improvements and basic science. The fact that grapevine is grafted in most countries, its major economical impact, and the strong impact of the environment and viticultural practices on the final quality of the berries, the complexity of berry "quality" and the diversity of grape genomes probably makes grapevine one of the most appropriate plant species to develop integrative and systems biology.

The present book illustrates some of the recent progress made in ecophysiology, molecular biology, cell biology and pathology of grapevine, as well as in precision viticulture and aroma characterization. We hope that this multidisciplinary contributions will be of interest for students, scientists and professionals involved in grapevine and wine activities.

Villenave d'Ornon, France Serge Delrot

Contents

Contributors

Atiako Kwame Acheampong Institute of Plant Sciences, A.R.O. Volcani Center, Bet Dagan 50250, Israel, vhettior@agri.gov.il

Jone Aguirreolea Facultades de Ciencias y Farmacia, Departamento Biología Vegetal, Sección Biología Vegetal Unidad Asociada al CSIC (EEAD, Zaragoza; ICVV, Logroño), Universidad de Navarra, Pamplona 31008, Spain

Elisa Angelini C.R.A. VIT Centro di Ricerca per la Viticoltura, Conegliano (TV) 31015, Italy, elisa.angelini@entecra.it

Xavier Aranda IRTA- Torre Marimon, Caldes de Montbui (Barcelona) 08140, Spain, xavier.aranda@irta.cat

Marouen Ayari Facultades de Ciencias y Farmacia, Departamento Biología Vegetal, Sección Biología Vegetal Unidad Asociada al CSIC (EEAD, Zaragoza; ICVV, Logroño), Universidad de Navarra, Pamplona 31008, Spain

Pilar Baeza Departamento de Producción Vegetal: Fitotecnia, Grupo de investigación en Viticultura. ETSI Agrónomos, Universidad Politécnica de Madrid, Madrid, 28040, Spain, pilar.baeza@upm.es

Fabienne Baillieul Laboratoire de Stress, Défenses et Reproduction des Plantes, Unité de Recherche Vigne et Vins de Champagne, Stress et Environnement, UPRES EA 2069, UFR Sciences Exactes et Naturelles, Université de Reims, Champagne-Ardenne Reims 51687, France, fabienne.baillieul@univ-reims.fr

Giorgia Batelli Plant Stress Genomics Research Center, 4700 King Abdullah University of Science and Technology, Thuwal 23955-6900, Kingdom of Saudi Arabia. Department of Botany and Plant Sciences, University of California at Riverside, Riverside, California 92521, USA, giorgia.batelli@kaust.edu.sa

Luigi Bavaresco CRA-Research Centre for Viticulture, Viale XXVIII Aprile, 26, 31015 Conegliano (TV), Italy; Istituto di Frutti-Viticoltura, Università Cattolica S. Cuore, Piacenza, I-29100, Italy, luigi.bavaresco@unicatt.it; luigi.bavaresco@entecra.it

Carme Biel IRTA- Torre Marimon, Caldes de Montbui (Barcelona) 08140, Spain, carmen.biel@irta.cat

Jana Boben Department of Biotechnology and Systems Biology, National Institute of Biology, Ljubljana 1000, Slovenia, jana.boben@mf.uni-lj.si

Sophie Bordiec Laboratoire de Stress, Défenses et Reproduction des Plantes, Unité de Recherche Vigne et Vins de Champagne, Stress et Environnement, UPRES EA 2069, UFR Sciences Exactes et Naturelles, Université de Reims, Champagne-Ardenne Reims 51687, France

Marjana Camloh Department of Biotechnology and Systems Biology, National Institute of Biology, Ljubljana 1000, Slovenia, marjana.camloh@nib.si

M. Carmen Antolín Facultades de Ciencias y Farmacia, Departamento Biología Vegetal, Sección Biología Vegetal Unidad Asociada al CSIC (EEAD, Zaragoza; ICVV, Logroño), Universidad de Navarra, Pamplona, 31008, Spain, cantolin@unav.es

Andrea Carra Department of Arboriculture and Pomology, University of Torino, Grugliasco 10095, Italy, andrea.carra@unito.it; andrea.m.carra@gmail.com

M.M. Chaves CBAA, Instituto Superior de Agronomia, Universidade Técnica de Lisboa, Tapada da Ajuda, Lisboa 1349-017, Portugal; LEM, Instituto de Tecnologia Química e Biológica, Universidade Nova de Lisboa, Oeiras 2780-901, Portugal, mchaves@isa.utl.pt

Christophe Clément Laboratoire de Stress, Défenses et Reproduction des Plantes, Unité de Recherche Vigne et Vins de Champagne, Stress et Environnement, UPRES EA 2069, UFR Sciences Exactes et Naturelles, Université de Reims, Champagne-Ardenne Reims 51687, France

Geneviève Conéjéro UMR 5004 Biochimie et Physiologie Moléculaire des Plantes, Plate forme d'Histocytologie et Imagerie Cellulaire Végétale, CIRAD TA A96/02 34398 Montpellier Cedex 5, France

Grant R. Cramer Department of Biochemistry and Molecular Biology, University of Nevada, Reno, NV 89557, USA, cramer@unr.edu

Omer Crane Institute of Plant Sciences, A.R.O. Volcani Center, Bet Dagan 50250, Israel

Steven Colas Laboratoire de Stress, Défenses et Reproduction des Plantes, URVVC-SE EA 2069, Université de Reims Champagne-Ardenne, UFR Sciences Exactes et Naturelles, Reims 51687, France

John Cushman Department of Biochemistry and Molecular Biology, University of Nevada, Reno, NV 89557, USA

Felicidad de Herralde IRTA-Torre Marimon, Caldes de Montbui (Barcelona) 08140, Spain, felicidad.deherralde@irta.cat

Serge Delrot Université de Bordeaux, UMR Ecophysiologie et Génomique Fonction nelle de la Vigne, Institut des Sciences de la Vigne et du Vin, Chemin de Leysotte, CS 50008, 33882 Villenave d'Ornon, France, serge.delrot@bordeaux.inra.fr

Marina Dermastia Department of Biotechnology and Systems Biology, National Institute of Biology, Ljubljana 1000, Slovenia, marina.dermastia@nib.si

Vincent de Rudnicki Cemagref, UMR ITAP, Montpellier 34196, France, vincent.derudnicki@cemagref.fr

Jérome Devy Laboratoire Signalisation et Récepteurs Matriciels, CNRS UMR MEOyC 6237, Université de Reims Champagne-Ardenne, UFR Sciences Exactes et Naturelles, Reims 51687, France

Tanja Dreo Department of Biotechnology and Systems Biology, National Institute of Biology, Ljubljana 1000, Slovenia, tanja.dreo@nib.si

Stefan Ebert Julius Kühn-Institut, Federal Research Centre for Cultivated Plants, Institute for Grapevine Breeding Geilweilerhof, Siebeldingen 76833, Germany

Jana Erjavec Department of Biotechnology and Systems Biology, National Institute of Biology, Ljubljana 1000, Slovenia, jana.erjavec@nib.si

J.M. Escalona Grup de Recerca en Biologia de les Plantes en Condicions Mediterrànies Universitat de les Illes Balears, Palma de Mallorca 07122, Spain, jose.escalona@uib.es

Jaume Flexas Departamento de Biologia, Grup de Recerca en Biologia de les Plantes en Condicions Mediterrànies, Universitat de les Illes Balears, Palma de Mallorca 07122, Spain, jaume.flexas@uib.es

Natacha Fontes Departamento de Biologia, Centro de Investigação e de Tecnologias Agro-Ambientais e Biológicas (CITAB), Universidade do Minho, Braga 4710-057, Portugal

Mario Fregoni Istituto di Frutti-Viticoltura, Università Cattolica S. Cuore, Piacenza I-29100, Italy

Wolfgang Friedt Department of Plant Breeding, Institute of Crop Science & Plant Breeding I, Justus-Liebig Universität Giessen, Giessen 35392, Germany

Alexander Gallé Grup de Recerca en Biologia de les Plantes en Condicions Mediterrànies Universitat de les Illes Balears, Palma de Mallorca 07122, Spain, alexander.galle@uib.es

Giorgio Gambino Plant Virology Institute, Grugliasco Unit, National Research Council, via L. da Vinci 44, Grugliasco 10095, Italy

Matteo Gatti Istituto di Frutti-Viticoltura, Università Cattolica S. Cuore, Piacenza I-29100, Italy

Hernâni Gerós Departamento de Biologia, Centro de Investigação e de Tecnologias Agro-Ambientais e Biológicas (CITAB), Universidade do Minho, Braga 4710-057, Portugal, geros@bio.uminho.pt

O.M. Grant East Malling Research, East Malling, UK

Ivana Gribaudo Istituto Virologia Vegetale CNR, U.O.S. Grugliasco, Via L. da Vinci 44, Grugliasco, TO 10095, Italy, i.gribaudo@ivv.cnr.it

Jerome Grimplet Horticulture, Forestry, Landscape and Parks Department, South Dakota State University, Brookings, SD 57006, USA

Kristina Gruden Department of Biotechnology and Systems Biology, National Institute of Biology, Ljubljana 1000, Slovenia, kristina.gruden@nib.si

J. Gulías Departmento de Biologia, Grup de Recerca de Biologia de les Plantes en condicions Mediterrànies, Universitat de les Illes Balears, Palma de Mallorca 07122, Spain, dbajg14@uib.es

Tamar Halaly Institute of Plant Sciences, A.R.O. Volcani Center, Bet Dagan 50250, Israel

Matjaž Hren National Institute of Biology, Department of Biotechnology and Systems Biology, Ljubljana 1000, Slovenia, matjaz.hren@nib.si

Lucile Jacquens Laboratoire de Stress, Défenses et Reproduction des Plantes, URVVC-SE EA 2069, Université de Reims Champagne-Ardenne, UFR Sciences Exactes et Naturelles, Reims 51687, France

Pedro Junquera Departamento de Producción Vegetal: Fitotecnia, Grupo de Investigación en Viticultura. ETSI Agrónomos, Universidad Politécnica de Madrid, Madrid 28040, Spain, pedro.junquera@upm.es

Alexandra Keren Institute of Plant Sciences, A.R.O. Volcani Center, Bet Dagan 50250, Israel

José Ramón Lissarrague Departamento de Producción Vegetal: Fitotecnia, Grupo de Investigación en Viticultura. ETSI Agrónomos, Universidad Politécnica de Madrid, Madrid 28040, Spain, joseramon.lissarrague@upm.es

Claudio Lovisolo DCA, Plant Physiology, University of Turin, Grugliasco 10090, Italy, claudio.lovisolo@unito.it

Sébastien Manteau Laboratoire de Stress, Défenses et Reproduction des Plantes, URVVC-SE EA 2069, Université de Reims Champagne-Ardenne, UFR Sciences Exactes et Naturelles, Reims 51687, France

Florence Mazeyrat-Gourbeyre Laboratoire de Stress, Défenses et Reproduction des Plantes, URVVC-SE EA 2069, Université de Reims, Champagne-Ardenne, UFR Sciences Exactes et Naturelles, Reims, 51687, France, florence.gourbeyre@univ-reims.fr

Hipólito Medrano Departmento de Biologia, Grup de Recerca de Biologia de les Plantes en condicions Mediterrànies, Universitat de les Illes Balears, Palma de Mallorca 07122, Spain, hipolito.medrano@uib.es

Erica Mica Scuola Superiore Sant'Anna, Piazza Martiri della Libertà 33, Pisa 56127, Italy

J. Miguel Costa CBAA, Instituto Superior de Agronomia, Universidade Técnica de Lisboa, Tapada da Ajuda, Lisboa 1349-017, Portugal; LEM, Instituto de Tecnologia Química e Biológica, Universidade Nova de Lisboa, Oeiras 2780-901, Portugal, miguelc@itqb.unl.pt

C. Miranda Departamento de Produccion Agraria, Universidad Pública de Navarra, Pamplona (Navarra) 31006, Spain

Laurence Monti-Dedieu Laboratoire de Stress, Défenses et Reproduction des Plantes, URVVC-SE EA 2069, Université de Reims Champagne-Ardenne, UFR Sciences Exactes et Naturelles, Reims 51687, France

Panagiotis N. Moschou Department of Biology, University of Crete, Crete, Greece

Montserrat Nadal Facultat d'Enologia de Tarragona, Departmento Bioquímica i Biotecnologia, Universitat Rovira i Virgili. Campus Sescelades, Tarragona 43007, Spain, montserrat.nadal@urv.cat

Petra Nikolić Department of Biotechnology and Systems Biology, National Institute of Biology, Ljubljana 1000, Slovenia, petra.nikolic@nib.si

Aliza Ogrodovitch Institute of Plant Sciences, A.R.O. Volcani Center, Bet Dagan 50250, Israel

Etti Or Department of Horticulture, Agricultural Research Organisation, The Volcani Center, Department of Fruit Tree Sciences, Bet Dagan 50250, Israel, vhettior@agri.gov.il

Anastasia K. Papadakis Departamento de Biologia, Centro de Investigação e de Tecnologias Agro-Ambientais e Biológicas (CITAB), Universidade do Minho, Braga 4710-057, Portugal

Philippe Pieri UMR 1287 Ecophysiologie et Génomique Fonctionnelle de la Vigne, Institut des Sciences de la Vigne et du Vin (ISVV Bordeaux), Villenave d'Ornon 33882, France, pieri@bordeaux.inra.fr

Fanja Rabenoelina Laboratoire de Stress, Défenses et Reproduction des Plantes, Unité de Recherche Vigne et Vins de Champagne, Stress et Environnement, UPRES EA 2069, UFR Sciences Exactes et Naturelles, Université de Reims, Champagne-Ardenne Reims 51687, France

Maja Ravnikar Department of Biotechnology and Systems Biology, National Institute of Biology, Ljubljana 1000, Slovenia, maja.ravnikar@nib.si

Martina Rex Julius Kühn-Institut, Federal Research Centre for Cultivated Plants, Institute for Grapevine Breeding Geilweilerhof, Siebeldingen 76833, Germany

M. Ribas-Carbó Departamento de Biologia, Grup de Recerca en Biologia de les Plantes en Condicions Mediterrànies, Universitat de les Illes Balears, Palma de Mallorca 07122, Spain, mribas@uib.cat

Charles Romieu Equipe Vigne, UMR DIAPC, INRA-SupAgro, Montpellier 34060, France, romieu@supagro.inra.fr

Ariel Rotman Institute of Plant Sciences, A.R.O. Volcani Center, Bet Dagan 50250, Israel

Ana Rotter Department of Biotechnology and Systems Biology, National Institute of Biology, Ljubljana 1000, Slovenia, ana.rotter@nib.si

Kalliopi A. Roubelakis-Angelakis Department of Biology, University of Crete, Crete, Greece, poproube@biology.uoc.gr

J.B. Royo Departamento de Produccion Agraria, Universidad Pública de Navarra, Pamplona (Navarra) 31006, Spain

Bernadette Ruelle Cemagref, UMR ITAP, Montpellier 34196, France

Denis Rusjan Biotechnical Faculty, University of Ljubljana, Ljubljana 1001, Slovenia, denis.rusjan@bf.uni-lj.si

Patricia Sánchez-De-Miguel Departamento de Producción Vegetal: Fitotecnia, Grupo de investigación en Viticultura. ETSI Agrónomos, Universidad Politécnica de Madrid, Madrid, 28040, Spain, patricia.sanchez@upm.es

Manuel Sánchez-Díaz Facultades de Ciencias y Farmacia, Departamento Biología Vegetal, Sección Biología Vegetal Unidad Asociada al CSIC (EEAD, Zaragoza; ICVV, Logroño), Universidad de Navarra, Pamplona 31008, Spain

Héctor Santesteban Facultades de Ciencias y Farmacia, Departamento Biología Vegetal, Sección Biología Vegetal Unidad Asociada al CSIC (EEAD, Zaragoza; ICVV, Logroño), Universidad de Navarra, Pamplona 31008, Spain

L.G. Santesteban Departamento de Produccion Agraria, Universidad Pública de Navarra, Pamplona (Navarra) 31006, Spain, gonzaga.santesteban@unavarra.es

Robert Savé IRTA- Torre Marimon, Caldes de Montbui (Barcelona) 08140, Spain, robert.save@irta.cat

Laurent Scheyer Cemagref, UMR ITAP, Montpellier 34196, France

Karen A. Schlauch Department of Biochemistry and Molecular Biology, University of Nevada, Reno, NV 89557, USA

Achim Schmitt Department of Plant Breeding, Institute of Crop Science & Plant Breeding I, Justus-Liebig Universität Gießen, Giessen 35392, Germany

Andrea Schubert Department of Arboriculture and Pomology, University of Torino, Grugliasco 10095, Italy, andrea.schubert@unito.it

Reinhard Töpfer Julius Kühn-Institut, Federal Research Centre for Cultivated Plants, Institute for Grapevine Breeding Geilweilerhof, Siebeldingen 76833, Germany, reinhard.toepfer@jki.bund.de

Sara Tramontini DCA, Plant Physiology, University of Turin, Grugliasco 10090, Italy

Björn Usadel Max Planck Institute of Molecular Plant Physiology, Am Muehlenberg 1, Potsdam-Golm 14474, Germany, usadel@mpimp-golm.mpg.de

Cornelis van Leeuwen UMR 1287 Ecophysiologie et Génomique Fonctionnelle de la Vigne, Institut des Sciences de la Vigne et du Vin (ISVV Bordeaux), Villenave d'Ornon 33882, France, k-van-leeuwen@enitab.fr

Philippe Vivin UMR 1287 Ecophysiologie et Génomique Fonctionnelle de la Vigne, Institut des Sciences de la Vigne et du Vin (ISVV Bordeaux), Villenave d'Ornon 33882, France, vivin@bordeaux.inra.fr

Chuanlin Zheng Institute of Plant Sciences, A.R.O. Volcani Center, Bet Dagan 50250, Israel

Eva Zyprian Julius Kühn-Institut, Federal Research Centre for Cultivated Plants, Institute for Grapevine Breeding Geilweilerhof, Siebeldingen 76833, Germany, eva.zyprian@jki.bund.de

Chapter 1
Grapevine Roots and Soil Environment: Growth, Distribution and Function

Felicidad de Herralde, Robert Savé, Xavier Aranda, and Carme Biel

Abstract As for many plants, belowground processes occurring in grapevine are much lesser well known than above ground processes. Root biomass, root distribution, root physiology need further study as well as their responses to the soil environment in relation to the performance of the grapevine (yield and quality). This chapter gives an overview of the methodologies used to study grapevine roots and their environment. First of all, root characteristics and soil environment must be described. Monitoring of soil water status is approached with different methods. Root sampling methods such as direct excavation, soil coring, root profiles, rhizotrons, minirhizotrons, and ingrowth cores are described and some results presented. Finally, ecophysiological measurements such as root biomass and distribution, root composition, hormones, and water uptake in roots are commented.

Contents

F. de Herralde (✉)
IRTA-Torre Marimon, Caldes de Montbui (Barcelona) 08140, Spain
e-mail: felicidad.deherralde@irta.cat

S. Delrot et al. (eds.), *Methodologies and Results in Grapevine Research*,
DOI 10.1007/978-90-481-9283-0_1, © Springer Science+Business Media B.V. 2010

Abbreviations

Ø	Diameter (mm)
Ψ_s	Soil water potential (MPa)
Θ	Soil water content (%)
SV	Soil volume (cm^3)
TFM	Total fresh mass (g)
SDM	Soil dry mass (g)
DMS	Dry mass of stones (g)
RDM	Roots dry mass (g)
DDM	Debris dry mass (g)
NI	Non irrigated treatment
I	Irrigated treatment
FDR	Frequency domain reflectrometry
TDR	Time domain reflectometry
ε	Dielectric constant of the soil matrix
ABA	Abscisic acid
ELISA	Enzyme-linked immunosorbent assay
HPFM	High-pressure flow meter
K_{LS}	Leaf specific hydraulic conductivity ($L{\cdot}m{\cdot}MPa^{-1}{\cdot}s^{-1}{\cdot}cm^{-2}$)
k	Hydraulic conductance ($L{\cdot}MPa^{-1}{\cdot}s^{-1}$)
R_R	Whole root system hydraulic resistance ($MPa{\cdot}s{\cdot}kg^{-1}$)
R_{RT}	Hydraulic resistance of trunk below the grafting point ($MPa{\cdot}s{\cdot}kg^{-1}$)
R_{GP}	Hydraulic resistance of the grafting point ($MPa{\cdot}s{\cdot}kg^{-1}$)
R_{CT}	Hydraulic resistance of cultivar trunk ($MPa{\cdot}s{\cdot}kg^{-1}$)
R_{CT}	Hydraulic resistance of the trunk ($MPa{\cdot}s{\cdot}kg^{-1}$)
Φ	Water flow (kg)
P	Pressure (MPa)
K	Hydraulic conductivity ($L{\cdot}m{\cdot}MPa^{-1}{\cdot}s^{-1}$)
l	Length (mm)
LPM	Low pressure method

1.1 Introduction

Belowground processes are much lesser well known than above ground processes. Root biomass, root distribution, root physiology need further study and its responses to the soil environment in relation to the performance of the grapevine (yield and quality). The main functions of roots are water and nutrient uptake, but their structural role should not be forgotten (Canadell et al. 1996, 1999).

In grapevines, the use of rootstocks not only prevents phylloxera infection effects, but also modulates the performance of the variety and its adaptation to the environmental conditions.

The study of root growth in grapevine requires a strong methodological and human effort, and that is why there are few quantitative studies on roots in comparison with aboveground processes.

Many studies have been carried out using belowground observation systems, to monitor root growth, root initiation and root turnover (Mullins et al. 1992). In those studies, it was determined that root growth occurs mainly between budbreak and anthesis and after harvest until leaf fall (Araujo and Williams 1988). More recent studies showed that root growth can also occur during summer (Comas et al. 2005, Eissenstat et al. 2006).

Different rootstocks are characterized by root growth, water uptake, water transport to shoots, metabolic activity and carbon storage in response to edaphoclimatic conditions (Savé et al. 2009).

Root distribution along soil profile is mainly affected by soil type, but its density in this profile is function of the genotypic characteristics of the combination rootstock-variety (Southey and Archer 1988, Williams and Smith 1991). Water availability and vineyard management also affect root growth and distribution.

Water uptake and transport is one of the major functions of roots. Root distribution in the soil and root turnover are key parameters for water uptake and they are also important for hydraulic redistribution (Bauerle et al. 2008) and partial root drying techniques (Dry et al. 2000, Stoll et al. 2000) as they affect the hormonal balance.

Since roots are the hidden half of plants, the study of roots and rhizosphere is difficult and time-costly and needs to be properly addressed and planned.

1.2 Methods and Results

1.2.1 Root Characteristics

When studying root systems, it is important to first define what we are looking for and for which reason.

One key is to define the studied organs or fractions (Fig. 1.1). In grafted grapevines, the grafting point defines the interface between above and belowground fractions, and between the genomes and connecting structures of the cultivar and the rootstock. Root trunk is the original structure, where all the roots sprout. Coarse roots (Ø > 2 mm) are the main roots and represent a high percentage of root biomass; they have structural, water and nutrient transport and storage functions. Fine roots (Ø < 2 mm) have water and nutrient foraging and uptake functions, and are frequently mycorrhized.

Root color is an indicator of root age from the youngest white root tips to the oldest black dead roots, through brown color range. It is well correlated with the metabolic activity, which decreases when the roots age (Comas et al. 2000).

1.2.2 Soil Environment

Measuring the soil or substrate environment is as important as defining the other experimental conditions when working with roots. Physicochemical characteristics

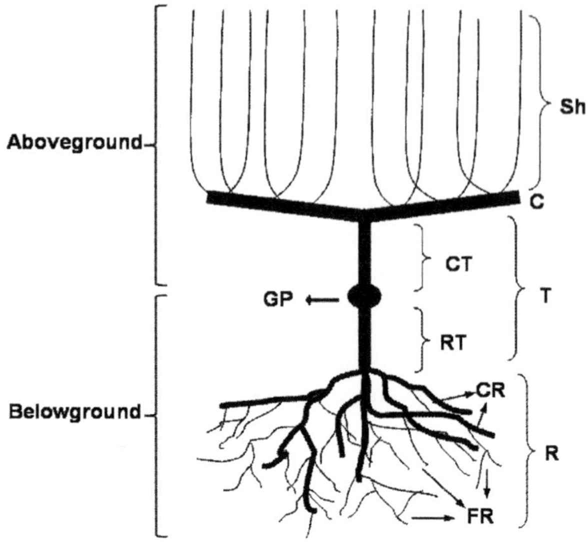

Fig. 1.1 Drawing representing the organs of vine *aboveground*: shoots (*Sh*), canes (*C*), cultivar trunk (*CT*), trunk (*T*); the interface as grafting point (*GP*); and *belowground*: rootstock trunk (*RT*), and roots (*R*), including coarse roots (*CR*) and fine roots (*FR*). Modified from de Herralde et al. (2006)

determine the results of the experiment and the possibility to compare and extrapolate them with data from other experiments.

Soil analysis of texture, nutrient content and water retention curves (Klute 1986) are needed for a basic characterization of the soil, and can be performed in any external analytical laboratory.

Water retention curves relate the soil water potential (Ψ_S, MPa) and soil water content (θ, %) (Fig. 1.2). Substrates are usually selected for having big porous spaces to prevent anoxia problems in containers. This promotes a wider θ in a narrower Ψ_S range. Wilting point (WP) is defined to occur at $\Psi_S = -0.1$ MPa. In soils,

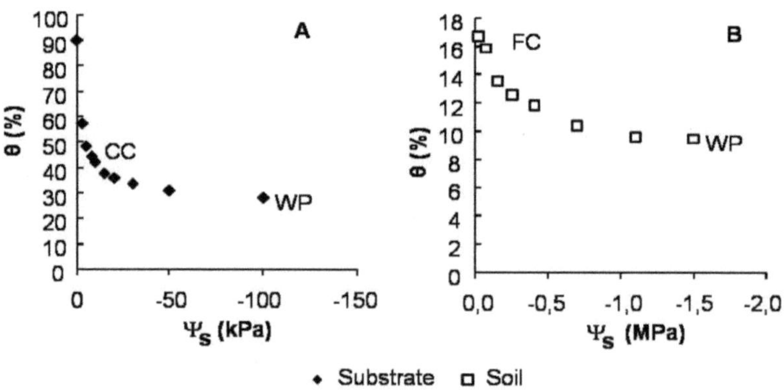

Fig. 1.2 Water retention curves in a substrate (**a**) and a soil (**b**) relating the soil water potential (Ψs, MPa) and soil water content (θ, %)

θ are usually lower and narrower, and wilting point is at $\Psi_S = -1.5$ MPa. This relationship is important for water management in the experiment. When this analysis is not available but texture is, an approximation can be calculated according to Gardner (1964).

There are many methods to measure and monitor soil water status. θ or $\Psi_S$can be either measured and afterwards related.

1.2.2.1 Gravimetric Method

This is a direct measurement of volumetric water content, which is the easiest and cheapest method. This measurement is not continuous and non destructive of a piece of soil and roots.

Extract an undisturbed soil sample of known volume (SV) (see Section 1.2.3.4). Preserve it from evaporation and take immediately to the lab. Weigh the sample to obtain the total fresh mass (TFM). Sieve the sample through a 2 mm mesh. Put the fraction below 2 mm in a tray and put in an oven at 105ºC until constant weight. Measure the soil dry mass (SDM). Weigh the fraction which is over 2 mm (2 mmFM). Separate stones, roots and other debris and wash them. Put these fractions in trays and put in an oven at 65ºC until constant weight. Measure the dry mass of stones (DMS), roots (RDM) and other debris (DMM)). Soil volumetric water content can be calculated as:

$$\theta\,(\%) = (\text{TFM} - 2\text{ mm FM} \;- \text{SDM}) * 100/\text{SV}$$

Root biomass is RDM. When measurements of root length and diameter are needed, they should be performed before drying them (see Section 1.2.3.1).

In a experiment with a non irrigated (NI) and irrigated treatment (I) in a soil comparing three rootstocks, θ (%) was measured. The data (Fig. 1.3) showed a pattern that changed along the season (being drier as season advanced), the soil profile (being drier in surface than in the deeper sections, except in the I plots at the harvest point) and the irrigation treatments, θ being reduced in the NI treatment, compared to the I treatment. Soil water content reduction during veraison was higher for both treatments at 30–50 cm depth, where most of the roots were found.

1.2.2.2 Frequency Domain Reflectometry (FDR) and Time Domain Reflectometry (TDR) Sensors

Both are indirect and non-destructive methods to measure θ. They are based on the relationship between the dielectric constant of the soil matrix (ε) and θ. The calculation of the dielectric constant of the material is determined from the frequency (FDR) or time (TDR) of an electromagnetic wave through a material. Both technologies have the advantage that θ can be continuously measured, providing daily and seasonal patterns of soil water dynamics and its relation to plant water uptake. The measurements are made with metal probes buried into the soil or external probes

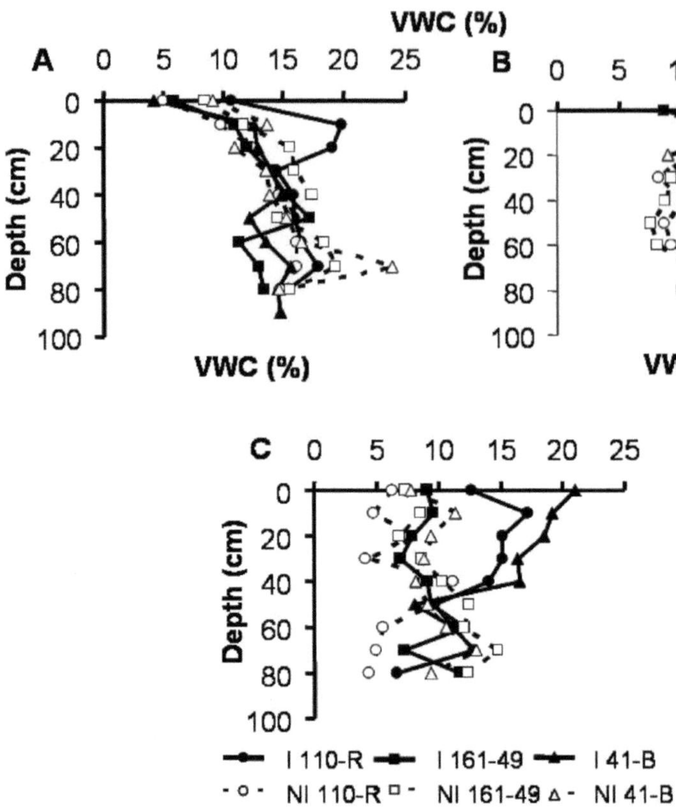

Fig. 1.3 Volumetric water content in soil profile (%), from surface to 1 m depth, at the phases of (**a**) pea size (Jul 18th), (**b**) veraison (Aug 10th) and (**c**) harvest (Sep 25th) 2008 for the two treatments I (*solid line*) and NI (*dotted line*), and the three rootstocks 110-R, 161-49 and 41-B

through accession tubes that read along the soil profiles. The different available devices give θ according to a default calibration. In order to obtain absolute values of θ, a calibration curve should be done for any particular soil along the soil profile. However, when working in very heterogeneous vineyards or in many different soil types, relative changes of θ can be very useful. Significant numbers of data sets, collected from various soil types and crops around the world, have shown that relative changes in volumetric soil water content based on the default calibration can be used to show the most important soil water trends in relation to optimum plant production (Alva and Fares 1998, Alva and Fares 1999, Paltineanu and Starr 1997, Starr and Paltineanu 1998a, b, Tomer and Anderson 1995). Figure 1.4 gives an example of small FDR probes inserted at 15, 30 and 45 cm in a schist soil containing 39.7% of coarse fraction (> 2 mm) and 60.3% of fine fraction (≤ 2 mm) and loam texture. After a very dry fall and winter 2007–2008, high rains in May recharged the profile. The summer reduced it to the minimum until September rains. These probes are easy to install and give good data in this type of soil.

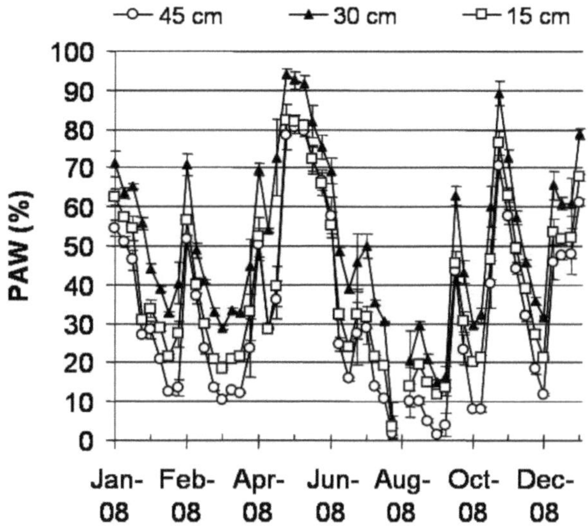

Fig. 1.4 Percentage of available water (PAW %) at 3 depths in a loam soil with a 40% of coarse elements in Priorat, Tarragona (Spain) during 2008. Data are weekly averages ± S.E

1.2.3 Root Sampling

Soil sampling can be done in many ways depending on the aim of the study and the available equipment. To quantify biomass, and to determine coarse roots structure and distribution in the soil, direct excavation and trenches and rhizotrons provide the best results. If the objective is the study of fine roots biomass, root turnover and sampling for physiological determinations, soil coring, ingrowth cores and minirhizotrons are more suitable.

1.2.3.1 Direct Excavation

When the main objective is to determine root biomass in the soil volume, this is the most recommended method (Böhm 1979). The minimum surface to explore is vine spacing in the row and interrows, assuming that horizontally the roots of the surrounding vines enter in the sampled area as much as the roots of the sampled vine go further. The depth to explore is determined by the root capacity to penetrate in the soil. There is a good relationship between the time invested in the excavation and the amount of biomass recovered, as deeper layers are dug. Stop digging when the effort is too much to retrieve a small amount of roots (Metcalfe et al. 2007). When bulk biomass in the explored volume is enough, an excavator equipped with a trencher or a backhoe is useful (Fig. 1.5). If that is not the objective or if the use of these equipments it is not possible, the excavation must be done manually with the help of tools such as hoes and shovels, and pressurized water or air (Fig. 1.5) to separate soil from roots.

If the roots structure should be preserved to process it, brushes may be needed. As fine and medium size roots can be broken during the excavation, all the soil

Fig. 1.5 Dry excavation with backhoe (**a**), with hand tools and pressurized air (**b**) and first coarse sieving of soil to separate roots (**c**)

extracted must be sieved with a battery of sieves of different mesh size, to retrieve as much as possible (Fig. 1.5).

From root trunk downwards, root structure and distribution can be measured. Main coarse roots are counted and diameter measured. Figures 1.6 and 1.7 provide examples of several parameters of root characteristics, describing and comparing three 1 year old rootstocks. Root geotropic angle is also frequently recorded. Once the root structure measurements are completed, roots can be cut at the sides of the dug hole, and the dimensions of the hole must be measured to determine the volume of rootzone explored. There are 3D digitalizing methods that can be used in situ and the root architecture can be modeled if needed (Danjon et al. 1999a, b, Tobin et al. 2007).

This is a destructive method and frequently difficult to perform in productive vines in commercial vineyards.

1.2.3.2 Trenches and Profile Root Distribution

There are several methods available for the study of roots distributions, but the most commonly used in viticulture is the profile wall method. The profile wall method typically consists in excavating a trench of 1- to 2-m depth at some predetermined location generally parallel to the vine row, establishing a grid of fixed subquadrat

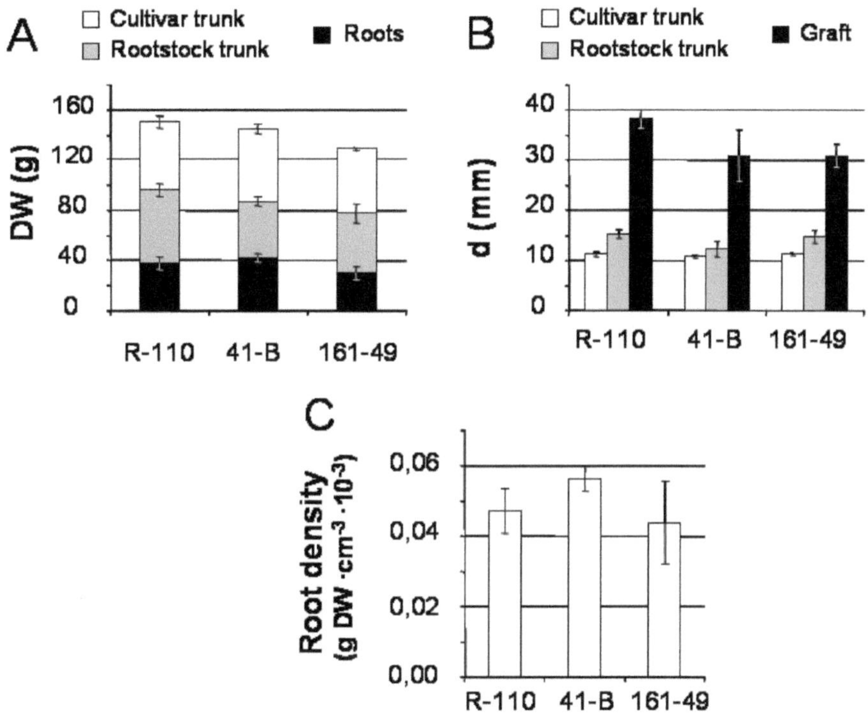

Fig. 1.6 Vine biomass of three rootstocks (R-110, 41B and 161-49) from dry excavation. (**a**) wood dry weight of variety trunk, rootstock trunk and roots (g). (**b**) Diameter of woody parts: variety trunk, rootstock trunk and graft (mm). (**c**) Root density (g DW roots cm^{-3} soil 10^{-3}). Data are mean of n=3 ± S.E

areas on a wall of the trench, and then recording root-wall intercepts. Smart et al. (2006) provided a thorough review of this method and results.

1.2.3.3 Rhizotrons

Rhizotrons are either structures built into the soil or big containers to make roots easily visible. These facilities have one or more glass walls to allow the direct observation of root growth. As in the case of trenches they lack the tridimensional study of root distribution, but offer the fourth-dimensional aspect, provided it can be followed in time. Growth can be monitored and measured directly on the glass wall or be photographed and the digital images analyzed using software such as WinRhizo^{TM}. An example obtained with our rhizotron compares root growth of three rootstocks (Fig. 1.8).

Differences in depth distribution between rootstocks, root type and phenological state are evident from Fig. 1.9. Root depth profiles were different for the three

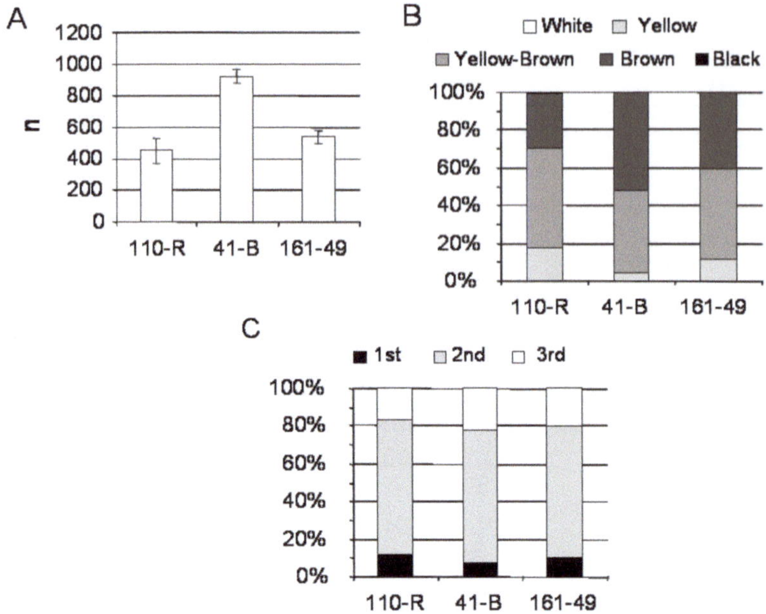

Fig. 1.7 Root traits of three rootstocks (R-110, 41B and 161-49) from dry excavation. (**a**) Root number (data are mean of n=3 ± S.E.). (**b**) Root color from white to black (%). (**c**) Root order (%)

rootstocks, showing a certain amount of growth after harvest and a clear ageing process shown by lignification (Fortea et al. 2009).

1.2.3.4 Soil Coring

As an alternative to destructive methods, or when the objective is to study fine roots, soil coring is an indirect way to assess root distribution in the soil profile, and

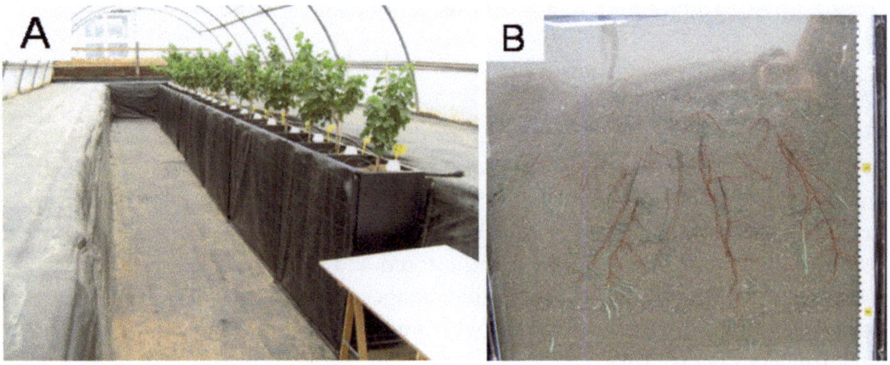

Fig. 1.8 Rhizotron facilities at IRTA-Torre Marimon. Each container is 1.2 m deep, 0.60 × 0.50 m surface. Soil is sandy loam. Frontal glass cover allows the view of growing roots that are marked periodically

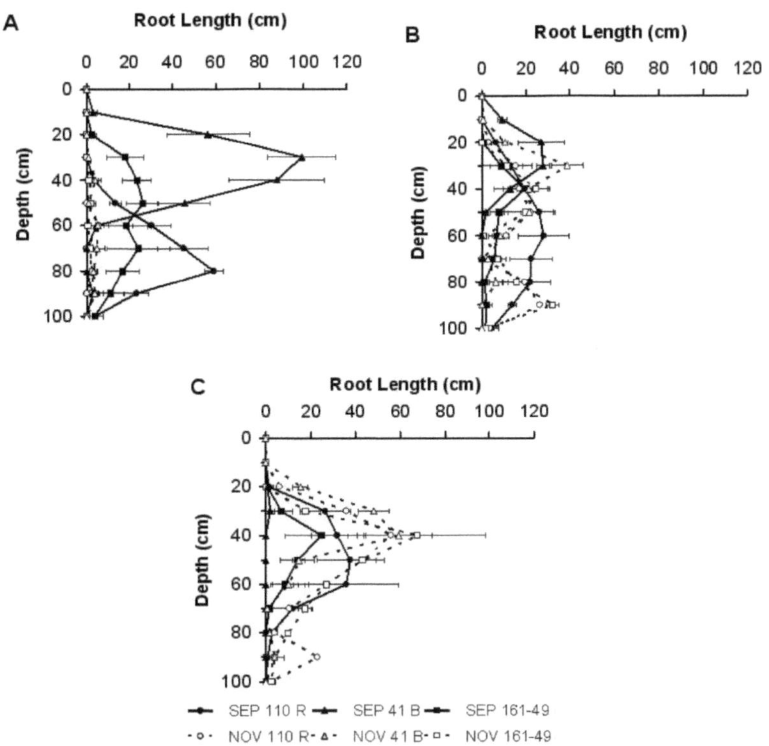

Fig. 1.9 Root length in soil profile for three rootstocks (110-R, 41-B and 161-49) in September and November 2008. (**a**) New roots (*white*). (**b**) Growing roots (*pale color*). (**c**) Lignified roots (*dark color*)

provides at the same time soil samples (Böhm 1979). Soil cores are to be performed using soil augers or probes manually or mechanically introduced and extracted from the soil (Fig. 1.10). The extracted cylinders can be processed in depth sections to develop a soil profile and the root distribution along it. Mechanical coring is recommended for being easier and more useful, especially in hard or stony soils, but the drawback is soil compaction in the soil sample. To obtain a good measurement of root distribution and any estimate of root biomass, many cores surrounding every vine must be done. This procedure can be more effort and time costly than a complete excavation.

1.2.3.5 Ingrowth Cores

They are a special type of core, used to study the root colonization of soil. A soil core is performed as described above. The extracted soil is sieved and all the roots, stones and other debris eliminated. A synthetic fabric mesh bag of the same size of the core and mesh size 3 mm is put in the hole, and refilled with the sieved soil. Let time to the roots grow inwards and extract the core to count and measure the roots.

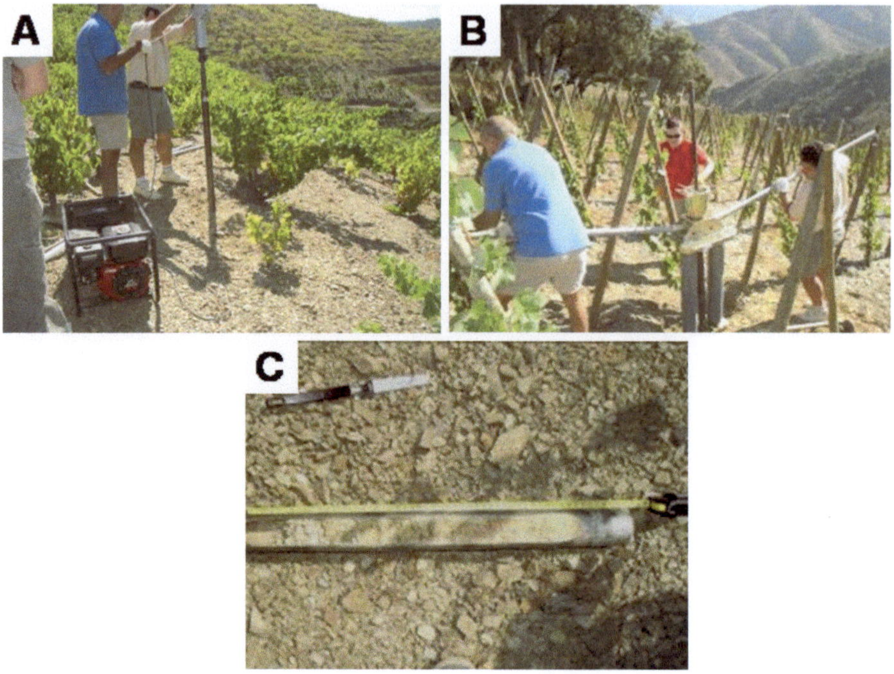

Fig. 1.10 Soil coring with mechanical auger (**a**), extraction (**b**) and sample processing (**c**)

If a time course study must be made, many ingrowth cores should be installed from the very beginning (Godbold et al. 2003).

1.2.3.6 Minirhizotrons

They are accession tubes enabling direct observations of root growth in its natural media. Their installation is less disturbing for the soil than that of trenches or rhizotrons. The soil is drilled in angle to the vertical to prevent direct percolation of water and light. The soil is filled with a transparent plastic tub, being careful not to break or scratch the surface of the tube. Observation is made by photograph, video or scanning. Digital images can be treated with adequate software, e.g. WinRizho^{TM}. This method is especially suitable to study fine roots turnover, root demography and root interactions with mycorrhiza or microfauna (Eissenstat et al. 2006). Installing access tubes in stony soils can be difficult. López et al. (1996) developed an inflatable minirhizotron especially for highly stony soils.

1.2.4 Ecophysiological Measurements

In order to describe root function, pot experiments provide fresh root material much more easily than field experiments, but the extrapolation of the results obtained is difficult because growth conditions in pots can be far from natural. Field sampling of roots can be performed as described above by direct excavation of soil coring. Roots must be properly stored in dry conditions, in ice-filled coolers or even quickly frozen in liquid nitrogen, depending on later analysis. Root function and activity can be determined in many ways. A few ones are described below.

1.2.4.1 Root Biomass

Excavated roots or roots extracted by soil coring are processed measuring length, diameter and color. This can be done by either computer image analysis or manually using callipers and color charts. After measurements have been taken, roots are classified in diametrical classes and dried in an oven at 65°C until constant weight. Big root trunks can be sectioned or let dry under semicontrolled environments like closed empty greenhouses where summer temperatures can be high. (See Figs. 1.6 and 1.7 for examples).

1.2.4.2 Root Composition

Many components of roots can be easily analyzed and provide estimates of root activity and function.

In carbon balance, dry matter partitioning and sink studies, the carbon and nitrogen composition of roots can be useful. The major storage of carbon in vine perennial tissues (roots, trunk, and canes) tissues is starch, which is accumulated in the ray parenchyma of the wood (Zapata et al. 2004). Roots usually contain the highest starch concentrations, followed by the trunk, cordon, canes and root-stock trunk. Starch contents of above and belowground plant parts are similar. Canopy management and vine spacing affect starch concentration and metabolism in roots (Hunter 1998, Hunter 2000a, b). Starch is most commonly analyzed by spectrophotometry (Zapata et al. 2004).

1.2.4.3 Hormonal Production in Roots

Hormonal activity plays an important role in the growth and development of new roots, as a detector of water stress and as long distance signal. Abscisic acid (ABA) plays a central role in the long distance drought signalling process in many plants (Gowing et al. 1990, Davies and Zhang 1991). Under certain circumstances, the ABA content of the root or the ABA concentration in the xylem can be related to soil water status (Zhang and Davies 1989). As such they can provide an estimate of the degree of soil drying. Several stresses such as water deficit, salinity or mineral nutrition deficits can induce ABA accumulation in roots. ABA concentration in root tissues depends on the anabolic and catabolic processes, on recirculation and exudation. This accumulation is regulated by ABA-metabolism (biosynthesis and degradation), recirculation, and exudation. Hartung et al. (2005) showed that endogenous ABA, radial water flows, and radial ABA flows are tightly linked. Under stress, longitudinal transport of ABA in the xylem acts as a long-distance signal regulating the water relations of the canopy, stomata and meristems. At the same time, ABA plays a direct role in roots, modulating root hydraulic conductance, root growth or being involved in the mechanisms related to dessication tolerance. In some crops, such as grapevine, ABA effects on berries improve the quality of the fruit. Measurements of ABA concentrations in roots, leaves, berries and xylem sap should be done as a whole. It has been described that stomatal behavior is more correlated with xylem sap ABA than with the concentration of ABA in leaves (total ABA) or in roots. Root ABA and leaf ABA are correlated only under water stress conditions. The water stress root signal (xylem sap ABA increase) seems to be important in the early stages of water stress. Coupling of the hydraulic and hormonal signals plays a key role in controlling vine water use (Peterlunger et al. 2000, Ren et al. 2007).

ABA is the major plant regulator, but its balance with other hormones such as cytokinins (Bravdo 2005), auxins (Jeong et al. 2004, Lovisolo et al. 2002), gibberellins (Bravdo 2005), polyamines (Antolin et al. 2008) has a big influence on root development and growth, water use regulation and berry ripening. ABA can be measured either by enzyme-linked immunosorbent assay (ELISA) (Rodrigues et al. 2008, Correia et al. 1995) or a combination of chromatography and spectrometry techniques (Müller et al. 2002, Schmelz et al. 2003, Vilaro et al. 2006). These techniques would deserve a specific chapter by themselves and further reading is recommended. Figure 1.11 shows the concentration of ABA in leaves and roots and their relation with stomatal conductance. The increase of ABA in leaves during ripening is proportional to its reduction in roots and may indicate the transport from roots to leaves, which would reduce stomatal conductance.

1.2.4.4 Root Water Uptake

Plant conductivity is the result of xylem anatomy and a combination of three different water pathways: apoplastic transport (Tyree 2003), symplastic transport, and the transcellular pathway through aquaporins (Tyerman et al. 2002). In some experiments, the conductance of the whole root system in vines (Peterlunger et al. 1990) or other species (Ramos and Kaufmann 1979) have been measured with a pressure chamber method or high-pressure flow meters (HPFM) (Tyree et al. 1995, Bogeat-Triboulot et al. 2002). The latter devices offer several advantages compared to the pressure chamber method. With the HPFM, whole root systems can be measured faster, within the soil, without disturbing them, and independently on their size. It also minimizes the effects of shoots and roots capacitance and the effects of the accumulation of solutes in root tips, changing osmotic and water potential in roots (Tsuda and Tyree 2000). In the last years, some knowledge about water transport has been obtained for grapevines (Sperry et al. 1988, Lovisolo et al. 2002, Tyree 2003, Chouzouri and Schultz 2005). In particular root hydraulic conductance has been correlated with vigor, growth and water use efficiency (Peterlunger et al. 1990) Xylem anatomy, vessel length, diameter and distribution are key parameters for hydraulic conductance (Tyree and Zimmermann 2002). In grapevines, leaf specific conductivity (K_{LS}) is lower in nodal segments than in internodal segments

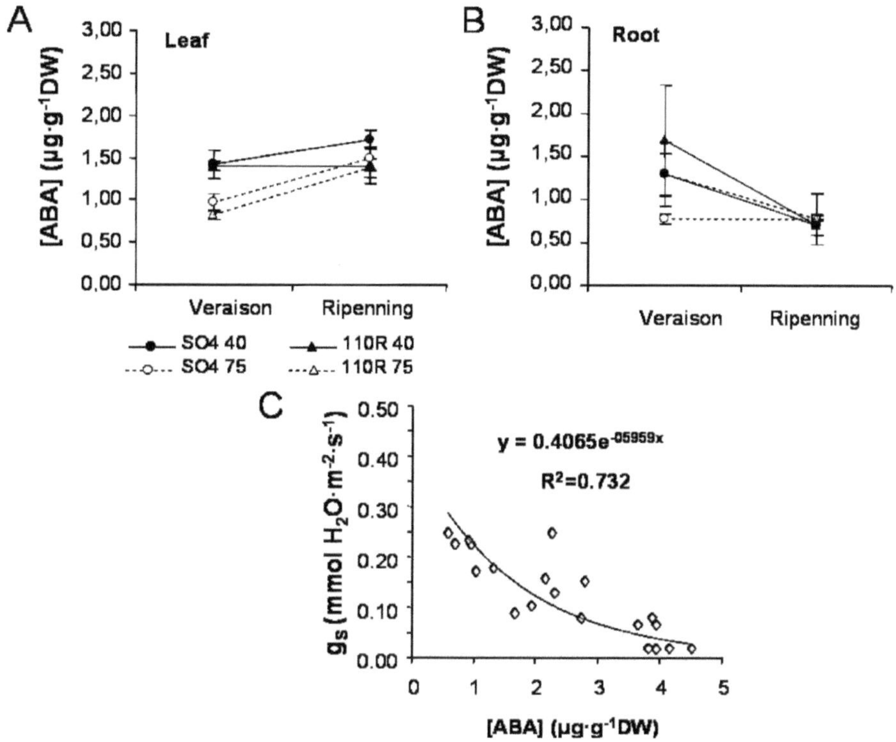

Fig. 1.11 ABA concentration (µg·g^{-1} DW) in leaves (**a**) and roots (**b**) of Tempranillo grafted onto SO4 and R-110, submitted to two irrigation treatments 40 and 75% of evapotranspiration, in veraison and ripening. Relationship between stomatal conductance (g_S) and ABA concentration (**c**)

of shoots. The nodes could thus be regarded as hydraulic constriction zones in the stem (Salleo et al. 1982). Grafts can often have lower hydraulic conductance (k) than the cultivar, thus reducing the whole hydraulic conductance, as it happens in grapevine (Bavaresco and Lovisolo 2000) or apple (Atkinson et al. 2003). Grafts have also been used to compare rootstocks and the effect of grafting on their hydraulic characteristics (De Herralde et al. 2006).

The method consists in perfusing water through the cut trunk of the vine from 0 to 500 kPa at a constant rate of 0.5 kPa·s^{-1}. In order to measure trunk and roots hydraulic conductance (k), the vine is cut 10–15 cm above the graft. Immediately, root system is connected to HPFM (Dynamax Inc. Houston TX, USA). Then, perform a transient measurement of whole root system hydraulic resistance (R_R).

Cut the trunk 10–15 cm below the graft and perform transient measurements of the whole trunk hydraulic resistance, which is the serial sum of the resistances of trunk below the grafting point (R_{RT}), the grafting point (R_{GP}), and the segment of cultivar trunk over the grafting point (R_{CT}). Cut again just below the grafting point and make another transient measuremet to obtain the resistance of the grafting point plus the cultivar trunk ($R_{GP} + R_{CT}$). Finally, cut just above and perform transient measurement of the resistance of the cultivar trunk R_{CT}).

Trunk resistance is the sum of the different segments resistance in series ($R_T = R_{RT} + R_{GP} + R_{CT}$), and trunk resistance is in series with root resistance (R_R). With HPFM, hydraulic conductance (k) is calculated as the slope of the plot of the water flow (Φ) versus the pressure (P).

$$k = d\Phi/dP = 1/R[L \cdot MPa^{-1} \cdot s^{-1}]$$

being each conductance the inverse of each resistance.

In order to compare the results, hydraulic conductances are often transformed into hydraulic conductivity (K)

$$K = k \cdot l[L \cdot m \cdot MPa^{-1} \cdot s^{-1}]$$

where l is the total length of the plant segment. Specific hydraulic conductances (k_S) and conductivities (K_S) can be calculated by dividing either by the conducting section, the diameter of the segment or other available data as leaf area supplied by the measured xylem or root biomass (Tyree and Zimmermann 2002).

HPFM and other equipments such as HCFM (Dynamax Inc.) and Xy'lem (Bronkhorst) provide a quick measurement of the hydraulic conductance and allow to monitor whole root systems or whole plants, which is not possible with the Low pressure method (LPM) (Sperry et al. 1988). LPM is more suitable for measuring embolism. In the case of trunk or root segments, both methods can be used and then compared.

Hydraulic resistance of root systems was measured in two rootstocks (110-R and SO4) submitted to two irrigation treatments (100% and 50% ET_0). Hydraulic resistance is higher in 110-R, more drought resistant, than in SO4, and it is well correlated with root biomass (Fig. 1.12) (de Herralde et al. 2006).

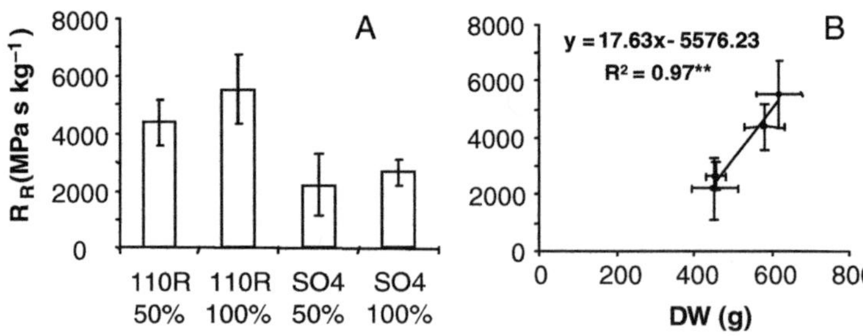

Fig. 1.12 Root hydraulic resitance of whole roots systems R_R ($MPa \cdot s \cdot kg^{-1}$) (**a**) and correlation of R_R and root biomass DW (g) (**b**). Data are mean of n = 7–8. Vertical and horizontal bars represent S.E

1.3 General Considerations and Further Research

This chapter intends to give an overview of techniques used in root research. It is not exhaustive but general enough to have a first approach to the "hidden half", as was called the belowground part of plants (Waisel et al. 2002).

Root research is needed to better understand the whole plant physiology. Root growth and biomass and is contribution to carbon sinks can be important, given the world vineyard surface. The hydraulic conductivity of roots, its response to the environmental conditions and how it is controlled and influences the whole plant water use efficiency are key points in the context of water shortages predicted in the Mediterranean climate areas. The influence of rootstock on vine growth and yield is well known, but much research remains to be done on the effect of rootstock on the quality of must and wine.

1.4 Disclaimer

All the commercial names and devices cited in this chapter are solely used as examples and the authors do not intend to recommend their specific use.

Acknowledgments Authors acknowledge the partial financial support of Spanish Ministry of Science and Education through the projects AGL2001-1285-CO3-02, AGL2005-06927-CO2-02 and AGL2008-04525-CO2-02, to the project CDTI IQAC-CSIC i Ferrer Bobet S.L. "Monitorització Hormonal en Vinya" and the Project ACCUA of Fundació Obra Social de la Caixa de Catalunya. Predoctoral Fellowship granted to M.Mar Alsina by IRTA. We thank the scientific collaboration of the Vitiviniculture Research Group (URV), leaded by Montse Nadal. We thank the collaboration of Eudald Crivillés from Miguel Torres S.A. We thank José Montero, Cristian Morales, Beatriz Grau, Oleguer Amoròs, Julián Cazaña, Marc Ferrer, Eulàlia Serra and Gemma Fortea for field and lab work.

References

Alva AK, Fares A (1998) A new technique for continuous monitoring of soil moisture content to improve citrus irrigation. Proc Fla State Hortic Soc 111:113–117

Alva AK, Fares A (1999) Precision scheduling of irrigation in sandy soils using capacitance probes. Dahlia Greidinger Int Symp Nutr Manag Salinit Water Stress. Technion IIT, Haifa Israel

Antolin MC, Santesteban H, Maria ES, Aguirreolea J, Sanchez-Diaz M (2008) Involvement of abscisic acid and polyamines in berry ripening of *Vitis vinifera* (L.) subjected to water deficit irrigation. Aust J Grape Wine Res 14:123–133

Araujo FJ, Williams, LE (1988) Dry-matter and nitrogen partitioning and root-growth of young field-grown Thompson Seedless grapevines. Vitis 27:21–32

Atkinson CJ, Else MA, Taylor L, Dover CJ (2003) Root and stem hydraulic conductivity as determinants of growth potential in grafted trees of apple (*Malus pumila* Mill.). J Exp Bot:54:1221–1229

Bauerle TL, Richards JH, Smart DR, Eissenstat, DM (2008) Importance of internal hydraulic redistribution for prolonging the lifespan of roots in dry soil. Plant Cell Environ 31:177–186

Bavaresco L, Lovisolo, C (2000) Effect of grafting on grapevine chlorosis and hydraulic conductivity. Vitis 39:89–92

Bogeat-Triboulot MB, Martin R, Chatelet D, Cochard H (2002) Hydraulic conductance of root and shoot measured with the thansient and dynamic modes of the high-pressure flowmeter. Ann For Sci 59:389–396

Böhm W (1979) Methods of studying root systems. Springer, New York

Bravdo BA (2005) Physiological mechanisms involved in the production of non-hydraulic root signals by partial rootzone drying – A review. In: Proc 7th Int Symp Grapevine Physiol Biotech. Acta Horticulturae 689:267–276

Canadell J, Djema A, Lopez B, Lloret F, Sabate S, Siscart D, Gracia CA (1999) Structure and dynamics of the root system. In: F Roda, J Retana, CA Gracia, J Bellot (eds) Ecology of mediterranean evergreen oak forests. Series: Ecological studies: analysis and synthesis 37:47–59

Canadell J, Jackson RB, Ehleringer JR, Mooney HA, Sala OE, Schulze ED (1996) Maximum rooting depth of vegetation types at the global scale. Oecol 108:583–595

Chouzouri A, Schultz HR (2005) Hydraulic anatomy, cavitation susceptibility and gas-exchange of several grapevine cultivars of different geographic origin. In: Proc 7th Int Symp Grapevine Physiol Biotech. Acta Hortic 689:325–331

Comas LH, Eissenstat DM, Lakso AN (2000) Assessing root death and root system dynamics in a study of grape canopy pruning. New Phytol 147:171–178

Comas LH, Anderson LJ, Dunst RM, Lakso AN, Eissenstat, DM (2005) Canopy and environmental control of root dynamics in a long-term study of Concord grape. New Phytol 167: 829–840

Correia MJ, Pereira JS, Chaves MM, Rodrigues M, Pacheco CA (1995) ABA xylem concentrations determine maximum daily leaf conductance of field-grown *Vitis vinifera* L. plants. Plant Cell Environ 18:511–521

Danjon F, Bert D, Godin C, Trichet P (1999a) Structural root architecture of 5-year-old *Pinus pinaster* measured by 3D digitising and analysed with AMAPmod. Plant Soil 217:49–63

Danjon F, Sinoquet H, Godin C, Colin F, Drexhage M (1999b) Characterisation of structural tree root architecture using 3D digitising and AMAPmod software. Plant Soil 211: 241–258

Davies WJ, Zhang JH (1991): Root signals and the regulation of growth and development of plants in drying soil. Annu Rev Plant Physiol Plant Mol Biol 42:55–76

de Herralde F, Alsina MM, Aranda X, Savé R, Biel C. 2006. Effects of rootstock and irrigation regime on hydraulic architecture of *Vitis vinifera* L. cv. Tempranillo. 2006. Int J Vine Wine Sci 40 133–139

Dry PR, Loveys BR, During H (2000) Partial drying of rootzone of grape. I. Transient changes in shoot growth and gas exchange. Vitis 39:3–7

Eissenstat DM, Lakso AN, Neilsen D, Smart DR, Bauerle TL, Comas LH, Neilsen GH (2006) Seasonal patterns of root growth in relation to shoot phenology in grape and apple. Acta Hortic 721:21–26

Fortea G, Savé R, Biel C, de Herralde F, Aranda X (2009) Late season root profile development of two contrasting vine rootstocks. In: Proc 7th ISRR Symp Root Res Appl (RootRAP). Vienna

Gardner WR (1964) Relation of root distribution to water uptake and availability. Agron J 56:41–45

Godbold DL, Fritz HW, Jentschke G, Meesenburg H, Rademacher P (2003) Root turnover and root necromass accumulation of Norway spruce (*Picea abies*) are affected by soil acidity. Tree Physiol 23:915–921

Gowing DJG, Davies WJ, Jones HG (1990) A positive root-sourced signal as an indicator of soil drying in apple, *Malus x domestica*-Borkh. J Exp Bot 41:1535–1540

Hartung W, Schraut D, Jiang F (2005) Physiology of abscisic acid (ABA) in roots under stress – a review of the relationship between root ABA and radial water and ABA flows. Aust J Agric Res 56:1253–1259

Hunter JJ (1998) Plant spacing implications for grafted grapevine. I. Soil characteristics, root growth, dry matter partitioning, dry matter composition and soil utilisation. S Afr J Enol Vitic 19:25–34

Hunter JJ (2000a) Implications of seasonal canopy management and growth compensation in grapevine. S Afr J Enol Vitic 21:81–91

Hunter JJ (2000b) Plant spacing effects on root growth and dry matter partitioning of *Vitis vinifera* L. cv. Pinot noir/99 Richter and implications for soil utilisation. Acta Hortic 526:63–74

Jeong ST, Goto-Yamamoto N, Kobayashi S, Esaka M (2004) Effects of plant hormones and shading on the accumulation of anthocyanins and the expression of anthocyanin biosynthetic genes in grape berry skins. Plant Sci 167:247–252

Klute A (1986) Methods of soil analysis. Part 1. Physical and mineralogical methods, 2nd edn. Am Soc Agron Inc., Madison, Wisconsin, USA

López B, Sabaté S, Gracia CA (1996) An inflatable minirhizotron system for stony soils. Plant Soil 179:255–260

Lovisolo C, Hartung W, Schubert A (2002) Whole-plant hydraulic conductance and root-to-shoot flow of abscisic acid are independently affected by water stress in grapevines Funct Plant Biol 29:1349–1356

Lovisolo C, Schubert A, Sorce, C (2002) Are xylem radial development and hydraulic conductivity in downwardly-growing grapevine shoots influenced by perturbed auxin metabolism? New Phytol 156:65–74

Metcalfe DB, Williams M, Aragão LEOC, da Costa ACL, de Almeida SS, Braga AP, Gonçalves PHL, de Athaydes Silva Junior J, Malhi Y, Meir P (2007) A method for extracting plant roots from soil which facilitates rapid sample processing without compromising measurement accuracy. New Phytol 174:697–703

Müller A, Düchting P, Weiler EW (2002) A multiplex GC-MS/MS technique for the sensitive and quantitative single-run análisis of acidic phytohormones and related compounds, and its application to *Arabidopsis thaliana*. Planta 216:44–56

Mullins MG, Bouquet A, Williams LE (1992) Biology of the grapevine. Cambridge University Press, UK, pp 239

Paltineanu IC, Starr JL (1997) Real-time soil water dynamics using multisensor capacitance probes: laboratory calibration. Soil Sci Soc Am J 61:1576–1585.

Peterlunger E, Marangoni B, Cipriani G (1990) Root hydraulic conductivity of grapevine rootstocks. Vignevini 17:43–46.

Peterlunger E, Buccella A, Iacono F (2000) Physiological changes in grapevine during adjustment to water stress: ABA, leaf gas exchanges and root hydraulic conductivity. Proc Fifth Int Symp Grapevine Physiol Acta Hort 526:201–206

Ramos C, Kaufmann MR (1979) Hydraulic resistance of rough lemon roots. Physiol Plant 45: 311–314

Ren H, Wei K, Jia W, Davies WJ, Zhang J (2007). Modulation of root signals in relation to stomatal sensitivity to root-sourced abscisic acid in drought-affected plants. J. Integr. Plant Biol. 49:1410–1420

Rodrigues ML, Santos TP, Rodrigues AP, de Souza CR, Lopes CM, Maroco JP, Pereira JS, Chaves MM (2008) Hydraulic and chemical signalling in the regulation of stomatal conductance and plant water use in field grapevines growing under deficit irrigation. Funct Plant Biol 35: 565–579

Salleo S, Rosso R, Lo Gullo MA (1982) Hydraulic architecture of *Vitis vinifera* L. and *Populus deltoides* Bartr. 1-year-old twigs: II. The nodal regions as "constriction zones" of the xylem system. G Bot Ital 116:29–40

Savé R, Sabaté S, de Herralde F, Biel, C, Miguel C, Alsina MM, Fortea G, Grau B, Vilanova A, Tomàs E, Aletà N, Aranda, X. (2009) Could be the root systems of cultured plants an important carbon sink under global change conditions? In: Proc 8th Int CO_2 Conf. Jena, France

Schmelz EA, Engelberth J, Alborn HT, O'Donnell P, Sammons M, Toshima H, Tumlinson III JH (2003) Simultaneous analysis of phytohormones, phytotoxins, and volatile organic compounds in plants. PNAS 100:10552–10557

Smart DR, Schwass E, Morano L, Lakso AN (2006) Grapevine root distributions: a comprehensive analysis and a review. Am J Enol Vitic 56:157–168

Southey JM, Archer E (1988) The effect of rootstock cultivar on grapevine root distribution and density. In: Van Zyl JL (ed.) The grapevine root and its environment. Tech comm. Dept Agric Water Supply. Pretoria, SA, 215:57–63

Sperry JS, Donelly JR, Tyree MT (1988) A method for measuring hydraulic conductivity and embolism in xylem. Plant Cell Environ 11:35–40

Starr JL, Paltineanu IC (1998a) Real-time soil water dynamics over large areas using multisensor capacitance probes and monitoring system. Soil Tillage Res 47:43–49

Starr JL, Paltineanu IC (1998b) Soil water dynamics using multisensor capacitance probes in nontraffic interrows of corn. Soil Sci Soc Am J 62:114–122

Stoll M, Loveys B, Dry P (2000) Hormonal changes induced by partial rootzone drying of irrigated grapevine. J Exp Bot 51:1627–1634

Tobin B, Čermák J, Chiatante D, Danjon F, Di Iorio A, Dupuy L, Eshel A, Jourdan C, Kalliokoski T, Laiho R, Nadezhdina N, Nicoll B, Pagès L, Silva J, Spanos I (2007) Towards developmental modelling of tree root systems. Plant Biosyst 141:481–501

Tomer MD, Anderson JL (1995) Field evaluation of a soil water-capacitance probe in a fine sand. Soil Sci 159:90–98

Tsuda M, Tyree MT (2000) Plant hydraulic conductance measured by the high pressure flow meter in crop plants. J Exp Bot 51:823–828

Tyerman SD, Niemietz CM, Bramley H (2002) Plant aquaporins: multifunctional water and solute channels with expanding roles. Plant Cell Environ 25:173–194

Tyree MT (2003) Hydraulic properties of roots. Root Ecol 168:125–150

Tyree MT, Zimmermann MH (2002) Xylem structure and the ascent of sap, 2nd edn. Springer, New York

Tyree MT, Patiño S, Benink J, Alexander J (1995) Dynamic measurements of root hydraulic conductance using a high pressure flowmeter in the laboratory and field. J Exp Bot 46:83–94

Vilaro F, Canela-Xandri A, Canela R (2006) Quantification of abscisic acid in grapevine leaf (*Vitis vinifera*) by isotope-dilution liquid chromatography-mass spectrometry. Analyt Bioanalyt Chem 386:306–312

Waisel Y, Eshel A, Kafkafi U (eds) (2002) Plant roots: the hidden half, 3rd edn. Marcel Dekker Inc., New York

Williams LE, Smith RJ (1991) The effect of rootstock on the partitioning of dry-weight, nitrogen and potassium, and root distribution of Cabernet-Sauvignon grapevines. Am J Enol Vitic 42:118–122

Zapata C, Deleens E, Chaillou S, Magne C (2004) Partitioning and mobilization of starch and N reserves in grapevine (*Vitis vinifera* L.). J Plant Physiol 161:1031–1040

Zhang J, Davies WJ (1989) Abscisic-acid produced in dehydrating roots may enable the plant to measure the water status of the soil. Plant Cell Environ 12:73–81

Chapter 2
Radiation Balance in Vineyards

Pilar Baeza, Patricia Sánchez-De-Miguel, and José Ramón Lissarrague

Abstract Light intercepted by grapevines is linearly related to vineyard productivity. Radiation intercepted is also linked to vegetative development and in turn with water consume by plants. Thus, potential yield and crop coefficient – kc – for irrigation management can indirectly be predicted based on canopy light interception and vegetative development. The radiation balance of a crop can be calculated as absorbed radiation (Ra) or intercepted radiation (Rit). Ra is the balance between the input of radiation in the vine-soil group (incident radiation –Ri- and reflected radiation by the soil –Rrs-) and the output of radiation from the vine-soil (transmitted radiation –Rt- and reflected by soil and plant together –Rrsp-). In crops in rows, the soil between rows must be taken into account especially in those periods when the canopy does not cover it completely, representing a drain of energy. But once the development of the canopy is complete, Rit can be used as a simplification of Ra. Rit is the difference between Ri and Rt, without considering the reflected radiation because its value is very small.

Keywords Photosynthetically active radiation · Ceptometer · PAR balance · Light interception · Grapevine

Contents

P. Baeza (✉)
Departamento de Producción Vegetal: Fitotecnia, Grupo de investigación en Viticultura. ETSI Agrónomos, Universidad Politécnica de Madrid, Madrid, 28040, Spain
e-mail: pilar.baeza@upm.es

S. Delrot et al. (eds.), *Methodologies and Results in Grapevine Research*,
DOI 10.1007/978-90-481-9283-0_2,

Abbreviations

A	Photosynthesis (e.g. μmol CO_2 m^{-2} s^{-1})
Ah	Area of the shade cast by the canopy over an horizontal surface (e.g. m^2)
Ap	External leaf area (e.g. m^2)
E	External leaves
Ea	Absortion efficiency (e.g. %)
Ei	Interception efficiency (e.g. %)
hs	Solar time
I	Internal leaves
K	Extinction coefficient (e.g. 0/1)
kc	Crop coefficient
LAI	Leaf area index (e.g. m^2 m^{-2})
PAR	Photosynthetic active radiation (e.g.μmol photons m^{-2} s^{-1})
R	Plant and soil reflexion coefficient
Ra	Absorbed radiation by the vine (e.g. μmol photons m^{-2} s^{-1})
Ri	Incident radiation (e.g. μmol photons $m^{-2}s^{-1}$)
Rit	Intercepted radiation by vines (e.g. μmol photons m^{-2} s^{-1})
Rrs	Radiation reflected by the soil (e.g. μmol photons $m^{-2}s^{-1}$)
Rrsp	Radiation reflected from soil and plant (e.g. μmol photons $m^{-2}s^{-1}$)
Rs	Coefficient of reflexion of soil on the crop
Rt	Radiation transmitted (e.g. μmol photons $m^{-2}s^{-1}$)
SA	External leaf area (e.g. m^2 m^{-2})
T	Transmission coefficient
VSP	Vertical shoot position
%Ra/Ri	Percentage of the incident radiation which is absorbed by the plant (e.g. %)
%Ras/Ri	Percentage of the incident radiation which is absorbed by the soil (e.g. %)
%Rit/Ri	Percentage of the incident radiation which is intercepted by the plant (e.g. %)

2.1 Introduction: Principles of Radiation Balance

The radiation balance is a global tool that characterizes the quantitative and qualitative production of the vineyard. On one hand, it integrates the leaf area – structure and geometry of the canopy – and the radiation intercepted by the vine (Fig. 2.1). On the other hand, it evaluates the light, heat and humidity microclimate of the vine. Many studies have found a close relationship between the balance of Photosynthetic Active Radiation (PAR) and total dry matter as well as the qualitative characteristics of in terms of sugar concentration and must and wine colour mainly (Carbonneau 1980, Robinson and Lakso 1991, Dokoozlian and Kliewer 1995, Schultz 1995,

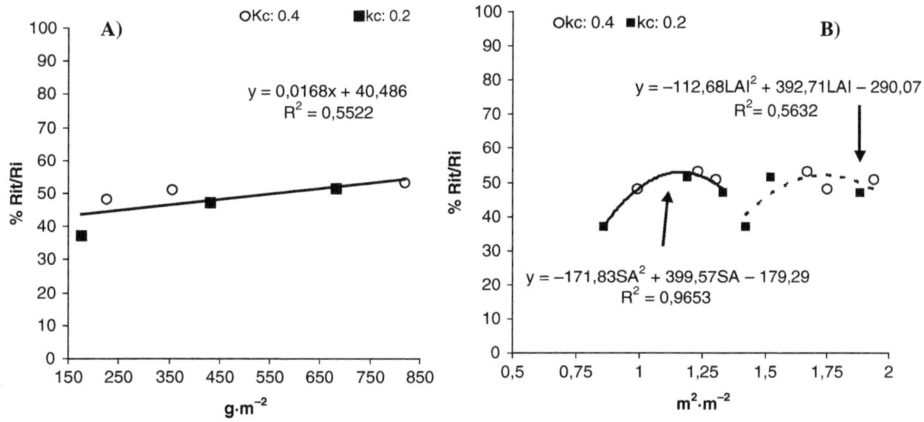

Fig. 2.1 Relationships between the percentage of incident radiation intercepted by the vine and: (**a**) the aboveground bio-mass produced ($g{\cdot}m^{-2}$), and (**b**) the leaf area index (LAI, $m^2{\cdot}m^{-2}$) and external leaf area (SA, $m^2{\cdot}m^{-2}$). Data from cv. Tempranillo and cv. Cabernet Sauvignon, Madrid, 2003–2005

Zufferey and Murisier 1997, Intrieri et al. 1998, Bergqvist et al. 2001, Spayd et al. 2002, Poni et al. 2003).

Radiation balance is the result of the interaction between the leaf area and the radiation that crosses it, and it is conditioned by various factors that affect one another, e.g. astronomical factors (the position of the sun, which is determined by its height and azimuth which in turn depend on latitude, day of year and time of day), meteorological factors (cloud cover affects the amount of available radiation and alters the ratio between direct and diffuse radiation), geometrical factors (the shape and dimensions of the canopy) orientation and separation between the rows, external and total leaf area, albedo of the soil and of the leaves, cultivation techniques ...

Methods for calculating the daily PAR balance in heterogeneous crops (those grown in rows) such as vines use an approximation to the canopy geometry and, around that approximation, the balance of homogeneous crops (those that only differ in the horizontal plane) is applied. For example, Riou et al. (1989) and Smart (1973) identify the perimeter of the canopy to the nearest geometric shape – e.g. a parallelepiped in a vertical shoot positioned (VSP) trellis – and assume that it is regularly distributed in the space. In these shapes, light interception model for homogeneous canopies is applied. Monsi and Saeki (1953) observed that light gradually diminished as solar bean passed through leaf layers across the canopy. The analytical expression of this fact is the extinction coefficient (k) in the Lambert-Beer law (Monsi and Saeki 1953).

In a vine or in a vineyard we find, on the one hand, exposed leaves that directly intercept solar radiation and, on the other, leaves not directly exposed to radiation but which receive light energy from diffuse radiation, that is the one reflected by the soil and by the vegetation itself, and also the radiation transmitted through the directly illumineted leaves. Sometimes, for specific locations, computer models are used to simulate the behavior of direct and diffuse radiation within the canopy and within the training system applying the laws of radiation (Monsi and Saeki 1953, McAdams 1954, Nilson 1971). However, on clear days, the main components of the daily PAR balance are the direct radiation that is absorbed by the vertical and horizontal planes of the canopy, the components of the diffuse solar radiation being less important (Smart 1973).

Absorption of radiation by grapevine canopy increases with leaf area up to a value where the shading of some leaves by others reduces the interception (Marcelis et al. 1998). Assuming that leaf area is randomly distributed within the canopy, the extinction of radiation can be expressed by the Lambert-Beer law. Russell et al. (1989) express this law as: $Rt/Ri = e^{(-K \cdot LAI)}$, where Rt is transmitted radiation, Ri is incident radiation, K is the extinction coefficient and LAI the Leaf Area Index. This coefficient K mainly depends on the optical and geometrical properties of the canopy and can be calculated through the average PAR daily balance or with data taken at a specific hour of the day. Oliveira and Santos (1995) provide an easy measure of this coefficient: $K = 2\,(Ah/Ap)$, where Ah is the area of the shade cast by the canopy over a horizontal surface and Ap is the external leaf area or surface perimeter of the canopy.

In order to measure the components of the PAR balance in the field, single radiation sensors, lineal sensors or ceptometers, pyranometers, hemispheric photography, thermopiles are used or, simply by measuring of the shadow cast by the canopy.

If no other factor is limiting, as PAR increases photosynthesis increases too until it reaches the saturation point (Fig. 2.2). In the curves describing the relationship between photosynthesis (A) and light (PAR), three different regions can be seen. The first is the light-dependent region, where the limiting factor is the light and the correspondence between PAR and A is linear. Second, a transition region appears

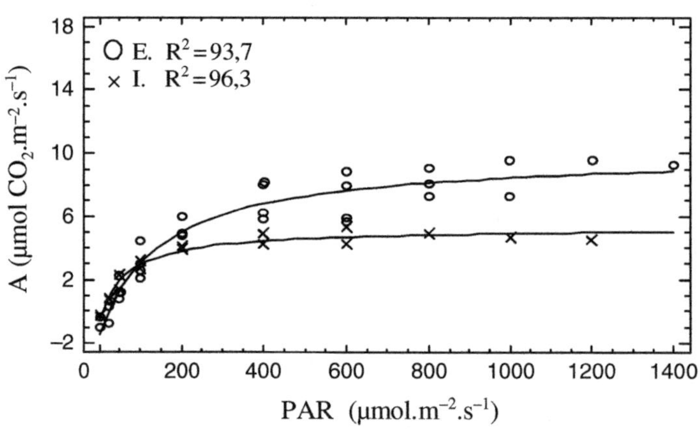

Fig. 2.2 Photosynthetic response curves (A, μmol $CO_2 \cdot m^{-2} \cdot s^{-1}$) to light (PAR, μmol$\cdot m^{-2} \cdot s^{-1}$) of cv. Cabernet Sauvignon leaves, external (E) and internal (I), at half position of canopy (Sánchez-de-Miguel 2007)

where there is a gradual curvature in the relation A/PAR. The last region corresponds to light-saturation, when the internal concentration of CO_2 in the leaves is the limiting factor and when a large increase in PAR is accompanied by a small increase in A. Therefore, the higher the Ra or Rit, the higher the A. However, two aspects must be taken into account: the photoinhibition and the absorption spectrum of the leaf. When PAR is absorbed above the light-saturation point (Is), A cannot handle all the energy absorbed by chlorophyll and other photosynthetic pigments, and there is an excess of energy that can trigger oxidative reactions of these pigments, which can destroy them (photooxidation), or can cause an excessive increase in the leaf temperature. In order to reduce these collateral effects, the plants decrease their photosynthetic potential by photoinhibition.

Regardless of the value of Ri, when radiation is intercepted, the leaves reflect a little more than 10%, transmit 9% and absorb the remaining 81% (Champagnol 1984). In this way, if the first layer of leaves receives 100% of Ri, the second layer receives between 5 and 10% and the third layer between 1 and 3%. Of radiation absorbed by leaves, 25% is emitted as infrared radiation and 75% warms the leaves and then is dissipated. The spectra of absorption, reflection and transmission of radiation varies with leaf thickness, age, water content, surface morphology and orientation, and only 1% of the energy received is used in photosynthesis.

The fraction of Ri absorbed per unit of leaf area mainly depends on the optical and geometrical properties of the canopy: number of leaves, position, angle and how the leaves cover the space leaving gaps between them. For example, the radiation throughout the day on an horizontal surface (e.g. overhead canopy or the soil surface in a flat land) follows a curve with a maximum at midday, as in the traditional grapevine bushes. However, a vertical surface (e.g. a VSP trellis) with North–South orientation shows an absorption curve with a relative minimum at noon, when the height of the sun is maximal, and two maxima (one before and another after midday) that coincide with the maximual value of the cosine of the angle of incidence of direct radiation on both vertical sides of the canopy. The orientation of the rows affects the interception of radiation only in the narrow and vertical geometries. Thus, the curve of Ra for a VSP trellis orientated east–west has the same shape as that of an horizontal surface (Fig. 2.3). In trellis systems such as VSP, the extinction coefficient is higher due to a compression of vegetation by the wires of the trellis which increases leaf density. Row spacing determines the proportion of bare soil and therefore the amount of radiation that is lost by soil absorption. Without changing the density of plantation, Rit increases when row spacing decreases and the distance between plants within the row increases. The height of the canopy improves the interception of radiation. In N–S orientated VSP trellis systems, decreasing row spacing and increasing the height of the canopy are complementary actions. Wide alleys are inefficient from the point of view of radiation interception, but the very narrow ones, as in the case of high canopies, can cause shade from one line to the other. Smart (1987) suggested a relationship between canopy height and row spacing of 1:1 and Schneider (1992) recommends a 0.8 ratio. The division of the canopy (e.g. Geneve Double Cordon – GDC –, Lyra) improves radiation interception, due to the increase of direct exposure and the thinning of the canopy density.

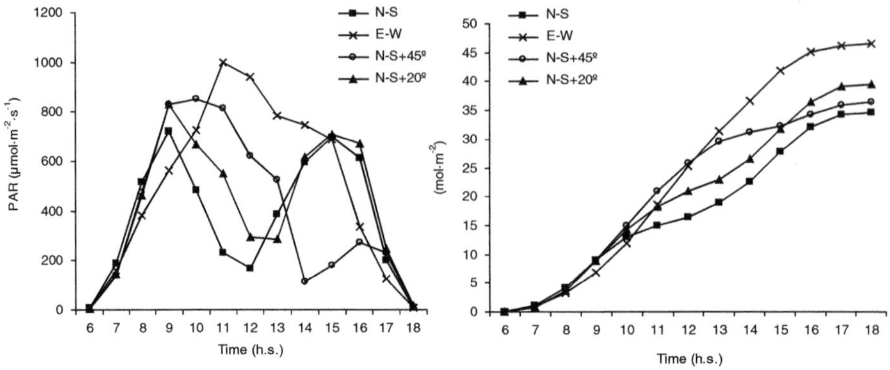

Fig. 2.3 Daily evolution of hourly (*left*) and accumulated (*right*) Rit in cv. Cabernet Franc according to the row orientation: North–South (NS), East–West (EW), Northeast–Southwest (N+45) and North–South plus 20 degrees to the east (N+20) (Jiménez et al. 2007)

Lastly, crop management (irrigation, soil management, canopy management) affects Ra and/or Rit due to its influence on leaf development and in the geometrical and optical characteristics of the leaves.

2.2 Materials and Methods

To determine the balance of radiation in our vineyard we must take data in 3 or 4 zones of the plot chosen at random (which we will call replications). The optimal number of replications depends on the uniformity of the plot (the less uniform the more number of replications), but in any case the balance of radiation will be carried out sampling the space between two vines of adjacent rows. The selected vines will present a uniform development that is representative of the repetition (i.e., the most common), where the chosen training system is maintained (goblet, sprawl, VSP, etc.).

In order to take the measurements a linear PAR sensor – or ceptometer – 1 m length (LI-191SA, LI-Cor, Lincoln, Neb.), a datalogger (LI-1000) and a fixed and leveled metallic structure as shown in Fig. 2.4, are used. However, if the ground is flat, we can miss out the fixed structures (taking care, in the case of lanes with cover crops, of the shadow cast on the ceptometer by elements that are not part of the grapevine canopy).

The methodology is based on Varlet-Grancher et al. (1989), considering a the following series of approximations and assumptions: (1) measurements are taken on clear days, therefore diffuse radiation is not considered; (2) leaves are distributed randomly within the canopy so that the Lambert-Beer law can be applied (Rt / Ri $= e^{(-K \cdot LAI)}$); (3) space between the soil level and the top of the vine is divided into three horizontal planes: one at the top of the canopy (the vine top), another at its lowest point, and the third at ground surface.

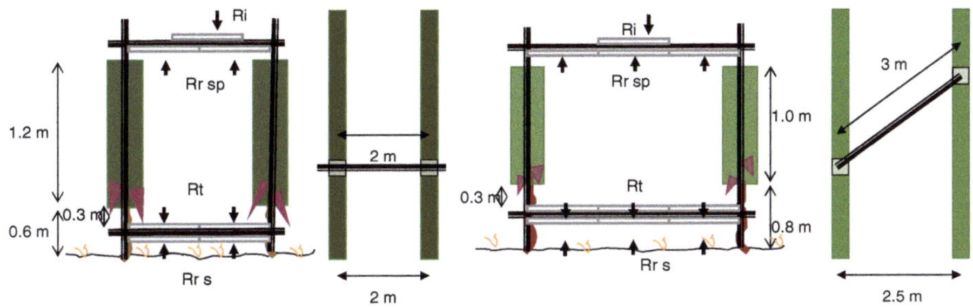

Fig. 2.4 Front and lateral view of the fixed metal structures and ceptometer positions for the calculation of the PAR balance in rows with a width multiple (e.g. 2 m) and not multiple (e.g. 2.5 m) to the length of the ceptometer (1 m). Ri: intercepted radiation; Rt: transmited radiation; Rrs: radiation reflected by soil; Rrsp: radiation reflected by soil and plant

There are four components of the balance of radiation in the canopy:

Ri, incident radiation
Rt, radiation transmitted
Rrs, radiation reflected by the soil
Rrsp, radiation reflected from soil and plant

These are determined by placing the ceptometer at the top of the vines upwards (Ri), downwards (Rrsp), at the lowest point of the vine – 30 cm – upwards (Rt) or downwards (Rrs) as shown in Fig. 2.4.

Before collecting the data the measurement path and the order in which the ceptometer will be positioned in the structures has to be established. For example, to

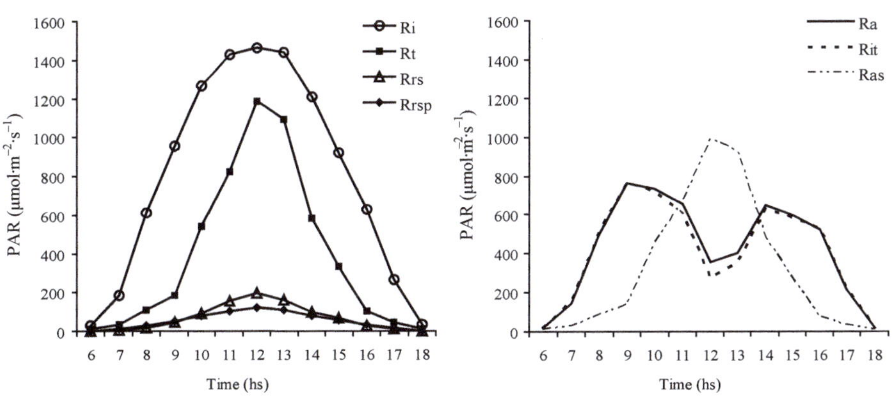

Fig. 2.5 Example of the daily evolution of the four components of absorbed daily PAR balance: Ri, Rt, Rrs, Rrsp, and the evolution of absorbed (Ra) and intercepted (Rit) PAR balance by the vine (Ra and Rit) and the absorbed by soil (Ras). Data from cv. Tempranillo in mid-maturation stage trained on a VSP trellis orientated North to South (Sánchez-de-Miguel 2007)

determine the absorbed radiation: firstly Ri, secondly Rrsp, thirdly Rt and finally Rrs, and, in each of these components, sampling from the left hand side of the structure to the right hand side. If we want to determine the intercepted radiation, then firstly Ri and secondly Rt sampling from left to right. In this case, instead of measuring Ri in each replication it can be measured 3 or 4 times hourly, placing the ceptometer on a flat surface clear of shadows (having previously confirmed that it is levelled).

After collecting all the data, the averages of Rt, Rrs, Rrsp are calculated, which in the field measurements are taken for 2 or more samples per position (Fig. 2.5). With these averages from the repetitions we obtain the following balances (Fig. 2.5) and coefficients:

Radiation absorbed by the vine	$Ra = (Ri + Rrs) - (Rt + Rrsp)$
Radiation intercepted by the vine	$Rit = Ri - Rt$
Radiation absorbed by the soil	$Ras = Rt - Rrs$
Transmission coefficient	$T = Rt/Ri$
Plant and Soil reflexion coefficient	$R = Rr(sp)/Ri$
Coefficient of reflexion of soil on the crop	$Rs = Rr(s)/Rt$
Absorption efficiency (%)	$Ea = (1 - T - R + Rs) \times 100$
Interception efficiency (%)	$Ei = (1 - T) \times 100$
Extinction coefficient, K	$T(\%) = e^{(-K\,LAI)}$

And so the percentages of the incident radiation which is absorbed or intercepted by the plant (%Ra/Ri; %Rit/Ri) or soil (%Ras/Ri).

These balances and ratios can be calculated by the hourly or accumulated data. The accumulated data throughout the day is obtained by integrating the daily evolution curves of the balance of radiation.

To determine the daily PAR balance on a clear day, measurements are taken every hour from sunrise to sunset.

The method herein proposed is up to all canopy shapes and alley spacing. It is very fast to perform and accurate. The simplest data collecting method just requires Rt to be measured as Ri can be download from a weather station. Undirected practical uses that outcome from the PAR balance are irrigation management and estimation of potential yield.

References

Bergqvist J, Dokoozlian N, Ebisuda N (2001) Sunlight exposure temperature effects on berry growth and composition of Cabernet-Sauvignon and Grenache in the Central San Joaquin Valley of California. Am J Enol Vitic 52:1–7

Carbonneau A (1980) Recherche sur les systèmes de conduite de la vigne: essai de maitrise du microclimat et de la plante entiere pour produire économiquement du raisin de qualité. I.N.R.A. Bordeaux I.N.R.A: 234

Champagnol F (1984) Photosynthèse. Eléments de physiologie de la vigne et de viticulture générale. Dehan, Montpellier, pp 40–66

Dokoozlian N, Kliewer WM (1995) The light environment within grapevines canopies. I. Description and seasonal changes during fruit development. Am J Enol Vitic 46:209–218
Intrieri C, Poni S, Rebucci B, Magnanini E (1998) Row orientation effects on whole canopy gas exchange of potted and field grown grapevines. Vitis 37:147–154
Jiménez L, Sánchez-de-Miguel P, Pedrosa R, de la Fuente M, Baeza P (2007) Influencia de la orientación de las filas en el microclima térmico de las hojas y la radiación fotosintética-mente activa interceptada por la vid (*Vitis vinifera* L. cv Cabernet franc) durante el periodo de maduración. Actas Congreso Clima y Viticultura. Conclivit I:119–125
Marcelis LFM, Heuvelink E, Goudriaan J (1998) Modelling biomass production and yield of horticultural crops: a review. Sci Hort 74:83–111
McAdams WH (1954) Heat transmission. McGraw-Hill, New York, London, p 532
Monsi M, Saeki T (1953) Über den Lichtfaktor in den Pflanzengesellschaften und seine Bedeutung für die Stoffproduktion. Jpn J Botany 14:22–52
Nilson T (1971) A theoretical analysis of the frequency of gaps in plant stands. Agric. Meteorol. 8: 25–38
Oliveira M, Santos M (1995) A semi-empirical method to estimate canopy leaf area of vineyards. Am J Enol Vitic 46:389–391
Poni S, Magnanini E, Bernizzoni F (2003). Degree of correlation between total light interception and whole-canopy net CO_2 exchange rate in two grapevine growth systems. Aus J Grape Wine Res 9:2–11
Riou C, Valancogne C, Pieri P (1989) Un modèle simple di'nterception du rayonement solaire par la vigne-vérification expérimentale. Agronomie 9:441–450
Robinson TL, Lakso AM (1991) Bases of yield and production efficiency in apple orchard systems. J Am Soc Hort Sci 116:188–194
Russell G, Jarvis P, Monteith J (1989) Absortion of radiation by canopies and stand growth. In: Russell G, Marshall B, Jarvis P (eds) Plant canopies: their growth, form and function. Cambridge University, New York, pp 21–39
Sanchez-de-Miguel P (2007) Producción y distribución de fotoasimilados en la vid (*Vitis vinifera* L.) durante el periodo de maduración. Cambios en la respuesta fotosintética a la luz de las hojas por factores biológicos, ambientales y culturales. Producción Vegetal: Fitotecnia. Madrid, Universidad Politécnica de Madrid, p 415
Schneider C (1992) Quelles techniques de conduite adopter pour favoriser la qualité et maitriser la production. Sitevinitech, Bordeaux
Schultz HR (1995) Grape canopy structure, light microclimate and photosynthesis I. A two dimensional model of the spatial distribution of surface area densities and leaf ages in two canopy systems. Vitis 34:211–215
Smart RE (1973) Sunlight interception by vineyards. Am J Enol Vitic 24:141–147
Smart RE (1987) The light quality environment of vineyards. OIV 3° Symposium International sur la Physiologie de la Vigne. Bordeaux, pp 370–373
Spayd S, Tarara J, Mee DL, Ferguson JC (2002). Separation of sunlight and temperature effects on the composition of *Vitis vinifera* cv. Merlot berries. Am J Enol Vitic 53:171–182
Varlet-Grancher C, Gosse G, Chartier M, Sinoquet H, Bonhomme, Allirand JM (1989). Mise au point: Rayonnement solaire absorbé ou intercepté par un couvert vegétal. Agronomie 9:1–21
Zufferey V, Murisier F (1997). Orientation des rangs de vigne et interception de l'energie lumineuse par le feuillage. Revue suisse de viticulture, arboriculture and horticulture 29:239–243

Chapter 3
Vegetative Development: Total Leaf Area and Surface Area Indexes

Patricia Sánchez-de-Miguel, Pilar Baeza, Pedro Junquera, and José Ramón Lissarrague

Abstract Canopy management determines canopy shape and spatial leaf area distribution which in turn determines vineyard productivity. There are two indexes evaluating vineyard productivity which involve leaf development: total leaf area – LAI – and external leaf area – SA –. The first one refers to total leaf area developed per m^2 of soil while SA refers to the external leaves, assuming that most of photosynthesis – 90% – is carried out by those leaves. This chapter aims to provide a feasible methodology to calculate both LAI and SA under different training systems and cultivars in order to predict vineyard productivity or/and to make decisions along the season. Relations between main leaf nerve length (cm) and leaf surface area (cm^2) are given for Airén, Albariño, Barbera, Cabernet franc, Cabernet sauvignon, Chardonnay, Godello, Garnacha, Graciano, Mencía, Merlot, Petit verdot, Pinot noir, Semillon, Syrah, Tempranillo, Treixadura, Verdejo and Viognier.

Keywords Leaf area index · Surface area · Potential yield · Training system

Contents

P. Sánchez-de-Miguel (✉)
Departamento de Producción Vegetal: Fitotecnia, Grupo de investigación en Viticultura. ETSI Agrónomos, Universidad Politécnica de Madrid, Madrid, 28040, Spain
e-mail: patricia.sanchez@upm.es

S. Delrot et al. (eds.), *Methodologies and Results in Grapevine Research*,
DOI 10.1007/978-90-481-9283-0_3,

Abbreviations

A	Area of the average leaf (e.g. cm^2)
ALA	Average leaf area per shoot (e.g. cm^2 $shoot^{-1}$)
Area	Individual leaf area (e.g. cm^2)
C	Average contour canopy length (e.g. m)
CW	Average canopy width (e.g. m)
D	Distance between vines along the row (e.g. m)
ET_0	Reference evapotranspiration (e.g. mm day^{-1})
g	Generatrix line, which is the length between the vertex (trunk at ground level) and the perimeter point of the circumference of the base in a inverted cone shaped canopy bush (e.g. m)
G	Length of gaps between plants (e.g. m)
GDC	Geneva double-curtain
H	Height of the cap head-trained bush or average canopy height of VSP (m)
L	Area of the largest leaf (e.g. cm^2)
LN	Main nerve length (e.g. cm)
NL	Number of leaves
R	Radius of the sphere canopy bush or radius at base of inverted cone shaped canopy bush (e.g. m)
S	Area of the smallest leaf (e.g. cm^2)
Suffix 1	Main leaf
Suffix 2	Lateral leaf
VSP	Vertical shoot-positioned
W	Row spacing (e.g. m)

3.1 Introduction

Until the early 80s the response of the vineyard (yield, pruning weight, ºBrix, etc.) was studied using simple factors (load, plant density, vine spacing, etc.), as if their effects were independent. Later, some researchers (Smart et al. 1982, Carbonneau and Huglin 1980, Intrieri 1987), in the light of previous work (Shaulis et al. 1966, Shaulis and May 1971, Smart 1973), rethought the study of the response of the vineyard as the effect of the canopy management as a whole. Thus, it arise the need to define new parameters that characterize grapevine canopy shape and could be used to explain vineyard capacity and its relationship with potential yield and must composition.

Among these parameters, there are two indexes explaining the vineyard potential productivity, the Leaf Area Index (LAI) and the Surface Area (SA). These indicators refer to leaves on which the net photosynthesis of the plant depends and, thus, the overall productivity of the vineyard (Schneider 1992, Smart 1973). On the other hand, they indicate the spatial distribution, which conditions the bunch microclimate, a factor that determines the quality characteristics of the harvest.

When the canopy is very porous – more than 50%, at the beginning of the growth cycle, plant photosynthetic activity mainly depends, all other factors being equal, on total leaf area (LAI). As the growing cycle progresses, the system becomes less porous, leaves crowd together and competition among them, for space and light arises. As a consequence, the overall photosynthesis of the plant depends, between 80 and 95%, on the outer leaves, i.e. the surface leaf area (SA) (Smart 1973, Williams et al. 1987, Carbonneau 1991). The overlapping of leaves begins when the number of shoots per meter of cordon is higher than 15 in cv. Gewurztraminer (Smart 1987) and between 10 and 12 shoots per meter of row in varieties with both medium sized (180–200 g/bunch) and large bunches (> 200 g/bunch), such as Tempranillo, Garnacha, Bobal, Graciano, Viura, Cabernet sauvignon, etc.

Total leaf area developed by the vine is defined as the Leaf Area Index (LAI) which is the ratio between total leaf area of the plant and vine spacing (Champagnol 1984, Carbonneau 1989). Surface Area (SA) is the canopy surface that defines the contours of the plant. Both parameters are expressed as $m^2\ m^{-2}$ (m^2 leaf area m^{-2} soil).

The leaf area follows an annual growth model similar to that of the shoots, because the vine is a deciduous plant (Johnson and Lakso 1985, Schultz 1992, Singh et al. 1994). At the beginning of the cycle, the shoots sprout from axillary buds formed during the previous season which contains a certain number of nodes, inter-nodes and preformed inflorescence primordia. Shoot growth in the current season is a combination between fixed growth and free growth. Fixed growth refers to the elongation of internodes and the expansion of leaves which were pre-formed leaves in the dormant bud, it accounts for up to 12 of the first-produced nodes. Free growth is the result of the elongation and production of new leaf primordia in the apical meristem activity (Mullins et al. 1992). Growth is maintained while the terminal meristem is active. In turn, the prompt buds of the afore mentioned shoots can develop generating new leaves in summer or lateral shoots.

The main causes of differences in final leaf area development are water availability (Hsiao 1973, Matthews et al. 1987, Williams and Matthews 1990, Schultz and Matthews 1993) and the length of the growth period (Schultz 1992).

The unfold of the leaves from the main shoot usually ends before flowering or at around 300 degree-days after budding (Williams 1987, Wermelinger and Koblet 1990, Miller and Howell 1996, Sipiora 2005); after flowering, average leaf area (cm^2) of those leaves undergoes little change.

Vine leaves are sinks of carbohydrates until they reach 50–80% of their final size (Koblet 1969); in this way, the shoots are "parasites" until they reach a leaf area of 50 cm^2 (Buttrose 1966). When the leaves are 95% of their final size 30–35 days after their appearance, they reach maximal photosynthetic activity which is maintained for 40–45 days, after which it decreases smoothly (Poni et al. 1992). However, photosynthesis is still high in the 100–140 days that follow leaf expansion (Intrieri et al. 1992, Schultz 1996, Zufferey 2000).

For optimal quantitative and qualitative ripening, the canopy should develop its leaf area quickly to reach the optimal values (balanced) avoiding competition between shoot growth and berry ripening. The values reported in the literature are highly variable and depend on many factors: climate, soil, variety, rootstock, planting density, canopy height, fertilization, water regime, etc. (Tables 3.1 and 3.2).

There are two main periods where the role of the leaf area is a key factor: fruit set and ripening of skin-pulp-seed. During the cycle, the health and efficiency of the canopy must be ensured, particularly during berry development, avoiding early ageing which jeopardizes ripening.

SA and LAI values reported largely range depending on training system and vineyard management, primarily the irrigation and trimming. Intrieri (1987) obtained 1.8 m^2 m^{-2} in GDC (Geneve Double Curtain) trained cv Trebbiano in the Po Valley (Italy). Martínez de Toda et al. (1991) reported 0.92–1.53 m^2 m^{-2} in bush-trained, unirrigated cv Garnacha in La Rioja (Spain), while on a VSP trellis, SA reached values of 1.08 m^2 m^{-2}. Schneider (1992) determined the SA of various training systems ranging between 0.67 m^2 m^{-2} for a VSP with a row spacing of 3.0 m and 1.16 $m^2 m^{-2}$ for a lyre, or 1.2 m^2 m^{-2} in a GDC of 3.5 m row spacing and 1.8 m canopy height. Baeza et al. (2005) obtained a SA of 1.2 m^2 m^{-2} in bushes with 3.0 m row spacing and, 1.7 $m^2 m^{-2}$ in a VSP with 2.0 m row spacing, both irrigated.

Table 3.1 Examples of optimal leaf area values reported in the literature

	Reference	Observations
LAI ($m^2 \cdot m^{-2}$)		
2 .0– 4.0	Champagnol (1984)	In cultivated or spontaneous vegetation. Sometimes up to 6 – 8 m^2 m^{-2}
2.0 – 6.0	Reynier (2001)	In vineyard
2.0	Smart (1987) and Champagnol (1993)	Regular layout, without large areas of bare soil nor leaf crowding, in particular in the area of renovation
Leaf area/yield ($m^2 \cdot kg^{-1}$)		
1.2	Hunter (2000)	
0.8 – 1.2	Smart and Robinson (1991)	< 0.5 m^2 kg^{-1} if vigour is low y > 2 m^2 kg^{-1} if vigour is high
0.8 – 1.5	Baeza and Lissarrague (2000)	
0.7 – 1.4	Dokoozlian and Hirschfelt (1995), Kliewer and Antcliff (1970), Kliewer and Weaver (1971)	
0.8 – 1.0	Jackson and Lombard (1993)	
Approximations		
Length of shoot = 1.2 – 1.6 m	Cloete et al. (2006)	It has sufficient leaves for the correct ripening of bunches, if said leaves are well exposed to sunlight
10 – 14 leaves per 200 g of yield	Koblet (1975)	

Table 3.2 Optimum values in relation to leaf area grouped into three levels of vigor (Smart and Robinson 1991)

	Vigor		
	Low	Moderate	High
Main shoot leaf area (cm^2)	< 80	20–160	> 180
Lateral shoot leaf area (cm^2)	< 25	30–40	> 50
Length of shoot without topping (cm)	50	100	> 200
Number of nodes main shoot (without topping)	< 0	15–20	> 25
Internode length (cm)	< 5	6–8	> 8
Number of lateral shoots per main shoot	< 3	3–5	> 8

Sánchez-de-Miguel (2007) obtained SA values of 0.99 m^2 m^{-2} in cv. Cabernet Sauvignon with a 2.5 m spacing between rows, and between 1.2 and 1.4 m^2 m^{-2} in cv Tempranillo with 2.0 m spacing between rows.

Various trials (Fernández et al. 1977, Smart et al. 1982, Carbonneau 1989, Kliewer and Dokoozlian 2005) have shown that 0.9–1.5 m^2 of exposed leaf area are necessary to ensure proper ripening for 1 kg of grapes. This value depends on water availability, practices, bunch microclimate, etc. An increase in this ratio does not necessarily result in a better quality harvest (Kliewer and Dokoozlian 2005, Murisier et al. 2007).

3.2 Practical Uses of LAI and SA

The main aim in determining LAI and SA is to find out the potential productivity of a training system. Other objectives are:

- To determine the potential maximal harvest that can be obtained under given conditions, and ensuring a proper maturation of the grape. In situations of excess fertility it would be possible to predict, with sufficient time, the need for bunch thinning at flowering, and in unusually cool years, it would be possible to determine the need for removal of bunches at veraison.
- Knowledge of the porosity (LAI/SA). This is an index of susceptibility to diseases such as powdery and downy mildew and Botrytis Bunch Rot. Depending on the measured value of this index and the climatic situation, the need for leaf removal or fungicide treatment can be predicted.
- Before planting, when setting performance targets per meter of row or hectare of the cultivar, it obliges the design of the plantation with an adequate SA, thus fixing the row spacing and canopy height.

3.3 Materials and Methods

3.3.1 *Leaf Area Index*

To calculate the LAI of a vineyard, data need to be recorded on a representative sample of shoots. Recommended sample size depends on vine load; it is recommended to sample at least 30% of the load of two vines in 3 or 4 areas of the plot randomly chosen. For example, for a load of 14 shoots, 4 shoots of each plant for each designated area or single plot are sampled. Eight shoots in each area of the plot are thus measured, which makes a total of 24–32 shoots per plot. These shoots should represent the average observed development in the plot.

If the evolution of the LAI must be followed, the shoots must to be tagged in the field with a resistant material, from the point when they reach 10–15 cm in length.

In these shoots, we take 3 different measurements in the main shoot (1) and another three in its laterals (2): main nerve length in the largest leaf (L) and the smallest leaf (S) of the main shoot ($NerveLL_1$; $NerveLS_1$) and those of the laterals ($NerveLL_2$; $NerveLS_2$) and the number of leaves (NL_1 and NL_2). Altogether we obtain 6 parameters per shoot.

For a given shoot, whether it is a main or lateral, the data regarding the length of the main nerve (LN, cm) is converted to an area (cm^2) by using the relationships between both variables (Table 3.3) obtained when the shoots were completely developed.

Table 3.3 Relationship between leaf area (cm^2) and main leaf length (LN, cm)

Cultivar	$Area_{leaf} = f$ (LNleaf)	R^2
Airén	$Area_{AI} = -10.23 + 2.85 \times LN + 1 \times LN^2$	0.95***
Albariño	$Area_A = 18.9 \times LN - 62.22$	0.90***
Barbera	$Area_B = 23.75 \times LN - 110.3$	0.88***
Cabernet Franc	$Area_{CF} = 20.10 \times LN - 72.07$	0.93***
Cabernet Sauvignon	$Area_{CS} = 0.38 + 1.21 \times LN^2$	0.93***
Chardonnay	$Area_{CH} = -0.07 - 1.73 \times LN + 1.39 \times LN^2$	0.98***
Garnacha	$Area_G = -2.74 + 0.99 \times LN + 1.04 \times LN^2$	0.87***
Godello	$Area_{GO} = 19 \times LN - 59.16$	0.88***
Graciano	$Area_{GR} = 25.66 \times LN - 114.02$	0.91***
Mencía	$Area_{MN} = 20.07 \times LN - 69.71$	0.88***
Merlot	$Area_M = 18.291 \times LN - 58.452$	0.86***
Petit Verdot	$Area_{PV} = 18.615 \times LN - 80.167$	0.94***
Pinot Noir	$Area_{PN} = 21.06 \times LN - 69.26$	0.91***
Semillon	$Area_{SE} = 26.05 \times LN - 106.11$	0.90***
Syrah	$Area_S = 21.41 \times LN - 75.409$	0.94***
Tempranillo	$Area_T = 20.306 \times LN - 69.302$	0.93***
Treixadura	$Area_{TR} = 22.13 \times LN - 83.39$	0.91***
Verdejo	$Area_{VJ} = 1.55 \times LN^2 - 2.81 \times LN + 7.01$	0.92***
Viognier	$Area_V = 20.36 \times LN - 71.94$	0.93***

R^2: correlation coefficient. ***: Significant at $p < 0.001$. Sample size was always more than 50 leaves.

In this way, the area of the largest (L_1) and of the smallest (S_1) main leaf as well as the area of the largest (L_2) and of the smallest (S_2) lateral leaf may be calculated. Through these measurements, the average leaf area are obtained as $A = (L + S)/2$, both in the main leaf (A_1) and the lateral leaf (A_2). Units are cm^2 $leaf^{-1}$.

Through the number of leaves (NL) the average leaf area per shoot is obtained as $ALA = A \cdot NL$, so that the average leaf area due to the main shoots (ALA_1) and laterals (ALA_2) can be differentiated. Separately or together, their units are cm^2 $shoot^{-1}$. Multiplying by shoot load per vine gives the leaf area per vine (cm^2 $vine^{-1}$). By dividing this value by 10,000 we obtained $m^2 \cdot vine^{-1}$. Finally, LAI is obtained by dividing the figure obtained by the vine spacing: m^2 m^{-2}.

3.3.1.1 Calculations

Largest (L) and smallest (S) leaf area (cm^2)	$= f(NerveL_{leaf})$ (see Table 3.3)
Area of the average leaf (A, cm^2)	$= (L + S)/2$
Average leaf area per shoot (ALA, cm^2 $shoot^{-1}$)	$= A \times NL$
Average leaf area per vine (cm^2 $vine^{-1}$)	$= ALA \times Load$
Leaf Area Index (LAI, $m^2 \cdot m^{-2}$)	$= (ALA \times Load)/(10.000 \times$ vine spacing)
Percentage of main shoot leaf area (%)	$= ALA_1(ALA_1 + ALA_2) \times 100$
Percentage of lateral shoot leaf area (%)	$= ALA_2/(ALA_1 + ALA_2) \times 100$

3.3.1.2 Example of Calculation

Cabernet Sauvignon/SO4 in Madrid (Colmenar de Oreja) (Fig. 3.1).
6 year-old vines trained to unilateral cordon Royat, North–South orientated.
80 cm cordon height and 115–130 cm of canopy height.
Vine spacing: 2.5×1.1 m^2 (3636 vines ha^{-1}).
Phenology: ripening (early September).

Bud Load: 14 Buds $Vine^{-1}$
Working with average values of the variables measured over 96 shoots across 24 vines.

Nerve L of L_1 (cm)	9.7
Nerve L of L_2 (cm)	7.1
Nerve L of S_1 (cm)	5.4
Nerve L of S_2 (cm)	3.6
L_1 (cm^2)	115
L_2 (cm^2)	62
S_1 (cm^2)	36

S_2 (cm^2)	16
A_1 (cm^2)	75
A_2 (cm^2)	39
NL_1	19
NL_2	29
ALA_1 (cm^2 shoot^{-1})	1,438
ALA_2 (cm^2 shoot^{-1})	1,383
ALA_{1+2} (cm^2 shoot^{-1})	2,821
Average leaf area per vine (m^2 vine^{-1})	3.95
LAI (m^2 m^{-2})	1.43
LAI_1 (m^2 m^{-2})	0.73
LAI_2 (m^2 m^{-2})	0.70
Percentage of main leaf area (%)	51
Percentage of lateral leaf area (%)	49

Fig. 3.1 Leaf development of cv Cabernet Sauvignon in maturation under two water regimes (unirrigated –**A**– and irrigated 0.45·ET_0 –**B**–). ET_0: reference evapotranspiration (mm·day^{-1}) Madrid, 2005

3.3.2 Surface Area

Surface Area (SA) should be determined in the same plants and at the same phenological stage at which total leaf area was taken. In order to take these measurements a metal or a flexible measuring tape, depending on the training system, is all one needs.

3.3.2.1 Measurement of SA in Bush Vines

Bushes can take many shapes, from a hemisphere or a spherical cap shape to an inverted cone in the case of erected cultivars. In the case of a hemisphere, the average radius is calculated (R) after the measurement, with a tape measure, of 4 diameters on the soil (Fig. 3.2) and also the height of the bush which is also considered a radius. SA is calculated as the surface of a hemisphere (Eq. 1).

$$\mathrm{SA} = \frac{2 \cdot \pi \cdot R^2}{(D \cdot W)} \tag{1}$$

where:

SA: surface area ($m^2\ m^{-2}$)
D: distance between vines along the row (m)
W: row spacing (m)
R: radius of the sphere (m).

If shape resembles a spherical cap of radius R and height H (Fig. 3.3) SA is given by using Eq. (2). The radius can be measured by using the shadow cast by the canopy on the soil and H, which is the maximum height of vegetation measured at the trunk.

$$\mathrm{SA} = \frac{2 \cdot \pi \cdot R \cdot H}{(D \cdot W)} \tag{2}$$

where:

SA: surface area ($m^2\ m^{-2}$)
D: distance between vines along the row (m)
W: row spacing (m)
R: radius of the sphere
H: height of the cap (m).

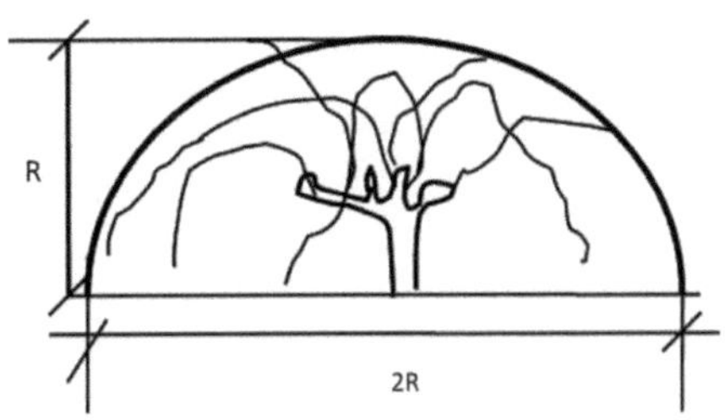

Fig. 3.2 Scheme of a head-trained, hemispherical canopy bush

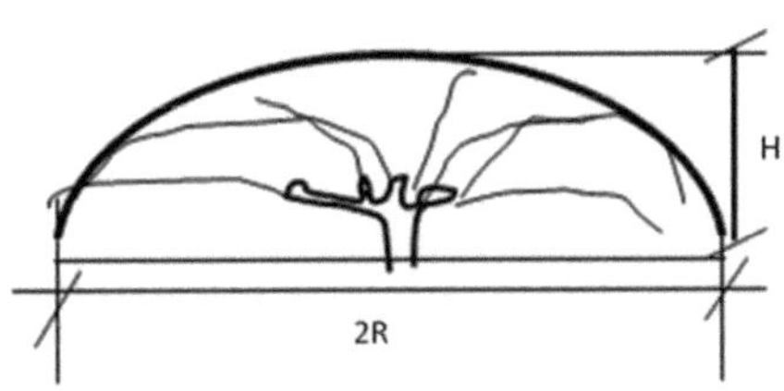

Fig. 3.3 Scheme of a trail growing habit, head-trained bush

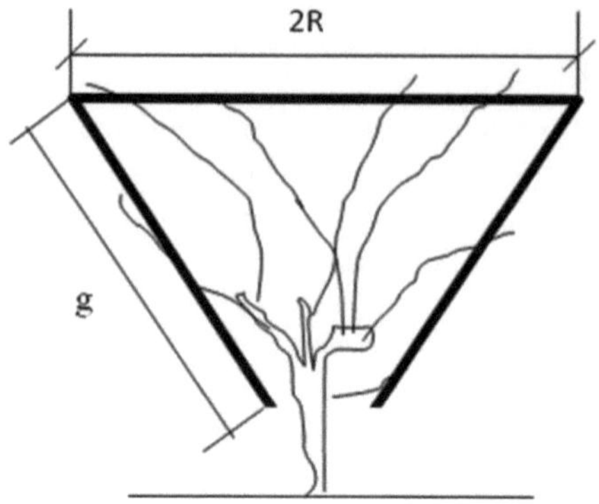

Fig. 3.4 Scheme of an inverted cone shaped canopy bush

In the case of an inverted cone (Fig. 3.4), SA is given by Eq. (3).

$$SA = \frac{(\pi \cdot R^2 + \pi \cdot R \cdot g)}{(D \cdot W)} \tag{3}$$

where:

SA: surface area ($m^2\ m^{-2}$)
D: distance between vines along the row (m)
W: row spacing (m)
R: radius at base of cone
g: generatrix line, which is the length between the vertex (trunk at ground level) and a point at the perimeter of the circumference of the base (m).

3.3.2.2 Measurement of SA in a VSP Trellis

Surface Area of a VSP is calculated as that of a parallelepiped (Fig. 3.5, Eq. 4) in which the photosynthetically active external surfaces are the two lateral sides and the upper part of the canopy. The lower side, although it is external, only receives diffuse radiation. Therefore its photosynthesis is small, and its contribution to the whole of the plant can be disregarded.

In each plant, 5 measurements of width of the upper part (CW) and 5 measurements of canopy height (H) are taken. In some trellises under unirrigated conditions, the canopy is not continuous along the row and so the length of these gaps or windows between plants along the row should be noted (G). We then obtain the average of W, F and G for each plant and calculate SA:

$$SA = \frac{(D - G) \cdot (2H + CW)}{(D \cdot W)} \tag{4}$$

where:

SA: surface area ($m^2\ m^{-2}$)
D: distance between vines along the row (m)
W: row spacing (m)
G: length of gaps between plants (m)
H: average canopy height (m)
CW: average canopy width (m).

It is possible to calculate the row spacing (W) of a vineyard with a given yield target. Assuming we need 1–1.5 m^2 SA/kg of grapes and that G = 0, we can apply the following equation in which D does not appear:

$$\mathrm{SA} = \frac{(2H + CW)10000}{W} \tag{5}$$

We can consider that CW ranges between 0.25 and 0.35 m in a conventional VSP trellis.

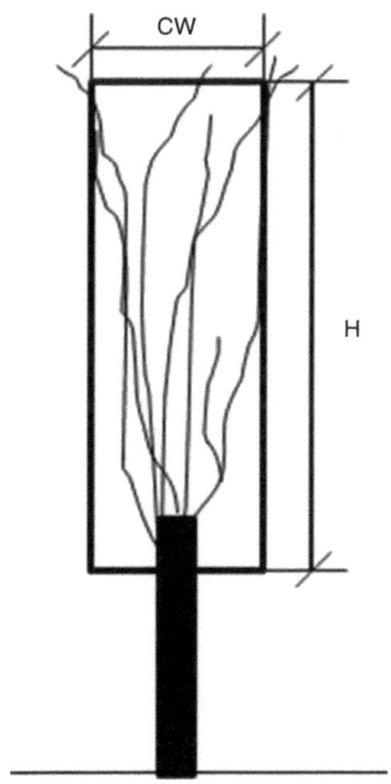

Fig. 3.5 Scheme of a vertical shoot positioned trellis

3.3.2.3 Measurement of SA in Curtains or Sprawl Trellis Systems

In the case of curtains or sprawls (Fig. 3.6) the canopy is not a wall of vegetation easily compared to a geometric shape. Thus a flexible tape is used to measure H. The tape is used to measure the plant contour from one side to the other one. The

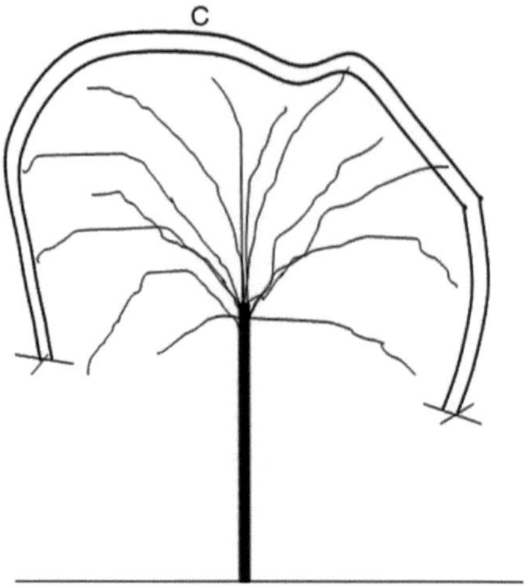

Fig. 3.6 Scheme of a sprawl canopy

measurements are made 4 or 5 times for each plant, and averaged. SA is obtained using Eq. (6).

$$SA = \frac{(D - G) \cdot C}{(D \cdot W)} \tag{6}$$

where:

SA: surface area ($m^2\ m^{-2}$)
D: distance between vines along the row (m)
W: row spacing (m)
G: length of gaps between plants (m)
C: contour canopy length (m).

At least, one leaf area characterization should be made during phase III of berry development, as it is the time when the clusters must ripen.

References

Baeza P, Lissarrague JR (2000) Definición y evaluación de los sistemas de conducción del viñedo. In: La conducción de la vid. Ed. Gobierno de la Rioja. Consejería de Agricultura, Ganadería y Desarrollo Rural

Baeza P, Ruiz C, Cuevas E, Lissarrague JR (2005) Ecophysiological and Agronomic Response of Tempranillo. Am J Enol Vitic 56:129–138

Buttrose M S (1966) The effect of reducing leaf area on the growth of roots, stems and berries of Gordo grapevines. Vitis 5:455–464

Carbonneau A (1989) L'exposition utile du feuillage: définition du potentiel du système de conduite. GESCO N°4, pp 25–47

Carbonneau A (1991) Observation sur vigne: codification des donnés agronomiques. Riv de Vitic et Enol 4:37–45

Carbonneau A, Huglin P (1980) Adaption of training systems to French regions. International Symposium of grapes and wines. U. Davis, California, pp 376–385

Champagnol F (1984). Le mode de conduite. Eléments de physiologie de la vigne et de viticulture generale. Dehan, Montpellier, pp 245–258

Champagnol F (1993) La dimension des baies, facteur de qualité de la vendange. Progrès Agricole et Viticole 110(1):11–16

Cloete H, Archer E, Hunter J J (2006). Shoot heterogeneity effects on Shiraz/Richter 99 grapevines. I. vegetative growth. S Afr J Enol Vitic 27:68–75

Dokoozlian N, Hirschfelt DJ (1995) The influence of cluster thinning at various stages of fruit development on Flame seedless table grapes. Am J Enol Vitic 46:429–436

Fernández J, Balkar J, Meyer LH (1977) Influencia de la iluminación sobre la actividad fotosintética de las hojas de vid cultivada en espaldera. Turrialba 27:3–6

Hsiao TC (1973) Plant responses to water stress. Ann Rev Plant Physiol 24:519–570

Hunter JJ (2000) Implications of seasonal canopy management and growth compensation in grapevine. S.Afr J Enol Vitic 21:81–91

Intrieri C (1987) Experiences on the effect of vine spacing and trellis-training system on canopy micro-climate, vine performance and grape quality. Acta Horticulturae 206:69–87

Intrieri C, Poni S, Silvestroni O, Filippetti I (1992) Leaf age, leaf position and photosynthesis in potted grapevines. Adv Hort Sci 1:23–27

Jackson DI, Lombard PB (1993) Environmental and management practices affecting grape composition and wine quality – A review. Am J Enol Vitic 44:409–430

Johnson RS, Lakso AN (1985) Relationships between stem length, leaf area, stem weight, and accumulated growing degree-days in apple shoots. J Am Soc Hortic Sci 110:586–590

Kliewer WM, Antcliff AJ (1970) Influence of defoliation, leaf darkening, and cluster shading on the growth and composition of Sultana grapes. Am J Enol Vitic 21:26–36

Kliewer WM, Dokoozlian NK (2005). Leaf Area/Crop Weight ratios of grapevines: Influence on fruit composition and Wine Quality. Am J Enol Vitic 56:170–181

Kliewer WM, Weaver RJ (1971) Effect of crop level and leaf area on growth, composition and coloration of Tokay grapes. Annual meeting of the American Society of Enologist, pp 172–177

Koblet W (1969) Migration of assimilates in vine shoots and influence of the leaf surface on grape yield and quality. Wein-Wiss 24:277–319

Koblet W (1975) Wanderung von Assimilaten aus verschiedenen Rebenblättern während der reifephase der Trauben. Wein-Wiss 30:241–249

Martínez de Toda F, Sancha JC, Llop E (1991) Estudio comparado del microclima luminoso en los sistemas de conducción en vaso y espaldera en Rioja. Riv Vitic Enol 4:149–156

Matthews M, Anderson M, Schultz H (1987) Phenologic and growth responses to early and late season water deficits in Cabernet franc. Vitis 26:147–160

Miller DP, Howell GS (1996) Effect of shoot number on potted grapevines. I. Canopy development and morphology. Am J Enol Vitic 47:244–250

Mullins MG, Bouquet A, Williams LE (1992) Developmental physiology: the vegetative grapevine. In: Mullins MG, Bouquet B, Williams LE (eds) Biology of the grapevine. Cambridge University Press, Cambridge, pp 80–111

Murisier F, Zufferey V, Triacca M (2007) Influence de l′ecartement des rangs et de la hauteur de la haie foliaire sur le comportement agronomique et le développement racinaire de la vigne. Revue Suisse Vitic Arboric Hortic 39:361–364

Poni S, Intrieri C, Silvestroni O (1992) Interactions of leaf age, fruiting and exogenous cytokinins in the gas-exchange of Sangiovese grapevines. Quad Vitic Enol Univ Torino 16:233–238

Reynier A (2001) Manual de Viticultura. Ed Mundiprensa (6ª Edición) 497 pp

Sanchez de Miguel 2007. Producción y distribución de fotoasimilados en la vid (*Vitis vinifera* L.) durante el periodo de maduración. Cambios en la respuesta fotosintética a la luz de las hojas por factores biológicos, ambientales y culturales. Doctoral Thesis at Universidad Politécnica de Madrid

Schneider C (1992) Quelles techniques de conduite adopter pour favoriser la qualité et maitriser la production. Sitevinitech, Bordeaux, pp 265–276

Schultz HR (1992) An empirical model for the simulation of leaf appearence and leaf area development of primary shoots of several gravevine (*V. vinifera* L.) canopy-systems. Sci Hortic 52:179–200

Schultz HR (1996) Leaf absorptance of visible radiation in *Vitis vinifera* L: estimates of age and shade effects with a simple field method. Sci Hortic 66:93–102

Schultz HR, Matthews MA (1993) Growth, osmotic adjustment, and cell-wall mechanics of expanding grape leaves during water deficits. Crop Sci 33:287–294

Shaulis NJ, Amberg H, Crowe D (1966) Response of concord grapes to light, exposure and geneva double curtain training. Am J Enol Vitic 89:268–280

Shaulis N, May P (1971) Response of Sultana vines to training on a divided canopy and to shoot crowding. Am J Enol Vitic 22: 215–222

Singh BP, McKinion JM, Sequeira RA, Reddy VR, Pachepsky LB, Acock B, Boote KJ, Pickering NB, DeJong TM, Grossman YL (1994) Modeling photosyntehesis and carbon partitioning. Hortic Sci 29(12):1411–1442

Sipiora MJ (2005) Influencia de la densidad de pámpanos por metro de cordón sobre la variedad de vid (*Vitis vinifera* L) Tempranillo en secano y regadío. Comportamiento agronómico, fisiológico y modelos descriptivos del desarrollo de la vid. Doctoral Thesis at Universidad Politécnica de Madrid, p 256

Smart RE (1973) Sunlight interception by vineyards. Am J Enol Vitic 24:141–147

Smart RE (1987) The light quality environment of vineyards. OIV – 3° Symposium International sur la Physiologie de la Vigne, Bordeaux, pp 370–373

Smart RE, Shaulis MJ, Lemom ER (1982) The effect of concord vineyard microclimate on yield I. The effects of pruning, training and shoot positioning on radiation microclimate. Am J Enol Vitic 33:99–108

Smart RE, Robinson M (1991) Sunlight into wine. A handbook for winegrape canopy management. Adelaide, Winetitles, pp 88

Wermelinger B, Koblet N (1990) Seasonal growth and nitrogen distribution in grapevine leaves, shoots and grapes. Vitis 29:15–26

Williams LE (1987) Growth of Thompson seedless grapevines: I – Leaf area development and dry weight distribution. J Am Soc Hortic Sci 112:325–330

Williams LE, Biscay PJ, Smith RJ (1987) Effect of interior canopy defoliation on berry composition and potassium distribution in Thompson Seedless grapevines. Am J Enol Vitic 38(4):287–292

Williams LE, Matthews M A (1990) Grapevine. Irrigation of Agricultural Crops. Agronomy monograph. 30:1019–1055

Zufferey V (2000) Echanges gazeux des feuilles chez Vitis vinifera L. (cv. Chasselas) en fonction des paramètres climatiques et physiologiques et des modes de conduite de la vigne. Doctoral Thesis at Zurich, Ecole Polytechnique Federale

Chapter 4
Vegetative Growth, Reproductive Development and Vineyard Balance

L.G. Santesteban, C. Miranda, and J.B. Royo

Abstract In the last decades, viticultural research has made important progress due to the implementation of many technical advances to this field. In addition to "classical" viticultural research, it is now possible to analyze vine behaviour with many different approaches, such as plant physiology, genomics, proteomics, precision viticulture and so on. These changes extend our understanding of viticulture, but they may also lead to a certain heterogeneity in the methodology used to estimate the basic characteristics of the vineyards (e.g.: vegetative and reproductive growth). Even worse, measurements of those parameters are sometimes neglected. In order to optimize the relevance and inter-disciplinarity of our work, and its applicability to the vineyard, we should tend toward protocol standardization, the most suitable measurements and ways of expressing results should be determined for each circumstance. This chapter attempts to shed some light on some methodological aspects related to the measurement of vine vegetative and reproductive growth and vineyard balance. We have tried to be comprehensive, but not exhaustive, since some topics are also the subject of other chapters in this book whereas others are beyond its scope.

Contents

L.G. Santesteban (✉)
Departamento de Produccion Agraria, Universidad Pública de Navarra, Pamplona (Navarra) 31006, Spain
e-mail: gonzaga.santesteban@unavarra.es

S. Delrot et al. (eds.), *Methodologies and Results in Grapevine Research*,
DOI 10.1007/978-90-481-9283-0_4,

Abbreviations

LAI	Leaf Area Index
NIR	Near Infrared
SA	External leaf area
TCSA	Trunk cross-sectional area
VPI	Vineyard Potentiality Index

4.1 Measuring Vegetative Growth

4.1.1 Estimation of Annual Shoot Growth

Vegetative growth is one of the main factors conditioning vineyard yield and grape quality, as it limits the photosynthetic potential of a vineyard. Leaf area index (LAI) and external leaf area (SA) are usually measured when this potential needs to be estimated. In Chapter 3, Sánchez-de-Miguel et al. (2010) give a comprehensive review of the methodology that can be used to measure those parameters. Apart from those direct methods, there is a great interest in estimating leaf area through indirect measurements. On one hand, simple parameters such as shoot length, only if secondary shoot length is included (Fig. 4.1a, b), and shoot basal cross-sectional area (Fig. 4.1) allow a cultivar-specific but relatively good estimation of leaf area. Shoot length allows to better estimate the leaf area for a single shoot, but as shoot basal cross sectional area can be measured very quickly over all the shoots in a vine, it might be a better tool to estimate vine leaf area.

On the other hand, there are several methods of LAI estimation based on remote sensing, that use either ground-level field measurements in the visible (Ollat et al. 1998, Johnson and Pierce 2004) and visible + NIR spectra (Drissi et al. 2009), or airborne cameras that take images in the visible + NIR spectra (Johnson et al. 2003, Hall et al. 2008). A more detailed description of the principles and use of these methods is far beyond the scope of this chapter; for more detailed reviews, see Hall et al. (2002) and Proffitt (2006). Pruning weight is also a widely used method for shoot growth estimation since it has been shown to be correlated to leaf area (Weaver and McCune 1960). However, this relation should be used carefully, due to the non-synchronic dynamics of leaf area formation and carbohydrate migration toward the shoots. This is especially true in warm areas, where shoot leaf area is built very early in the season but carbohydrate transport (from and to the shoots) keeps on

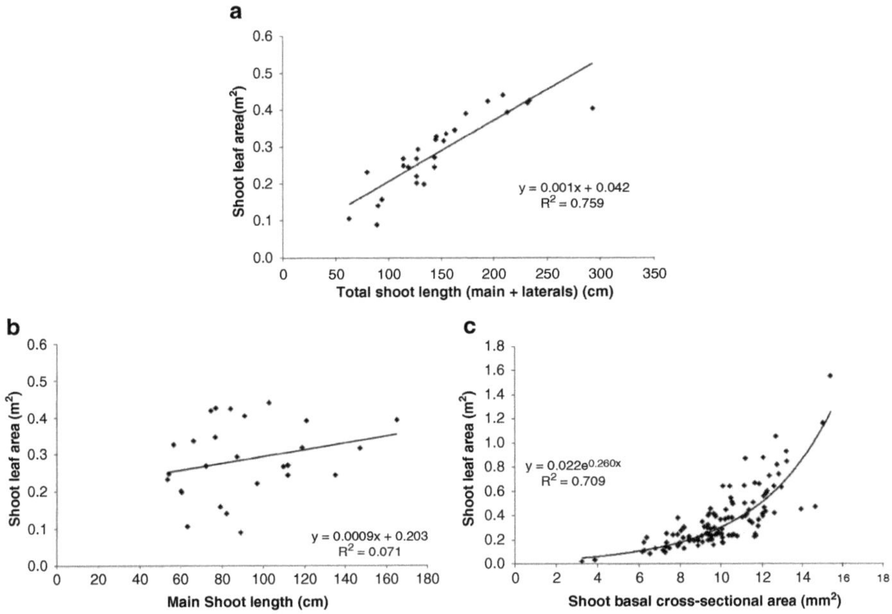

Fig. 4.1 Relation between shoot length and basal cross-sectional area and shoot leaf area in cv. "Tempranillo". (**a**) Relation between total shoot leaf area and main shoot length ($P = 0.046$). (**b**) Relation between total shoot leaf area and total shoot length ($P < 0.001$). (**c**) Relation between total shoot leaf area and shoot basal cross-sectional area ($P < 0.001$). Adapted from Santesteban (2003)

throughout summer and even early autumn, weeks or even months after harvest (Holzapfel and Smith 2007).

4.1.2 Estimation of Annual Vine Growth

Surprisingly, viticultural research has seldom taken trunk size and growth into account. An extremely simple measurement, assessing the width of 10–20 trunks per plot (depending on variability) in two orthogonal directions with a Vernier calliper, to estimate trunk cross-sectional area (TCSA), allows an integration of vineyard carbohydrate balance. As shown in Fig. 4.2, when measured in a single vineyard (>80 ha, highly variable), TCSA is well related to those factors limiting vine photosynthetic activity (in this case, soil conductivity) and, when several vineyards of a single variety are compared, trunk average annual growth ($cm^2 \cdot year^{-1}$) is also well related to average yield and sugar production per vine (Fig. 4.3). In our opinion, TCSA measurement should be included in any experimental protocol attempting to understand vine behaviour, due to its simplicity and physiological significance; and TCSA related to vineyard age (TCSA$\cdot$ year^{-1}) may be used as a very useful Vineyard Potentiality Index (VPI) to characterize vineyards.

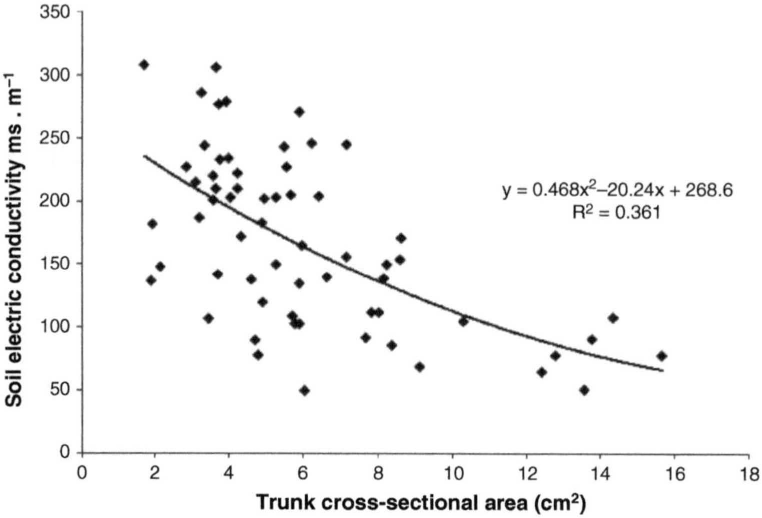

Fig. 4.2 Relation between trunk cross-sectional area (TCSA) and soil conductivity (*P<0.001*). Data from 64 points at a 7-year-old cv. "Tempranillo" vineyard in Navarra (Spain). Soil electric conductivity was measured in 2008 spring with an EM-38 conductivity meter (Geonics, Ontario, Canada). Source: Santesteban and Tisseyre, unpublished data

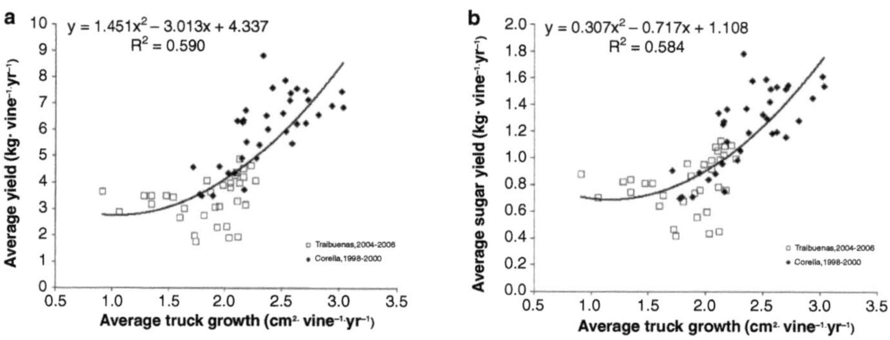

Fig. 4.3 Relation between annual trunk growth and average yield expressed as (**a**) kg · vine^{-1} and (**b**) sugar kg· vine^{-1} (*P < 0.001*). Average yield and trunk growth data from several cv. "Tempranillo" vineyards at two locations in Southern Navarra (Spain) have been pooled for regression analysis (3-year series)

4.2 Measuring Reproductive Growth

4.2.1 Estimation of Inflorescence and Flower Number

Reproductive structure development in grapevine is known to be a relatively complex process that comprises three stages: (i) formation of the anlagen,

(ii) differentiation of the anlagen to form inflorescence primordia and (iii) flower differentiation (Srinivasan and Mullins 1981). The first two stages occur inside the dormant buds in the season preceding inflorescence appearance, starting in basal buds about 2 weeks prior to anthesis and ending 3 weeks later (Swanepoel and Archer 1988), and determine the number of inflorescences per bud. The third stage, flower differentiation, takes place in the next season, in a short period of time around budburst (Srinivasan and Mullins 1981, May 2000). Therefore, flower number per bud in one season depends on the number and degree of development of inflorescence primordia, formed during the previous season, and on the conditions around budburst.

Inflorescence or cluster number can be easily determined in grapes. However, in spite of being a noteworthy parameter in viticulture because it is the first variable indicative of the productive potential in a vineyard, flower number is seldom determined in viticultural research. Indeed, its estimation is very time-consuming. Attempting to overcome this problem, some authors estimate flower number after establishing a linear relationship between inflorescence size or degree of branching and flower number (Bessis 1960, Casteran et al. 1981, Maigre 1996, Dunn and Martin 2000). When both estimations of flower number are compared, rachis length, including that of the main branch (also known as wing or shoulder), appears to be more accurate than those performed comparing branch number.

Although there is a quite close relationship between rachis length or rachis branching degree and flower number, it varies highly depending on the year, experimental conditions and even inflorescence position in the shoot (Santesteban 2003, Dunn and Martin 2007). Therefore, this relationship should be calculated separately for each experimental condition (year, site, irrigation treatment…) and inflorescence insertion range. The great interest of these relations notwithstanding, they are very lengthy to obtain, so it is necessary to develop simpler and more general methods to estimate flower number. Some authors have successfully compared the number of flowers that can be counted in a picture of one inflorescence to its real flower number (Intrieri et al. 2008).Therefore, once the relationship is determined in adjacent vines, flower number can be estimated non-destructively. However, due to tediousness of flower counting, further research should explore the possibilities given by image analysis (Schneider 1992, Duchêne et al. 2001) which might not require major adjustments for different varieties, years or locations.

4.2.2 Estimation of Fruit-Set Rate

Fruit-set can be defined as the conversion of floral organs into fruits, and fruit-set rate is defined as the proportion of flowers that become fruit. Regular fruit set ratios in grapevine can range from 20 to 50% (Huglin and Balthazard 1975), although for some varieties such as Muscat of Alexandria, set rates of about 5% can be satisfactory enough (Coombe 1973). When fruit set rate stands below what is considered normal, it is said that *coulure* or poor set has happened. Nevertheless, apart from few varieties such as Grenache (Bernard 1986) and Merlot (Zapata et al. 1999), for which fruit set problems are quite frequent, very bad weather is required to diminish fruit set rate to an extent that affects normal yield. Non-set ovaries fall from the cluster during the 3 weeks after flowering and maximal falling rate occurs about 12 days after flowering (Bessis and Fournioux 1992). Unlike for other fruit-crops, in grapevine there is no second fall event afterwards (Bernard 1986, Bessis and Fournioux 1992).

Estimation of fruit-set rate requires knowledge of the number of flowers at bloom and of the number of retained berries. As stated above, flower number estimation is a time-consuming task, so it is not usually done in viticultural research. Bagging of the inflorescences retained on the vine allows an estimation of flower number and fruit set rate by counting the number of caps and ovaries dropped, and the number of retained berries. The bags are assumed not to affect pollination and fertilisation (Scholefield et al. 1977, May 2000, May 2004, Chkhartishvili et al. 2006).

4.2.3 Estimation of Fruit Load

Fruit load is usually expressed as cluster number or as berry number. In Fig. 4.4, a comparison has been drawn between the effect of fruit load on berry size expressed as either cluster number or berry number. As can be seen, the latter parameter allows a better estimation of the effect of fruit load than the former one and, therefore, it should be preferred when attempting to evaluate fruit load effects in experiments dealing with practices such as thinning, pruning, deficit irrigation, modelling and so on.

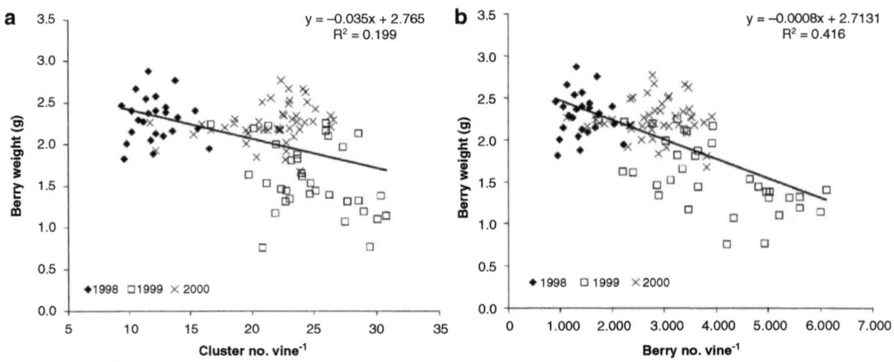

Fig. 4.4 Effect of fruit load expressed as cluster no. (**a**) and as berry no. (**b**) per vine ($P<0.001$). Data from several cv. "Tempranillo" vineyards in Southern Navarra (Spain) with varying pruning levels (from 12 to 20 buds per vine) and irrigation schedule. Data from 1998, 1999 and 2000 have been pooled for regression analysis. (**a**) Effect of cluster number on berry weight. (**b**) Effect of berry number in berry weight

When berry number per vine needs to be estimated early, it can be done through sampling 20–30 clusters in adjacent vines. Once average berry number per cluster is calculated, berry number per vine is estimated by counting cluster number. However, the most usual way to estimate berry number is to calculate it at harvest as:

$$\text{Berry number} = (1 - K_{\text{rachis}})\frac{\text{Yield}\,(\text{g}\cdot\text{vine}^{-1})}{\text{Berry weight}\,(\text{g})}$$

where K_{rachis} is the proportion of cluster weight for which the rachis stands for (rachis weight/cluster weight). Although some authors do not take rachis weight into account when estimating berry number (De la Hera et al. 2007), it has to be considered, since K_{rachis} can vary from 0.06 to 0.10 in varieties with small clusters such as Chardonnay (Lavín et al. 2001) to smaller values in varieties with bigger clusters as 0.05–0.07 in Shiraz (Vallone et al. 2004) or 0.04 in Tempranillo and Grenache (Hidalgo Togores 2003).

4.3 Estimating Vineyard Balance

Getting a "balanced" vineyard is one of the most frequently mentioned topics when vinegrowers and viticulturists are asked about their goals. However, when further asked, they (we) fail to define precisely what a balanced vineyard is. Classical works define it as the one that brings its fruit to a given sugar content, depending on the use fruit will be given (Winkler 1930, Winkler 1954, 1958). This definition, although somehow simplistic since it only refers to sugar accumulation, may be useful enough to define what a balanced vineyard is (or, at least, to define what an unbalanced vineyard is).

Several indices have been classically used to estimate vine balance; mainly pruning wood to yield and leaf area to yield ratios. The former index was first used by Ravaz in 1907 (quoted by Maccarrone et al. (1996)), and after that it is usually named Ravaz Index (RI). Balanced vines are usually said to have RI values between 5 and 10 (Bravdo et al. 1985, Kliewer and Dokoozlian 2005), although for high-yielding varieties with thin shoots such as Cinsaut, these values may range from 4 to 15, and for varieties with long or thick shoots such as Grenache and Syrah, they range from 3 to 8 (Champagnol 1984). In a similar way, several authors have established that leaf: fruit ratios of about 0.8–1.4 $m^{-2}kg^{-1}$ are necessary for proper fruit ripening (Kliewer and Dokoozlian 2005), although some research states that these values range from scarcely 0.5–1.0 $m^{-2}kg^{-1}$ (Kliewer and Antcliff 1970) up to 1.5–1.7 $m^{-2}kg^{-1}$ (Winkler 1930, Buttrose 1966).

Nevertheless, under water-limiting conditions, the interest of these ratios is less clear, since overcropping is more related to water conditions than to light

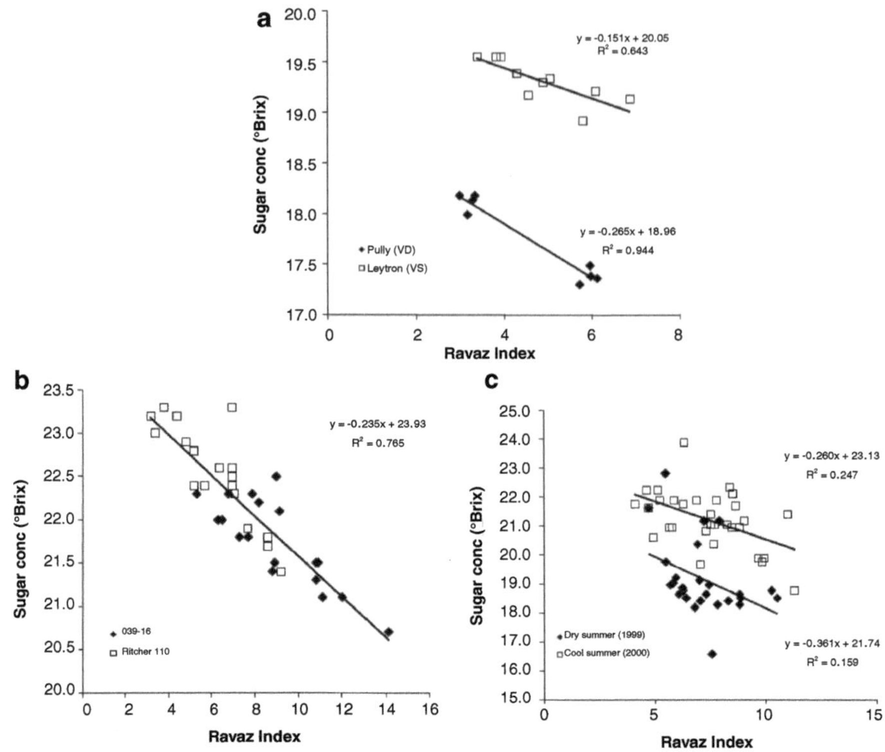

Fig. 4.5 Relationships between pruning weight:yield ratio (Ravaz Index) and grape sugar content at harvest under different growing conditions. (**a**) Switzerland – sunlight limiting conditions; (**b**) California (USA) – fully irrigated; (**c**) Southern Navarra (Spain) non-fully irrigated. Redrawn after (Kliewer and Dokoozlian 2005, Murisier and Zufferey 2006). (**a**) Murisier et al. 2006, 2007, Switzerland. (**b**) Kliewer and Dookozlian, 2005, California, USA. (**c**) Santesteban and Royo; Spain, unpublished data

interception. In fact, in an earlier work with cv. "Tempranillo" we showed that the relative importance of leaf to fruit ratio was much smaller than that of plant water status to determine sugar content (Santesteban and Royo 2006). In Fig. 4.5, it can be seen that Ravaz index is well related to sugar content under circumstances at which fruit:leaf ratio constitutes the limiting factor, either in cool climates where light interception has to be maximized (Fig. 4.5a) or in warm climates where high yields are sought through a relatively non-limiting watering (Fig. 4.5b). However, under typical Mediterranean circumstances of Southern Europe, where sugar accumulation is mainly limited by water deficit (Fig. 4.5c), this relation is not found. Under those circumstances, other indexes had to be proposed, trying to better integrate the circumstances that allow considering a vineyard as balanced. Score cards similar to those designed by Smart and Robinson (1991) and by Tardaguila

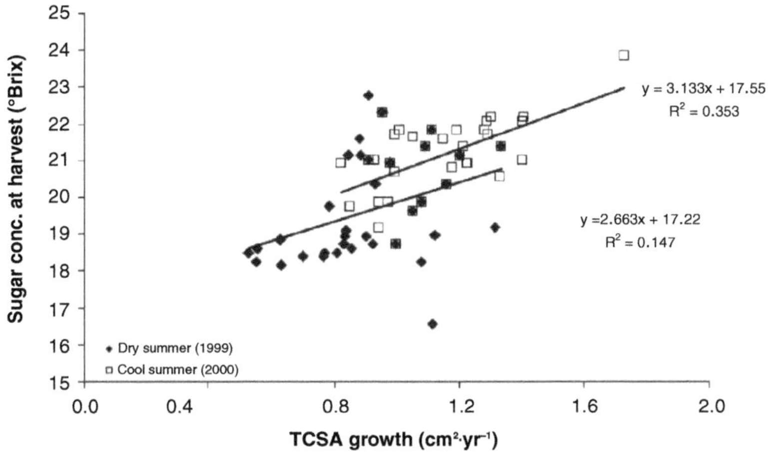

Fig. 4.6 Relation between annual trunk growth and sugar concentration at harvest ($P < 0.001$). Data from several cv. "Tempranillo" vineyards in Southern Navarra (Spain) with varying pruning levels (from 12 to 20 buds per vine) and irrigation scheduling

and De Toda (2008), complemented with other simple indices such as annual trunk growth, related to the total carbohydrate balance (Fig. 4.6), could be a useful tool to estimate vineyard balance.

4.4 How to Express Results: Per Vine? Per Square Meter? Per Meter of Row?

Finally, we would like to face one problem that sometimes is found by researchers. In viticultural research reports, results are frequently expressed on a per vine basis. When extrapolation of the obtained information to other vineyards is intended, or when data from several vineyards are analysed as a whole, some authors keep on using per vine values (Considine 2004, Folwell et al. 1994), whereas other researchers prefer to do it on a square meter (Murisier et al. 2007, Yuste et al. 2005) or on a per meter of row basis (Reynolds et al. 1995). We have not been able to find any analysis in bibliography that compares these three ways of analysing data for modelling. In order to cast some light on this problem, we have pooled data from several experiments of ours, and compared the relations obtained when attempting to explain yield using fruit load and Vineyard Potentiality Index (VPI, TCSA·year^{-1}) data.

As it can be seen in Table 4.1, expressing results per vine provided a better estimation of the yield from bunch number and VPI data, at least under Mediterranean conditions and for the analysed range of vine spacing. Nevertheless, further research is needed to determine which would be the best mode to express results to understand vine behaviour under different climatic and viticultural circumstances.

Table 4.1 Comparison of three ways of data expression: per vine, per square meter and per meter of row. The ability of bunch number and VPI to forecast yield has been compared with data from 77C. Sauvignon vineyards monitored in 2008 and with annual data from about 90 Tempranillo plots, where experimental data have been gathered between 1998 and 2007. Planting densities range from 1980 to 3800 vines ha^{-1}

					$BN \cdot vine^{-1}$			$VPI \cdot vine^{-1}$		
	R^2	df	*P*	Intercept	b	β	*P*	b	β	*P*
C. sauvignon	0.644	74	<0.001	−482.1	103.9	0.702	<0.001	345.4	0.185	0.028
Tempranillo	0.661	290	<0.001	−1798.0	220.1	0.623	<0.001	1896	0.300	<0.001
					$BN \cdot m^{-2}$			$VPI \cdot m^{-2}$		
	R^2	df	*P*	Intercept	b	β	*P*	b	β	*P*
C. sauvignon	0.535	74	<0.001	−337.0	116.7	0.684	<0.001	646.6	0.241	0.003
Tempranillo	0.349	290	<0.001	−13.17	178.5	0.550	<0.001	797.5	0.146	0.002
					$BN \cdot m^{-1}$			$VPI \cdot m^{-1}$		
	R^2	df	*P*	Intercept	b	β	*P*	b	β	*P*
C. sauvignon	0.545	74	<0.001	−763.9	114.7	0.691	<0.001	519.9	0.205	0.011
Tempranillo	0.444	290	<0.001	−1185.6	203.9	0.602	<0.001	1701	0.245	<0.001

BN: bunch number; VPI: vineyard potentiality index (TCSA/vineyard age), $cm^2 \cdot year^{-1}$; b: non-standardized coefficient. β: standardized coefficient. In all cases, Variance Inflation Factor values were close to 1, far from those that can be considered troublesome due to colineality (Neter 1996).

Acknowledgments Some of the results included are part of several research projects funded by *Gobierno de Navarra* (refs. IIM11879.RI1; PA03029; PA058; PA057), Fundación Fuentes Dutor, Verdtech Nuevo Campo SA and Bodegas Ochoa. Besides, the authors would like to thank the long list of vineyard owners and managers that have gently allowed us prowling around their vineyards, and also to all the students and staff that contributed to this research with their time and effort.

References

Bernard AC (1986) Une approche de la coulure du Grenache. Progrès Agricole et Viticole 103:395–399

Bessis R (1960) Deux méthodes rapides d'apreciation du nombre de fleurs dans les grappes de la vigne. Comptes Rendus des Sciences de l' Academie d'Agriculture de la France 14:823–828

Bessis R, Fournioux JC (1992) Zone d'abscission et coloure de la vigne. Vitis 31:9–21

Bravdo B, Hepner Y, Loinger C, Cohen S, Tabacman H (1985) Effect of irrigation and crop level on growth, yield and wine quality of Cabernet Sauvignon. Am J Enol Vitic 36:132–139

Buttrose MS (1966) The effect of reducing leaf area on the growth of roots, stems and berries of Gordo grape vines. Vitis 5:455–464

Casteran P, Raynier A, Rivet P (1981) Evaluation du nombre de fleurs des bourgeons de quelques cepages de *Vitis vinifera* L. Progrès Agricole et Viticole 98:595–599

Champagnol F (1984) Elements de physiologie de la vigne et de viticulture general. Dehan Imprimerie, Montpelier, France

Considine JA (2004) Grapevine productivity and yield components: A case study using field vines of Zante currant. Aust J Grape Wine Res 10:108–115

Coombe BG (1973) The regulation of set and development of the grape berry. Acta Horticulturae 34:261–273

Chkhartishvili N, Vashakidze L, Gurasashvili V, Maghradze D (2006) Type of pollination and indices of fruit set of some Georgian grapevine varieties. Vitis 45:153–156

De la Hera ML, Romero P, Gomez-Plaza E, Martinez A (2007) Is partial root-zone drying an effective irrigation technique to improve water use efficiency and fruit quality in field-grown wine grapes under semiarid conditions? Agric Water Manag 87:261–274

Drissi R, Goutouly JP, Forget D, Gaudillere JP (2009) Nondestructive measurement of grapevine leaf area by ground normalized difference vegetation index. Agron J 101: 226–231

Duchêne E, Meluc D, Panigai L et al. (2001) Élaboration du nombre de baies par m^2 pour le Pinot Noir et le Chardonnay in Alsace, Bourgogne and Champagne. J Int Sci Vigne Vin 35: 215–224

Dunn GM, Martin SR (2000) Do temperature conditions at budburst affect flower number in *Vitis vinifera* L. cv. 'Cabernet Sauvignon'? Aust J Grape Wine Res 6:116–124

Dunn GM, Martin SR (2007) A functional association in Vitis vinifera L. cv. Cabernet Sauvignon between the extent of primary branching and the number of flowers formed per inflorescence. Aust J Grape Wine Res 13:95–100

Folwell RJ, Santos DE, Spayd SE, Porter LH, Wells DS (1994) Statistical technique for forecasting concord grape production. Am J Enol Vitic 45:63–70

Hall A, Lamb DW, Holzapfel BP, Louis J (2002) Optimal remote sensing applications in viticulture – a review. Aust J Grape Wine Res 8:36–47

Hall A, Louis JP, Lamb DW (2008) Low-resolution remotely sensed images of winegrape vineyards map spatial variability in planimetric canopy area instead of leaf area index. Aust J Grape Wine Res 14:9–17

Hidalgo Togores J (2003) Tratado de enología. Mundi-Prensa, Madrid

Holzapfel B, Smith J (2007) Carbohydrate reserves at harvest and leaf fall impact on vine productivity in the next season. Bulletin de l'OIV 80:17–29

Huglin P, Balthazard J (1975) Variabilité et fluctuation de la composition des inflorescences et des grappes chez quelques variétés de *Vitis vinifera*. Vitis 14:6–13

Intrieri C, Filippetti I, Allegro G, Centinari M, Poni S (2008) Early defoliation (hand vs mechanical) for improved crop control and grape composition in Sangiovese (*Vitis vinifera* L.). Aust J Grape Wine Res 14:25–32

Johnson LF, Roczen DE, Youkhana SK, Nemani RR, Bosch DF (2003) Mapping vineyard leaf area with multispectral satellite imagery. Comput Electron Agric 38:33–44

Johnson LF, Pierce LL (2004) Indirect measurement of leaf area index in California North Coast vineyards. Hortscience 39:236–238

Kliewer WM, Antcliff AJ (1970) Influence of defoliation, leaf darkening, and cluster shading on the growth and composition of Sultana grapes. Am J Enol Vitic 21:26–36

Kliewer WM, Dokoozlian NK (2005) Leaf area/crop weight ratios of grapevines: Influence on fruit composition and wine quality. Am J Enol Vitic 56:170–181

Lavín A, Gutierrez A, Rojas MS (2001) Niveles de carga en viñedos jóvenes cv. Chardonnay y sus efectos sobre producción y calidad del vino. Agricultura Técnica 61:26–34

Maccarrone G, Bogoni M, Scienza A (1996) Assessment of source-sink relationships with simple indices in grapevines. Acta Horticulturae 427:177–186

Maigre D. (1996) Comportement viticole et physiologique de la vigne soumise à différents modes d'entretien du sol. Revue Suisse Viticulture, Arboricolture et Horticulture 28: 303–312

May P (2000) From bud to berry, with special reference to inflorescence and bunch morphology in *Vitis vinifera* L. Aust J Grape Wine Res 6:82–98

May P (2004) Flowering and fruitset in grapevines. Phylloxera and Grape Industry Board of South Australia in association with Lythrum Press, Adelaide

Murisier F, Zufferey V (2006) Influences of plantation density and height of foliage on the quality of grapes and wines. Trials using Chasselas a Leytron (VS). Revue Suisse de Viticulture, Arboriculture, Horticulture 38:271–276

Murisier F, Zufferey V, Triacca M (2007) Influence of row spacing and hedging height on yield, quality and root development of vines. Revue Suisse de Viticulture, Arboriculture et Horticulture 39:361–364

Neter J (1996) Applied linear regression models, 3rd edn. Irwin, Chicago

Ollat N, Fermaud M, Tandonnet JP, Neveux M (1998) Evaluation of an indirect method for leaf area index determination in the vineyard: combined effects of cultivar, year and training system. Vitis 37:73–78

Proffitt T (2006) Precision viticulture: a new era in vineyard management and wine production. Winetitles. Ashford, South Australia

Reynolds AG, Wardle DA, Naylor AP. (1995) Impact of training system and vine spacing on vine performance and berry composition of Chancellor. Am J Enol Vitic 46:88–97

Sánchez-de-Miguel P, Baeza P, Junquera P, Lissarrague JR (2010) Vegetative growth, total leaf area and surface indexes. In: Delrot S, et al. (eds) Methodologies and results in grapevine research. Springer, pp 31–44

Santesteban G (2003) Evaluación del efecto del estrés hídrico sobre el crecimiento, la fertilidad y la producción de la vid (*Vitis vinifera* L.) cv. 'Tempranillo'. *In* Dpt Producción Agraria. Universidad Pública de Navarra, Pamplona

Santesteban LG, Royo JB (2006) Water status, leaf area and fruit load influence on berry weight and sugar accumulation of cv. 'Tempranillo' under semiarid conditions. Scientia Horticulturae 109:60–65

Schneider C (1992) Analyse des composants de la production viticole alsaciane. Recherche d'une méthode fiable de mesure du nombre de fleurs/inflorescences. In: Comptes Rendus 4e Symposium Internationale Physiologie de la Vigne. Turin (Italia): Fondazione Giovanni Dalmaso

Scholefield PB, May P, Neales TF (1977) Harvest pruning and trellising of Sultana vines. II. Effects on early spring development. Scientia Horticulturae 7:123–132

Smart RE, Robinson M (1991) Sunlight into wine. A handbook for winegrape canopy management. Winetitles, Adelaide, Australia

Srinivasan C, Mullins MG (1981) Physiology of flowering in the grapevine – a review. Am J Enol Vitic 32:47–63

Swanepoel JJ, Archer P (1988) The ontogeny and development of *Vitis vinifera* L. cv. Chenin Blanc inflorescence in relation to phenological stages. Vitis 27:133–141

Tardaguila J, de Toda FM (2008) Assessment of Tempranillo grapes quality in the vineyard by Vitur score-sheet. J Int Sci Vigne Vin 42:59–65

Vallone RC, Nijensohn L, Cavagnaro JB, Merlo EE, Maffei JA (2004) Validation of a simple biological indicator of plant water status to determine irrigation timing of grapevines (Vitis vinifera, L). Acta Horticulturae 646:77–89

Weaver RJ, McCune SB (1960) Effect of overcropping Alicante Bouchet grapevines in relation to carbohydrate nutrition and development of the vine. Proc Am Soc of Hort Sci 75:341–353

Winkler AJ (1930) The relation of number of leaves to quality and size of table grapes. Proc Am Soc Hortl Sci 27:158–160

Winkler AJ (1954) Effects of overcropping. Am J Enol Vitic 5:4–12

Winkler AJ (1958) The relation of leaf are and climate to vine performance and grape quality. Am J Enol Vitic 9:10–23

Yuste J, Asenjo JL, Alburquerque MV, Rubio JA (2005) Relationships among physiology, growth and production as affected by water regime and vine spacing of 'Tempranillo' grapevines. Proceedings of the 7th International Symposium on Grapevine Physiology and Biotechnology, pp 343–348

Zapata C, Magne C, Brun O et al. (1999) La coulure chez la vigne. Rôle des réserves carbonées et azotées. Le Vigneron Champanois 5:43–54

Chapter 5
Methodologies for the Measurement of Water Flow in Grapevines

J.M. Escalona and M. Ribas-Carbó

Abstract Many methods are available to calculate mass flow of water in the transpiration stream by using heat as a tracer. Measurements can be taken in herbaceous and woody tissues, and in any conductive organ including roots. Depending of the method, measurements are taken either in the part of the conductive organ where the sensors are located, or in the whole perimeter of the conductive organ. Some methods integrate the sap flow in the whole sapwood, while others give information to calculate sap flow at different depths below the cambium. Calibration is convenient in all cases, being compulsory for the invasive methods, since probe insertion alters the xylem characteristics. This chapter describes two main groups of methods, invasive and non-invasive. For each method, brief theoretical information and practical considerations are given. All the information help the users to choose the most suitable method for his/her own purposes.

Contents

J.M. Escalona (✉)
Grup de Recerca en Biologia de les Plantes en Condicions Mediterrànies Universitat de les Illes Balears, Palma de Mallorca 07122, Spain
e-mail: jose.escalona@uib.es

S. Delrot et al. (eds.), *Methodologies and Results in Grapevine Research*,
DOI 10.1007/978-90-481-9283-0_5, © Springer Science+Business Media B.V. 2010

Abbreviations

HPV	Heat pulse velocity
HP	Heat pulse method
CHP	Compensation heat pulse
HD	Heat dissipation
HFD	Heat field deformation
HRM	Heat ratio method
SHB	Stem heat balance
THB	Trunk heat balance

5.1 Introduction

Grapevines as well as other vines present several morphological characteristics that suggest a high water use compared to other broadleaved species. Large xylem vessels allowing for high flow rates (Zimmermann 1983), rapid shoot and leaf development even under dry conditions, and an extended root system capable of exploring large soil volumes for water (Richards 1983), are some of the typical characteristics to all *Vitis* species. Calculations of seasonal water use are obtained from changes in soil water content over time. However, their low time resolution as well as sampling problems due to the extensive root system results in guidelines rather than precise numbers which would be necessary for describing and modeling the reaction of a plant to environmental changes. Moreover, the contribution of different vegetation types like green cover crops or the vines themselves cannot be separated.

During the last decade, sap flow measurements have become increasingly popular, especially after several systems have been marketed. Sap flow methods are easily automated, allowing continuous records of plant water use with high time resolution. Understanding of the theoretical basis of the techniques of sap flow measurement is needed in order to select the most appropriate methods for each application and to avoid and decrease potential errors. Previous reviews (Smith and Allen 1996), describe the most important methods for measuring sap flow in plant stems. This chapter, update previous knowledge of sap flow techniques and describes several aspects (basis, technical specifications, advantages and problems) of sap flow systems used in grapevines.

5.2 Invasive Methodologies

5.2.1 Granier Heat Dissipation Technique (hd)

The method of Granier based on the heated probe technology, developed by Vieweg and Ziegler (1960), independently also by Ittner (1968), Balek and Pavlik (1977) and better quantified by Granier (1985, 1987). It is based on the detection of convective

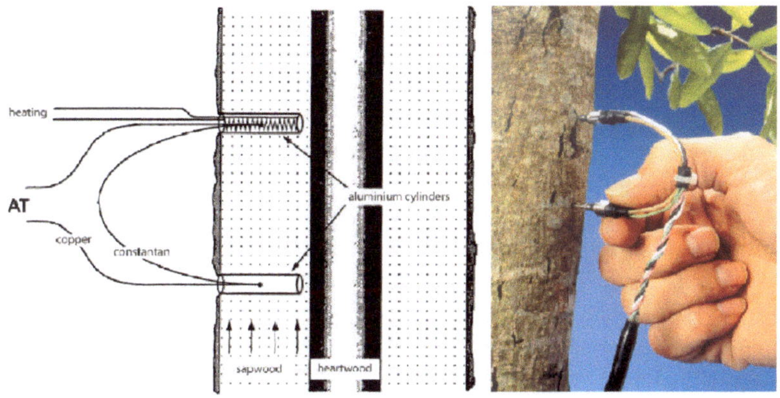

Fig. 5.1 Granier technique (scheme and picture from Dynamax Inc.)

heat transport (heat carried with the sap stream). Two cylindrical probes with a diameter of 2 mm and length of 20 mm are inserted radially into the stem in a vertical distance of 100–150 mm. The upper probe is heated constantly and the temperature difference between the two probes is measured (Fig. 5.1). Under no flow conditions, the temperature in a zone around the heater is increased up to the point where the heat conduction through the wood is in equilibrium with the heat energy supplied by the heater. Then, the temperature differential is at maximum, decreasing rapidly when the heat energy around the upper probe is transported away with the xylem stream. Granier (1985) developed mean sap flow velocity and measured temperature differences that are valid for a number of species:

$$U = 119 * 10^{-6}\ K\ 1.231\ (m^3 m^{-2} s^{-1})$$

where $U =$ sap flux density

$K = (\Delta T_{max} - \Delta T) * \Delta T^{-1}$ (dimensionless)

$\Delta T_{max} =$ temperature difference under no flow conditions

$\Delta T =$ measured temperature difference.

To calculate mass flow, sap flux density (U), is then multiplied by the conducting sapwood area at the height of the heated probe.

Braun and Schmid (1999b) developed some adaptation of this system for sap flow measurements in grapevine, modifying the probe length, and demonstrated that grapevine stems do not develop heartwood and that the whole cross section area excluding the bark can be regarded as sapwood.

5.2.1.1 Practical Considerations

One drawback of the Granier heat dissipation method (HD) is that requires calibration during a period of zero sap flow (Granier 1985). A period of zero sap flow is usually assumed to take place at night-time, and a maximum temperature difference (ΔT_{max}) between a heated and a reference unheated probe is recorder predawn. However, sap flow does not necessarily cease in branches and roots at night; thus, estimates of sap flux density based on such principle may be incorrect. Sap may continue to flow during the first hours in the night because of water movement for new growth, water redistribution in roots (Sakuratani 1981) and refilling of stems tissue after prolonged drought (Escalona et al. 2002). Far from being an exception, night-time sap flow is widespread among species and range from 5 to 30% of daily water loss. The zero flux assumption is not restricted to HD, but also applies to the heat balance methods of Cermak et al. (1972) and Sakuratani (1981) and the T_{max} heat pulse method of Green et al. (2003). These methods are described later. Using the Granier system, Tatarinov et al. (2005) have shown that for needle type probes, the ratio of the temperatures differences between the heater and the unheated probe inserted in the trunk (ΔT), depends of the heat conductivity of the wood. Because ΔTmax depends on wood heat conductivity (0.15–0.40 $Wm^{-1}K^{-1}$), changes in wood water content are expected to produce significant variations in ΔT_{max}. The scenario is further complicated because Granier formula is sensitive to the choice of ΔT_{max}, so that undetected small amounts of sap flow during the night may lead to large errors in sap flux density computations. In order to resolve this problem, Regalado and Ritter (2007), propose a relationship between ETP obtained by Penman Monteith equation and ΔT_{max} for the estimation of zero flow considering ΔT and introducing a constant in the formula.

5.2.2 Heat Pulse Systems

Heat pulse method permit measurements of sap flow rate by determining the velocity of a short pulse of heat carried by the moving sap stream. Heating and sensor probes must be installed by drilling holes into sap wood, so this method is suitable only for use on woody stems like vines.

Heat pulse methods date back some 70 years to the work of Huber (1932), who first conceived the idea of using heat as a tracer of sap flow. Almost 25 years later, Marshall (1958) developed a theoretical framework for heat pulse based on a set of analytical solutions.

Swanson (1962) was one of the first to use Marshall's analytical solution in his analysis of the compensation heat pulse method (CHP method). The velocity of sap ascending a stem is determined by compensation of the measured velocity of a heat pulse for the dissipation of heat by conduction through the matrix of wood fibres, water and gas within the stem (Swanson 1994). Two temperature sensors are placed asymmetrically on either side of a line of heat source. Immediately after release of a pulse of heat of 1–2 s duration, temperature increases more in the upstream sensor than at the downstream sensor because of conduction. Nevertheless, the heat carried by the moving sap quickly warms the downstream sensor, so the temperature of the two sensors equalizes after some time (t_e) approximately 60 s. This is the time required for convection in the moving sap stream to move the peak of the heat pulse from the heater to the point midway between the two temperature sensors, so t_e decreases as sap velocity increases. The velocity of the heat pulse (v_h) is this given by the following equation (Swanson and Whitfield 1981)

$$V_h = (x_d - x_u)/2t_e$$

Where the (t_e) is the time (s) delay for the temperatures at points x_u and x_d to become equal. And x_d and x_u are the distances (cm) between the heater and the upstream and downstream sensors, respectively.

Because sap velocities in woody stems normally vary with radial depth, sensors are usually implanted at several depths below the cambium of the stem, so that the radial profile of sap flux density across the sapwood can be determined. Mass flow rates of sap through the stem (F_m) are then calculated from the integral of the sap flux profile over the cross-sectional area of the sapwood. For species with sapwood considered to be thermally homogeneous, like vines, the heat pulse technique can be used to measure transpiration almost without calibration. Nevertheless, validation of this technique and derivation of any required calibration functions can be accomplished by comparing rates of sap flow determined by the heat pulse method with rates measured by an independent method (gravimetrically if possible).

Cohen et al. (1981) developed an alternative improved heat pulse method that relies on measuring the time t_m recorder to reach a maximum temperature in a single sensor located downstream at distance x_d from the lime heater. This method is referred as the T_{max} method. The heat pulse velocity V_m (ms^{-1}) is calculated from

$$V_m = (x_d^2 - 4Kt_m/t_m)^{0.5}$$

The only other factor required to determine V_m is the thermal diffusivity K, which is determined from the following equation:

$$K = xd^2/4t_m$$

That is calculated assuming zero sap flow.

Calculations of V_z and V_m assume that the heater and temperature probes have no effect on the measured heat flow. In reality, convection of the heat pulse is disturbed by the presence of the heater and the temperature probes and also by the disruption of the xylem tissue with their placement. These perturbations produce a systematic underestimation in the measured heat pulse velocity (Cohen et al. 1981, Green et al. 2003).

Recently, Intrigliolo et al. (2009), used this system for measuring sap flow in grapevines with small trunks and under cold winters that might have caused xylcm injuries. They use two thermocouples at a radial distance of 5 and 12 mm from the surface of the trunk (without bark). One probe measures the maximum temperature rise at distance of 10 mm downstream from the heater. The second reference probe, which is located 40 mm below the heater measures changes in ambient stem temperature (Green et al. 2003). Heat pulse velocity (V_c) is converted into sap flow (V_s) using $V_s = V_c(0.441\ F_{wood} + F_{water})$, where F_{wood} and F_{water} are calculated from the volume fractions of wood and water respectively (Becker and Edwards 1999). These authors estimate the portion of sapwood (tissue actively involved in transporting water in the trunk cross sectional area) at the end of the season by cutting trunk sections at the height of the gauges insertion and analysing it by binocular microscopy. The determination of volume fraction of wood and water can be made by Archimede's principle by coring a sample from different vines in the same vineyard and assuming that no air is present in the sapwood tissue. The results of their experiments show some deviations and errors associated to the technique. Comparison of sap flux data measured by T_{max}-heat pulse method with other transpirations determinations (gas exchange) showed that five out of six heat pulse gauges clearly underestimated canopy transpiration using the standard assumptions. In addition, the deviations from the actual values were considerable and varied from vine to vine. This suggests that there was not a systematic error in sap flow readings, but rather a seemingly random deviation from the actual values. Water transport in the sapwood is generally not uniform across the trunk section (Braun and Schmid 1999b). Hence, there might have been differences between vines in the location of the probes with respect to the portion of the trunk's sapwood that transport the majority of the water (Fig. 5.2).

The heat ratio method (HRM), is a heat-pulse method developed by Burguess et al. (2001), able to accurately measure low rates of sap flow. The HRM method measures the ratio of the increase in temperature, following the delivery of a heat pulse, at points equidistant downstream and upstream from the linear heater. HRM configuration, correction for wounding and other operational details are given by Burguess et al. (2001). The HRM method is sensitive to the direction of sap flow, being able to measure reverse flow in roots and other conductive organs.

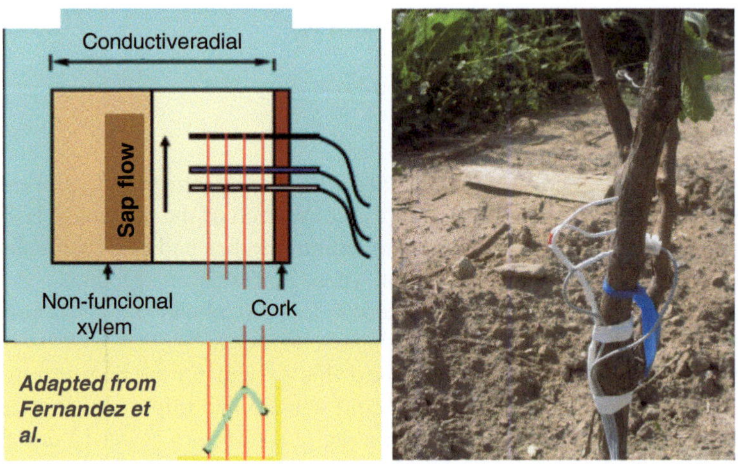

Fig. 5.2 Scheme and picture of HPV sensors installation (with permission of Fernández and Intrigliolo respectively)

5.2.2.1 Practical Considerations

It is essential to position the heater and sensor probes correctly. Accurate spacing between the probes is achieved by using a guide jig when drilling the holes, which must be done carefully to prevent excessive damage to the stem. The exact position of each probe should be carefully measured, so that errors caused by misalignment of probes can be corrected. In general, probe sets should be moved regularly to new stems because wound reactions in the woody tissue of the stem often develop 14–21 days after implanting the probes. These reactions are thought to be caused by the deposition of resin in the xylem vessels or tracheas surrounding the implantation site (conifers and other species), or possibly cavitations (vines), with the result that sap flow moves away from the sensor probes as the wound reaction develops, seriously decreasing the accuracy of the technique (Fig. 5.3).

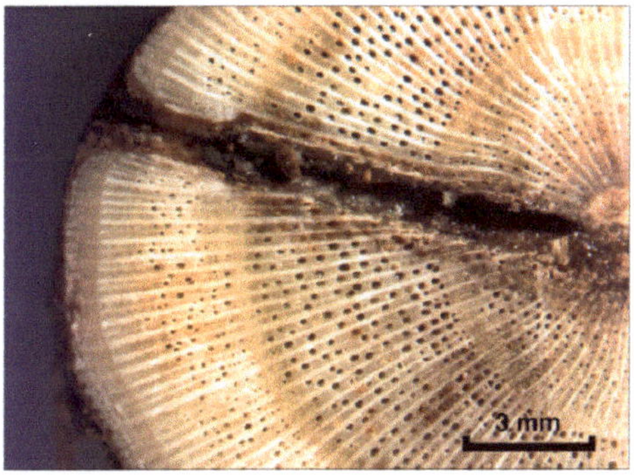

Fig. 5.3 Damage produced by probes insertion in. sap stream section

Unusually, wound reactions can develop in much less than 14 days and so it is essential that users of the heat-pulse technique monitor their data for changes in sensitivity to sap flow and move the probes more frequently when necessary.

5.2.3 Other Invasive Methodologies

Heat field deformation (HFD). This method, developed by Nadezhdina et al. (1998), is based on the analysis of temperature differences around a linear heater inserted in the sapwood. These temperature differences characterise the deformation of the heat field around the heater caused by the ascent of sap. The HFD method can measure wide-sized stems and trunks in a broad range of flow rates including low, zero and reverse flows. The HFD sensors consist of a linear heater and two pairs of differential thermocouples (Symmetrical and asymmetrical) measuring the temperature in axial (dT_{sym}) and tangential (dT_{asym}) directions around the heater (Nadezhdina et al. 2002) as raw data. Sap flow is then calculated from the mentioned temperature differences. The multi-pint sensor has several thermocouples along each needle and allows measurements of sap flow radial profile. dT_{sym} is also known as the sap flow index (SFI), which can be used as a stress indicator (Nadezhdina 1999).

The HFD system is available from ICT International (see Table 5.1)

Table 5.1 List of main companies that manufacture sap flow equipment and types of system

Manufactures of sap flow equipments	Sap flow techniques
Dynamax, Inc. 10808 Fallstone Road Suite 350, Huston USA www.dynamax.com; www.fruitionsciences.com	TDP sap velocity (HD) Dynagage (HB, Sap flow sensors from 2.1 to 165 mm stem diameter) Flow 4 (data analysis system)
East 30 Sensors 1610 Kitzmiller Road Pullman, Washington USA support@east30sensors.com	Sap flow sensors (HP)
Ecomatik Muenchener str.22. D85221 Munich Germany www.ecomatik.de	SF-L sensor (HD sap flow sensor)
EMS Brno Turisticka 5 62100 Brno Czech Republic www.emsbrno.cz	Sap flow meter T4.2 (HB external heating 6–20 mm stems and branches) Sap flow EMS 51 (THB for large stems (from 12 mm)
ICT International PO Box 503. Armidales NSW 2350 Australia www.ictinternational.com	Sap flow systems (HFD, HRM methods)
Phytech Ltd. Kibbutz Yad Mordechai, 79145 Israel www.phytech.com	SF4 & SF5 sap flow sensors (HB from 1 to 10 mm diameter) SF8 sap flow sensor (Granier type above 15 mm diameter)
Tranzflo NZ Ltd. 15 Parata St. Palmerston North 4410. New Zealand www.tranzflow.co.nz	The Green's HPV system
UP-GmbH Bahnhofstrasse D-03046 Cottbus Germany www.upgmbh.com	Ex618 M1 and BAS (HD sap flow system)

Abbreviations: HPV: heat pulse velocity; HP: heat pulse method; CHP: compensation heat pulse HD: heat dissipation; HFD: heat field deformation; HRM: heat ratio method; SHB: stem heat balance; THB: trunk heat balance

5.3 Non-invasive Methodologies

5.3.1 Stem Heat Balance Method (SHB)

Heat is applied to the entire circumference of the stem encircled by the heater and the mass flow of sap is obtained from the balance of the fluxes of heat into and out of the heated section of stem (Sakuratani 1981, Baker and van Bavel 1987). The foam insulation and weather shield surrounding the stem extend above and below the heater sufficiently to minimize extraneous thermal gradients across the heater section of stem and reduce solar heating of the stem to a negligible level. Heat input to the stem section is thus limited to the electrical power supplied to the heater (P).

The heat balance of the stem is as follows:

$$P = Q_v + Q_r + Q_f$$

where Q_v is the rate of vertical heat loss by conduction in the stem, Q_r is radial heat loss by conduction and Q_f is heat uptake by the moving sap stream. The value of Q_f is determined by subtracting Q_v and Q_r from P. The value of P is calculated from the electrical resistance and voltage across the heater, while Q_v and Q_r are determined from measurements of dT_A and dT_B and dT_r. Finally, Q_f is converted to the mass flow rate of sap (Fig. 5.4).

The value of Q_v is calculated using Fourier's law for one-dimensional heat flow from the upward and downward temperature gradients away from the heater (Sakuratani 1981, Baker and van Bavel 1987). The radial component of the stem heat balance Qr directly depend on the effective thermal conductance (K_{sh}) of the sheath of materials surrounding the heater. The value of K_{sh} is unknown and depends of the thermal conductivity of the insulating sheath and stem diameter. This component should to be calculated during periods when no sap flows are known to be zero.

Once all other components of the stem heat balance are known, Q_f is determined by difference and mass flow rate of sap (F_m) using:

$$F_m = Q_f * c_s^{-1} * dT_{AB}^{-1}$$

where c_s is the specific heat capacity of sap and dT_{AB} is the sap temperature gradient across the heater assuming that heating of the sap is uniform in radius.

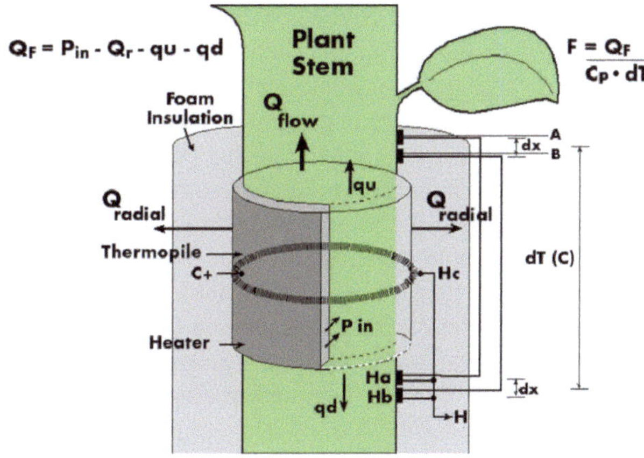

Fig. 5.4 Scheme of heat balance method (from Dynamax Inc.)

Stem heat balance gauge includes a flexible heater, typically a few cm in width, which is wrapped around the stem and enclosed in a layer of cork, a layer of foam insulation and aluminum coated PVC weather shield. Pairs of thermocouple junctions connected in series are embedded in the cork band to form a thermopile; one junction from each pair is positioned on the inner surface of the cork and the other on the outer surface, so that the thermopile measures the radial temperature gradient away from the heater (dT_r). Gauges also contain another set of thermocouples composed of two pairs of thermocouple junctions. These thermocouples are positioned against the surface of the stem and are aligned axially along the stem, with one junction from each pair above the heater and one below, in a staggered arrangement. The two thermocouple pairs the temperature gradients dT_A and dT_B which are used to calculate components in the heat balance of the stem (Fig. 5.4).

5.3.1.1 Practical Considerations

Stem heat balance method can be used to measure sap flow in both woody (Steinberg et al. 1989) and herbaceous (Baker and van Bavel 1987) stems. Stem heat balance gauges to fit stems with diameter ranging from 2 to 125 mm are commercially available (see Table 5.1). Individual gauges can only be used on stems with diameter ranging within relatively narrow limits, for example 2–3.5 mm for the smallest gauges and 100–125 for the largest. As a consequence, a number of different size gauges are required when sap flow measurements must be made on stems of different sizes.

Gauges should be installed on straight sections of stem without swellings or lumps that could cause poor contact between the stem surface and the heater or thermocouples. Loose bark and any small branches or leaves sprouting from the section to be enclosed by the gauge should be carefully removed. Application of silicone-grease-based electrical insulating chemical to the stem surface prior to installation of the gauge is usually recommended for several reasons: to ensure good thermal contact between the gauge and stem, to allow slippage of the gauge during installation, to prevent ingress of water and condensation, to prevent sensor corrosion, and to allow movement of the gauge during contraction and expansion of the stem. After preparation of the surface, the gauge is positioned on the stem and fixed using Velcro straps. It is important that water entry is prevented, as it can cause erratic measurements and damage the electrical components of the gauge. This can be completely prevented by attaching a conical collar made of heavy-duty polythene to the stem just above the gauge and sealing the joint with grafting wax.

In the commercially available systems, constant power is supplied to the gauge heater. The voltage across the heater must be adjustable so that it can be set between 3 and 10 V, depending on the size of the gauge and sap flow rates. The voltage should be selected to maintain a measurable increase in sap temperature (> 1ºC) during periods of high sap flow, but without heating the stem to damaging temperatures when flow is low.

Systems with a variable power supply that is controlled to maintain a constant increment in stem temperature have also been constructed (Grime et al. 1995,

Weibel and Boersma 1995), but the power control requires either complex datalogger programs or additional circuitry. Nevertheless, variable power systems have important advantages over the conventional constant power systems: overheating of the stems at low flow rates is avoided; power consumption is lower, an important consideration when operating sap flow gauges at remote sites away from mains power; the dynamic response is improved, and the heat storage term is reduced. Also, the advantage of variable power control for heat balance sap flow gauges are evident under high flow rates (> 600 g h^{-1}), where high rates of power must be applied. Grapevine is a good example of this fact. Recently, Tarara and Fergurson (2006) have developed algorithms (open-loop algorithms) based on day length and the theoretical diurnal course of irradiance, that reduce the possible errors in Q_v and Q_r calculation.

Accurate determination of rates of sap flow using stem heat balance gauges critically depends on the correct evaluation of the effective thermal conductance (K_{sh}), which must be done when sap flow is zero. For this propose, it is often assumed that there is no sap during the hours before dawn (Steinberg et al. 1989). However, sap flow can occur at night particularity under conditions of dry air advection (Green et al. 1989), and so it can be necessary to determine K_{sh} in other ways, for example, from data obtained after heave rainfall, or after all foliage (Steinberg et al. 1989). A commercial equipment by Environmental Systems inc. provide very simple gauges containing a heater system with electrical resistances connected in series and two thermocouples inserted in the stem, that measure the temperature up and down from the heater. The gauges easy to install and the software permit to obtain directly the mass flow of the stem subtracting day by day the sap flow measured during the night.

Different equipments based on heat balance methodologies have been used to measure sap flow in grapevines (Eastham and Gray 1988, Lascano et al. 1992, Braun and Schmid 1999a, Calo et al. 1999, Escalona et al. 2002). In general, data obtained from heat balance gauges provide a good estimation of plant water consumption. However, the literature also provides examples of deviations between the water use measured with the stem heat balance and that determined by the transpiration method with canopy gas exchange equipments (Dragoni et al. 2006). Also, it is a very interesting tool for ecophysiological studies and agronomical applications.

5.3.2 Other Non-invasive Methods of Sap Flow Measures

Nuclear magnetic resonance has been used to measure phloem and xylem sap flow velocities non-invasively (Xia et al. 1993, Peuke et al. 2001). A pulse magnetic field gradient is applied along the direction of flow causing proton spin at frequency directly proportional to the intensity of the applied magnetic field. An inverse magnetic field gradient is applied after a certain time, turning all spins back to a net phase shift of zero. However, if the protons have moved between the respective applications of the magnetic field gradients, the phase shifts will not return to zero. These phase shifts are proportional to flow velocity (Peuke et al. 2001). Short time resolutions and precise quantitative velocity measurements can be achieved with NMR. However, the cost and dimensions of the equipment and the complexity of data processing render NMR unsuitable for field applications.

Pulsed-laser system is based to apply a heat pulse to the surface of the stem with near-infrared laser source. The heat propagation is monitored externally by means of an infrared camera. Heat pulse velocities are determined from the thermometric data and related to the more useful quantity, mass flow rate. Based on this technique, Helfter el al. (2007) developed a compact stand-alone and non-invasive system allowing for direct detection of phloem and xylem sap movements and allows for precise measurements of phloem flow velocities.

Acknowledgment The authors wish to acknowledge Mike van Bavel (Dynamax), Diego Intrigliolo and Enrique Fernández for their contributions on this chapter as specialists in sap flow techniques and for allowing use of pictures and schemes. Funding was provided by the Spanish Ministry of Education (project AGL 2008-04525-02-01). J. Escalona benefit from Balearic Government and Balearic University.

References

Baker JM., Van Bavel CHM (1987) Measurement of mass flow of water in the stems of herbaceous plants. Plant Cell Environ 10:777–782

Balek J, Pavlik O (1977) Sap stream velocity as an indicator of the transpirational process. J Hydrol 34:193–200

Becker P, Edwards WRN (1999) Corrected heat capacity of wood for sap flow calculations. Tree Physiol 19:767–768

Braun P, Schmid J (1999a) Sap flow measurements in grapevine (*Vitis vinifera* L.) 1. Stem morphology and use of the heat balance method. Plant Soil 215:39–45

Braun P, Schmid J (1999b) Sap flow measurements in grapevine (*Vitis vinifera* L.) 2. Granier measurements. Plant Soil 215:47–55

Burguess SSO, Adams MA, Turner NC, Beverly CR, Ong CK, Khan AAH, Bleby TM (2001) An improved heat pulse method to measure low and reverse rates of sap flow in woody plants. Tree Physiol 21:589–598

Calo A, Giorgessi F, Sansone L, Tomasi D, Zerbi G (1999) Recherches sur le rapport entre le flux de seve, la transpiration et la vigueur dans la vigne selon le mode de conduite. Vitis 38:7–13

Cermak J, Deml M, Penka M (1972) A new method of sap flow rate determination in trees. Biol Plantarum 15:171–178

Cohen Y. Fuchs M. and Green G.C. 1981. Improvement of the heat-pulse method for determining sal flow in tress. Plant Cell Environ 4:391–397

Dragoni D, Lakso AN, Piccione RM, Tarara JM (2006) Transpiration in grapevines in the humid. Northeastern United States. Am J Enol Vitic 57(4) :460–467

Eastham J. and Gray S. 1988. A preliminary evaluation of the suitability of sap flow sensors for use in scheduling vineyard irrigation. Am J Enol Vitic 49:171–176

Escalona J, Flexas J, Medrano H (2002) Drought effects on water flow, photosynthesis and growth of potted grapevines. Vitis 41:57–62

Granier A (1985) Une nouvelle méthode pour la mesure du flux de sève brute dans le tronc des arbres. Ann Sci Forest 42:193–200

Granier A (1987) Evaluation of transpiration in a Douglas-fir stand by means of sap flow measurements. Tree Physiol 3:171–176

Green SR, Mcnaughton KG, Clothier BE (1989) Nocturnal water use by kiwifruit and apples. Agric Forest Meteor 48:251–261

Green S, Clothier B, Jardine B (2003) Theory and practical application of heat pulse to measure sap flow. Agron J 95:1371–1379
Grime VL, Morison JIL, Simmonds LP (1995) Sap flow measurements from stem heat balances: a comparison of constant with variable power methods. Agric Forest Meteor 74:27–40
Helfter C, Shephard JD, Martínez-Vilalta J, Mencuccini M, Hand DA (2007) Noninvasive optical system for the measurement of xylem and phloem sap flow in woody plants of small stem size. Tree Physiol 27:169–179
Huber B (1932) Beobachtung und messung pflanzicher sartströme. Ber Deuch Bot Ges 50:89–109
Intrigliolo DS, Lasko AN, Piccioni RM (2009) Grapevine cv. Riesling water use in the northeast United States. Irrig Sci 27:253–262
Ittner E (1968) Der Tagesgang der Geschwindigkeit des Transpirationsstromes im Stamm einer 75-jahringer Fichte. Oec Plant 3:177–183
Lascano RG, Baumhardt RL, Lipe WN (1992) Measurement of water flow in young grapevines using the stem heat balance method. Am J Enol Vitic 49:693–700
Marshall DC (1958) Measurement of sap flow in conifers by heat transport. Plant Physiol 33: 385–396
Nadezhdina N (1999) Sap flow index as an indicator of plant water status. Tree Physiol 19:885–891
Nadezhdina N, Cermák J, Nadezhdin V (1998) Heat field deformation method for sap flow measurements. In: Cermák J, Nadezhdina N (eds) Measuring sap flow in intact plants. IUFRO Publications, Publishing House of Mendel University, Brno, Czech Republic, pp 72–92
Nadezhdina N, Cermák J, Ceulemans R (2002) Radial pattern of sap flow in woody stems of dominant and understory species: scaling errors associated with positioning of sensors. Tree Physiol 22:907–918
Peuke AD, Rokitta M, Zimmermann U, Schreiber L, Haase A (2001) Simultaneous measurement of water flow velocity and solute transport in xylem and phloem of adult plants of ricinus communis over a daily time course by nuclear magnetic resonance spectrometry. Plant Cell Environ 24:491–503
Regalado CM, Ritter A (2007) An alternative method to estimate zero flow temperature differences for Granier's thermal dissipation technique. Tree Physiol 27:1093–1102
Richards D (1983) The grape root system. . Hort Rev 5:127–168
Sakuratani T (1981) A heat balance method for measuring water flux in the stem of intact plants. J Agric Meteor 37:9–17
Smith DM, Allen SJ (1996) Measurement of sap flow in plant stems. J Exp Bot 47:1833–1844
Steinberg S, Bavel CHM, McFarland MJ (1989) A gauge to measure mass flow rate of sap in stems and trunks of woody plants. J Am Soc Hort Sci 117:466–472
Swanson RH (1962) An instrument for detecting sap movement in woody plants. Rochy Mountains Forest and Range Experiment Station, Fort Collings CO paper no 68 16p
Swanson RH (1994) Significant historical developments in thermal methods for measuring sap flow in trees. Agric Forest Meteor 72:113–132
Swanson RH, Whitfield DWA (1981) A numerical-analysis of heat pulse velocity theory and practice. J Exp Bot 32:221–239
Tarara JM., Fergurson JC (2006) Two algorithms for variable power control of heat- balance sap flow gauges under high flow rates. Agr J 98:830–838
Tatarinov F, Kucera J, Cienciala E (2005) The analysis of physical background of tree sap flow measurements based on thermal methods. Measur Sci Technol 16:1157–1169
Vieweg GH, Ziegler H (1960) Thermoelektrische Registrierung der Gerchwindikkeit des Transpirationsstromes I. Deutsch Botan Ges Ber 73:221–226
Xia YE, Sarafis E, Campbell E, Callaghan P (1993). Non invasive imaging of water flow in plants by NMR microscopy. Protoplasma 173:170–176
Weibel FP, Boersma K (1995) An improved stem heat balance method using analogue heat control. Agric Forest Meteor 75:191–208
Zimmermann MH (1983) Xylem Structure and the ascent of sap. Springer Verlag, Heidelberg. New York

Chapter 6
Methods for Assessment of Hydraulic Conductance and Embolism Extent in Grapevine Organs

Claudio Lovisolo and Sara Tramontini

Abstract The aim of this chapter is to assess the interplay of hydraulic conductance and xylem embolism (cavitation) in root, in shoot, in leaf and in the whole grapevine. To measure hydraulic conductance, three main methods are available, which are based on evaporating (EFM), pulling (VPM) or pushing (HPM) water out of the plant organ. The three methods are expected to give similar results upon plant water status. Under drought, hydraulic conductance assessment must take into account the extent of embolism. EFM does not modify xylem cavitation and gives good estimates of hydraulic conductance, even if it does not give direct evidence of the embolism phenomenon. VPM involves pulling water through the organ using a vacuum pump, but it is limited by the atmospheric pressure (about 0.1 MPa); when the pulling-tension does not exceed the organ water potential, presence of embolism is not perturbed. The HPM is the easiest method to modulate a wide range of pressures, forcing flows into the sample; HPM measurements can displace native embolism. Extent of embolisms is represented as Percent Loss of Conductivity (PLC) of organ-segments. By imposing an appropriate pressure to HPM systems, it is possible to assess organ-segment PLC.

Contents

C. Lovisolo (✉)
DCA, Plant Physiology, University of Turin, Grugliasco 10090, Italy
e-mail: claudio.lovisolo@unito.it

S. Delrot et al. (eds.), *Methodologies and Results in Grapevine Research*,
DOI 10.1007/978-90-481-9283-0_6, © Springer Science+Business Media B.V. 2010

Abbreviations

E	Steady state transpirational flow rate
EFM	Evaprative flux method
E_{plant}	Plant transpiration
G_{plant}	plant hydraulic conductance
HD	Heat dissipation
HPV	Heat pulse velocity
K_{leaf}	Leaf hydraulic conductance
K_{root}	Root hydraulic conductance
PAR	Photosynthetic active radiation
RKM	Rehydration kinetics method
RWU	Relative water uptake
SHB	Stem Heat Balance
THB	Trunk Heat Balance

6.1 Introduction on Available Methods

Water transport in *Vitis vinifera* is strongly influenced by extent of embolism. Under high levels of tension, induced by drought or freezing (Tyree and Sperry 1989), gas-filled xylem vessels may become disrupted by breakage of water columns, which causes embolism formation. This process is not unidirectional and repaired cavitated vessels have been already observed even when water in neighbouring conduits is under tension (Holbrook et al. 2001). Given the importance of this phenomenon for grapevine species (Lovisolo et al. 2010), it cannot be delinked from hydraulic conductance measurements. To induce air embolism experimentally without modifying xylem development (Lovisolo and Schubert 1998), plants can be submitted to a quick and severe water stress (Schultz and Matthews 1988). On the contrary, when a non-hydraulic stress signal from root to shoot is intended to be studied, a moderate water stress has to be designed, to induce neither embolisms nor modifications of xylem anatomy to the plants (Sauter et al. 2001).

When discussing hydraulic conductance, mention should be made on the role of water channels (aquaporins). Aquaporins are ubiquitous in plant organs, but they are particularly abundant in the plasma membranes of root cells and in xylem parenchyma cells. They are thought to control the radial movement of water through roots and in living tissues adjacent to xylem vessels, therefore aquaporins are likely involved in embolism formation or repair (Kaldenhoff et al. 2008).

Grapevine aquaporins have been cloned and/or characterized by Baiges et al. (2001), Picaud et al. (2003), Perrone et al. (2006), Reid et al. (2006), Galmés et al. (2007), Fouquet et al. (2008), Glissant et al. (2008), Schlosser et al. (2008), Shelden et al. (2009), Vandeleur et al. (2009).

Today, the availability of the grapevine genomic sequence (Jaillón et al. 2007, Zharkikh et al. 2008) provides a powerful molecular data base to aquaporin studies.

The role of aquaporins in water transport may become crucial under water stress, when the proportions of water flowing through different pathways, apoplastic, symplastic and transmembrane, may change (Steudle 2000), when water movements between leaf cells and veins may regulate stomatal function (Cochard et al. 2002), when the frequency of cycles of embolization and refilling increases (Sperry et al. 2002), and when conditions of reduced transpiration do not create high driving forces derived from significant water potential gradients (Javot et al. 2003). Cochard et al. (2007) confirmed the importance of aquaporins in response to light in walnut (*Juglans regia*) leaves, and showed a direct correlation between the transcript abundance of two different aquaporins and the pattern of leaf hydraulic conductance.

6.2 Entire Plant

Methods for measuring hydraulic conductance on the whole plant are generally not invasive. For this reason, in most cases the results are obtained from data on water potential and transpiration of the plant (EFM).

In potted plants, the amount of water flowing through the plant (E_{plant}) can be estimated by extrapolating the leaf transpiration rate, measured by gas exchange, to the whole canopy surface, and validating gravimetrically by pot weighing of daily amounts of transpired water (Lovisolo and Schubert 1998, 2000). Plant hydraulic conductance (g_{plant}) is calculated in this way *in vivo* by the ratio between E_{plant} and the water potential difference from the soil to the stem (Lovisolo et al. 2002).

Other techniques for measuring sap flow, based on thermodynamic principles and requiring specific instrumentation (as more deeply described in Chapter 5 of this book), enable measurements of sap flux in the field. The main methods are: heat pulse velocity (HPV), trunk segment heat balance (THB), stem heat balance (SHB), heat dissipation (HD, better known as Granier measurements) and heat field deformation (Čermák et al. 2004).

THB methods are characterized by direct electric heating delivered continuously to a volume of xylem tissue surrounding a needle-type sensor inserted radially into the stem or a better defined volume of tissue included between heating plates or several heating elements placed in parallel into the xylem. With this method, which was initially optimized for forest trees, the heat balance of a defined heated space is calculated, splitting the input energy between the conductive heat losses and the warming of water flow (Lundblad et al. 2001, Čermák et al. 2004).

HPV methods measure sap velocity by delivering heat pulses from an active electrode and registering temperature increase by thermocouples shortly above and below the pulsing electrode (Köstner et al. 1998).

In general, THB methods are more quantitative than HPV, but they require more complex instrumentation than the second ones and, perhaps more importantly, a considerable power supply, not simple to obtain in field measurements. Swanson (1994) suggests a combination of THB and HPV methods: the quantity of sap flow determined with a THB method can be used to calibrate simultaneous velocity measurements made with HPV apparatus.

The SHB method (Sakuratani 1981, Baker and van Bavel 1987) adapted to grapevines was used to measure xylem sap flow by Lovisolo and Schubert (1998). It needs daily recalculations of the sheath conductance of the gauge using minimum pre-dawn values obtained after night calibration loops, and to fix a stem thermal conductance value (i.e. 0.54 W m^{-1} $°C^{-1}$ for 4-year old plants of cv. Freisa, according to Lovisolo and Schubert 1998).

Granier measurements are more invasive but the method is relatively simple and continuous, compared with the semi-continuous HPV. It is based on the measurement of the temperature differential between two probes inserted radially into the stem. This method has been calibrated for mature plants of *Vitis vinifera* by Braun and Schmid (1999).

The plant hydraulic conductance (g_{plant}) can be round down (leaf hydraulic conductance excluded) as the ratio between the plant transpiration (E_{plant}) and the water potential drop along the sap pathway (i.e. the difference of water potential between soil and stem) (Améglio et al. 1999).

$$g_{plant} = E_{plant}/(\Psi_{soil} - \Psi_{stem})$$

Ψ_{soil} is measured by tensiometry, or by Watermark technology (Irrometer company, Riverside, CA, USA). It can also be assumed as being equal to the leaf water potential measured pre-dawn. Alternatively, it can be assessed by means of soil moisture/water potential curves previously assessed for the pot substrate (Lovisolo and Schubert 1998) or by measuring gravimetrically soil moisture. Ψ_{stem} can be measured using a pressure chamber (Scholander et al. 1965) on several leaves along the stem (see par 6.3) (Lovisolo et al. 2002), wrapped in the evening before measurement with a double layer (inside plastic and outside aluminium) bag and assumed to represent the water potential of the corresponding stem xylem (Liu et al. 1978).

6.3 Leaves

Leaf water potential (Ψ_{leaf}) can be measured using a Scholander-type pressure chamber (Soil Moisture Equipment Corp., Santa Barbara, CA, USA). This simple and accurate equipment is a helpful tool for field measurements. The value obtained for the water potential (Ψ, MPa) can be used as approximates of the average Ψ of the concerning organ, since:

- in xylem tissue, the osmotic pressure component at apoplastic level is much lower, and therefore negligible, than the hydrostatic pressure
- xylem is in close relation with most of the cells composing an organ. It is therefore acceptable to estimate the obtained hydrostatic pressure as the average Ψ equilibrium point between the apoplast (xylem) and the symplast of the organ

When using a Scholander chamber, it is important to select the right moment of the day, Ψ being an inconstant value. For Ψ_{leaf} e Ψ_{stem} 12.00 am is a commonly accepted standard condition, corresponding with the time of maximal transpiration in the plant.

When measuring the most basal leaf, wrapped in the evening before with a double layer bag, it is acceptable to assume that the data obtained represent the water potential of the corresponding root stalk xylem ($\Psi_{root\text{–}stalk}$) (Liu et al. 1978). Following the same approximation, Lovisolo and Schubert (2006) proposed to measure Ψ_{shoot} as $\Psi_{root\text{–}stalk}$ on bagged leaves at internode 5th, 10th and 15th from the shoot base, representing the Ψ_{shoot} of the corresponding xylem, and to measure Ψ_{leaf} on transpiring leaves at internode 8th and13th from the shoot base.

To minimize the reduction in vine leaf area due to sampling, only one replicate for pressure and water potential is advisable to be performed each time, if dynamics over the year are requested (Cochard et al. 2002).

Leaf hydraulic conductance on a surface area basis (K_{leaf}, mmol s^{-1} m^{-2} MPa^{-1}) is obtained from

$$K_{leaf} = F/(P \times LA)$$

Water flow (F; mmol s^{-1}) entering the petiole of a cut leaf is measured when exposed to positive pressure (+P) or allowed to transpire and exposed to negative leaf pressure ($-P$, Ψ_{leaf}) as determined by a Scholander chamber, considering the total leaf area (LA, m^2). Cochard et al. (2007) used this system at 0.2 MPa (+P) and 0 MPa ($-P$) in order to determine the light dependence of K_{leaf}. To do so, after rehydrating in full darkness and connecting the detached leaves to a tension/pressure controlling flow apparatus (XYL'EM, Bronkhorst, France), the flow value is measured in full darkness/light conditions, with 1 h duration cycles.

Measurements on leaf material regarding the degree of xylem embolism due to the presence of air in the vessels have been described rigourously by Cochard et al. (2002) in walnut (*Juglans regia* × *nigra*) trees, but only partially in grapevine (Lovisolo et al. 2010).

The determination of percent loss of conductivity (PLC) is performed with a high-resolution liquid mass flowmeter following a procedure introduced by Sperry et al. (1988) and further refined by Cochard et al. (2000). The leaf is cut from the plant, and six segments 15–20 mm long excised from the leaf rachis, and four segments detached from the midribs of four leaflets. The whole operation has to be carried out under water. The segments are attached to the tubing of the PLC apparatus and their initial hydraulic conductance (K_{init}) is determined with a hydrostatic pressure gradient of approximately 3 kPa. The samples are then flushed with water pressurized to 0.1 MPa, defining the maximum conductance (K_{max}). The PLC is extrapolated as $PLC = 100 \times (1 - K_{init}/K_{max})$.

In leaves infiltrated under pressure, stomata may represent extra resistance in series with leaf mesophyll hydraulic resistance. Cochard et al. (2007) measured the dimension of the stomata pore (with a cryogenic scanning electron microscope) to quantify their putative effect on total leaf hydraulic conductance. Following Tyree et al. (2005), the hydraulic conductance K of all stomatal pores (presupposing them as 15 μm deep cylinders) in parallel on the leaf blade is determined in agreement with the following equation:

$$K = \frac{\Pi}{64\,N\eta}\sum_{1}^{N}\frac{10^{15}DL^{3}\times l^{3}}{18e\ \ L^{2}+l^{2}}$$

where N is the number of stomata measured, η is the water viscosity at 20°C (10^{-3} Pa s), L and l are the pore major and minor axes (m), e is the stomatal pore depth (15×10^{-6} m), D is the stomatal density (m^{-2}), and $10^{15}/18$ is a conversion factor between $m^3\ s^{-1}\ Pa^{-1}\ m^{-2}$ to mmol s^{-1} MPa^{-1} m^{-2}.

An indirect value of the hydraulic conductance at the leaf level (Ding et al. 2004) is provided through measurement of relative water uptake (RWU) and index of the transmembrane transport (I_{ATC}) in leaf tissue. Leaf discs, 12 mm in diameter, are cut from leaves using a leaf punch. Discs are weighed (initial fresh weight FWi), and immersed in Petri dishes filled with 40 mL deionized water for 15 min. They are then placed between two pieces of dry filter paper for 5 min and re-weighed (wet fresh weight, FWw).

$$I_{ATC} = (RWU_{control} - RWU_{treated})/RWU_{control} * 100$$

This unpublished method, designed and tested by the authors on several grape genotypes, can be used to quantify differences in hydraulic conductance between watered and drought conditions, and to estimate aquaporin activity. In the second case, leaf tissue is imbibed with substances able to inhibit aquaporins. At the present time, specific inhibitors of aquaporins are not available (Kaldenhoff et al. 2008). To measure aquaporin activity, before cutting leaf discs and assessing FWi, detached leaves are treated by immerging the cut end of their petioles in a solution containing the non-specific inhibitor (e.g. 0.05 mM $HgCl_2$) for 1 h, and both mercury-treated and controls are immersed in 15 mM KCl. During this 1-h period, leaves should remain exposed to the light in order to absorb solutions from the cut petiole by transpiration.

Recently, Scoffoni et al. (2008) observed that heterobaric species, i.e. those with bundle sheath extensions being *V. vinifera* one of them, have a twofold greater K_{leaf} light response than homobaric species. When using HPM on leaf material with this scope of analysis, possible artifacts and secondary effects must be considered. First, flooding the airspaces opens new flow pathways, also through stomata, influencing the obtained values of K_{leaf} light response. Second, the water solution forced in the leaf modifies both apoplastic ABA concentration and aquaporin activation (Cochard et al. 2007) causing variations on the obtained data (Scoffoni et al. 2008).

For these reasons, two alternative methods were proposed: the rehydration kinetics method (RKM; Brodribb and Holbrook 2003) and the evaporative flux method (EFM; Sack et al. 2002).

The first one is based on an analogy between the rehydration of desiccated leaves and the charging of a capacitor in series with a resistor. A leaf from a dehydrated shoot (whose Ψ_o, MPa is defined as the initial low water potential) is rehydrated for a given time period (t, s), and the new, higher water potential (Ψ_f, MPa) measured. K_{leaf} is determined from the following equation:

$$\tau = t/\ln\left(\frac{\psi_f}{\psi_o}\right)$$

$$K_{leaf} = \frac{C \times LMA \times SWC}{\tau}$$

where C is the change in relative water content per change in water potential (MPa^{-1}), LMA is the leaf dry mass per area ($g\ m^{-2}$), SWC is the saturated water content (mass of water per dry mass in hydrated leaf) and τ is the rehydration time constant (s). Desiccating shoots below -0.2 MPa removes the problem of negative K_{leaf} values and by less than -0.8 MPa minimizes cavitation (Brodribb and Holbrook 2003). A more accurate description of the RKM is given in the article.

In the second method, EFM, K_{leaf} is obtained from the steady-state transpirational flow rate (E, mmol $m^{-2}\ s^{-1}$) divided by the water potential driving force ($\Delta\Psi_{leaf}$, MPa). Leaves, after excision under perfusing solution, are supported abaxial surface down above a box fan, where they are either illuminated by ambient irradiance, or by a light source suspended above a Pyrex container filled with water (1,000–1,500 μmol photons $m^{-2}\ s^{-1}$ PAR). E is calculated every 30 s by a computer connected with the petiole. Leaves are left to transpire at least 30 min and until flow rate stabilizes with a coefficient of variation $< 5\%$ for at least 3–5 min (times ranged up to 2 h). After data recording, the leaves are rapidly removed and placed into a Whirl-Pak bag that has been previously exhaled in, and Ψ_{leaf} is measured after at least 10 min equilibration. K_{leaf} is obtained from $E/\Delta\Psi_{leaf}$ (where $\Delta\Psi_{leaf} = -\Psi_{leaf}$), normalized by leaf area and standardized to 25°C to correct for the temperature dependence of water viscosity (Sack et al. 2002).

6.4 Roots

Water flow at root level is detected by pulling (VPM) or pushing (HPM) methods. The VPM involves pulling water through the root system using a vacuum pump, but it is limited by the atmospheric pressure (about 0.1 MPa). The HPM is the easiest method to modulate a wide range of pressures, forcing flows into the root. HPM methods often refer to High Pressure Flow Meter (HPFM) technique. A detailed description of this technique can be found in Tyree et al. (1995). This instrument has been experimentally used to measure the hydraulic properties of field-growing root systems (Nardini et al. 2006, Lovisolo et al. 2007), besides leaves, petioles and stems.

It is advisable to start measuring hydraulic conductance after or contemporarily with leaf gas exchange and water potential measurements.

The use of a modified water pressure chamber allows the operator to put thc sample under pulling and pushing conditions during the same measurement. Lovisolo et al. (2002) described a controlled tension-pressure apparatus, able to receive the entire root system. The roots are entirely inserted into the apparatus, filled with water, and the shoot is cut 2 cm below the apex (Schubert et al. 1995). This method allows to modify the composition of the water solution inside the root chamber, through changes of osmolarity and/or addition of aquaporin inhibitors to discriminate between apoplastic and transcellular contributions to root water pathways. Moreover, by changing the osmolarity of the solution, it is possible to influence lateral water flow from vessels into adjacent xylem cells, and swelling of pectin hydrogels in inter-vessel pit membranes (van Ieperen 2007).

Interesting examples of the effects of different solutions are given by Lovisolo and Schubert (2006), van Ieperen and van Gelder (2006), Cochard et al. (2007), Nardini et al. (2007), Lovisolo et al. (2008b).

Measurements are taken applying either an apical suction (−0.07 MPa) through the sleeve to the cut apical internode (Kolb et al. 1996), or a basal pressure (+0.2 MPa) to the water in the chamber (Sperry et al. 1988). After collecting and weighing water spilling from the cut apex, the hydraulic conductance is calculated, following an Ohm's law analogy from tension (or pressure) gradient and flow measurements (Lovisolo and Schubert 1998). This method assesses the hydraulic conductance of the entire plant, independently from grafting interferences between root and shoot, which decrease the hydraulic conductance of grafted vines as compared with own-rooted ones (Bavaresco and Lovisolo 2000).

The HPFM allows to measure the hydraulic properties in field-growing root systems. Several experiments using HPFM for measuring root hydraulic conductance have been carried out with success on *Olea europaea* L. (olive) plants, which may forecast possible satisfactory results also in *Vitis* spp.

Two essential precautions: (i) to make the measurements during the morning (minimizing in this way the potential impact of diurnal periodicity on root hydraulic conductance, as indicated by Tsuda and Tyree 2000) (ii) and to cut the stem, below the graft union, under water. The use of a watertight container around the base of the plant enables the operator to cut off the stem under water, about 5.0 cm below the graft union (Lovisolo et al. 2007).

The HPFM, connected to the root system, is run by increasing the pressure (P) up to 0.4 MPa at a rate of 5–8 kPa s^{-1} and by measuring the instantaneous flow every 2 s.

Considering that during HPFM measurements of root systems, the direction of the flow is opposite to the physiological direction of transpiration, solutes are expected to be pushed toward the root tips. Here, they would concentrate driving osmotic water counterflows, therefore leading to underestimates of the true root hydraulic resistance (R_{root}) (Tyree et al. 1994, Gascó et al. 2007). In order to minimize the change in solute concentration and to provide good estimates of root hydraulic resistance, rapid transient measurements are preferred to the quasi-steady state mode in the case of root systems (Fig. 6.1) (Tyree et al. 1995, Bogeat-Triboulot et al. 2002). The root hydraulic conductance (K_{root}) is obtained as the slope of the linear regression of flow data versus pressure data (Fig. 6.2) (Tyree et al. 1995). R_{root} is calculated as the inverse of K_{root}. All measurements are corrected at a temperature of 22°C (calibration temperature for the HPFM).

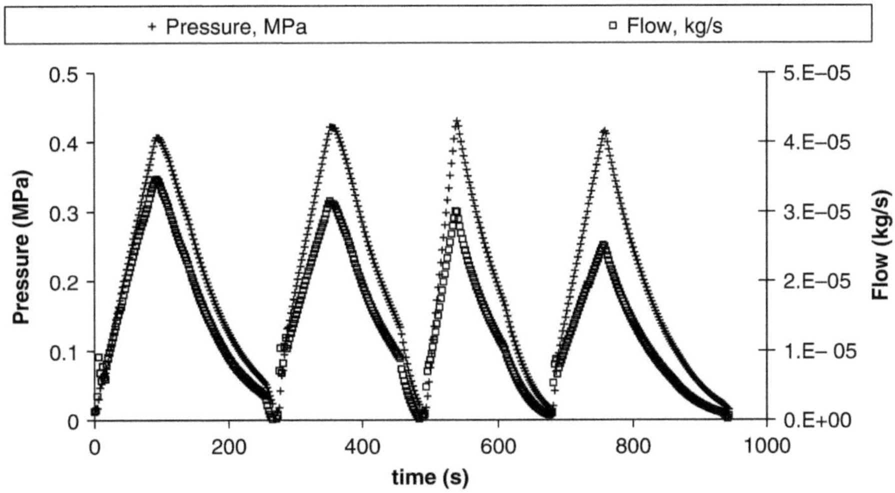

Fig. 6.1 The HPFM, connected to the root system, is run increasing four times the pressure (P) up to 0.4 MPa at a rate of about 5 kPa s^{-1} and measuring the instantaneous flow every 2 s. Considering that, during HPFM measurements of root systems, the direction of the flow is opposite to the physiological direction of transpiration, solutes are expected to be pushed toward the root tips. Here, they would concentrate driving osmotic water counterflows, therefore leading to overestimates of the true root hydraulic conductance. In order to minimize the change in solute concentration and provide good estimates of root hydraulic conductance, rapid transient measurements are preferred to the quasi-steady state mode in the case of root systems

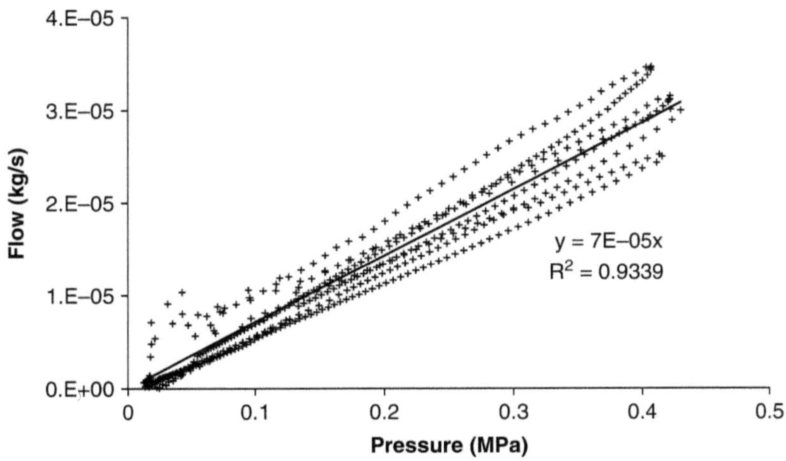

Fig 6.2 The root hydraulic conductance ($K_{root} = 7 \cdot 10^{-5}$ kg s^{-1} MPa^{-1}) is obtained as the slope of the linear regression of flow data versus pressure data ($R^2 = 0.9339$)

6.5 Measurements on Xylem Segments

6.5.1 *Xylem Portions*

When hydraulic conductance of a plant segment is related to its length, it is named hydraulic conductivity.

When measuring the hydraulic conductivity of plant portions with HPFM, an initial low pressure (Pi) is applied. If the aim of the analysis includes presence and effects of embolisms (i.e. embolism extent), its value has to be previously set for each plant organ in order not to displace air embolisms from open xylem vessels (Table 6.1). The theoretical initial pressure is calculated as $4 \times \tau/D$ (Yang and Tyree 1992). τ is the surface tension of water (0.0728 Nm^{-1}) and D is the vessel diameter, assuming that the maximum diameter for embolism is equal to vessel diameter (Salleo et al. 2004). A transient pressure increase (5 kPa s^{-1}) is applied until the final pressure is reached, and then maintained for 3 min. The difference between the initial and the final hydraulic conductivity, after transient water flushing, represents the degree of embolism, which can be quantified as the percent loss of conductivity: PLC (%) $= 100 * (K_{hf} - K_{hi})/(K_{hf})$ (Lovisolo et al. 2008a) (Fig. 6.3).

Detailed work has shown that the HPFM light response may arise from aquaporin expression and/or activation (Nardini et al. 2005, Cochard et al. 2007, Voicu et al. 2008), even not in grapevines.

Measurements on apical twigs can be carried out with the HPFM immediately after the cut, keeping it immersed in water to prevent transpiration, by increasing P to 0.3 MPa, in the steady-state mode, i.e. measuring at the pre-established pressure (Tyree et al. 1995, Lovisolo et al. 2007). Within 10–15 min, the flow becomes stable and the hydraulic conductance of the whole leafy twig (K_{twig}) can be calculated as the ratio between the recorded flow (F) and the pressure difference applied (ΔP).

The hydraulic conductance of leaves (K_{leaf}), can be measured using the HPFM, connecting it to an apical leafy twig. Measurements of K_{leaf} are carried out in the open under a light shading net, at a photosynthetically active radiation at the shoot level between 600 and 700 μmol m^{-2} s^{-1}, as measured with a quantum sensor (Lovisolo et al. 2007). The reason for this precaution is that, in some tree

Table 6.1 Set up for measurements of hydraulic conductivity of both initial and final pressure (Pi and Pf), designed for each grapevine organ in order not to displace air embolisms from open xylem vessels (Pi), or to displace them (Pf) (rewritten after Lovisolo et al. 2008a)

	Position on the plant	Cut (under water)	Pi (kPa)	Pf (kPa)
Petiole	Central region of the shoot	1 cm from the nodes leaf cut 1 cm above petiole insertion after a few seconds from start of measurement	40	600
Shoot	Apical portion	40 cm long	20	300
Roots	Central half of the main root	–	10	300

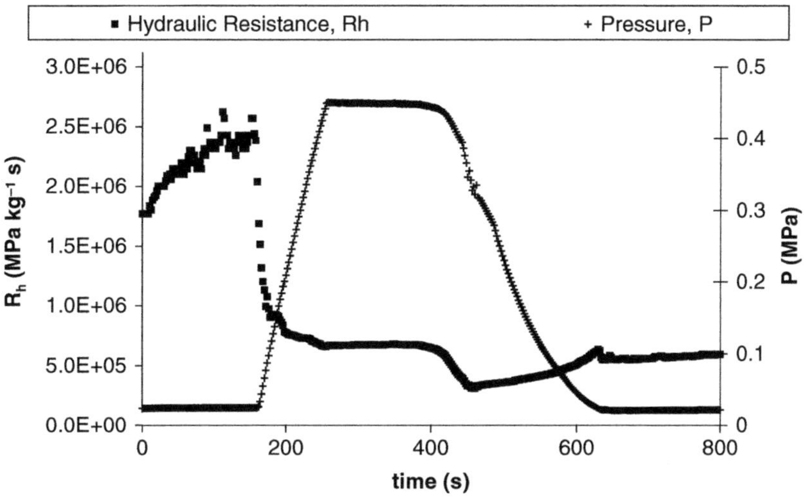

Fig. 6.3 Time course of a measurement of hydraulic resistance (R_h) according to an imposed pressure setup (P). After 3 min at constant low pressure, a transient pressure increase (5 kPa s^{-1}) is applied until a final pressure, which is maintained for about 3 min; afterwards the initial pressure is restored. The difference between the initial (K_{hi}) and the final hydraulic conductivity (K_{hf}) (hydraulic conductivity is the inverse of hydraulic resistivity, i.e. hydraulic resistance (R_h) divided by the length of the sample), after transient water flushing, represents the degree of embolism, which can be quantified as the percent loss of conductivity: PLC (%) $= 100 * (K_{hf} - K_{hi})/(K_{hf})$ (Lovisolo et al. 2008a)

species, the leaves hydraulic conductance is strongly downregulated in dark conditions (Tyree et al. 2005). After measuring the whole leaf/stem system, the leaf blades are removed, and the new hydraulic conductance of the leafless stem measured again. Because leaves can be considered as having resistances in series with stems (Gascó et al. 2004), the inverse conductances are additive in series; thus, K_{leaf} is calculated as the difference in resistance between leafy twig and leafless stem (Nardini 2001). After each experiment, the total leaf surface area of the twig ($A_{L\ twig}$) should be measured using a leaf area meter and scale K_{leaf} by $A_{L\ twig}$ (Lovisolo et al. 2007).

For measurements of the hydraulic conductance with the pulling method, Muramatsu and Hiraoka (2008) propose to apply a vacuum pressure of −0.07 MPa. The liquid volume (V) flowing through the stem per unit time (t) can be used as an index of hydraulic conductance, being proportional to the applied pressure gradient per unit length (dP / dl), predefined as a constant.

$$DV/dt = -K_{capillary}\ dP/dl$$

They proposed to put the resulting values of hydraulic conductance in relation with the water flow indcx and the vessel area index, derived as:

$$\text{Water flow index} = \sum_{i}^{n} r_i^4/S \times (\text{xylem ratio})$$

where r is the radius of each vessel element and S is the xylem area. Both data are obtained from microscopic observations.

Xylem ratio = xylem area / stem section area

The calculated vessel area index is defined as total vessel area per mm^2 cross-section area of the sample.

6.5.2 By Nuclear Magnetic Resonance

Holbrook et al. (2001) described a non-invasive method based on magnetic resonance to monitor the functional status of xylem units in living plants. These measurements showed that grapevines repaircavitated vessels only during periods of reduced transpiration.

References

Améglio T, Archer P, Cohen M, Valancogne C, Daudet FA, Dayau S, Cruiziat P (1999) Significance and limits in the use of predawn leaf water potential for tree irrigation. Plant Soil 207:155–167
Baker JM, van Bavel CHM (1987) Measurement of mass flow of water in the stems of herbaceous plants. Plant Cell Environ 10:777–782
Baiges I, Schaffner AR, Mas A (2001) Eight cDNA encoding putative aquaporins in *Vitis* hybrid Richter-110 and their differential expression. J Exp Bot 52:1949–1951
Bavaresco L, Lovisolo C (2000) Effect of grafting on grapevine chlorosis and hydraulic conductivity. Vitis 39:89–92
Bogeat-Triboulot MB, Martin R, Chatelet D, Cochard H (2002) Hydraulic conductance of root and shoot measured with the transient and dynamic modes of the high-pressure flowmeter. Ann Sci Forest 59:389–396
Braun P, Schmid J (1999) Sap flow measurements in grapevines (*Vitis vinifera* L.) 2. Granier measurements. Plant Soil 215:47–55
Brodribb TJ, Holbrook NM (2003) Stomatal closure during leaf dehydration, correlation with other leaf physiological traits. Plant Physiol 132:2166–2173
Čermák J, Kučera J, Nadezhdina N (2004) Sap flow measurements with some thermodynamic methods, flow integration within trees and scaling up from sample trees to entire forest stands. Trees 18:529–546
Cochard H, Bodet C, Améglio T, Cruiziat P (2000) Cryoscanning electron microscopy observations of vessel content during transpiration in walnut petioles: facts or artifacts? Plant Physiol 124:1191–1202
Cochard H, Coll L, Le Roux X, Améglio T (2002) Unraveling the effects of plant hydraulics on stomatal closure during water stress in walnut. Plant Physiol 128:282–290
Cochard H, Venisse JS, Barigah TS, Brunel N, Herbette S, Guilliot A, Tyree Mt, Sakr S (2007) Putative role of aquaporins in variable hydraulic conductance of leaves in response to light. Plant Physiol 143:122–133
Ding X, Iwasaki I, Kitagawa Y (2004) Overexpression of a lily PIP1 gene in tobacco increased the osmotic water permeability of leaf cells. Plant Cell Environ 27:177–186
Fouquet R, Leon C, Ollat N, Barrieu F (2008) Identification of grapevine aquaporins and expression analysis in developing berries. Plant Cell Rep 27:1541–1550
Galmés J, Pou A, Alsina MM, Tomàs M, Medrano H, Flexas J (2007) Aquaporin expression in response to different water stress intensities and recovery in Richter-110 (*Vitis* sp.): relationship with ecophysiological status. Planta 226:671–681

Gascó A, Nardini A, Raimondo F, Gortan E, Motisi A, Lo Gullo MA, Salleo S (2007) Hydraulic kinetics of the graft union in different *Olea europaea* L. scion/rootstock combinations. Environ Exp Bot 60:245–250

Gascó A, Nardini A, Salleo S (2004) Resistance to water flow through leaves of *Coffea arabica* is dominated by extra-vascular tissues. Funct Plant Biol 31:1161–1168

Glissant D, Dedaldechamp F, Delrot S (2008) Transcriptomic analysis of grape berry softening during ripening. J Int Sc Vigne Vin 42:1–13

Holbrook NM, Ahrens ET, Burns MJ, Zwieniecki MA (2001) In vivo observation of cavitation and embolism repair using magnetic resonance imaging. Plant Physiol 126:27–31

Jaillón O, Aury JM, Noel B, Policriti A, Clepet C, Casagrande A, Choisne N, Aubourg S, Vitulo N, Jubin C, Vezzi A, Legeai F, Hugueney P, Dasilva C, Horner D, Mica E, Jublot D, Poulain J, Bruyère C, Billault A, Segurens B, Gouyvenoux M, Ugarte E, Cattonaro F, Anthouard V, Vico V, Del Fabbro C, Alaux M, Di Gaspero G, Dumas V, Felice N, Paillard S, Juman I, Moroldo M, Scalabrin S, Canaguier A, Le Clainche I, Malacrida G, Durand E, Pesole G, Laucou V, Chatelet P, Merdinoglu D, Delledonne M, Pezzotti M, Lecharny A, Scarpelli C, Artiguenave F, Pè ME, Valle G, Morgante M, Caboche M, Adam-Blondon AF, Weissenbach J, Quétier F, Wincker P (2007) The grapevine genome sequence suggests ancestral hexaploidization in major angiosperm *phyla*. Nature 449:463–467

Javot H, Lauvergeat V, Santoni V, Martin-Laurent F, Guclu J, Vinh J, Heyes J, Franck KI, Schaffner AR, Bouchez D, Maurel C (2003) Role of a single aquaporin isoform in root water uptake. Plant Cell 15:509–522

Kaldenhoff R, Ribas-Carbo M, Flexas J, Lovisolo C, Heckwolf M, Uehlein N (2008) Aquaporins and Plant Water Balance. Plant Cell Environ 31:658–666

Kolb KJ, Sperry JS, Lamont BB (1996) A method for measuring xylem hydraulic conductance and embolism in entire root and shoot systems. J Exp Bot 47:1805–1810

Köstner B, Granier A, Cermák J (1998) Sapflow measurements in forest stands: methods and uncertainties. Ann Sci For 55:13–27

Liu WT, Wenkert W, Allen LH, Lemon ER (1978) Soil-plant water relations in a New York vineyard: resistances to water movement. J Am Soc Hortic Sci 103:226–230

Lovisolo C, Hartung W, Schubert A (2002) Whole-plant hydraulic conductance and root-to-shoot flow of abscisic acid are independently affected by water stress in grapevines. Funct Plant Biol 29:1349–1356

Lovisolo C, Perrone I, Carra A, Ferrandino A, Flexas J, Medrano H, Schubert A (2010) Drought-induced changes in development and function of grapevine (*Vitis* spp.) organs and in their hydraulic and non hydraulic interactions at the wholeplant level: a physiological and molecular update. Funct Plant Biol 37: in press. doi: 10.1071/FP09191

Lovisolo C, Perrone I, Hartung W, Schubert A (2008a) An abscisic acid-related reduced transpiration promotes gradual embolism repair when grapevines are rehydrated after drought. New Phytol 180:642–651

Lovisolo C, Schubert A (1998) Effects of water stress on vessel size and xylem hydraulic conductivity in *Vitis vinifera* L. J Exp Bot 49:693–700

Lovisolo C, Schubert A (2000) Downward shoot positioning affects water transport in field-grown grapevines. Vitis 39:49–53

Lovisolo C, Schubert A (2006) Mercury hinders recovery of shoot hydraulic conductivity during grapevine rehydration: evidence from a whole-plant approach. New Phytologist 172: 469–478

Lovisolo C, Secchi F, Nardini A, Salleo S, Buffa R, Schubert A (2007) Expression of PIP1 and PIP2 aquaporins is enhanced in olive dwarf genotypes and is related to root and leaf hydraulic conductance. Physiol Plant 130:543–551

Lovisolo C, Tramontini S, Flexas J, Schubert A (2008b) Mercurial inhibition of root hydraulic conductance in *Vitis* spp. rootstocks under water stress. Environ Exp Bot 63:178–182

Lundblad M, Lagergren F, Lindroth A (2001) Evaluation of heat balance and heat dissipation methods for sapflow measurements in pine and spruce. Ann For Sci 58:625–638

Muramatsu N, Hiraoka K (2008) Hydraulic conductance and xylem anatomy in fruit tree shoots. J Japan Soc Hort Sci 77:122–127

Nardini A (2001) Are sclerophylls and malacophylls hydraulically different? Biol Plant 44: 239–245

Nardini A, Gascò A, Raimondo F, Gortan E, Lo Gullo MA, Caruso T, Salleo S (2006) Is rootstock-induced dwarfing in olive an effect of reduced plant hydraulic efficiency? Tree Physiol 26:1137–1144

Nardini A, Gascò A, Trifilò P, Lo Gullo MA, Salleo S (2007) Ion-mediated enhancement of xylem hydraulic conductivity is not always suppressed by the presence of Ca^{2+} in the sap. J Exp Bot 58:2609–2615

Nardini A, Salleo S, Andri S (2005) Circadian regulation of leaf hydraulic conductance in sunflower (*Helianthus annuus* L. cv Margot). Plant Cell Environ 28:750–759

Perrone I, Carra A, Lovisolo C, Schubert (2006) *Vitis vinifera* aquaporin PIP2 (pip2) mRNA, complete cds. NCBI GenBank: DQ358107.1

Picaud S, Becq F, Dedaldechamp F, Ageorges A, Delrot S (2003) Cloning and expression of two plasma membrane aquaporins expressed during the ripening of grape berry. Funct Plant Biol 30:621–630

Reid KE, Olsson N, Schlosser J, Peng F, Lund ST (2006) An optimized grapevine RNA isolation procedure and statistical determination of reference genes for real-time RT-PCR during berry development, BMC Plant Biol 6:27. doi: 10.1186/1471-2229-6-27

Sack L, Melcher PJ, Zwieniecki MA, Holbrook NM (2002) The hydraulic conductance of the angiosperm leaf lamina: a comparison of three measurement methods. J Exp Bot 53: 2177–2184

Sakuratani T (1981) A heat balance method for measuring water flux in the stem of intact plants. J Agric Met 37:9–17

Salleo S, Lo Gullo MA, Trifilò P, Nardini A (2004) New evidence for a role of vessel-associated cells and phloem in the rapid xylem refilling of cavitated stems of *Laurus nobilis* L. Plant Cell Environ 27:1065–1076

Sauter A, Davies WJ, Hartung W (2001) The long-distance abscisic acid signal in the droughted plant: the fate of the hormone on its way from root to shoot. J Exp Bot 52:1991–1997

Schlosser J, Olsson N, Weis M, Reid K, Peng F, Lund S, Bowen P (2008) Cellular expansion and gene expression in the developing grape (*Vitis vinifera* L.). Protoplasma 232:255–265

Scholander PF, Hammel HT, Bradstreet ED, Hemmingsen EA (1965) Sap pressure in vascular plants. Science 148:339–346

Schubert A, Restagno M, Novello V, Peterlunger E (1995) Effects of shoot orientation on growth, net photosynthesis, and hydraulic conductivity of *Vitis vinifera* L. cv Cortese. Am J Enol Vitic 46:324–328

Schultz HR, Matthews MA (1988) Resistance to water transport in shoots of *Vitis vinifera* L. Plant Physiol 88:718–724

Scoffoni C, Pou A, Aasamaa K, Sack L (2008) The rapid light response of leaf hydraulic conductance: new evidence from two experimental methods. Plant Cell Environ 31: 1803–1812

Shelden MC, Howitt SM, Kaiser BN, Tyerman SD (2009) Identification and functional characterisation of aquaporins in the grapevine, *Vitis vinifera*. Funct Plant Biol 36:1065–1078

Sperry JS, Donnelly JR, Tyree MT (1988) A method for measuring hydraulic conductivity and embolism in xylem. Plant Cell Environ 11:35–40

Sperry JS, Hacke UG, Oren R, Comstock JP (2002) Water deficits and hydraulic limits to leaf water supply. Plant Cell Environ 25:251–263

Steudle E (2000) Water uptake by roots: effects of water deficit. J Exp Bot 51:1531–1542

Swanson RH (1994) Significant historical developments in thermal methods for measuring sap flow in trees. Agric For Meteorol 72:113–132

Tsuda M, Tyree MT (2000) Plant hydraulic conductance measured by the high pressure flow meter in crop plants. J Exp Bot 51 (345):823–828

Tyree MT, Nardini A, Salleo S, Sack L, El Omari B (2005) The dependence of leaf hydraulic conductance on irradiance during HPFM measurements: any role for stomatal response? J Exp Bot 56:737–744

Tyree MT, Patiño S, Bennink J, Alexander J (1995) Dynamic measurements of root hydraulic conductance using a high-pressure flowmeter in the laboratory and field. J Exp Bot 46:83–94

Tyree MT, Sperry JS (1989) Vulnerability of xylem to cavitation and embolism. Ann Rev Plant Physiol Plant Mol Biol 40:19–38

Tyree MT, Yang S, Cruiziat P, Sinclair B (1994) Novel methods of measuring hydraulic conductivity of tree root systems and interpretation using AMAIZED, a maize root dynamic model for water and solute transport. Plant Physiol 104:189–199

Vandeleur RK, Mayo G, Shelden MC, Gilliham M, Kaiser BN, Tyerman SD (2009) The role of plasma membrane intrinsic protein aquaporins in water transport through roots: diurnal and drought stress responses reveal different strategies between isohydric and anisohydric cultivars of grapevine. Plant Physiol 149:445–460

van Ieperen W (2007) Ion-mediated changes of xylem hydraulic resistance in *planta*: fact or fiction? Trends in Plant Sci 12:137–142

van Ieperen W, van Gelder A (2006) Ion-mediated flow changes suppressed by minimal calcium presence in xylem sap in *Chrysanthemum* and *Prunus laurocerasus*. J Exp Bot 57:2743–2750

Voicu MC, Zwiazek JJ, Tyree MT (2008) Light response of hydraulic conductance in bur oak (*Quercus macrocarpa*) leaves. Tree Phsysol 28:1007–1015

Yang S, Tyree MT (1992) A theoretical model of hydraulic conductivity recovery from embolism with comparison to experimental data on *Acer saccharum*. Plant Cell Environ 15:633–643

Zharkikh A, Troggio M, Pruss D, Cestaro A, Eldrdge G, Pindo M, Mitchell JT, Vezzulli S, Bhatnagar S, Fontana P, Viola R, Gutin A, Salamini F, Skolnick M, Velasco R (2008) Sequencing and assembly of highly heterozygous genome of *Vitis vinifera* L. cv Pinot Noir: Problems and solutions. J Biotech 136:38–43

Chapter 7
Comparison of Three Operational Tools for the Assessment of Vine Water Status: Stem Water Potential, Carbon Isotope Discrimination Measured on Grape Sugar and Water Balance

Cornelis van Leeuwen, Philippe Pieri, and Philippe Vivin

Abstract Yield, grape composition and wine sensory attributes tightly depend on vine water status. Hence, the measurement of vine water uptake is important for research purposes as well as for practical vineyard management. Many techniques have been developed over the past decades. Among them, three are of particular interest, because they are easy to implement, robust and because their utilization is complementary: stem water potential, carbon isotope discrimination measured on grape sugar and water balance. The present chapter describes and compares these three methods. It also indicates in which situation each of them will be most useful for researchers and vineyard managers.

Contents

C. van Leeuwen (✉)
UMR 1287 Ecophysiologie et Génomique Fonctionnelle de la Vigne, Institut des Sciences de la Vigne et du Vin (ISVV Bordeaux), Villenave d'Ornon 33882, France
e-mail: k-van-leeuwen@enitab.fr

S. Delrot et al. (eds.), *Methodologies and Results in Grapevine Research*,
DOI 10.1007/978-90-481-9283-0_7, © Springer Science+Business Media B.V. 2010

Abbreviations

ET_0	Daily reference evapotranspiration
FTSW	Fraction of transpirable soil water
SI	Stress index indicator
TDR	Time Domain Reflectometry
TTSW	Total transpirable soil water

7.1 Introduction

Given the importance of vine water status on yield and quality parameters, its measurement is very important for vineyard management and research in both irrigated and non-irrigated vineyards. Many methods have been developed over the last five decades. Most of them were first implemented on annual crops or fruit tree crops, and then adapted to vineyards. Measurements of vine water uptake can be grouped according to three different approaches:

- measurements of soil water,
- physiological indicators or
- water balance modelling.

Some of these methods are more suitable for research purposes, others for practical vineyard management. The grapevine is a deep rooting species that is often cultivated in stony soils. Hence, measurements of soil water potential by means of tensiometers or soil water content by neutron moisture probes or Time Domain Reflectometry (TDR) devices are difficult to implement. Vine water status can also be monitored through the use of physiological indicators (Cifre et al. 2005). These indicators are based on the principle that vine physiology is modified by water deficit. Many physiological indicators have been developed over the last four decades: (1) transpiration (Hsiao 1973), (2) water potentials (Begg and Turner 1970), (3) microvariations in stem or berry diameter (Garnier and Berger 1986, Greenspan et al. 1994), (4) differences between leaf and air temperatures (Jones 1999), (5) carbon isotope discrimination measured on grape sugars (Farquhar et al. 1989, van Leeuwen et al. 2001, Gaudillère et al. 2002), (6) sap flow measurements (Escalona et al. 2002) and (7) growth parameters (Pellegrino et al. 2005a). Among these accurate physiological indicators, two practical tools for determining vine water status, stem water potential and carbon isotope discrimination measured on grape sugars, will be described in this chapter.

Water potential can be measured on vine organs (generally leaves) with a pressure chamber (Scholander et al. 1965). Water potentials in vine organs are closely related to vine water status. Several applications of water potentials have been developed (Begg and Turner 1970). Water potentials represent instantaneous vine water status. Several measurements must be carried out with regular intervals to follow the dynamic evolution of vine water status during the season.

Ambient atmospheric CO_2 contains approximately 1.1% of the stable isotope ^{13}C. Plants incorporate preferentially ^{12}C and thus "discriminate" ^{13}C. In water stress conditions, isotope discrimination is less effective. Hence, the $^{13}C/^{12}C$ ratio (also called $\delta^{13}C$) measured on sugars produced by photosynthesis changes with vine water uptake conditions and can be used as an indicator of vine water status (Farquhar 1989). When measured on grape sugar at ripeness, this ratio can be used as an integrative measurement of average water uptake conditions from veraison through ripeness (Gaudillère et al. 2002). The measurement can be outsourced to specialised laboratories and need no other field work than grape sampling.

Water balance can be calculated on a daily time step when soil water holding capacity, soil surface management (bare soil or cover crop), canopy architecture, daily rainfall, daily mean temperature, daily irrigation and daily reference evaporation (ET_o) are known (Lebon et al. 2003, Pieri and Gaudillère 2005). Among these input data, soil water holding capacity is the most difficult to measure accurately because of the depth of vine rooting. It can be derived from soil texture or measured by means of a neutron moisture probe. When soil water balance is correctly monitored, it yields the remaining amount of water in soil all through the season without measurements in the field.

In this chapter, the above mentioned tools for assessment of vine water status are described. Examples of possible applications for vineyard management are provided.

7.2 Stem Water Potential

Water potentials in vascular plants can be measured by means of a pressure chamber (Scholander et al. 1965). These potentials are measured on plant organs, generally leaves. The pressure chamber technique can be used to measure:

- leaf water potential during the day,
- pre-dawn leaf water potential and
- stem water potential.

Among these three applications of the pressure chamber technique, stem water potential provides an accurate and robust estimation of vine water status (Choné et al. 2001).

Stem water potential is measured during the day on a leaf that is bagged with an opaque plastic bag at least 1 h prior to measurement (Begg and Turner 1970). Stem water potential values reach a minimum in the early afternoon. This moment is generally chosen for comparing measurements among sites. Six to ten measurements on separate vines are necessary to represent water status in a vineyard block, depending on the size of the block and the intra-block variability of the vine water status. To carry out the measurement, the leaf is cut from the vine and the petiole is re-cut with a sharp cutter. The leaf is introduced into the pressure chamber with the petiole coming out of a hole in the cap. The pressure is progressively increased inside the pressure chamber by means of a bottle of compressed nitrogen. When a drop of sap appears on the section of the petiole, no more nitrogen is introduced inside the pressure chamber. The pressure inside the chamber is indicated on the manometer. This pressure corresponds to the stem water potential. Stem water potential are always negative values and should be expressed in MPa (1 MPa = 10 bar). The opaque bag prevents the leaf from transpiring. During the hour between bagging and taking the water potential measurement, the water potential in the leaf balances with the water potential in the stem xylem. Although the measurement is carried out on a single leaf, the obtained value represents whole vine water potential. Hence, when measurements are carried out on several leaves of the same vine, the coefficient of

Table 7.1 Stem, leaf and pre-dawn leaf water potentials measured on 6 vines with 6 replicates per vine (*Vitis vinifera* Cabernet-Sauvignon, 5 September 2006, Margaux, Bordeaux, France)

Midday stem water potential (MPa)							
	Vine 1	Vine 2	Vine 3	Vine 4	Vine 5	Vine 6	Mean 6 Vines
	−1.08	−1.22	−1.10	−0.90	−0.93	−1.11	
	−1.10	−1.24	−1.13	−0.92	−0.97	−1.12	
	−1.11	−1.28	−1.16	−0.99	−0.98	−1.13	
	−0.96	−1.18	−0.98	−0.9	−0.94	−0.88	
	−1.01	−1.19	−1.06	−0.95	−0.98	−0.89	
	−1.08	−1.21	−1.09	−0.96	−0.98	−1.00	
Mean	−1.06	−1.22	−1.09	−0.94	−0.96	−1.02	−1.05
Standard deviation	0.06	0.04	0.06	0.04	0.02	0.12	0.06
Variation coefficient %	−5.5%	−3.0%	−5.9%	−3.9%	−2.1%	−11.4	−5.3%
Midday leaf water potential (MPa)							
	Vine 1	Vine2	Vine 3	Vine 4	Vine 5	Vine 6	Mean 6 vines
	−1.32	−1.40	−1.37	−1.23	−1.17	−1.37	
	−1.33	−1.41	−1.40	−1.26	−1.23	−1.38	
	−1.35	−1.43	−1.43	−1.28	−1.24	−1.39	
	−1.13	−1.33	−1.26	−1.03	−0.94	−1.02	
	−1.16	−1.33	−1.28	−1.04	−0.96	−1.03	
	−1.20	−1.34	−1.29	−1.04	−0.96	−1.03	
Mean	−1.25	−1.37	−1.34	−1.14	−1.08	−1.20	−1.23
Standard deviation	0.10	0.04	0.07	0.12	0.14	0.19	0.11
Variation coefficient %	−7.9%	−3.2%	−5.1%	−10.6%	−13.2%	−15.9%	-9.3%
Pre-dawn leaf water potential (MPa)							
	Vine 1	Vine 2	Vine 3	Vine 4	Vine 5	Vine 6	Mean 6 vines
	−0.23	−0.32	−0.25	−0.15	−0.13	−0.10	
	−0.23	−0.32	−0.26	−0.16	−0.13	−0.12	
	−0.24	−0.33	−0.27	−0.17	−0.13	−0.12	
	−0.26	−0.34	−0.27	−0.17	−0.14	−0.13	
	−0.32	−0.34	−0.28	−0.18	−0.14	−0.13	
	−0.33	−0.36	−0.34	−0.18	0.15	0.14	
Mean	−0.27	−0.33	−0.28	−0.17	−0.14	−0.12	−0.22
Standard deviation	0.04	0.01	0.03	0.01	0.01	0.01	0.02
Variation coefficient %	−16.4%	−4.0%	−10.5%	−7.3%	−5.8%	−8.98%	−8.8%

variation (%) of stem water potential is consistently lower than pre-dawn leaf water potentials or leaf water potentials (Table 7.1).

7.2.1 Impact of Soil Water Availability and Climatic Conditions on Stem Water Potential

Stem water potential values reflect soil water availability, but they also depend on climatic parameters of the day of measurement (temperature, solar radiation, Vapour Pressure Deficit). In order to assess the impact of soil water availability and climatic conditions on stem water potential values, stem water potential was measured 2 days in a row, in cool and cloudy conditions (7 September 2004) and in warm and sunny conditions (8 September 2004) on 46 plots of three vines inside a vineyard block located in the Bordeaux area (*Vitis vinifera* cv. Merlot). Soil water content varied highly among the 46 plots, but can be considered stable from one day to another on a given plot. Stem water potential values obtained on 7 and 8 September were highly correlated (Fig. 7.1). They ranged from −0.42 MPa to −1.36 MPa depending on soil water availability (Table 7.2). The effect of climate on stem water potential is reflected by the difference of the average stem water potential value measured on the two days: –0.94 on 7 September and –1.02 on 8 September. It can thus be concluded that the effect of climate on stem water potential is limited compared to the effect of soil water availability. However, comparisons of soil water availability through stem water potential readings should preferably be carried out in similar climatic conditions, for example on sunny days without extreme temperatures.

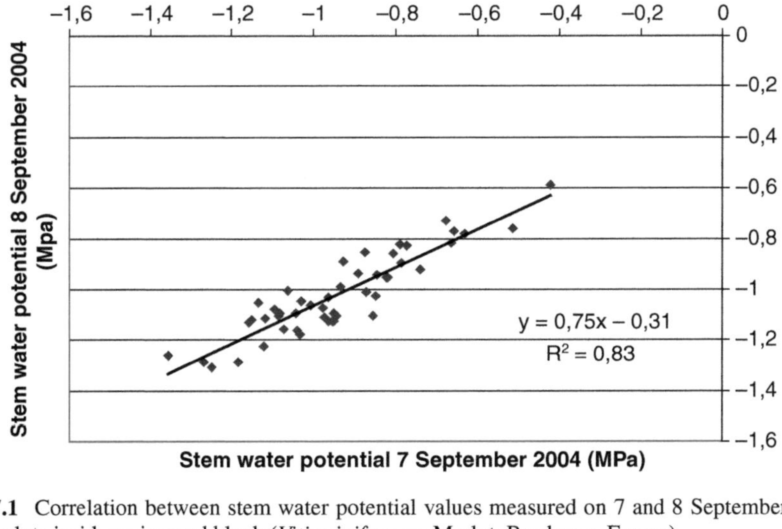

Fig. 7.1 Correlation between stem water potential values measured on 7 and 8 September 2004 on 46 plots inside a vineyard block (*Vitis vinifera* cv. Merlot, Bordeaux, France)

Table 7.2 Stem water potential measured on 46 plots inside a vineyard block (*Vitis vinifera* cv. Merlot, Bordeaux, France)

	7 Sept. 2004	8 Sept. 2004
Miminum ψ stem (MPa)	−1.36	−1.31
Maximum ψ stem (MPa)	−0.42	−0.59
Average ψ stem (MPa)	−0.94	−1.02

7.2.2 Possible Applications of Stem Water Potential

7.2.2.1 Assessment of Vine Water Status in Relation to Soil Water Availability

Vine water status is highly dependant on soil water availability (Fraction of Transpirable Soil Water or FTSW). Hence, vine water status can vary from one block to another depending on soil characteristics. Vine water status was monitored on three blocks of a Bordeaux wine estate in 2005, which was a dry vintage. One block was planted on a gravely soil ("G", low soils water holding capacity), another block was planted on a clayey soil ("C", medium soil water holding capacity) and the third block was planted on a sandy soil with a water table within the reach of the roots ("S"). Differences in vine water status depending on soil water availability can easily be shown by seasonal dynamics of stem water potential values (Fig. 7.2).

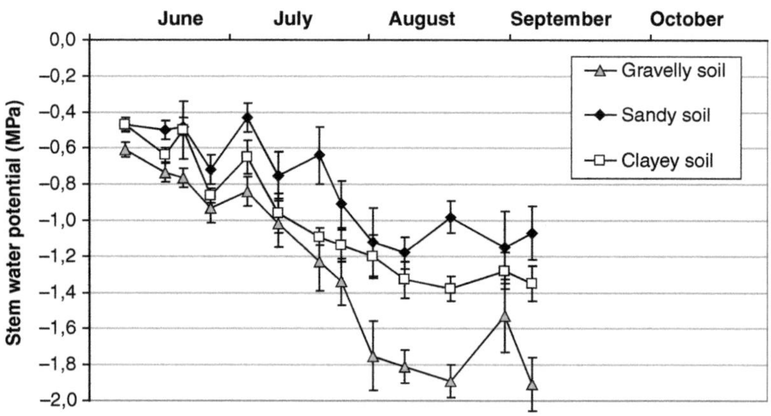

Fig. 7.2 Comparison of seasonal stem water potential on three soils with various soil water holding capacities in the Saint-Emilion region in 2005 (*Vitis vinifera* L. cv. Merlot, Bordeaux). The more negative the values are, the greater the water deficit stress. Error bars indicate standard deviation (n = 6)

7.2.2.2 Assessment of Seasonal Vine Water Status Dynamics Among Vintages

Vine water status is related to the fraction of transpirable soil water, which varies during the season depending on the climatic conditions of the vintage (rainfall amount and distribution, Potential Evaporation or ET_0). Stem water potential can be used to assess seasonal dynamics of vine water status on a vineyard block and to compare this evolution among vintages. In Fig. 7.3, this evolution is shown for a Bordeaux vineyard block planted on a gravely soil in the Saint-Emilion region during the years 2004 (moderately dry vintage), 2005 (very dry vintage) and 2007 (wet vintage).

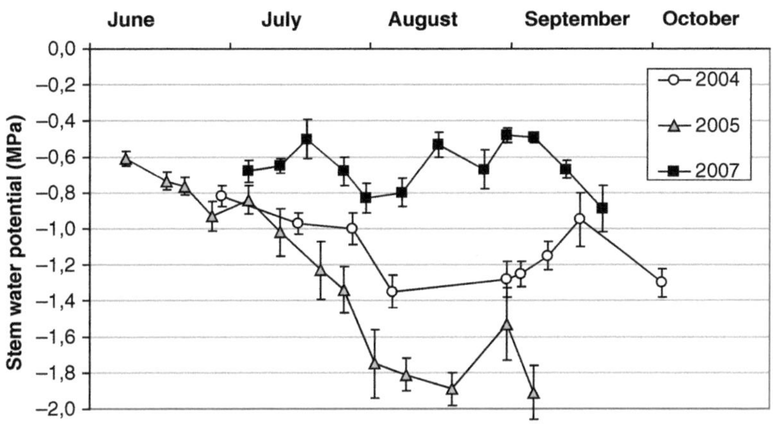

Fig. 7.3 Comparison of seasonal stem water potential on a gravely soil in the Saint-Emilion region in three vintages (*Vitis vinifera* L. cv. Merlot, Bordeaux). The more negative the values are, the greater is the water deficit stress. Error bars indicate standard deviation (n = 6)

7.2.2.3 The Use of Stem Water Potential in Irrigation Management

Because stem water potential represents whole vine water status during the day, it is a particularly useful tool for irrigation management. It accurately represents vine water status, even if soil water content is heterogeneous, which is the case in irrigated vineyards (Shackel 2006). However, the specific threshold of stem water potential that causes irreversible damage on canopy and grapes varies with the vigor of the vines. Water deficit damage is caused by vascular embolism (Schultz and Matthews 1988). Xylem vessels are small in vines that have been exposed to early water deficit stress and large in vines that developed under unlimited water uptake conditions (Lovisolo and Schubert 1998). On vigorous vines with large xylem vessels that are suddenly exposed to excessive water deficit, irreversible embolism might occur at stem water potential readings of −1.2 MPa. Low vigor vines that are progressively exposed to water deficits might resist to stem water potential levels of –1.6 MPa (no excessive leaf necrosis, no berry shrivel). Pre-dawn leaf water potentials are less useful in irrigation management, because they underestimates vines water stress when soil water is heterogeneous, which is the case in irrigated vineyards (Améglio et al. 1999).

7.3 Carbon Isotope Discrimination Measured on Grape Sugar

Ambient atmospheric CO_2 contains 98.9% of ^{12}C isotope and 1.1% of ^{13}C isotope. ^{12}C is more easily used by the enzymes of photosynthesis for hexose production. Therefore, the sugar produced by photosynthesis contains a higher proportion of the ^{12}C isotope than ambient CO_2. This process is called "isotope discrimination". When plants face water deficit conditions, isotope discrimination is reduced because of stomatal closure (Farquhar et al. 1989). Therefore, the $^{13}C/^{12}C$ ratio in photoassimilates provides a signature of plant water status over the period in which they were synthesised. When measured on grape sugar at ripeness, the $^{13}C/^{12}C$ ratio (so-called $\delta^{13}C$) indicates average vine water status during grape ripening (van Leeuwen et al. 2001, Gaudillère et al. 2002).

$\delta^{13}C$ can easily be measured by mass spectrometry in specialized laboratories. Grape juice is obtained from grapes sampled at ripeness or close to ripeness. Freezing of grape juice does not alter $^{13}C/^{12}C$ ratio. Two mL of juice are introduced into an Eppendorf tube and centrifuged at 10,000 RPM. Tin capsules (TIN 6∗4 mm) are delicately introduced into a 96-well (8 mm) microplate (SARSTEDT n° 83.1835). Five μL of grape juice is introduced in each tin capsule by means of a micropipette P10. The location of the samples must be carefully registered. The microplate is placed in a non-ventilated stove at 60°C during 24 h. Tin capsules are compressed and turned into small balls without any remaining air. Any contact with the hands or other carbon containing material must be avoided. Samples should be preferable double encapsulated to avoid problems with "sticky samples". The microplate is then closed and sent to a laboratory specialized in stable isotope analysis.

The $^{13}C/^{12}C$ ratio in the sample is compared to that in an international standard, the so-called PDB standard which is a rock in which this ratio is particularly stable:

$$\delta^{13}C = \frac{(^{13}C/^{12}C\ \text{sample} - ^{13}C/^{12}C\ \text{PDB standard})}{^{13}C/^{12}C\ \text{PDB standard}} * 1000$$

It is expressed in p. 1000

The results vary from −20 p. 1000 (severe water deficit stress) to –27 p. 1000 (no water deficit stress). Repeatability is excellent: the error of measurement is 0.12 and the variation coefficient is 0.2% (Table 7.3). $\delta^{13}C$ is well-correlated to stem water potential values measured between veraison and ripeness (Fig. 7.4).

Table 7.3 Repetability of $\delta^{13}C$ measurements on grape sugar at ripeness

Replicate	δ13C
1	−22.88
2	−22.87
3	−22.87
3	−22.87
4	−22.95
5	−23.02
6	−22.89
7	−22.93
8	−22.97
mean:	−22.92
standard deviation:	0.051
coefficient of variation:	−0.2%

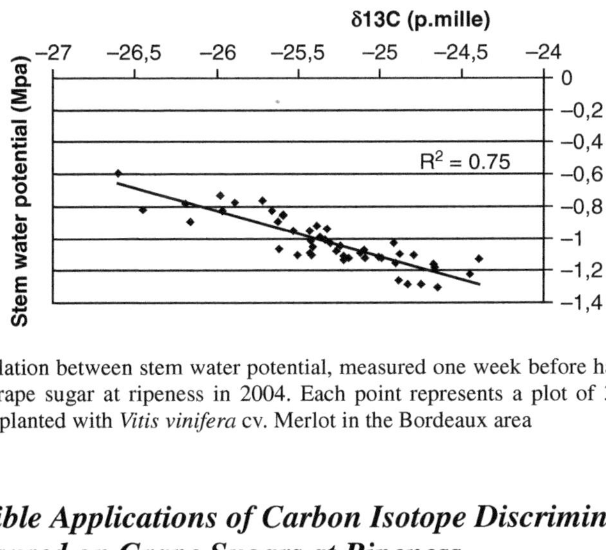

Fig. 7.4 Correlation between stem water potential, measured one week before harvest, and $\delta^{13}C$ measured on grape sugar at ripeness in 2004. Each point represents a plot of 3 vines inside a vineyard block planted with *Vitis vinifera* cv. Merlot in the Bordeaux area

7.3.1 Possible Applications of Carbon Isotope Discrimination Measured on Grape Sugars at Ripeness

7.3.1.1 Assessment of Vine Water Status in Relation to the Climate of the Vintage, Soil Water Availability and Grapevine Variety

$\delta^{13}C$ was measured on grape sugars at ripeness in three grapevine varieties planted on three different soil types during 4 vintages. Grapevine varieties were *Vitis vinifera* cv. Merlot, *V. vinfera* cv. Cabernet franc and *V. vinfera* cv. Cabernet-Sauvignon. Soils were gravely (low soil water holding capacity), clayey (medium soil water holding capacity) and sandy. On the sandy soil, a water table was present within the reach of the roots. Among the four vintages, 1997 and 1999 were rainy, 2000 was dry and 1998 was very dry. Three factor analysis of variance showed a significant effect of vintage, soil and cultivar on $\delta^{13}C$ (Table 7.4). The soil effect was greatest and accounted for 57% of the total variance. $\delta^{13}C$ values showed that vines faced

Table 7.4 Three-factor analysis of variance of the effect of soil, vintage and grapevine variety on $\delta^{13}C$ measured on grape sugars at ripeness. Homogeneous groups are determined with Newman-Keuls test. NS = not significant

	P	% of variance explained by each factor		
Factor vintage (F1)	< 0.00001	33%		
Factor soil (F2)	< 0.00001	57%		
Factor grapevine variety (F3)	0.0003	2%		
Inter F1/F2	0.00001	6%		
Inter F1/F3	0.0067	1%		
Inter F2/F3	NS	< 0.3%		
Factor vintage				
	Means		Homogeneous groups	
1998	−22.20	A		
2000	−23.18		B	
1999	−24.05			C
1997	−24.14			C
Factor soil				
Gravel	−22.05	A		
Clay	−23.60		B	
Sand	−24.54			C
Factor grapevine variety				
Cabernet-sauvignon	−23.25	A		
Cabernet franc	−23.29	A		
Merlot	−23.63		B	

greatest water deficit stress on the gravely soil, followed by the clayey soil. Water deficit stress was weak on the sandy soil with a water table. The climate of the vintage accounted for 33% of the total variance. $\delta^{13}C$ values showed that water deficit stress was greatest in 1998, followed by 2000. Water deficit stress was mild in 1999 and 1997. A weak but significant effect of the grapevine variety on $\delta^{13}C$ values was shown (2% of the total variance).

7.3.1.2 Spatialization of Vine Water Status at Intra-Block Scale

The advantage of the carbon isotope discrimination method lies in the fact that it does not require any field measurement other than grape sampling at ripeness. Hence, many measurements can easily be carried out. This is not the case with a pressure chamber, which is a more time consuming and labour intensive tool. In dry farmed vineyards, $\delta^{13}C$ is a valuable tool for determining water deficits at block scale or even at intra-block scale. In 2005, $\delta^{13}C$ measurements were carried out on 92 plots (composed of 3 vines each) inside a 0.3 ha vineyard block in the Bordeaux area. These point measurements were transformed into a map of intra-block vine water status variations (Fig. 7.5a). For the validation of this map, stem water potential measurements were carried out on every vine of every plot on 7 September 2005 between 1:30 and 4 PM. Stem water potential values were transformed into

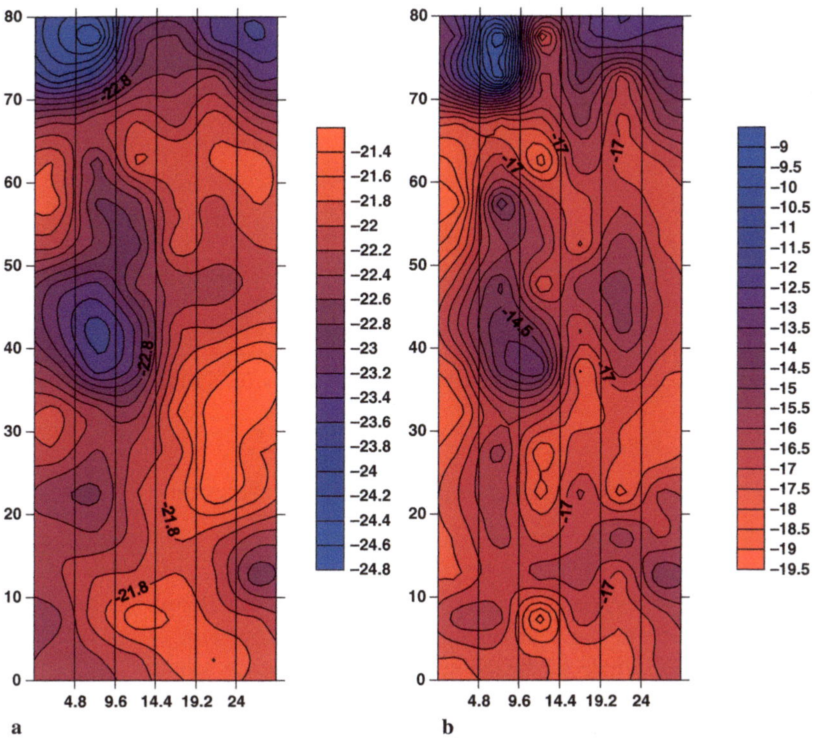

Fig. 7.5 Spatial distribution of vine water status in a vineyard block assessed with (**a**) $\delta^{13}C$ measured on grape sugar at ripeness (expressed in p. 1000) and (**b**) stem water potential (expressed in bar) measured one week prior to harvest (*Vitis vinifera* cv. Merlot, Bordeaux 2005)

a map of intra-block vine water status variations (Fig. 7.5b). Both maps are highly comparable. However, the map based on stem water potential readings was much more difficult to obtain: 12 persons and 6 pressure chambers were mobilized to carry out the measurements.

7.3.1.3 Spatialization of Vine Water Status at Estate Level

At the estate level, it is practically impossible to produce vine water status maps with the pressure chamber technique. It is possible to produce such maps by measuring carbon isotope discrimination on grape sugar because only grape sampling at ripeness is required. On the Bordeaux estate presented in Fig. 7.6a, dry blocks (high $\delta^{13}C$ values) correspond to gravely soils with low water-holding capacity. Blocks with higher water-holding capacities (due to a higher clay content in the soil) show more negative $\delta^{13}C$ values. In this study, the $\delta^{13}C$ method was validated with soil resistivity measurements. Soil resistivity is high when soils contain low clay and high gravel content (Fig. 7.6b). Soil resistivity is low on blocks with high clay content, resulting in high water holding capacity.

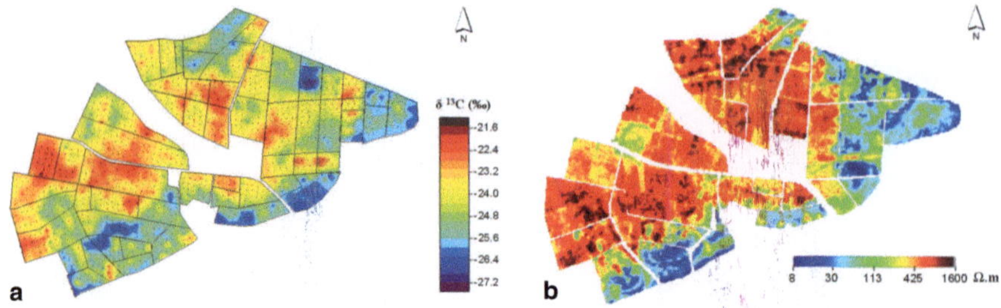

Fig. 7.6 Spatial representation of vine water status, assessed by $\delta^{13}C$ measurements on grape sugar (**a**), in a Bordeaux estate in 2006 compared with soil resistivity mapping carried out in the same estate (**b**). Cold colours represent low water deficit (**a**) or low soil resistivity (**b**) while warm colours represent high water deficit (**a**) and high soil resistivity (**b**). Data: SOVIVINS, F 33650 Martillac

7.3.1.4 The Use of $\delta^{13}C$ to Assess the Necessity of Irrigation

When a dry-farming grower wants to know if some of his vineyards block would benefit from irrigation, $\delta^{13}C$ can be used as an objective tool to assess the level of vine water deficit stress on each vineyard block of his estate. However, because $\delta^{13}C$ is measured on grape sugar at ripeness, it allows assessment of the water deficit experienced by the vines when the season is over. Hence, contrary to stem water potential, this technique cannot be used for day-to-day irrigation management but it can be used to validate irrigation strategies afterwards.

7.3.2 Thresholds for $\delta^{13}C$ Values with Respect to Vine Water Deficit

The range of $\delta^{13}C$ values with regard to vine water deficit are presented in Table 7.5. For comparison, thresholds for midday leaf water potential values, pre-dawn leaf water potential values and midday stem water potential values with regard to vine water deficit are also included.

Table 7.5 $\delta^{13}C$ and water potential values with respect to vine water deficit thresholds

	$\delta^{13}C$	Midday stem water potential (MPa)	Midday leaf water potential (MPa)	Pre-dawn leaf water potential (MPa)
No water deficit	<-26	>-0.6	>-0.9	>-0.2
Weak water deficit	-24.5 to -26	-0.6 to -0.9	-0.9 to -1.1	-0.2 to -0.3
Moderate to weak water deficit	-23 to -24.5	-0.9 to -1.1	-1.1 to -1.3	-0.3 to -0.5
Moderate to severe water deficit	-21.5 to -23	-1.1 to -1.4	-1.3 to -1.4	-0.5 to -0.8
Severe water deficit	>-21.5	<-1.4	<-1.4	<-0.8

7.4 Water Balance Modelling

Plant water status in vineyards is effectively assessed in field conditions by classical techniques on the basis of indicators such as pre-dawn leaf water potential or stem water potential or stomatal conductance, which all characterize the plant reaction to the highly variable soil water availability (van Leeuwen and Seguin 1994, Schultz 1996, Pellegrino et al. 2005b). However, the soil water availability simply results from the dynamical equilibrium of water exchanges with the atmosphere or the deeper soil layers. Even if the soil surface evaporation and the vine transpiration are both down-regulated when soil water availability itself decreases, all these water exchanges may be derived from climatic variables, mainly rainfall and reference evapotranspiration (ETo). The soil is then considered as a water reservoir of limited capacity that is filled or emptied mechanically, according to the balance of input and output water fluxes. The main advantages of this approach are (i) to provide a sound insight of water status at any time, from a limited set of climatic variables and system parameters; (ii) the ability to simulate any conditions, past or future, in any vineyard, merely by changing the input variables or the system parameters, and therefore (iii) the quantification of sensitivity to input parameters (Pieri and Gaudillère 2005).

As was demonstrated by several studies, a dynamical soil water balance model is therefore useful in forecasting the soil available water and the plant reaction at any time during the growing season (Trambouze and Voltz 2000, Lebon et al. 2003, Pellegrino et al. 2005b). A sensible soil water balance model may therefore be useful for both evaluating the qualitative potential of a vine growing area and optimizing the management techniques such as row azimut, row height/spacing ratio, row porosity (related to vine vigor), irrigation, soil surface management (Smart 1990). Moreover, like any other model, a soil water balance model may be used to simulate conditions that are not encountered in the reality. In that way, water balance modelling is very useful to predict the impact of climate change, especially if water shortage is expected to become widespread.

7.4.1 The Stomatal Regulation Function

The soil water balance model presented here is based on energy partitioning between the vine rows and the soil surface, calculated geometrically from the solar radiation interception ratio (k). The main parameters are therefore the soil water holding capacity, or total transpirable soil water (TTSW) and the geometrical descriptors: row azimut (az), row foliage height/ distance ratio (Hf/D) and row porosity (p), this later parameter may be linked to vine vigor. The model includes a stomatal regulation function that simulates the effect of decreased soil water availability upon vine transpiration and therefore feedback on soil water content evolution. The model functioning was validated in various soil and climate conditions (Lebon et al. 2003, Pellegrino et al. 2005b). If the root system is assumed in equilibrium and well adapted to the soil, the major impact of the fraction of transpirable soil water (FTSW) upon plant transpiration has been extensively demonstrated and is therefore used in modeling (Ritchie 1981, Trambouze and Voltz 2000, Lebon et al. 2003, Pellegrino et al. 2005b). Taking advantage of a nearly universal and close relationship between soil volumetric water content (θ) and matrix water potential (ψ) even with changing soil texture (Saxton et al. 1986, Saxton and Rawls 2006) (Fig. 7.7a, b), the bi-linear feed-back function previously describing plant stomatal regulation was replaced by a unique non-linear function of FTSW (Fig. 7.7c). This simplification actually assumes that the equivalent daily relative conductance (more exactly the daily relative transpiration TR/TRmax) is linearly related to the soil water potential. From fitting the water potential – FTSW relationships in 12 soil types (Fig. 7.7b) we found (Eq. 1):

$$\mathrm{TR/TRmax} = 1 - \exp(-5\ \mathrm{FTSW}) \tag{1}$$

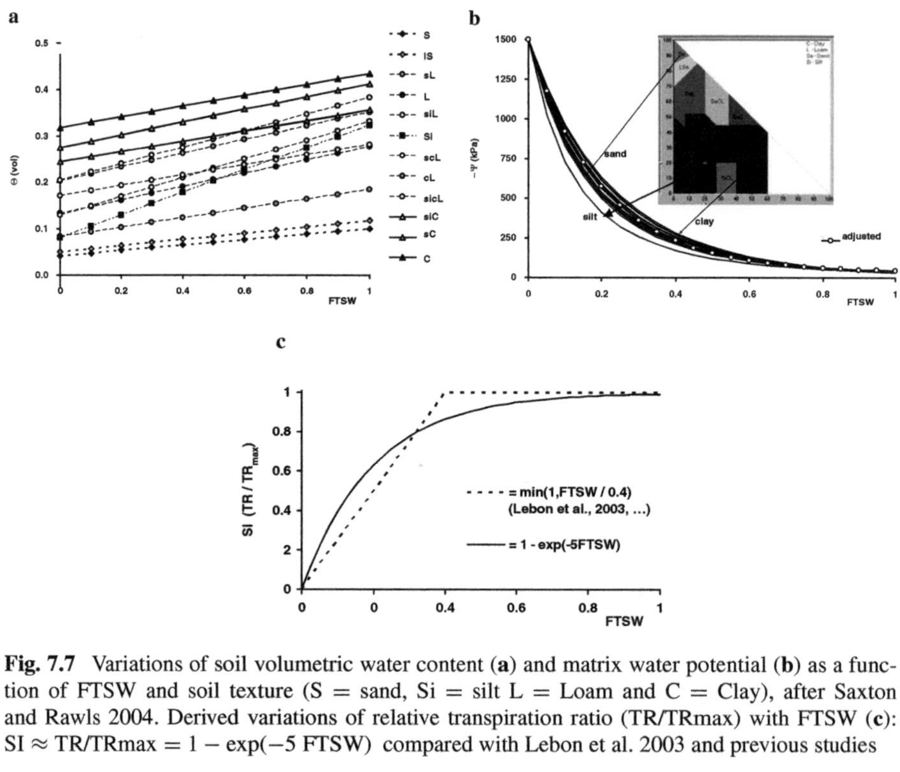

Fig. 7.7 Variations of soil volumetric water content (**a**) and matrix water potential (**b**) as a function of FTSW and soil texture (S = sand, Si = silt L = Loam and C = Clay), after Saxton and Rawls 2004. Derived variations of relative transpiration ratio (TR/TRmax) with FTSW (**c**): SI ≈ TR/TRmax = 1 − exp(−5 FTSW) compared with Lebon et al. 2003 and previous studies

7.4.2 Water Fluxes and Relations of Computed Water Stress Index with $\delta^{13}C$

The model can efficiently calculate the water fluxes and the evolution of soil water content as well as vine water stress all along the growing period (Fig. 7.8).

Several different water stress indicators may be derived from the model calculations. Here, the mean (.m) or minimum (.min) relative transpiration prior to veraison (ver) or from veraison to maturity (mat) was considered as a meaningful stress index indicator (SI). On the other hand, $^{13}C/^{12}C$ discrimination measurements ($\delta^{13}C$) of sugars in mature berries provided an independent and integrated estimation of water stress during the maturation period (Gaudillère et al. 2002). Measurements of $\delta^{13}C$ from different years (1997–2004) in 3 different soils of the same estate and therefore with similar training system characteristics (Château Cheval-Blanc near Bordeaux, vertical shoot positioning trellis) could be linearly related to the calculated mean relative vine transpiration rate during the maturation period, "SI.m mat" (Fig. 7.9). As expected, the different soils with contrasted holding capacities resulted in very different ranges of experienced vine water stress. Therefore, the gravely soil, with the lowest TTSW value, contributed most to the relationship.

One possible application of the model is historical retrospective study of water fluxes and water balance. For instance, simulations applied in a typical Bordeaux vineyard indicate variations of stress indicator SI (and correlated $\delta^{13}C$) were closely linked to wine vintage quality; 1995, 1998, 2000, 2001 and 2003 are considered very good to excellent vintages and were all associated with lower values of computed stress index (Fig. 7.10).

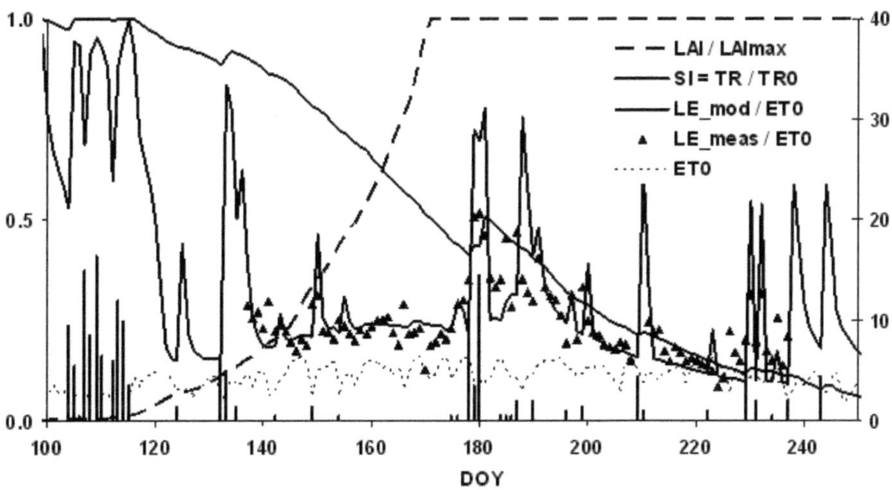

Fig. 7.8 Variations of water fluxes and vines water stress during year 2005 in a vineyard near Bordeaux. Rainfall (*histogram bars*) and reference evapotranspiration (ETo): right hand scale. *Left hand scale*: ratios of leaf area index (LAI/LAImax), actual evapotranspiration (LE/ETo), both modelled and measured and "stress index" (SI = TR/TR0, where TR is actual vine transpiration, and TR0 potential transpiration)

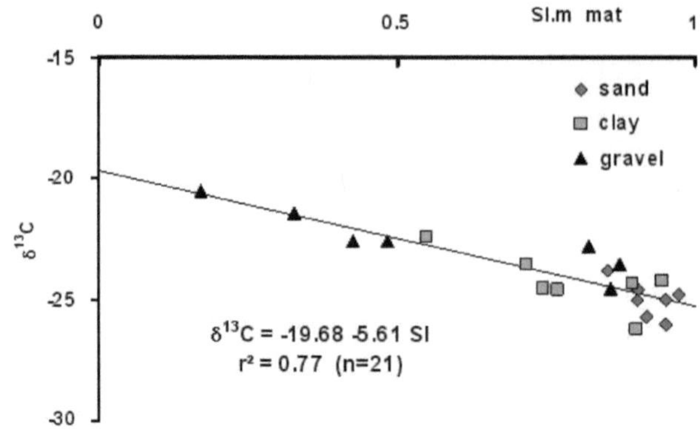

Fig. 7.9 Relationship of ^{13}C discrimination index ($\delta^{13}C$) with the mean water stress index calculated by the model between veraison and maturity (SI.m mat) in 3 different soils of water holding capacity: total transpirable soil water (TTSW) was 250, 170, 120 mm in soils labelled sand (with a water table within the reach of the roots, clay and gravel, respectively)

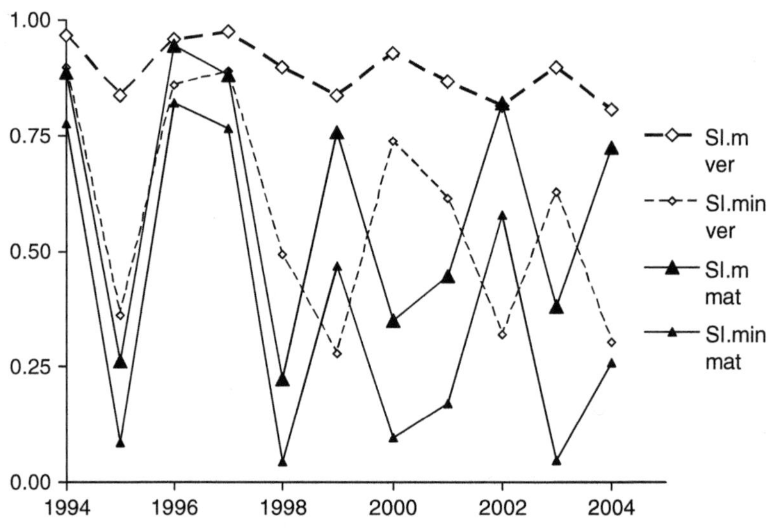

Fig. 7.10 Variations of mean (.m) and minimum (.min) values of stress index (SI) at veraison (ver) and maturity (mat). Meteorological data in Bordeaux 1994–2004. No cover crop, soil available water capacity (TTSW = 150 mm), row height/width ratio (Hf/W = 4), row height/spacing ratio (Hf/D = 0.8), row azimut (N–S rows), and row minimum porosity (p = 0.25) all constant

7.4.3 Sensitivity Analysis of the Water Balance Model

The water balance model represents a complex reality in a rather simplified way. However, it may be useful to unveil system properties that are not easily predictable,

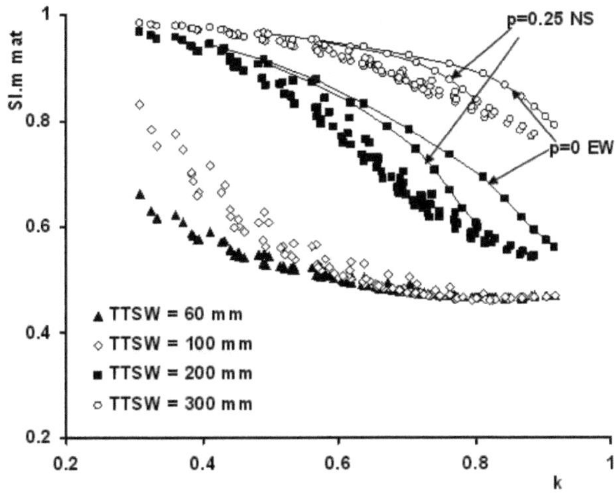

Fig. 7.11 Mean stress index from veraison to maturity (SI.m mat) as a function of mean ratio of radiation interception by the rows (k) and soil water holding capacity (TTSW). Climate data of Bordeaux 1994–2004. Row height/width ratio (Hf/w = 4) was kept constant. Different values of input parameters height/spacing ratio (Hf/D), row azimuth (N–S *vs*. E–W rows) and row porosity (p) were used to summarise the sensitivity to k at a given value of TTSW

especially if non-linear interactions are to be taken into account. For instance, the results of the water balance model are naturally highly sensitive to the TTSW value introduced as input parameter. However, the integrated effects of different soil (TTSW) and training system (radiation interception ratio k) parameters over different climates (years 1994–2004 in Bordeaux) would be entirely un-predictable without the help of such model (Fig. 7.11).

Lower values of TTSW or higher values of radiation interception by the vine rows (k) are linked to lower values of SI, indicating that the water stress experienced by the vines was stronger or longer. These integrated results of simulations confirm the water stress was mostly sensitive to the soil water holding capacity, especially if the more usual range of k [0.4–0.7] was considered. From the same data set, the overall sensitivity of average calculated water stress to variations of the training system parameters was estimated (Fig. 7.12). Training system parameters included the row azimuth (az = 0 for N–S, 90 for E–W rows) and the row minimum porosity (p = 0 and 0.25 represent vigorous and average vines, respectively).

Figure 7.12 summarizes the results of simulations at a constant row height/spacing ratio. Again, a major effect of TTSW is observed but the sensitivity to training system parameters is still not negligible at intermediate TTSW range, i.e. outside very dry and very wet conditions. In most common viticulture conditions however, higher Hf/D, lower row porosity and North–South rows all contribute to higher water stress intensity (lower SI) since they increase the k ratio and transpiration. In the case illustrated dealing with the last years in the Bordeaux area, the vine grower's choice of training system parameters could compensate for differences in soil water holding capacity up to 50 mm (Fig. 7.12).

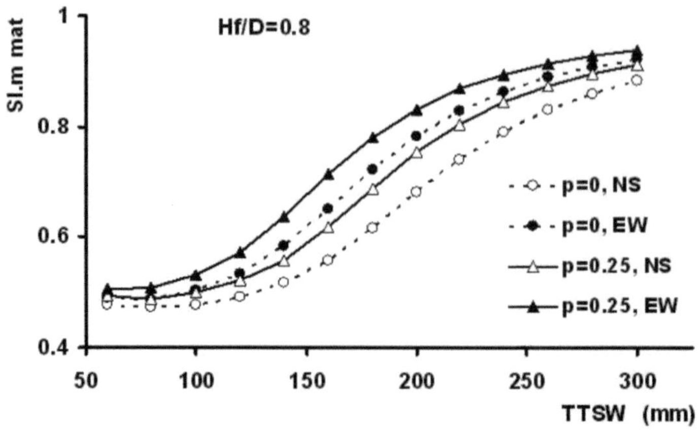

Fig. 7.12 Variations of mean water stress index at maturity (SI.m mat) with total transpirable soil water (TTSW) row azimuth (az, 0 for N–S, 90 for E–W rows), and row minimum porosity (p). Climate data of Bordeaux, 1994–2004. Here, the row height / spacing ratio was kept constant (Hf/D = 0.8)

7.4.4 Possible Applications of the Water Balance Model

With a given target of water stress index, that might result from a separate analysis of quality and/or economic performance goals, these results provide a basis for an optimisation of the training system, taking fully into account the climatic constraints and the climate variability.

The same model could also easily be used for near-real time vine water status estimation and irrigation management (decision making about irrigation investment could also be greatly improved by an analysis of sensitivity to irrigation parameters in any given climate context).

Since it is based on climate data and on a limited set of easily accessible parameters, the model can readily be applied to a changing climate. Short term weather forecasts may be included in input variables to improve risk analysis and real time vineyard management. Impact of long term climate change may be assessed from long term climate simulations derived from global atmosphere circulation models.

7.5 Conclusions

A large range of methods for assessing vine water uptake conditions have been developed over the years. Some of them are more accurate and more easily performed than others. Vine water status is accurately assessed by water balance modelling and stem water potential measurements. Stem water potential may be the most accurate tool for irrigation management. It can be used to quantify the degree of water stress experienced by the plant and to monitor changes in vine water status over the season. The principal drawback of this technique is that its implementation is time consuming. Carbon isotope discrimination ($\delta^{13}C$), measured on grape sugar at ripeness, indicates the average water deficit stress experienced by the vines during grape ripening. Because carbon isotope discrimination measurement allows the execution of many measurements, it is a useful tool for block characterisation (for example in terroir studies). Its measurement takes place at ripeness. Hence, it cannot be used for irrigation management during the growing season. However, it can be used at harvest to validate irrigation strategies. Water balance modelling allows water status assessment derived from easily available climatic data. It is useful for real-time vineyard monitoring, training system optimization as well as the study of climate change impact.

Acknowledgments Figure 7.5 was provided by Jean-Pascal Goutouly (UMR EGFV, ISVV, F-33882 Villenave d'Ornon). Figure 7.6 was provided by David Pernet (SOVIVINS, F33650 Martillac).

References

Améglio T, Archer P, Cohen M, Valancogne C, Daudet F.-A, Dayau S, Cruiziat P (1999) Significance and limits in the use of predawn leaf water potential for tree irrigation. Plant Soil 207:155–167.

Begg J, Turner N (1970) Water potential gradients in field tobacco. Plant Physiol 46: 343–346

Choné X, van Leeuwen C, Dubourdieu D, Gaudillère JP (2001) Stem water potential is a sensitive indicator for grapevine water status. Ann Bot-London 87:477–483

Cifre J, Bota J, Escalona J, Medrano H, Flexas J (2005) Physiological tools for irrigation scheduling in grapevine (Vitis vinifera L.): an open gate to improve water use efficiency? Agric Ecosyst Environ 106:159–170

Escalona J, Flexas J, Medrano H (2002) Drought effects on water flow, photosynthesis and growth of potted grapevines. Vitis 41:57–62

Farquhar G, Ehleringer J, Hubick K (1989) Carbon isotope discrimination and photosynthesis. Annual Rev Plant Physiol Plant Molecular Biol 40:503–537

Garnier E, Berger A (1986) Effects of water stress on stem diameter changes of peach trees growing in the field. J Appl Ecol 23:193–209

Gaudillère JP, van Leeuwen C, Ollat N (2002) Carbon isotope composition of sugars in grapevine, an integrated indicator of vineyard water status. J Exp Bot 53:757–763.

Greenspan M, Shackel K, Matthews M (1994) Developmental changes in the diurnal water budget of the grape berry exposed to water deficits. Plant Cell Environ 17:811–820

Hsiao T (1973) Plant responses to water stress. Annu Rev Plant Physiol 24: 519–570

Jones G (1999) Use of InfraRed thermometry for estimation of stomatal conductance as a possible aid for irrigation scheduling. Agric For Meteorol 95:139–149

Jones H (2004) Irrigation scheduling: advantages and pitfalls of plant-based methods. J Exp Bot 55:2427–2436

Lebon E, Dumas V, Pieri P, Schultz H (2003) Modelling the seasonal dynamics of the soil water balance of vineyards. Funct Plant Biol 30:699–710

Lovisolo C, Schubert A (1998) Effects of water stress on vessel size and xylem hydraulic conductivity in Vitis vinifera L. J Exp Bot 49:693–700

Pellegrino A, Lebon E, Simonneau T, Wery J (2005a) Toward a simple indicator of water stress in grapevine (Vitis vinifera L.) based on the differential sensitivities of vegetative growth components. Aust J Grape Wine Res 11:306–315

Pellegrino A, Lebon E, Voltz M, Wery J (2005b) Relationships between plant and soil water status in vine (Vitis vinifera L.). Plant Soil 266:129–142

Pieri P, Gaudillère JP (2005) Vines water stress derived from a soil water balance model – sensitivity to soil and training parameters. In: Schultz H (ed) Proceedings of the XIVth International Conference GESCO, 23–27 August 2005, Geisenheim, Germany, pp 457–463

Ritchie J. (1981) Water dynamics in the soil-plant-atmosphere system. Plant Soil 58:81–96

Saxton K, Rawls W, Romberger J, Papendick R (1986) Estimating generalized soil water characteristics from texture. Trans Am Soc Agric Eng 50:1031–1035

Saxton K, Rawls W (2006) Soil water characteristic estimates by texture and organic matter for hydrologic solutions. Soil Sci Soc Am J 70:1569–1578

Scholander P, Hammel H, Edda D, Bradstreet E, Hemmingsen E (1965) Sap pressure in vascular plants. Science 148:339–346

Schultz H (1996) Water relations and photosynthetic responses of two grapevine cultivars of different geographical origin during water stress. Acta Hort ISHS 427:251–266

Schultz H, Matthews M (1988) Resistance to water transport in shoots of Vitis vinifera L. Plant Physiol 88:718–724

Shackel K (2006) Water relations of woody perennial plant species, in: VAN LEEUWEN, C. et al. (ed) Proceedings of the VIth International Terroir Congress, 2–7 July 2006, Bordeaux: ENITA – Montpellier: Syndicat Viticole des Coteaux du Languedoc, France, pp 54–63

Smart R, Dick J, Gravett I, Fisher B (1990) Canopy management to improve grape yield and wine quality – Principles and practices. S Afr J Enol Vitic 11:3–17

Trambouze W, Voltz M (2000) Measurement and modelling of the transpiration of a Mediterranean vineyard. Agric For Meteorol 2913:1–14

Van Leeuwen C, Seguin G (1994) Incidences de l'alimentation en eau de la vigne, appréciée par l'état hydrique du feuillage, sur le développement de l'appareil végétatif et la maturation du raisin (Vitis vinifera variété Cabernet Franc, Saint Emilion 1990). J Int Sci Vigne Vin 28: 81–110

Van Leeuwen C, Gaudillère JP, Trégoat O (2001) Evaluation du régime hydrique de la vigne à partir du rapport isotopique $^{13}C/^{12}C$. J Int Sci Vigne Vin 35:195–205

Chapter 8
Gas-Exchange and Chlorophyll Fluorescence Measurements in Grapevine Leaves in the Field

Alexander Gallé and Jaume Flexas

Abstract The analysis of photosynthetic processes under field conditions, in particular under varying climatic conditions has become an important issue for plant scientists and agronomists. An easy and robust measuring technique is needed to assess the underlying biophysical and biochemical processes of photosynthesis. Advances in analysis of leaf traits, such as photosynthetic activity and limitations, have been made due to improvements of leaf gas exchange analysis and chlorophyll fluorimetry. In this chapter the basics of photo-biochemistry and physics, as well as the fundamental model of photosynthesis by Farquhar et al. are described. Recent methods on how to determine various photosynthetic parameters are discussed, including a section on potential errors and mistakes. Finally, the potential of combined measurements of leaf gas exchange and chlorophyll fluorescence is introduced, emphasizing the importance of limitations of CO_2 diffusion across a leaf ("mesophyll conductance").

Contents

A. Gallé (✉)
Grup de Recerca en Biologia de les Plantes en Condicions Mediterrànies Universitat de les Illes Balears, Palma de Mallorca 07122, Spain
e-mail: alexander.galle@uib.es

S. Delrot et al. (eds.), *Methodologies and Results in Grapevine Research*,
DOI 10.1007/978-90-481-9283-0_8,

Abbreviations

A_N	Net CO_2 assimilation rate (e.g. mmol CO_2 m^{-2} s^{-1})
C_a	Ambient CO_2 concentration (e.g. micro mol CO_2)
C_c	CO_2 concentration in the chloroplast (e.g. ppm CO_2)
C_i	Leaf internal CO_2 concentration, in the substomatal cavities (e.g. ppm CO_2)
E	Leaf transpiration rate (e.g. mol H_2O m^{-2} s^{-1})
Fm	Maximum chlorophyll *a* fluorescence
Fm′	Maximum chlorophyll *a* fluorescence in the light adapted state
Fs	Apparent chlorophyll *a* fluorescence in the light adapted state
Fv	Variable chlorophyll *a* fluorescence
Fo	Basal chlorophyll *a* fluorescence (in the dark)
Fo′	Basal chlorophyll *a* fluorescence (after light–dark transition)
Φ_{PSII}	Apparent efficiency of the PSII photochemistry
g_c	Cuticular conductance for water vapour (or CO_2)
g_m	Mesophyll conductance for CO_2 (e.g. mol CO_2 m^{-2} s^{-1})
g_s	Stomatal conductance for water vapour or CO_2 (e.g. mol H_2O m^{-2} s^{-1})
Γ^*	CO_2 compensation point between photosynthesis and photorespiration
IRGA	Infra red gas analyser
$J_{(max)}$	(maximum) Photosynthetic electron transport rate
NPQ	Non-photochemical quenching
PSII	Photosystem II
qP	Photochemical quenching
R_d	Rate of day respiration or respiration in the light (e.g. μmol O_2 m^{-2} s^{-1})
Rubisco	Ribulose-1,5-bisphosphate carboxylase/oxygenase
TPU	Triose-phosphate utilization
$V_{c,max}$	Apparent carboxylation rate of rubisco
VPD	Vapour pressure deficit

8.1 Introduction: Photosynthetic Primary Reactions Versus Gas Exchange

Photosynthesis is the fundamental process whereby plants capture and process sunlight and CO_2 to achieve growth and reproduction. Biophysical processes, which include CO_2 transport from the atmosphere through stomata to the sites of carboxylation in the chloroplast, and biochemical processes located in the chloroplast thylakoid membranes, stroma, mitochondria and cytosol of the cell, determine the rate of net CO_2 assimilation (A_N). All these processes as well as environmental variables like light intensity and temperature can have different effects on A_N. To understand the biology behind changes in A_N and to allow predictions of environmental and genetic influences on plant productivity, the dissection of biophysical and biochemical factors and the calculation of photosynthesis parameters is an important tool.

The uptake of CO_2 for photosynthetic carbon assimilation and the loss of water by leaf transpiration are tightly coupled, being influenced by external factors like changes in water availability or in light intensity, and by internal factors (e.g. C3, C4 or CAM photosynthesis) depending on the genomes. Hence, monitoring CO_2 and water vapour exchange across a leaf simultaneously with the now available portable infrared gas-exchange analysers (IRGA's) provides a very good measure of photosynthetic activity and its limitations.

8.2 The Basics of Gas-Exchange Analysis: Net Photosynthesis, Stomatal Conductance, Leaf Transpiration and the Sub-stomatal Concentration of CO_2

During the last 20 years, gas exchange analysis has become a common tool for measurements of photosynthetic CO_2 uptake in leaves, as a result of the concomitant availability of portable IRGA's for field measurements (Long et al. 1996). Among them, two major systems have evolved, i.e. a closed path and an open path gas-exchange system. In brief, in a closed system there is no net flow of air through the system, while in an open or differential system there is a net air flow. Nowadays, the open gas-exchange configuration is the most commonly used system, as in the CIRAS-1 (PP Systems Ltd., Hitchin, Herts, UK) and the LI-6400 (LI-COR Inc., Lincoln, Nebraska, USA).

As depicted in Fig. 8.1, the major parts of this system consist of (1) an air supply unit, (2) a precision flow-meter, (3) a transparent enclosure or leaf chamber and (4) IRGA(s). Air of known and controlled CO_2 (Ce) and water vapour (we) mole fraction is pumped from the air supply unit and split into two air streams. In order to provide a certain concentration of CO_2 and water vapour, they can be first removed from the air by using soda lime ($Ca(OH)_2$ and NaOH granulate) and Drierite ($CaSO_4$). Then disposable compressed CO_2 gas cylinders can be used to provide a controlled rate of CO_2 added back to the CO_2-free air, while water vapour can be regulated using the bypass loop or by changing the flow rate. One of the split air streams provides air for the reference cell of the IRGA(s) and the other one passes via flowmeter and the leaf chamber to the analysis cell of the IRGA(s).

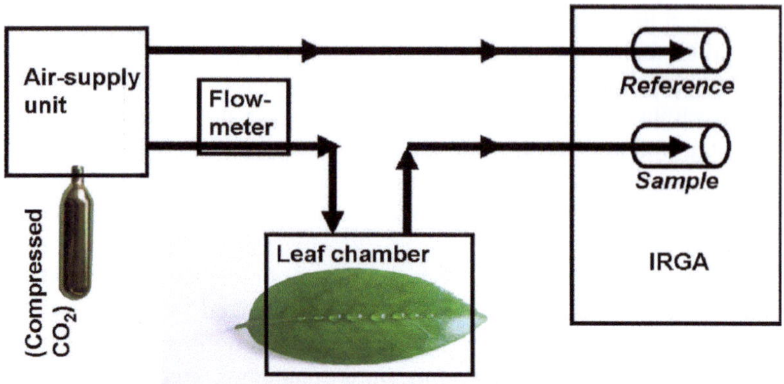

Fig. 8.1 Schematic layout of an open gas exchange system. Air from the outside is pumped into the air-supply unit and CO_2 and water vapour are removed by absorbents and then replaced at a controlled level. Before entering the leaf chamber, the air stream is split into a reference and an analyse stream, passing via a thermal mass flowmeter. Finally, both streams pass a dual beam IRGA and the air is released into the atmosphere again. Comparing reference and analysis beam provides information about changes in CO_2 and water vapour, hence leaf gas exchange rates can be determined (A_N, g_s, etc.)

The difference in mole fractions of CO_2 and water vapour between the chamber entrance (reference cell; C_e, w_e) and outlet (analysis cell; C_o, w_o) is recorded by the IRGA(s) and A_N may then be calculated:

$$A_N = \frac{u_e\left(c_e - \frac{(1-w_e)}{(1-w_o)}c_o\right)}{s} \qquad \text{or}$$

$$A_N = u_e\frac{c_e - c_o}{s} - c_oE$$

where '*ue*' is the flow of air passing through the chamber (mol s^{-1}), 's' the surface area of the enclosed leaf (m^2), and in the simplified version E is the transpiration rate (mol H_2O m^{-2} s^{-1}). From E stomatal conductance for water vapour (g_s) can be deducted, implementing leaf temperature and air pressure for the estimation of leaf water vapour ("total conductance"), and subtracting the boundary layer conductance. The only assumption done is that the relative humidity inside leaf substomatal chamber is 100%, which is close to the real value according to direct measurements. Inter-conversion of g_s for water vapour to g_s for CO_2 can be easily done by dividing with the factor 1.6, which is the ratio of H_2O/CO_2 diffusivities in air. Finally, the internal or sub-stomatal CO_2 concentration (C_i) can be derived from simultaneous measurements of A_N and g_s, according to Fick's first law of diffusion: $C_i = C_a - A_N/g_s$, where C_a is the CO_2 concentration around the leaf.

However, the CO_2 concentration within the mesophyll at the sites of carboxylation in chloroplasts (C_c) is of main interest to determine the real or apparent photosynthetic activity. This has become evident because of the draw-down of CO_2 from C_i to C_c, which may change, especially under stressful situations (Flexas et al. 2002, Flexas et al. 2008, Niinemets et al. 2009b). Thus, equally to the equation of

Fick's first law for C_i it can be written $C_c = C_i - A_N/g_m$, where g_m is the mesophyll conductance for CO_2. Determination of g_m can then be used to calculate C_c, which may help to better estimate the apparent or real photosynthetic activity (Ethier and Livingston 2004, Flexas et al. 2008, Long and Bernacchi 2003, Niinemets et al. 2009b). Using C_c instead of C_i has been shown to be a powerful tool for the analysis of A_N/C_i curves (the response of A_N to varying C_a) and thus, leading to more reliable results of photosynthetic activity in vivo (see Section 8.3). To assess g_m a combination of leaf gas-exchange and chlorophyll fluorescence analysis can be used, besides other more sophisticated techniques (i.e. isotope-ratio mass spectrometry), which is advantageous in the field when using a portable gas-exchange system equipped with a fluorimeter (see Section 8.6).

8.3 The Farquhar et al. Model of Photosynthesis and CO_2-Response Curves of Gas Exchange

Due to the path of CO_2 across a leaf through stomata and mesophyll to the sites of carboxylation, limitations of photosynthesis include stomatal and non-stomatal factors, which cannot be easily distinguished. A quantitative model has been established by Farquhar and co-workers in the early 80's (Farquhar et al. 1980), based on the idea that the response of A_N to C_i only depends on mesophyll processes, eliminating the effects of boundary layer and stomata ($A_N = g_s(C_a - C_i)$). Rapid variations of CO_2 concentrations around a leaf, so called "A_N/C_i curves" are therefore commonly used to determine the limitations of photosynthesis (Long and Bernacchi 2003, Sharkey et al. 2007), either by Rubisco activity, by RuBP-regeneration or by triose-phosphate utilization (TPU) (Fig. 8.2).

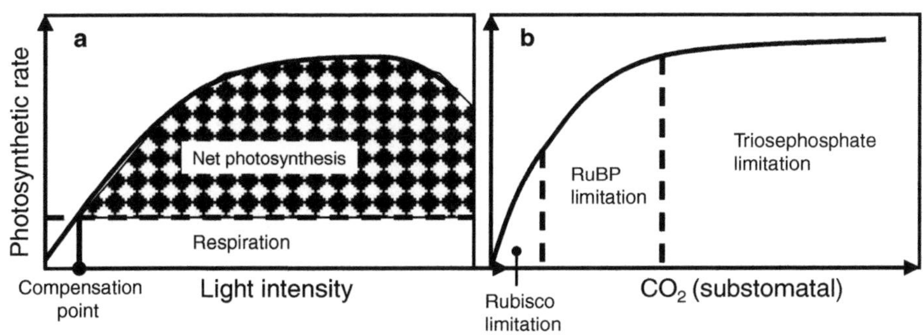

Fig. 8.2 The typical response of net CO_2 assimilation (A_N) to varying light intensities (**a**) and to internal CO_2 (a; "A_N/C_i curve"). A_N increases with increasing light intensity until a saturation plateau is reached, while under excessive light conditions A_N starts to decline again due to photoinhibition processes. A_N increases also with increasing CO_2 until a plateau is reached. The different parts of this response curve represent possible limitations to photosynthesis, wherein the initial linear part refers to the (maximal) Rubisco activity, the second part to the (maximal) electron transport rate and the last to the utilization of triose-phosphate

In brief, a common protocol for these A_N/C_i curves is as follows: the initial point of the curve is taken under ambient CO_2 (380 ppm) and steady state photosynthesis under saturating light. Thereafter, CO_2 concentration around the leaf is lowered stepwise (in 50 ppm intervals) to 50 ppm, providing about at least 2 min at each step for stabilization. Then, CO_2 concentration is increased in intervals of 150 ppm up to about 2,000 ppm (depending on the maximum values allowed by the IRGA system used). The initial slope of this A_N/C_i curve represents the apparent Rubisco activity ($V_{c,max}$) in vivo, while a decrease of dA_N/dC_i thereafter indicates a limitation by RuBP-generation and thus describes the maximum electron transport rate (J_{max}). A further increase of C_i may lead to a decrease of $A_N(dA_N/dC_i \leq 0)$ when TPU becomes limiting.

As CO_2 and O_2 compete for the Rubisco binding sites, the model has to account for that and A_N is therefore mathematically expressed as: $A_N = v_c - 0.5v_o - R_d = v_c(1 - \Gamma^*/C_i) - R_d$. The term $(1 - \Gamma^*/C_i)$ is used to account for recently assimilated carbon that is released in photorespiration, R_d denotes for respiration in the light without photo-respiration. Γ^* is the CO_2 concentration at the photosynthetic compensation point, when photorespiratory CO_2 efflux equals photosynthetic CO_2 uptake ($v_c = 2\ v_o$). This is different from the CO_2 compensation point Γ, where $v_c = 2\ v_o + R_d$. Several methods have been established to determine Γ^*. Different models for $V_{c,max}$, J_{max} and TPU determination have been proposed, being useful depending on the conditions of plant and environment, as well as on the scientific questions addressed (Evans et al. 1986, Ethier and Livingston 2004, Sharkey et al. 2007, Warren 2008, Niinemets et al. 2009b). An easy way to analyse A_N/C_i curves is to use the online program by Sharkey et al. (2007), which is available on: http://www.blackwellpublishing.com/PlantSci/pcecalculation/default.asp The Farquhar et al. model has been used successfully to model photosynthesis in grapevines (Maroco et al. 2002, Schultz 2003, Flexas et al. 2009; see Table 8.1).

8.4 Potential Errors: Influence of Leaf Temperature Recordings, Cuticular Conductance, and Heterogeneous Stomatal Closure

When conducting gas exchange measurements a number of precautions have to be taken to minimize or eliminate possible errors. First, a common problem during gas-exchange measurements is a leakage of CO_2 through the foam gaskets of the leaf chamber, which has its greatest effects during a run of an A_N/C_i curve when CO_2 differences between inside and outside the chamber become large (Flexas et al. 2007, Rodeghiero et al. 2007). One fast and easy method to correct for this CO_2 leakage is to do another A_N/C_i curve with a killed leaf (boiled in water) and to use this curve for recalculation of the in vivo data sets. Gradients of temperature and water vapour across and along the clamped leaf should be minimized and the environment inside the chamber maintained as homogeneous as possible, to ensure measurements of net CO_2 assimilation in a well-defined microclimate. Moreover,

Table 8.1 Overview of parameters that can be determined with IRGA's and/or fluorimeters, citing the most relevant literature. For details of abbreviations see text

Parameter	IRGA	Fluorimeter (PAM)	References (methodology)	References (grapevine)
A_N; g_s; E; C_i	√		Farquhar et al. (1980); Long et al. (1996); Long and Bernacchi (2003)	Chaves et al. (1987); Downton et al. (1987); Escalona et al. (1999); Medrano et al. (2003); Schultz (2003)
F_v/F_m, Φ_{PSII}; qP; qN (NPQ)		√	Genty et al. (1989); Horton et al. (1996); Maxwell and Johnson (2000); Krause and Jahns (2004)	Maroco et al. (2002); Medrano et al. (2002); Guan et al. (2004); Flexas et al. (1999, 2002);
ETR (J); J_{max}		√	Krall and Edwards (1992); Maxwell and Johnson (2000); Epron et al. (1995); Valentini et al. (1995)	Flexas et al. (1999, 2002, 2009); Souza et al. (2003)
$V_{c,max}$	√	(√)	Farquhar et al. (1980); Long and Bernacchi (2003); Ethier and Livingston (2004); Sharkey et al. (2007)	Schultz (2003); Souza et al. (2003); Flexas et al. (2009)
C_c; g_m	√	√	Harley et al. (1992); Evans et al. (1986); Ethier & Livingston (2004); Sharkey et al. (2007); Flexas et al. (2007); Niinemets et al. (2009 a,b,c)	Flexas et al. (2002, 2009); Sampol et al. (2003); Moutinho-Pereira et al. (2004)

valid measurements of g_s and C_i depend on correctly recorded leaf temperatures (see above), which seems trivial but not always easy to achieve.

Moreover, kinetic parameters of the enzyme Rubisco respond to changes in temperature and thus, a temperature response function has to be incorporated into the analysis of A_N/C_i curves, particularly for the estimation of $V_{c,max}$. The in vivo temperature response functions of the parameters based on the chloroplast CO_2 concentration have been derived from in vitro assays and are implemented in the current models, although these are for tobacco (Bernacchi et al. 2001, 2003) and no estimate is available for grapevines.

Measurements of g_s with an IRGA system in fact provide the sum of stomatal and cuticular conductance (g_c), the latter being usually low under standard conditions (relative high g_s) and therefore negligible. However, under stressful conditions such as drought g_s may become very low and thus g_c contributes to a larger extent to the measured g_s. To account for this possible overestimation of g_s due to g_c, the following relatively easy-to-conduct methods can be applied. First, in hypostomatal plants the lower side of the leaf, which commonly presents most or all stomata, can be sealed with Teflon or other water-impermeable lubricant and then the conductance of the sealed leaf is measured. The recorded conductance represents the upper-side leaf conductance (without stomata!), in particular the g_c (Boyer et al. 1997). Another way to determine g_c is to measure g_s continuously on a detached leaf until the decline of g_s flattens out and values become stable. This end point represents the remaining conductance (g_c) after stomata have been completely closed. More sophisticated methods can be found elsewhere (Hoad et al. 1996, Santrucek et al. 2004).

Heterogeneous stomatal closure or patchy stomatal response is a common phenomenon in leaves. However, it can be neglected when transpiration and g_s are high (non-stressed conditions) or when gas exchange measurements are carried out on a relatively large leaf area, covering a high number of stomata. In particular under drought stress, when stomata close und hence g_s decreases, the stomatal patchiness may impair the correct determination of C_i, as demonstrated in grapevines under certain conditions (Downton et al. 1988a, b) but not in others (Flexas et al. 2002, 2009). An easy way to check for patchiness is to do several measurements on different parts of the same leaf and compare/compile the data (Flexas et al. 2002), or to conduct A_N/C_i curves under increasing VPD and decreasing g_s (Grassi and Magnani 2005, Flexas et al. 2009).

8.5 The Basics of Chlorophyll Fluorescence Analysis: The Concept of Variable Fluorescence, the Kautsky Effect and Its Interpretation

Excitation of chlorophyll molecules by absorbed light energy can lead to three different processes: driving photosynthesis (photochemistry), dissipation of excess energy to heat and emission of chlorophyll fluorescence. As these three processes are in competition, a decrease in one results in an increase in the yield of the other two. Hence, information about changes in the efficiency of photochemistry and heat dissipation can be derived by measuring chlorophyll fluorescence (Maxwell and Johnson 2000). This is easy to measure, due to the fact that the spectrum of fluorescence differs from that of absorbed light, with a peak of emission of longer wavelength (>710 nm) than that of absorption (<680 nm). Therefore, fluorescence yield can be quantified by exposing a leaf to light of defined wavelength and measuring the amount of light re-emitted at longer wavelength. An important modification for the application of chlorophyll fluorescence has been undertaken by using a modulated measuring system, in which the light source for fluorescence measurement is switched on and off at high frequency and the detector is tuned to detect only the emitted fluorescence from this signal. Hence, background illumination and most importantly full sunlight in the field does not interfere with the fluorescence measurements.

Changes in the yield of chlorophyll fluorescence have been observed first by Kautsky and co-worker in the early 1960's (Kautsky et al. 1960), when transferring leaves form the dark into the light. They recorded first an increase in yield of chlorophyll fluorescence over a period of about 1 s, followed by a decline during several minutes. The rise of fluorescence yield has been explained by the consequence of reduction of electron acceptors downstream of PSII, namely plastoquinone (i.e. Q_A). After all Q_A are reduced, no more electrons are accepted from PSII (PSII reaction centers are 'closed'), until they are passed onto the next acceptor Q_B. Hence, during this time the yield of chlorophyll fluorescence becomes maximal (Fm). Thereafter, the yield of chlorophyll fluorescence drops to almost initial values during several minutes, the so-called fluorescence quenching, which can be explained in two ways. First, an increased rate of electrons might be transported away from PSII to the final acceptors in the carbon metabolism (activated by light), which is named "photochemical quenching". Besides that, there is an increase in thermal dissipation of excess energy, which is termed "non-photochemical quenching (NPQ)".

According to this so-called 'Kautsky effect', several photochemical and non-photochemical parameters can be derived from quenching analysis (Horton et al. 1996, Maxwell and Johnson 2000, Krause and Jahns 2004). To assess changes in the PSII photochemistry, the maximum efficiency of the PSII photochemistry (Fv/Fm), the (apparent) efficiency of the PSII photochemistry (Φ_{PSII}) and the fraction of open PSII reaction centers (qP) are the most useful quenching parameters. As depicted in Fig. 8.3, Fv/Fm is determined in a dark-adapted leaf as Fv/Fm = (Fm–Fo)/Fm, where Fo is the fluorescence signal after switching on the measuring light and Fm is the maximal fluorescence (see above) after applying a saturating flash (> 4,000 μmol photons m^{-2} s^{-1}). Under non-stressful conditions Fv/Fm ranges around 0.8, with some species-specific variations. In principle, a decrease in Fv/Fm results from an increase in thermal dissipation (non-photochemical quenching) at the expense of photochemical activity. Thus, a lowered Fv/Fm is a good indicator for sustained impaired photochemistry ("photoinhibition"), when it is measured after an appropriate period (usually 10–30 min) of dark adaptation. To assess the PSII photochemistry in a leaf under light-adapted conditions (e.g. in the field), Φ_{PSII} and qP can be determined (Genty et al. 1989). Φ_{PSII} is calculated as $\Phi_{PSII} = (Fm'-Ft)/Fm'$, where Ft is recorded under continuous light exposure by a fixed intensity ('actinic light', which can be given by the light source of an IRGA leaf chamber or fluorimeter) and Fm′ is obtained when superimposed with saturating light pulse(s). When the light is turned off after the saturating puls Fo′can be determined and thus qP is calculated as $qP = (Fm' - Ft)/(Fm' - Fo')$. Φ_{PSII} can give a measure of photosynthetic linear electron transport (ETR, *J*) and the photosynthetic capacity in vivo (Genty et al. 1989), when calculated as:

$J = \Phi_{PSII} \times PAR \times \alpha \times \beta$. PAR denotes for the incident light (μmol photons m^{-2} s^{-1}), α is the fraction of the absorbed light by the measured leaf and β the fraction of absorbed light energy distributed to PSII. For α and β general assumptions can be made, as for a mature non-succulent leaf α is usually around 0.84 and β = 0.5 under steady state photosynthesis.

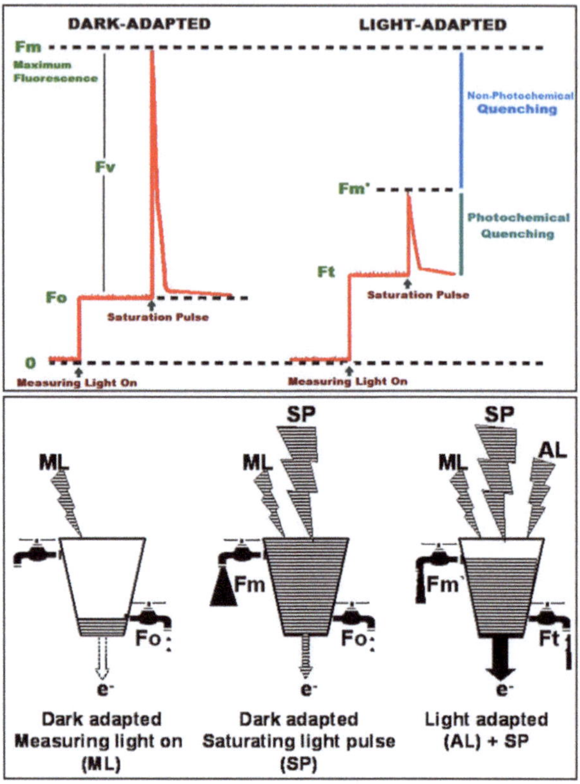

Fig. 8.3 Schematic overview of chlorophyll fluorescence analysis. See text for details

Φ_{PSII} and qP can be interrelated with Fv/Fm: Fv/Fm $= \Phi_{PSII}$/qP. In fact, qP and Fv/Fm provide information about the underlying processes which have altered efficiency, while Φ_{PSII} relates to achieved efficiency. Optimal utilization of photochemical energy in carbon metabolism (including photorespiration) is characterized by high qP values, while qP declines when light absorption exceeds requirements of carbon assimilation. Thus among other factors, qP serves as a valuable indicator of "light stress".

Quantification of non-photochemical quenching always requires measurements on dark-adapted leaves, in particular determination of Fm. Thus, the most commonly used parameter is NPQ, which is linearly related to thermal dissipation and covers a range from 0 to infinity: NPQ $=$ (Fm $-$ Fm$'$)/Fm$'$. Note that changes in NPQ reflect changes in heat dissipation relative to the dark adapted state. An increase in NPQ can result from either processes that protect the leaf against light-induced damage or damage itself. Here, various processes can be involved, as e.g. xanthophyll cycle activity (also shown in grapevines, Medrano et al. 2002) and phosphorylation/dephosphorylation of light harvesting complexes (Demmig-Adams and Adams III 1996, Horton et al. 1996, Krause and Jahns 2004, Szabo et al. 2005).

All the chlorophyll fluorescence parameters described here have been used in studies in grapevines (Flexas et al. 2002, Maroco et al 2002, Medrano et al. 2002; see Table 8.1).

8.6 The Application Potential of Simultaneous Gas Exchange and Chlorophyll Fluorescence Measurements

Combining gas exchange and chlorophyll fluorescence analysis has become a very powerful tool to investigate the relationship between CO_2 fixation, light use efficiency and photoinhibition. As CO_2 fixation and electron transport can be measured simultaneously, it is possible to estimate the extent of photorespiration in vivo. Here, A_N and Φ_{PSII} are measured under different light intensities ("light response curve") and under non-photorespiratory conditions (elevated CO_2 or 1–2% O_2). Hence, a linear plot of the quantum yield of CO_2 fixation (Φ_{CO2}) and PSII photochemistry (Φ_{PSII}) reveals the number of electrons required for one molecule of CO_2 to be fixed (Epron et al. 1995, Valentini et al. 1995). Assuming that this relationship holds true under photorespiratory conditions, the photorespiration can be estimated in vivo, which has been widely used to assess the photo-protective role of photorespiration under drought stress (Valentini et al. 1995, Niinemets et al. 1999, Flexas et al. 2009).

Moreover and as already mentioned above (see 8.2.), leaf intrinsic diffusion components (g_m, C_c) can be derived from these simultaneous measurements according to Fick's law of diffusion ($A_N = g_m(C_i - C_c)$) (Flexas et al. 2008, Niinemets et al. 2009b). g_m can be derived from fluorescence and gas exchange data, using the so called constant *J* or variable *J* method when photosynthesis is limited by the regeneration of RuBP or by Rubisco (Bernacchi et al. 2002, Epron et al. 1995, Harley et al. 1992). Most commonly and due to the fact that photosynthesis limitation by RuBP and TPU are not commonly observed, the variable *J* method according to Harley et al. (1992) is applied and calculated as:

$$g_m = \frac{A_N}{C_i - \dfrac{\Gamma^*(J + 8(A_N + R_d))}{J - 4(A_N + R_d)}}$$

For the non-photorespiratory CO_2 compensation point Γ^*, the value of 42.9 μmol mol^{-1} CO_2 (at 25°C) from Bernacchi et al. (2001) can be used, which is in good agreement with values for *Vitis* that range between 42 and 43 μmol mol^{-1} (Flexas et al. 2009). The respiration in the light or 'day'respiration (R_d) can be measured with an IRGA system according to Laisk (1977) or by dividing the measured dark respiration by two according to Niinemets and co-workers (Niinemets et al. 2005, Niinemets et al. 2009c). For the "Laisk method" A_N/C_i curves are conducted under three different light intensities (Von Caemmerer 2000), which are usually 800, 300 and 100 PAR. The intersection of these three curves gives an estimate of R_d (y-axis), as well as of Γ^* (x-axis) according to Warren and Dreyer (2006).

Among the different models for determination of g_m and C_c from gas exchange and fluorescence measurements (Loreto et al. 1992, Evans and von Caemmerer 1996, Ethier and Livingston 2004, Niinemets et al. 2009a), the freely available and easy-to-handle software of Sharkey et al. (2007) can be recommended to the novice: http://www.blackwellpublishing.com/PlantSci/pcecalculation/default.asp Although precautions and additional measurements should be taken when attempting the measurement of g_m and C_c (see details and discussion in Pons et al. 2009), this has been done successfully in grapevines under different conditions (Flexas et al. 2002, 2009, Sampol et al. 2003, Moutinho-Pereira et al. 2004).

Finally, by using simultaneous gas exchange and fluorescence measurements, a quantitative limitation analysis of photosynthesis can be conducted (Flexas et al. 2009, Grassi and Magnani 2005, Wilson et al. 2000), where the total limitation is divided in three components: stomatal limitation ("g_s"), mesophyll limitation ("g_m") and biochemical limitation ("$V_{c,max}$ or J_{max}"). Such limitation analysis can provide a powerful tool for studying the photosynthetic limitations under different stress situations, as well as during the photosynthetic recovery from stress (Flexas et al. 2009).

Acknowledgments The authors wish to acknowledge all other relevant studies in the field that could not be included due to limitation of space. A. Gallé benefited from a STSM of the COST 858 programme in 2007. Founding was provided by the Spanish Ministry of Education and Research (BFU2005-03102/BFI and BFU2008-01072/BFI) and the Swiss National Science Foundation (PBBEA-117524).

References

Bernacchi CJ, Singsaas EL, Pimentel C, Portis AR, Jr., Long SP (2001) In vivo temperature response functions for leaf steady-state photosynthesis models. Photosynth Res 69:235

Bernacchi CJ, Portis AR, Nakano H, von Caemmerer S, Long SP (2002) Temperature response of mesophyll conductance. Implications for the determination of Rubisco enzyme kinetics and for limitations to photosynthesis in vivo. Plant Physiol 130:1992–1998

Bernacchi CJ, Pimentel C, Long SP (2003) In vivo temperature response functions of parameters required to model RuBP-limited photosynthesis. Plant Cell Environ 26:1419–1430

Boyer JS, Wong SC, Farquhar GD (1997) CO_2 and water vapor exchange across leaf cuticle (epidermis) at various water potentials. Plant Physiol 114:185–191

Chaves MM, Harley PC, Tenhunen JD, Lange OL (1987) Gas exchange studies in two grapevine cultivars. Physiol Plant 70:639–647

Demmig-Adams B, Adams III WW (1996) The role of xanthophyll cycle carotenoids in the protection of photosynthesis. Trends Plant Sci 1:21–26

Downton WJS, Grant WJR, Loveys BR (1987) Diurnal changes in the photosynthesis of field-grown grape vines. New Phytol 105:71–80

Downton WJS, Loveys BR, Grant WJR (1988a) stomatal closure fully accounts for the inhibition of photosynthesis by abscisic acid. New Phytol 108:263–266

Downton WJS, Loveys BR, Grant WJR (1988b) non-uniform stomatal closure induced by water stress causes putative non-stomatal inhibition of photosynthesis. New Phytol 110:503–509

Epron D, Godard D, Cornic G, Genty B (1995) Limitation of net CO_2 assimilation rate by internal resistances to CO_2 transfer in the leaves of two tree species (*Fagus sylvatica* L and *Castanea sativa* Mill). Plant Cell Environ 18:43–51

Escalona JM, Flexas J, Medrano H (1999) Stomatal and non-stomatal limitations of photosynthesis under water stress in field-grown grapevines. Aust J Plant Physiol 26:421–433

Ethier GJ, Livingston NJ (2004) On the need to incorporate sensitivity to CO_2 transfer conductance into the Farquhar-von Caemmerer-Berry leaf photosynthesis model. Plant Cell Environ 27: 137–153

Evans JR, Sharkey TD, Berry JA, Farquhar GD (1986) Carbon isotope discrimination measured concurrently with gas-exchange to investigate CO_2 diffusion in leaves of higher-plants. Aust J Plant Physiol 13:281–292

Evans JR, von Caemmerer S (1996) Carbon dioxide diffusion inside leaves. Plant Physiol 110: 339–346

Farquhar GD, Caemmerer SV, Berry JA (1980) A biochemical-model of photosynthetic CO_2 assimilation in leaves of C-3 species. Planta 149:78–90

Flexas J, Escalona JM, Medrano H (1999) Water stress induces different levels of photosynthesis and electron transport rate regulation in grapevines. Plant Cell Environ 22:39–48

Flexas J, Bota J, Escalona JM, Sampol B, Medrano H (2002) Effects of drought on photosynthesis in grapevines under field conditions: an evaluation of stomatal and mesophyll limitations. Funct Plant Biol 29:461–471

Flexas J, Diaz-Espejo A, Berry J, Cifre J, Galmes J, Kaldenhoff R, Medrano H, Ribas-Carbo M (2007) Analysis of leakage in IRGA's leaf chambers of open gas exchange systems: quantification and its effects in photosynthesis parameterization. J Exp Bot 58:1533–1543

Flexas J, Ribas-Carbo M, Diaz-Espejo A, Galmes J, Medrano H (2008) Mesophyll conductance to CO_2: current knowledge and future prospects. Plant Cell Environ 31:602–621

Flexas J, Baron M, Bota J, Ducruet J-M, Galle A, Galmes J, Jimenez M, Pou A, Ribas-Carbo M, Sajnani C, Tomas M, Medrano H (2009) Photosynthesis limitations during water stress acclimation and recovery in the drought-adapted *Vitis* hybrid Richter-110 (*V. berlandieri* x *V. rupestris*). J Exp Bot 60:2361–2377

Genty B, Briantais JM, Baker NR (1989) The relationship between the quantum yield of photosynthetic electron-transport and quenching of chlorophyll fluorescence. Biochimica et Biophysica Acta 990:87–92

Grassi G, Magnani F (2005) Stomatal, mesophyll conductance and biochemical limitations to photosynthesis as affected by drought and leaf ontogeny in ash and oak trees. Plant Cell Environ 28:834–849

Guan XQ, Zhao SJ, Li DQ, Shu HR (2004) Photoprotective function of photorespiration in several grapevine cultivars under drought stress. Photosynthetica 42:31–36

Harley PC, Loreto F, Di Marco G, Sharkey TD (1992) Theoretical considerations when estimating the mesophyll conductance to CO_2 flux by analysis of the response of photosynthesis to CO_2. Plant Physiol 98:1429–1436

Hoad SP, Grace J, Jeffree CE (1996) A leaf disc method for measuring cuticular conductance. J Exp Bot 47:431–437

Horton P, Ruban AV, Walters RG (1996) Regulation of light harvesting in green plants. Ann Rev Plant Physiol Plant Mol Biol 47:655–684

Kautsky H, Appel W, Amann H (1960) Chlorophyllfluorescenz und Kohlensaureassimilation .13. Die Fluorescenzkurve und die Photochemie der Pflanze. Biochem Zeitschrift 332:277–292

Krall JP, Edwards GE. 1992. Relationship between photosystem II activity and CO_2 fixation in leaves. Physiol Plant 86:180–187

Krause GH, Jahns P (2004) Non-photochemical energy dissipation determined by chlorophyll fluorescence quenching: characterization and function. In: Chlorophyll a fluoerescence: signature of photosynthesis, vol 19. Springer, Dordrecht, pp 463–495

Laisk A (1977) Kinetics of photosynthesis and photorespiration in C3 plants. Nauka, Moscow

Long SP, Farage PK, Garcia RL (1996) Measurement of leaf and canopy photosynthetic CO_2 exchange in the field. J Exp Bot 47:1629–1642

Long SP, Bernacchi CJ (2003) Gas exchange measurements, what can they tell us about the underlying limitations to photosynthesis? Procedures and sources of error. J Exp Bot 54:2393–2401

Loreto F, Harley PC, Di Marco G, Sharkey TD (1992) Estimation of mesophyll conductance to CO_2 flux by three different methods. Plant Physiol 98:1437–1443

Maxwell K, Johnson GN (2000) Chlorophyll fluorescence – A practical guide. J Exp Bot 51: 659–668

Medrano H, Bota J, Abadía A, Sampol B, Escalona JM, Flexas J (2002) Drought effects on light energy dissipation in high light-acclimated, field-grown grapevines. Funct Plant Biol 29: 1197–1207

Medrano H, Escalona JM, Cifre J, Bota J, Flexas J (2003) A ten-year study on the physiology of two Spanish grapevine cultivars under field conditions: effects of water availability from leaf photosynthesis to grape yield and quality. Funct Plant Biol 30:607–619

Meyer S, Genty B (1998) Mapping intercellular CO_2 mole fraction (C_i) in *Rosa rubiginosa* leaves fed with abscisic acid by using chlorophyll fluorescence imaging – Significance of C_i estimated from leaf gas exchange. Plant Physiol 116:947–957

Maroco JP, Rodrigues ML, Lopes C, Chaves MM (2002) Limitations to leaf photosynthesis in field-grown grapevine under drought – Metabolic and modelling approaches. Funct Plant Biol 29:451–459

Moutinho-Pereira JM, Correia CM, Gonçalves BM, Bacelar EA, Torres-Pereira JM (2004) Leaf gas exchange and water relations of grapevines grown in three different conditions. Photosynthetica 42:81–86

Niinemets U, Bilger W, Kull O, Tenhunen JD (1999) Responses of foliar photosynthetic electron transport, pigment stoichiometry, and stomatal conductance to interacting environmental factors in a mixed species forest canopy. Tree Physiol 19:839–852

Niinemets U, Cescatti A, Rodeghiero M, Tosens T (2005) Leaf internal diffusion conductance limits photosynthesis more strongly in older leaves of Mediterranean evergreen broad-leaved species. Plant Cell Environ 28:1552–1566

Niinemets U, Diaz-Espejo A, Flexas J, Galmes J, Warren CR (2009a) Role of mesophyll diffusion conductance in constraining potential photosynthetic productivity in the field. J Exp Bot 60:2249–2270

Niinemets U, Diaz-Espejo A, Flexas J, Galmes J, Warren CR (2009b) Importance of mesophyll diffusion conductance in estimation of plant photosynthesis in the field. J Exp Bot 60: 2271–2282

Niinemets U, Wright IJ, Evans JR (2009c). Leaf mesophyll diffusion conductance in 35 Australian sclerophylls covering a broad range of foliage structural and physiological variation. J Exp Bot 60:2433–2449

Pons TL, Flexas J, von Caemmerer S, Evans JR, Genty B, Ribas-Carbo M, Brugnoli E (2009) Estimating mesophyl conductance to CO_2: methodology, potential errors and recommendations. J Exp Bot 60:2217–2234

Rodeghiero M, Niinemets U, Cescatti A (2007) Major diffusion leaks of clamp-on leaf cuvettes still unaccounted: how erroneous are the estimates of Farquhar et al. model parameters? Plant Cell Environ 30:1006–1022

Sampol B, Bota J, Riera D, Medrano H, Flexas J (2003) Analysis of the virus-induced inhibition of photosynthesis in malmsey grapevines. New Phytol 160:403–412

Santrucek J, Simanova E, Karbulkova J, Simkova M, Schreiber L (2004) A new technique for measurement of water permeability of stomatous cuticular membranes isolated from *Hedera helix* leaves. J Exp Bot 55:1411–1422

Schultz HR (2003) Extension of a Farquhar model for limitations of leaf photosynthesis induced by light environment, phenology and leaf age in grapevines (*Vitis vinifera* L. cvv. White Riesling and Zinfandel). Funct Plant Biol 30:673–687

Sharkey TD, Bernacchi CJ, Farquhar GD, Singsaas EL (2007) Fitting photosynthetic carbon dioxide response curves for C-3 leaves. Plant Cell Environ 30:1035–1040

Souza CR de, Maroco JP, dos Santos TP, Rodrigues ML, Lopes CM, Pereira JS, Chaves MM (2003) Partial root zone drying: regulation of stomatal aperture and carbon assimilation in field-grown grapevines (*Vitis vinifera* cv. Moscatel). Funct Plant Biol 30:653–662

Szabo I, Bergantino E, Giacometti GM (2005) Light and oxygenic photosynthesis: energy dissipation as a protection mechanism against photo-oxidation. EMBO Reports 6:629–634

Valentini R, Epron D, Deangelis P, Matteucci G, Dreyer E (1995) In-situ estimation of net CO_2 assimilation, photosynthetic electron flow and photorespiration in turkey oak (*Q. cerris* L) leaves – Diurnal cycles under different levels of water-supply. Plant Cell Environ 18:631–640

Von Caemmerer S (2000) Biochemical model of leaf photosynthesis. Techniques in plant sciences No. 2. CSIRO Publishing, Collingwood, Victoria, Australia, p 50

Warren CR, Dreyer E (2006) Temperature response of photosynthesis and internal conductance to CO_2: results from two independent approaches. J Expl Bot 57:3057–3067

Warren CR (2008) Stand aside stomata, another actor deserves centre stage: the forgotten role of the internal conductance to CO_2 transfer. J Exp Bot 59:1475–1487

Wilson KB, Baldocchi DD, Hanson PJ (2000) Quantifying stomatal and non-stomatal limitations to carbon assimilation resulting from leaf aging and drought in mature deciduous tree species. Tree Physiol 20:787–797

Chapter 9
Measuring Water Use Efficiency in Grapevines

Hipólito Medrano, J. Flexas, M. Ribas-Carbó, and J. Gulías

Abstract Viticulture is strongly conditioned by water availability in many areas and this limitation increases according to climatic change predictions. Maximizing grape water use efficiency (WUE) has become a key objective in management practices as well as a target for new environmental friendly cultivars. The goals of this chapter are to define the different spatial and temporal scales at which WUE can be measured and to describe the presently available methodologies to perform these determinations. Three different levels of WUE are distinguished: Vineyard or Crop (WUEc), Plant (WUEp) and Leaf (WUE_l). The meaningful of the different WUE parameters depends on the objective of the study and the methodology used. Positive and negative implications of the available methodologies to measure or estimate WUE at the different spatial and temporal scales are discussed, as well as the existing relationships between these different parameters to estimate WUE.

Contents

H. Medrano (✉)
Departamento de Biologia, Grup de Recerca de Biologia de les Plantes en condicions Mediterrànies, Universitat de les Illes Balears, Palma de Mallorca 07122, Spain
e-mail: hipolito.medrano@uib.es

S. Delrot et al. (eds.), *Methodologies and Results in Grapevine Research*,
DOI 10.1007/978-90-481-9283-0_9,

Abbreviations

A_N	Leaf net photosynthesis
A_N/E	Instantaneous leaf water use efficiency
A_N/g	Intrinsic leaf water use efficiency
C_a	CO_2 concentration in the atmosphere
C_i	CO_2 concentration at the substomatal cavity
$\Delta^{13}C$	^{13}C discrimination
E	Leaf transpiration
e_i	Water vapor pressure at the substomatal cavity
e_a	Water vapor pressure in the atmosphere
g_s	Stomatal conductance
TDM	Total plant dry matter
TWC	Total plant water consumption
VPD	Vapor pressure deficit
WUE	Water use efficiency
WUEc	Vineyard or Crop water use efficiency
WUEp	Plant water use efficiency
WUE_l	Leaf water use efficiency

9.1 Introduction

Water availability is an important factor for current and future viticulture because most of viticulture areas are under semiarid conditions, and because the grape growing and drought season are coincident forcing many grape growers to rely on irrigation. Moreover, for those areas, climatic change predictions agree on an increase of the aridity (Schultz 2000, Chaves et al. 2007). Water use by grapevines along the growth season is larger than for other crops (250–1,000 mm, Netzer et al. 2005, Williams and Ayars 2005, Zhang et al. 2007) and thus water availability is one of the major worldwide limitations to grape production in current viticulture areas. Consequently, there is an increasing concern about the optimization of water use to secure the sustainability of this crop, and water-use efficiency (WUE) is becoming a key parameter to qualify the sustainability of viticulture, similar to other irrigated crops because agriculture is the most water consuming sector. Water scarcity is a generalized condition for important agricultural areas and currently this problem is reflected as a priority for the United Nations policy in what has been called the "Blue Revolution" and summarised as *more crop per drop* (Annan 2000).

In addition to these social and environmental problems related with vineyard water use, grape quality is quite sensible to soil water availability showing, in general, the highest grape quality parameters under suboptimal soil water content (Matthews et al. 1990, Medrano et al. 2003, Salón et al. 2005). Therefore, improving water use efficiency in grapevines has become a matter of increasing interest for researchers, agronomists and farmers both to explore ways towards the highest quality of harvest and to increase crop sustainability (Chaves et al. 2007, Flexas et al. 2009).

9.2 Defining Water Use Efficiency: From Leaf to Plants and Vineyards

The concept of WUE derived from early field assays that measured "water requirements" or "transpiration coefficients" in different crops (Briggs and Shantz 1913). These ways to express WUE are still used in recent literature as i.e. "irrigation water productivity" (Sadras 2009), at time with others, thus the term "water use efficiency" is still a matter of debate (Tanner and Sinclair 1983, Jones 2004, Morison et al. 2008).

The concept of WUE always reflects a balance between gains (moles of carbon, crop yield) and costs (moles of water, volume of water used). This balance can be measured at different levels from instantaneous fluxes in the leaf (photosynthesis vs. transpiration rates) to values concerning whole plant or crop (Fig. 9.1), although in a wider context, this concept is also applied to whole agricultural system. Ideally, its magnitude should be without units because gains and costs should have the same units, even though gains are commonly thousand times lower (μmolCO_2/mmol H_2O; kilos dry mass/Tons of water).

WUE values can therefore be analyzed at different space and time scales (Fig. 9.1) and all of them are significant for the evaluation or optimization of water use. At the leaf and plant levels, it is common to use instantaneous leaf gas exchange rates, soil water content depletion, lysimeters, sap flow measurement techniques,

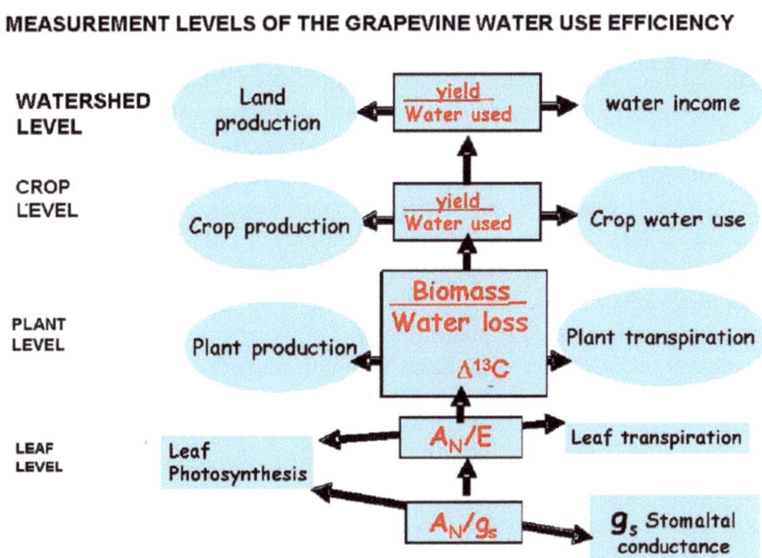

Fig. 9.1 Different levels for water use efficiency measurement. From leaf to watershed, (instantaneous to growth season or yearly basis) there is a progressive integration of different crop production processes and water expenses with different measurement techniques implicated to determine the final WUE balance

etc. which permit comparisons at the short time scale of environmental effects, ecophysiological responses or genetic capacities. To integrate these values over time, ^{13}C discrimination measurements are especially useful for a time integral of leaf or crop balances. For crop WUE, measurements of total water consumption during the growth season (soil evaporation, runoff, crop transpiration) and total dry mass or yield production are needed. Increasing the efficiency of the irrigation facilities and soil management practices is also needed in order to avoid non-transpired water losses.

Beyond these levels, sometimes there is a special interest to extra-crop, non-agronomic inefficiencies in water use. The important component of the whole system of water storage, conveyance, etc, which can cause losses as high as 60% of the total water consumed (Morison et al. 2008) needs to be evaluated.

The main difficulties for the determination of grapevine WUE come from the accurate determination of transpired water and plant dry mass accumulation under realistic (vineyard) conditions. In this respect, indirect calculations made on the basis of irrigation volume, aboveground biomass or harvest weight, even though they are actually neither the transpired water nor the total dry mass production, are quite convenient to compare environments (soil, climate), agronomical practices (irrigation dosage, trellis systems), and genetic components (rootstocks and varieties) (Sadras 2009).

The next sections will describe the techniques available to estimate and measure WUE at different levels, as well as their interest and limitations.

9.2.1 Determination at the Vineyard Level

Assessing WUE at the vineyard or wine region scale (WUEc) requires taking into account several parameters related to temporal and spatial scales, because the discontinuous canopy crop is complex. However, there is an increasing interest for those methodologies which are in part associated with regional planning, sustainable use of natural resources and vulnerability of agrosystems to climatic change (Jarosz et al. 2009). Among those methodologies, Eddy covariance was chosen as a standard reference for carbon and water fluxes in regional and local networks (Baldocci 2003). The method requires accurate and frequent measurements of different micrometeorological variables (as air temperature, wind speed, vapour pressure deficit and CO_2) in large areas of plain, uniform vineyards, showing reasonable estimates of energy balance, water and CO_2 fluxes on the canopy (Spano et al. 2004).

Direct measurement of vineyard water consumption under field conditions should also take into account the amount of water lost by runoff, percolation and direct soil evaporation. It is uneasy to determine the amount of water that is actually consumed by the vineyard, and the larger and deeper grape root system makes it more difficult to obtain an accurate measurement of total plant water use by soil humidity probes or the lysimeters. In addition, the determination of biomass production also presents several difficulties, like measuring root biomass accumulation and leaf biomass losses due to pests and diseases along their live. Total irrigation water together with yield (and market prices) can provide a rough estimate of vineyard WUEc, water necessities, and eventually can help to qualify the sustainability of vineyard region.

The accuracy of the methodology for vineyard assessment therefore depends on the crop characteristics and environmental variability, but, the accumulated evidence suggest that the Eddy covariance technique is suited to derive global carbon and water fluxes at the local scale (Trambouze et al. 1998, Valentini et al. 2000, Spano et al. 2004). Thus, it is a promising approach for the estimations of the CO_2 and water fluxes, enabling to provide gross values of vineyard water use, crop efficiency and WUE (Trambouze et al. 1998, Yunusa et al. 2004, Zang et al. 2007). Estimates of the vineyard evapotranspiration and its components from energy balance, Eddy covariance and Bowen ratio energy balance methods seems a good approach to assess water consumption (Trambouze et al. 1998, Kordova et al. 2000, Yunusa et al. 2004, Zhang et al. 2007).

9.2.2 Plant Water Use Efficiency

Water use efficiency at the plant level (WUEp) is a key parameter from a research point of view since it allows to deepen in the understanding of the physiological and ecophysiological mechanisms controlling the efficiency of plants to fix carbon. To measure WUEp accurately, it is useful to have a simplified system where water loss by soil water evaporation, percolation or runoff is minimized. In this context, to measure whole plant water transpiration by gravimetric methods in pots or containers (as individual lysimeters) sounds a reasonable approach. In addition to water consumption, accurate determination of biomass accumulation is a key point for WUEp. Both vegetative and fruit dry mass should be taken into account when a physiological and/or ecophysiological approach is developed. Moreover, root biomass weighing is particularly difficult and should also be considered in order to take the whole plant biomass production. The methodology to estimate root biomass in the field is not easy and, usually implies a rough approach. Studies on potted plants may provide a more accurate estimate of plant biomass production, including roots. In this context, WUEp can be determined as:

$$\text{WUEp} = \text{TDM/TWC}$$

where, TDM: Total plant dry matter production (including leaves, stems, bunches and roots) and TWC: Total plant water consumption.

Nevertheless, this simplified system has several flaws since the canopy is not representative of real vineyard canopies which largely determine total plant water consumption (Williams and Ayars 2005) and carbon assimilation (Escalona et al. 2003). Another limiting factor when using potted plants is that root development is severely constrained, which could modify the root/shoot ratio and constrain the plant development. Moreover, these kind of experiments are short and hardly last for the whole season, which implies that the selection of the experimental period is crucial

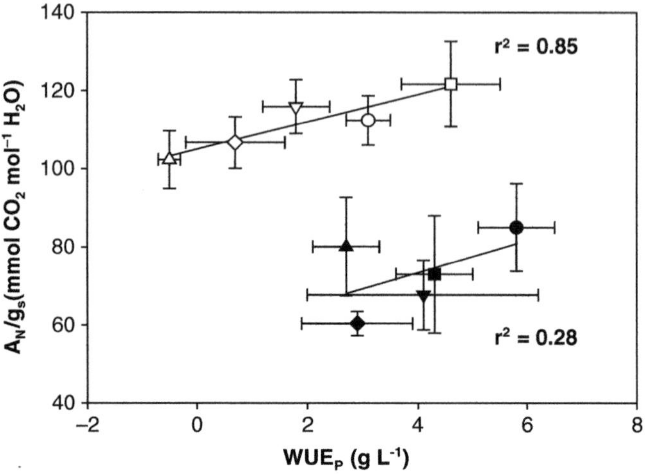

Fig. 9.2 The relationship between intrinsic WUE_{leaf} (A_N/g_s) and whole plant WUE_P (based on vegetative growth including roots, first-year plants not producing grapes) in 5 different cultivars of *Vitis vinifera* growing outdoors in pots during summer in Mallorca (Balearic Islands, Spain). Closed symbols represent irrigated treatment at field capacity and open symbols refer to non-irrigated treatment defined by the leaf maximum daily g_s (about 0.05 mol H_2O m^{-2} s^{-1}). Symbols for cultivars are: Grenache (circles), Callet (squares), Tempranillo (triangles up), Malvasia of Banyalbufar (triangles down) and Cabernet Sauvignon (diamonds). Values represent means ± SE of 6 (A_N/g_s) or 4 replicates (whole plant WUE). Regression lines are displayed with their r^2. (Flexas et al. 2009)

because WUEp depends on the seasonal variations of VPD and plant growth patterns. In this sense, to obtain meaningful results, the experiments must be carried out during the active growth period of grapevine. Because of these factors, the results obtained with potted plants cannot be directly extrapolated to the vineyard level. In spite of these limitations, this is the most accurate method to measure WUEp, and to compare environmental and genetic effects on the whole plant WUE (Fig. 9.2).

To avoid the difficulties encountered in extrapolating data from potted plants to the vineyards, several methods have been developed to estimate water use by field plants. In addition to those pointed out in the previous sections, sap flow measurements provide a reasonable good estimate of total plant water loss for most woody crops (Smith and Allen 1996), enabling daily and seasonal estimations of water expenses by the plant (Ginestar et al. 1998, Braun and Schmith 1999, Escalona et al. 2002). Nevertheless, the accuracy of this technique needs to be assessed for the particular measuring conditions. Moreover, other water losses like soil water evaporation percolation, and runoff are difficult to estimate accurately, which leads to misleading values of total crop water consumption by this methods. (See Chapter 5 on Sap Flow techniques for more details).

Lysimeters are the best (and most expensive) way to "weigh" the effective water loss by plants in comparative field conditions, but in grapevines as in other fruit tree crops the large and deep root system adds difficulties for a proper design of the lysimeter size. Nevertheless, lysimeters provide a unique way to estimate water use, calculate crop coefficients and compare some agronomic practices such as pruning or irrigation dosage (Williams et al. 2003, Netzer et al. 2005, Williams and Ayars 2005) that can ultimately be used as reference values for other experiments with more common facilities. The lysimeter design consist of a minimal soil depth taking into account the crop root system (deeper than 1 m), a minimal cropping area (more than 10 m^2, more than 4 vines), and a uniform environment representative of the typical vineyard conditions. This requires to build a complex and expensive equipment in which soil and plants are placed inside a big container in the middle of a standard crop. Such container should be weighed continuously or every day, thus providing the best and more realistic data on water consumption. The increase in total plant dry weight remains difficult to estimate, so that measurements of water use are more accurate than biomass increases of the crop.

9.2.3 Instantaneous Leaf Water Use Efficiency

Both CO_2 and H_2O absorb radiation at specific IR wavelength, for which both are measured simultaneously using IRGAs (see Chapter 8). The exchange of CO_2 by a leaf over a short time period is related to the instantaneous rate of net CO_2 assimilation (A_N), while the exchange of water is related to the instantaneous transpiration (E), from which it is possible to calculate stomatal conductance (g_s) provided that leaf temperature is determined and some assumptions are made (see Chapter 8 on leaf gas exchange). The ratio of A_N to leaf transpiration E (A_N/E) is called "instantaneous water-use-efficiency (WUE)", and reflects the mols of carbon fixed by the leaf per mol of water used in transpiration. Because E depends on the vapor pressure deficit, (VPD) the same leaf would give lower values of WUE_l when the environment is drier. To reduce this "interference", and better qualify the leaf control of water loss, the leaf-level instantaneous WUE is most often approached using the "intrinsic" WUE, i.e. the ratio of net assimilation (A_N) to stomatal conductance (g_s), which was introduced to compare photosynthetic properties at a common evaporative demand (Osmond et al. 1980). Differences between genotypes in A_N/g_s have been reported to have a genetic basis in grapevines (Bota et al. 2001, Gibberd et al. 2001, Soar et al. 2006), and breeding for high WUE has become a main objective for viticulture (Flexas et al. 2009). If the VPD is not substantially changed, both ratios should correlate (Fig. 9.3).

Despite reflecting an instantaneous value only, measuring A_N/E and, specially, A_N/g_s has the advantage of being rapid, non destructive and of reflecting the leaf capacity for WUE_l, making it an interesting tool to compare, for instance, the potential WUE_l of different treatments or cultivars. Moreover, and particularly in grapevines in particular, it has been shown that both instantaneous measurements of A_N and E in sun-exposed leaves strongly correlate with daily-weighed values (i.e., with the accumulated carbon gain and water loss by a given leaf during a whole day), for which they can be used to estimate daily values of WUEp (Medrano et al. 2003). This can be useful since measuring WUEp is time consuming. Nevertheless, this

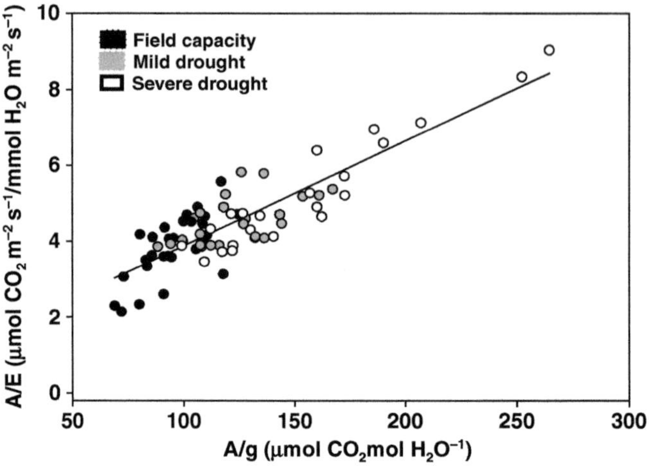

Fig. 9.3 Relationship between instantaneous leaf water use efficiency (A/E) and intrinsic leaf water use efficiency (A/g_s) in grapevines cv. Tempranillo grown outdoors in pots and under three different water regimes: Field capacity, Mild drought (70% of field capacity), Severe drought (50% of field capacity)

may require testing the correlation in more cultivars under different environmental conditions, since such a relationship is far from being universal (Fig. 9.2).

Concerning the IRGA measurements of A_N and E (or g_s) aimed at providing an estimate of WUE_l, it is important to perform them under conditions that are as close as possible to those experienced by the leaf in its natural environment. Leaf to air vapor pressure deficit (VPD) and leaf temperature are unavoidably modified once the leaf has been enclosed in the cuvette, but this can be minimized setting high air flow rates in the instrument. On the other hand, no regulation of temperature should be used, and natural light may be preferred over artificially-provided lights. When using natural light, particular attention should be paid to ensure that leaf orientation in the chamber is similar to natural conditions, including irradiance and leaf orientation. Shading by the cuvette should be avoided as it creates a heterogeneous light environment across the leaf. Photosynthesis responds rapidly to variations in light conditions, but stomata respond more slowly. Therefore, assuming that leaves are near steady-state conditions in the field, any significant change in A_N without a change in g_s might have implications for calculations of WUE_l.

9.2.4 *Stable Isotope Discrimination Techniques for WUE Estimation*

A technique based on stable isotope of carbon (^{13}C) has frequently been used for analysis of WUE at different scales of both time and space (Pate 2001).

This technique is based on the differential use of carbon isotopes by the enzymes that fix CO_2, RUBP carboxylase (Rubisco) and PEP carboxylase, as well as on the physical differences of the isotopes that induce different diffusivities (O'Leary et al. 1992). Considering both enzymatic and diffusive fractionation, total fractionation against the heavy isotope of carbon (^{13}C) can be estimated following a simplification of the equation of Farquhar et al. (1982)

$$\Delta^{13}\mathrm{C} = a + (b - a)\,\mathrm{C_i}/\mathrm{C_a}$$

where C_a and C_i are the CO_2 concentrations in the atmosphere and the substomatal cavity, respectively. *a* is the fractionation associated with diffusion of CO_2 in air (4.4‰) and *b* is the fractionation associated with the enzymatic fixation of CO_2 which accounts for the fractionation of both Rubisco (95%) and PEP carboxylase (5%) and the combination gives a fractionation of $b = 27‰$ (Brugnoli et al. 1998). This equation assumes that fractionation by photorespiration and respiration are small and almost unaccountable (Brugnoli et al. 1998). After considering mesophyll conductance as another barrier for CO_2 diffusion to the chloroplast (Pons et al. 2009), $\Delta^{13}C$ can be used as an estimate of the C_i/C_a ratio.

Water use efficiency can be considered from different points of view and different levels. At the leaf level, the ratio of photosynthesis (A_N) to transpiration (E) gives the instantaneous WUE of the leaf. This value can be derived from gas exchange measurements and is also related to the ratio C_i/C_a, as follows:

$$\mathrm{WUE} = (\mathrm{A_n}/\mathrm{E}) = 1/1.6^*[\mathrm{C_a^*}(1 - \mathrm{C_i}/\mathrm{C_a})/(\mathrm{e_i} - \mathrm{e_a})]$$

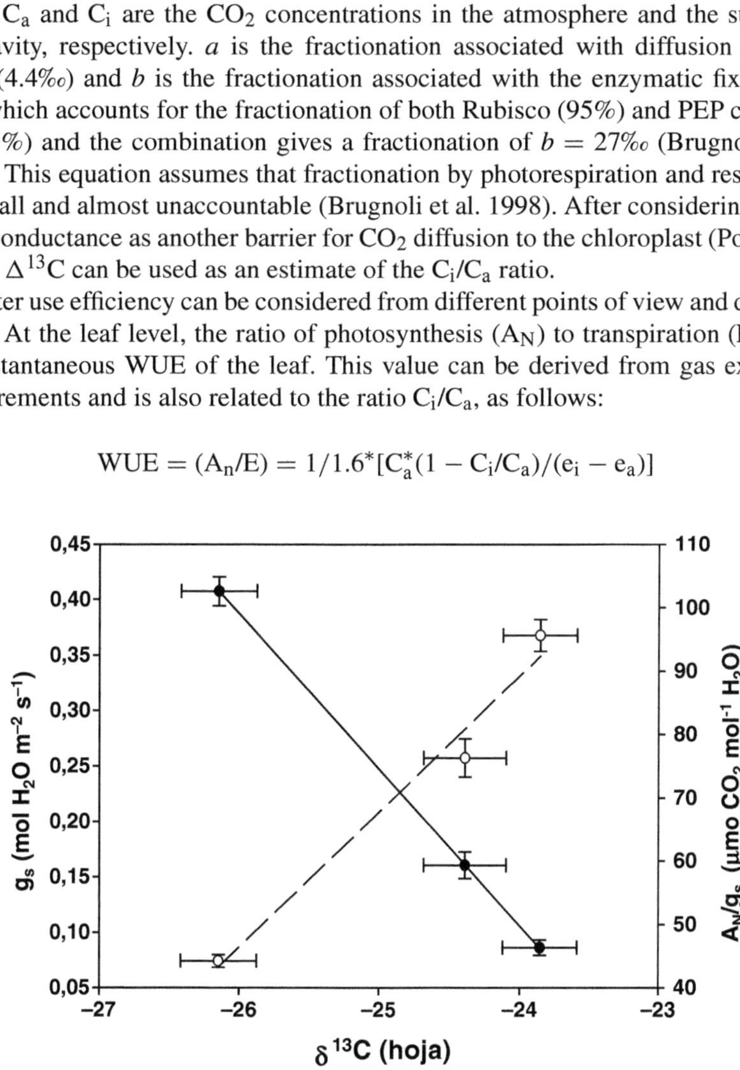

Fig. 9.4 The $\delta^{13}C$ of leaf biomass reflected the treatment conditions (totally correlated (R^2=1) with gmax). The $\delta^{13}C$ also reflect the average of A_N/gs of each treatment (R^2=0.97)

where e_a and e_i are the vapor pressures in the atmosphere and inside the leaf, respectively.

Therefore, the higher WUE, the lower the ratio C_i/C_a would be and, consequently, the lower $\Delta^{13}C$. The inverse relationship between WUE and $\Delta^{13}C$ has extensively been shown (Farquhar and Richards 1984, see Fig. 9.4). The relationship between WUE and $\Delta^{13}C$ can be observed at different integration times. Instantaneous WUE can be estimated with on-line carbon isotope fractionation during photosynthesis (Evans et al. 1986), short-term (1–2 days) WUE can be estimated by measurements of the isotopic composition of leaf soluble sugars (Brugnoli et al. 1988) or can be integrated on a much longer time when dry matter of plant tissue is assayed (Farquhar and Richards 1984). Plant material commonly analyzed for carbon isotope fractionation includes leaves, roots, shoots and wood. Technically, plant material is collected and dried in an oven at 60ºC for 48 h. Dry matter is homogenized in a ball milling and 0.2–2 mg of powder is placed in a small tin capsule and wrapped up (Dawson and Brooks 2001). The sample is placed in an elemental analyzer where it is combusted. The CO_2 generated is injected into a continuous flow isotope ratio mass analyzer (CF-IRMS).

9.3 General Consideration: Future Perspectives

The techniques used to determine the grapevine WUE are more and more difficult from the leaf to the vineyard level. At the leaf level, the main problem is the representativeness of those instantaneous, single, particular leaf values relative to the whole plant, but not the difficulties of the measurement itself. For the whole plant, the problem comes from errors related to the assessment of the carbon or biomass gain (mainly roots weight). In the vineyard, the particularities of the crop (discontinuous cover, trellis system) severely limit the accuracy of carbon and water fluxes determination. However, this topic is increasingly mentioned in recent literature. This reflects the need for rapid progress in evaluating agronomical practices and improving crop WUE. This is also needed for genetic improvement aiming at a more economic use of water, which seems to run in parallel with the improvement of grape quality. This subject is already, and will be even more in the near future, an important research field because of the necessity to improve grape WUE for a sustainable viticulture.

Acknowledgments The authors wish to acknowledge funding provided by the Spanish Ministry of Education and Research (AGL2008-04525-CO2-01/AGR).

References

Annan K (2000) UN Millenium Summit. http://www.un.org/millenium/summiet.htm. Accessed 20 May 2009

Baldocci DD (2003) Assessing the eddy covariance technique for evaluating carbon dioxide exchange rates of ecosystems: past, present and future. Glob Change Biol 9: 479–492

Bota J, Flexas J, Medrano H (2001) Genetic variability of photosynthesis and water use in Balearic grapevine cultivars. Ann App Biol 138:353–365
Braun P, Schmith J, (1999); Sap flow measurement in grapevines. Stem morphology and the use of gear balance method. Plant and Soil 215:39–45
Briggs LJ, Shantz HJL (1913). The water requeriments of plants. A review of the literature. USDA Plant Industries Bulletin 285
Brugnoli E, Hubick KT, von Caemmerer S, Wong SC, Farquhar GD (1988) Correlation between the carbon isotope discrimination in leaf starch and sugars of C_3 plants and the ratio of intercellular and atmospheric partial pressures of carbon dioxide. Plant Physiol 88:1418–1424
Brugnoli E, Scartazza A, Lauteri M, Monteverdi MC, Maguas C (1998) Carbon isotope discrimination in structural and non-structural carbohydrates in relation to productivity and adaptation to unfavourable conditions. In: Griffiths H (ed) Stable Isotopes integration of biological, ecological and geochemical processes. BIOS Scientific Publishers Limited, Oxford
Chaves MM, Santos TP, Souza CR, Ortuño MF, Rodrigues ML, Lopes CM, Maroco JP, Pereira JS (2007) Deficit irrigation in grapevine improves water-use-efficiency without controlling vigour and production quality. Ann App Biol 150:237–252
Dawson TE, Brooks PD (2001) Fundamentals of stable isotope chemistry and measurement. In: Unkovich M, Pate J, Mcneill A, Gibbs J (eds) Stable Isotope techniques in the study of biological processes and functioning of ecosystems. Kluwer Academic Publishers, Dordrecht
Escalona JM, Flexas J, Medrano H (2002) Drought effects on water flow, photosynthesis and growth of potted grapevines. Vitis 41:57–62
Escalona JM, Flexas J, Bota J, Medrano, H (2003) Distribution of leaf photosynthesis and transpiration within grapevine canopies under different drought conditions Vitis 42:57–64
Evans JR, Sharkey TD, Berry JA, Farquhar GD (1986) Carbon isotope discrimination measured concurrently with gas exchange to investigate CO_2 diffusion in higher plants. Aust J Plant Physiol 13:281–292
Farquhar GD, Richards RA (1984) Isotopic composition of plant carbon correlates with WUE of wheat genotypes. Aust J Plant Physiol 11:539–552
Farquhar GD, O'Leary MH, Berry JA (1982) On the relationship between carbon isotope discrimination and the intercellular carbon dioxide concentration in leaves. Aust J Plant Phsyiol 9:121–137
Flexas J, Galmés J, Gallé A, Gulías J, Pou A, Ribas-Carbo M, Tomàs M, Medrano H (2010) Improving water use efficiency in grapevines: potential physiological targets for biotechnological improvement. Aust J Grape and Wine Res 16(1):106–121
Gibberd MR, Walker RR, Blackmore DH, Condon AG (2001) Transpiration efficiency and carbon-isotope discrimination of grapevines grown under well-watered conditions in either glasshouse or vineyard. Aust J Grape and Wine Res 7:110–117
Ginestar C, Eastham J, Gray S, Iland P (1998) Use of sap-flow sensors to schedule vineyard irrigation . I. Effects of post-veraison water deficit on water relations, vine growth, and yield of Shiraz grapevines. Am J Enol Vitic 49:413–420
Jarosz N (2009) Effect of land use on carbon dioxide, water vapour and energy exchange over terrestrial ecosystems in Southwestern France during the CERES campaign. Biogeosciences Discuss, 6:2755–2784
Jones H G, (2004) Plants and microclimate. Cambridge University Press, Cambridge
Kordova L, Biezuner I, Mahner, Schwantz C (2000) Estimation of actual evapotranspiration from vineyard by utilizing eddy correlation method. Acta Hort 537, ISHS
Matthews MA, Ishii R, Anderson MM, O'Mahony M (1990) Dependence of wine sensory attributes on vine water status. J Sci Food and Agric 51:321–335
Medrano H, Escalona JM, Cifre J, Bota J, Flexas J (2003) A ten-year study on the physiology of two Spanish grapevine cultivars under field conditions: effects of water availability from leaf photosynthesis to grape yield and quality. Func Plant Biol 30:607–619
Morison JIL, Baker NR, Mullineaux PM, Davies WJ (2008) Improving water use in crop production. Philos Trans Royal Soc London series B 363:639–658

Netzer Y, Yao C, Shenker M, Bravdo B, Schwartz A, Cohen S (2005) Water consumption of 'Superior' Grapevines Grown in a Semiarid Region. Acta Hort. 689, ISHS 2005

O'Leary MH, Madhavan S, Apneth P (1992) Physical and chemical basis of carbon isotope fractionation in plants. Plant Cell Environ 15:1099–1104

Osmond CB, Björkman O, Anderson DJ (1980) Physiological processes in plant ecology. Springer, New York

Pate JS (2001) Carbon isotope discrimination and plant water-use efficiency. In: Unkovich M, Pate J, Mcneill A, Gibbs J (eds) Stable Isotope techniques in the study of biological processes and functioning of ecosystems. Kluwer Academic Publishers, Dordrecht

Pons TL, Flexas J, von Caemmerer S, Evans JR, Genty B, Ribas-Carbo M, Brugnoli E (2009) Estimating mesophyll conductance to CO_2: methodology, potential errors, and recommendations. J Exp Botany 60:2217–2234

Sadras VO (2009) Does partial root-zonedrying improve irrigation water productivity in the field? A meta-analysis. Irrig Sci 27:183–190

Salón JL, Chirivella C, Castel JR (2005) Response of cv. Bobal to timing of deficit irrigation in Requena, Spain: water relations, yield, and wine quality. Am J Enol Vitic 56:1–8

Schultz HR (2000) Climate change and viticulture: A European perspective on climatology, carbon dioxide and UV-B effects. Aust J Grape Wine Res 6:2–12

Smith DM, Allen SJ (1996) Measurement of sap flow in plant trunks. J Exp Bot 47:1833–1844

Soar CJ, Speirs J, Maffei SM, Penrose AB, McCarthy MG, Loveys BR (2006) Grape vine varieties Shiraz and Grenache differ in their stomatal response to VPD: apparent links with ABA physiology and gene expression in leaf tissue. Aust J Grape Wine Res 12:2–12

Spano D, Duce P, Snyder RL (2004) Estimate of Mass and energy fluxes over Grapevine Using Eddy Covariance Technique. Acta Hort 664:631–638

Tanner CB, TR Sinclair (1983) Efficient water use in crop production: Research or re-search? p. 1–27. *In* H.M. Taylor et al. (ed.) Limitations to efficient water use in crop production. ASA, CSSA, and SSSA, Madison, WI.

Trambouze W, Bertuzzi P, Voltz M (1998) Comparison of methods for estimating actual evapotranspiration in a row–cropped vineyard. Agric Forest Meteorol 91:193–208

Valentini R, Matteucci G, Dolman AJ, Schulze ED, Rebmann C, Moors E, Granier A, Gross P, Jensen NO, Pilegaard K, Lindroth A, Grelle A, Bernhofer C, Grünwald T, Aubinet M, Ceulemans R, Kowalski AS, Vesala T, Rannik U, Berbigier P, Loustau D, Guomundsson J, Thorgeirsson H, Ibrom A, Morgenstern K, Clement R, Moncrieff J, Montagnani L, Minerbi S, Jarvis, P G (2000) Respiration as the main determinant of carbon balance in European forests, Nature 404:861–865

Williams LE, JE Ayars (2005) Grapevine water use and the crop coefficient are linear functions of the shaded area measured beneath the canopy. Agric Forest Meteorol 132:201–211

Williams LE, Phene CJ, Grimes DW, Trout TJ (2003) Water use of young Thompson Seedles grapevines in California. Irrig Sci 22:1–9

Yunusa IAM, Walker RR, Lu P (2004) Evapotranspiration components from energy balance, sapflow and microlysimetry techniques for an irrigated vineyard in inland Australia. Agric Forest Meteorol 127:93–107

Zhang B Z, Kang SZ, Zhang L, Du TS, Li SE, Yang XY (2007) Estimation of seasonal crop consumption in a vineyard using Bowen ratio-energy balance method. Hydrol Proces 21: 3635–3641

Chapter 10
Use of Thermal Imaging in Viticulture: Current Application and Future Prospects

J. Miguel Costa, O.M. Grant, and M.M. Chaves

Abstract In order to optimise crop management, improve yield and quality, modern viticulture increasingly looks for reliable and fast methods to monitor plant physiology for early stress detection. Imaging technologies can be used for real time, non-invasive and non-destructive monitoring of grapevine physiological status. Thermal imaging allows us to measure the infrared (IR) radiation emitted by plants and to visualise spatial variation of their surface temperature. Recent developments extended the affordability and potential of thermal imaging in plant physiology and agronomy. Biological processes such as leaf transpiration can be monitored in real-time using thermal imaging. Thus, the technique can be used for continuous assessment of a plant's physiological condition, and indirectly to detect the impact of non-optimal growth conditions (e.g. drought stress) on plant performance. The present paper briefly describes the principles behind thermal imaging and shows how it can be used for early stress detection, crop management and breeding in viticulture.

Contents

J.M. Costa (✉)
CBAA, Instituto Superior de Agronomia, Universidade Técnica de Lisboa, Tapada da Ajuda, Lisboa 1349-017, Portugal; LEM, Instituto de Tecnologia Química e Biológica, Universidade Nova de Lisboa, Oeiras 2780-901, Portugal
e-mail: miguelc@itqb.unl.pt

S. Delrot et al. (eds.), *Methodologies and Results in Grapevine Research*,
DOI 10.1007/978-90-481-9283-0_10,

Abbreviations

IR	Infrared
W	Spectral radiant exitance (radiation)
ε	Emissivity
B	Stefan Boltzmann Constant (5.67×10^{-12} W cm^{-2} K^{-4})
T_{leaf}	Leaf temperature (ºC)
T_a	Air temperature (ºC)
T_{dry}	Temperature of the dry reference (ºC)
T_{wet}	Temperature of the wet reference (ºC)
r_{HR}	Parallel resistance to heat and radiative transfer (s m^{-1})
r_{aw}	Boundary layer resistance to water vapour (s m^{-1})
r_s	Leaf resistance to water vapour loss (s m^{-1})
γ	Psychrometer constant (J kg^{-1})
R_{ni}	Net isothermal radiation (Wm^{-2})
ρ	Air density (kg m^{-3})
c_p	Specific heat capacity of air (J kg^{-1} K^{-1})
D	Air vapour pressure deficit (Pa)
s	Slope of the curve relating saturating water vapour pressure to temperature (Pa K^{-1})
CWSI	Crop water stress index
I_G	Linear thermal index
Ψ_{pd}	Pre-dawn leaf water potential (MPa)
g_s	Stomatal conductance to water vapour (mol H_2O m^{-2} s^{-1})
A_n	Net assimilation (μmol CO_2 m^{-2} s^{-1})
WUE_i	Intrinsic water use efficiency (μmol CO_2 mol^{-1} H_2O)

10.1 Introduction

Environmental and biotic stresses often cause substantial reductions in crop yield and/or quality in the agricultural/horticultural sector. Major horticultural production areas worldwide are located in hot and dry regions like the Mediterranean area (Costa et al. 2007, Flexas et al. 2009) where high radiation, high temperatures and vapour pressure deficit (VPD) and soil water deficits often occur simultaneously and cause losses of production and quality. This is particularly true for the wine industry whose main production areas are located in hot and dry areas, including Southern Europe, California and Australia. Climate scenarios for arid areas predict increasing water shortage due to lower precipitation, and higher average air temperatures in combination with more extreme environmental phenomena (IPPC 2007). This will put increased pressure on water resources for irrigated horticulture. Climate change will also influence the range and the incidence of pests and diseases affecting grapevine. Moreover, there is a need to optimise the use of inputs such as water, biocides, and energy to increase the sustainability of the sector.

Approaches to minimise losses in yield and quality can be achieved either by crop breeding, optimisation of management, or a combination of both (Condon et al. 2004, Cifre et al. 2005, Costa et al. 2007). The use of methodologies that help the easy and accurate assessment of the plant physiological condition and support strategies of crop breeding and/or management is therefore of major interest for the wine industry.

In recent decades, the development of imaging technologies extended the possibility to study plant physiology and develop approaches for fast and accurate remote diagnosis of a wide range of plant stresses (Nilsson 1995, Chaerle and Van Der Straeten 2000, Jones and Schofield 2008, Liew et al. 2008, Sirault 2009). Among them, thermal imaging emerged as a fast and non-invasive method to assess plant transpiration and, indirectly, to assess the physiological condition of several crops including grapevine (Jones et al. 2002, 2009, Grant et al. 2006, 2007, Möller et al. 2007, Stoll and Jones 2007, Loveys 2008, Stoll et al. 2008a, b). Advances in thermal imaging technology and image processing enhanced the diagnostic power of this technique at different scales of study, from single leaves to whole vineyards. The present chapter describes the basic principles of IR imaging and points out the potential and the limitations of the technique for stress detection and breeding in viticulture.

10.2 Thermal Imaging Background

10.2.1 Infrared Radiation

Infrared (IR) radiation is electromagnetic radiation with a wavelength in the region of the spectrum between 0.75 and 1000 μm (Fig. 10.1). According to the Planck's radiation law, every object at a temperature above the absolute zero (0 K) emits electromagnetic radiation in the region between 0.75 and 1000 μm. The higher the temperature, the greater the energy radiated.

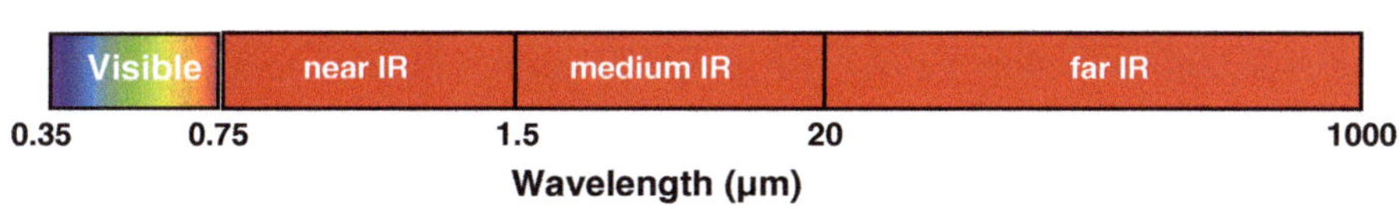

Fig. 10.1 Visible and thermal infrared spectrum (adapted from Gaussorgues 1999)

10.2.2 Thermal Imaging

Thermal imaging allows visualisation of the IR radiation leaving an object as the temperature across the object's surface. Instead of the 400–750 nm range corresponding to the common visible light camera, IR cameras operate between 1 and 14 μm. Within this interval, the most commonly used wavelengths for thermal imaging are 3–5 μm or 7–14 μm (Gaussorgues 1999, Kaplan 2007). Within these windows, IR atmospheric transmission is close to a maximum. Thermal sensitivity is higher at smaller wavelengths (3.5–5 μm) compared to higher ones (7–14 μm) because small wavelengths correspond to higher energy levels. However, for certain applications, the use of longer wavelengths may be advantageous. For observations of targets at long distance, and therefore, through longer atmospheric paths, the window of wavelengths between 8 and 12 μm would minimise errors due to atmospheric absorption of IR radiation.

In recent years infrared thermal technology has progressed rapidly in terms of both image resolution (quality) and speed of image capture. Detection technology developed fast and in the 1990s thermal cameras with very high thermal sensibility (of some milliKelvins) were made available (Gaussorgues 1999). The market is presently dominated by uncooled IR cameras whose market share keeps growing, supported by lower prices and more user friendly instruments and software (Kaplan 2007).

10.2.3 Important Concepts Related to Thermal Imaging

10.2.3.1 Emissivity and Black Body Radiation

One of the most important concepts to consider when working with IR radiation is emissivity (ε), which describes the efficiency with which a material radiates energy. In other words, the emissivity at a particular wavelength represents the amount of radiation emitted from an object as a fraction of that emitted by a blackbody. A blackbody consists in a theoretical object that absorbs 100% of the radiation that hits it, and reflects none, and therefore appears perfectly black. A blackbody has an $\varepsilon = 1$. Real-world objects are not blackbodies but grey bodies. They absorb a certain fraction of the incident radiation, and reflect and transmit the rest. Their emissivity will be greater than 0 and less than 1, varying with the type of the material (Table 10.1).

10.2.3.2 The Stefan-Boltzman Law

The radiation emitted by an object depends on several factors, among which two are of particular importance. One is the absolute temperature of the object (which defines the wavelength at which the maximum radiation occurs in addition to its influence over the amount of total radiation). The second one is its emissivity. As a result, the object's temperature and its emissivity will define the amount of IR energy emitted by the object. According to the Stefan–Boltzmann law, the total energy

Table 10.1 Emissivities of common materials, plant leaves and fruits

Material	Emissivity
Aluminium polished	0.10
Aluminium anodized	0.65
Glass	0.97
Ceramic	0.93
Water	0.95
Plastic	0.93
Cloth	0.95
Leaves	0.95–0.97
Fruits	0.90–0.97
Human skin	0.98

Sources: Hellebrand et al. (2001), Jones et al. (2002), Kaplan (2007), Bulanon et al. (2008).

radiated per surface area unit for a black body in unit time is directly proportional to the 4th power of its temperature:

$$W = \varepsilon^{*}B^{*}T^{4}(\mathrm{W\ cm^{-2}})$$

where

W = spectral exiting radiation
ε = Emissivity
B = Stefan Boltzmann Constant (5.67×10^{-12} W cm^{-2} K^{-4})
T = Temperature (K)

Therefore, when the total radiation emitted by an object and its emissivity are known, the temperature of that object can be determined.

10.2.4 How Does IR Radiation Become Visible to Human Eyes?

IR imaging devices make electronic images out of emissions of IR radiation (Fig. 10.2).

Early sensor technology typically used a mechanical scanning system to focus IR energy onto a single element detector which resulted in images with poor resolution and which often required the support of visible images to identify the areas of interest (Jones 2004). The early IR sensors needed to be cooled (by cryogenic fluids or compressed gases) which resulted in heavy (up to about 20 kg) and expensive equipment, with high maintenance costs and reduced portability. However, these sensors were extremely sensitive and accurate. As a result, they are still in use for high precision measurements. They operate in the 3–5 μm region which provides excellent thermal sensitivity.

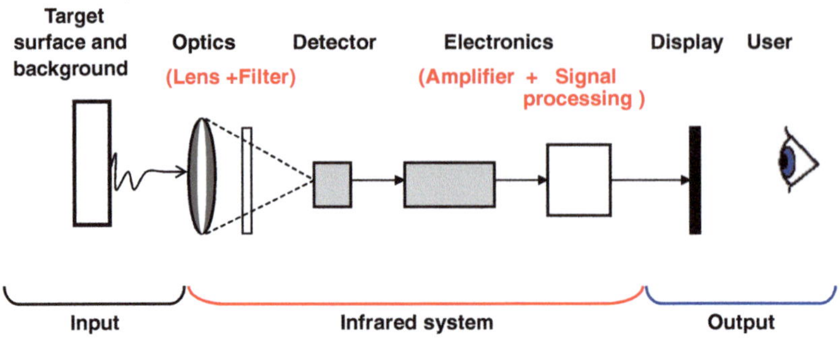

Fig. 10.2 Simple representation of the components of an IR thermal imaging device

Uncooled sensors, in turn, operate at or above room temperature. They represent the most common form of thermal imaging technology available today (Kaplan 2007). The reduced cost and weight and a simpler IR detector system allows for a more widespread usage. Uncooled detectors usually operate in the 7–14 μm range IR wavelengths.

10.2.5 Leaf Energy Balance and Thermal Imaging

Temperature balance in leaves/plants largely depends on the transpiration rate and consequently on leaf stomatal conductance. The use of IR imaging in plant eco-physiology and agronomic studies is based on the concept of leaf energy balance and its regulation by plants and environment (see Jones 2004 for a review).

10.2.5.1 Leaf Energy Balance

Leaves, like all objects, interact with their physical environment through energy-exchange processes. In order to optimize their metabolic functioning, plants attempt to maintain their energy balance (rate of energy absorbed = rate of energy loss) in equilibrium (Jones 1992). However, when energy components are not balanced, leaf temperature changes (increase or decrease) until a new equilibrium is reached (Lambers et al. 1998). The high surface to volume ratios that characterise most leaves facilitates a fast energy exchange, and equilibrium in energy exchange can be achieved within a matter of seconds (Ehleringer 1989).

Leaf temperature is the result of the interaction of several factors/processes: the temperature of the surrounding environment, the absorbed/re-emitted radiation and the processes of convection and transpiration (Ehleringer 1989). Therefore, the basic components of the leaf energy balance can be divided into:

(1) Absorption of shortwave radiation (solar radiation) by leaf tissues;
(2) Net absorption and emission of long wave radiation (sky and terrestrial IR radiation);

(3) Heat transfer by convection and conduction. Heat is exchanged between the leaf surface and the surrounding bulk air masses through physical contact. This occurs when there is a difference between the temperatures of the air and the leaf. Heat transfer occurs in the direction of the temperature gradient.
(4) Evaporative cooling related to leaf transpiration. Transpiration implies the movement of water by diffusion from the inner leaf surfaces through stomatal pores to the outside air. The phase transition from liquid to gaseous water (an endothermic process) causes leaf cooling.

10.2.5.2 Evaporative Energy Exchange

Heat loss associated with water evaporation is the result of the energy demand for the evaporative process. Water evaporation cools the leaf during the day time when the leaf transpires, but it can increase temperature when water condenses on the leaf (Lambers et al. 1998). The evaporative energy exchange due to transpiration is influenced by three major aspects: stomatal conductance to water vapour (g_S), the difference in water vapour pressure between the leaf and the air and the boundary layer conductance.

When temperature rises, the air expands and is able to contain more water vapour and the evaporation from the wet surfaces of the leaf cells raises the vapour pressure to saturation. Although at higher temperature the air surrounding the plants contains more water, the rise in the water vapour concentration of the air is usually less rapid than that of the leaf (Lambers et al. 1998).

10.2.6 Estimating Leaf Stomatal Conductance from Thermal Imaging Measurements

The most common application of leaf temperature measurement for plant physiologists and agronomists is the monitoring of stomatal opening. Direct estimation of leaf stomatal conductance to water vapour from temperature measurements avoids the need for multiple and time consuming leaf gas-exchange measurements and allows for assessment of stomatal conductance over large crop areas.

The existing methods used to estimate g_s from leaf temperature have been developed on the basis of the leaf energy balance (Jones 1992, Jones et al. 2002, Leinonen et al. 2006) which is presented below

$$T_{\text{leaf}} - T_a = \left[r_{HR}(r_{aw} + r_s)\gamma R_{ni} - \rho c_p r_{HR} D\right] / \{\rho c_p \left[\gamma(r_{aw} + r_s) + s r_{HR}\right]\},$$

where T_{leaf} and T_a (ºC) are leaf and air temperatures, respectively, r_{HR} is the parallel resistance to heat and radiative transfer (s m^{-1}), r_{aw} is the boundary layer resistance to water vapour (s m^{-1}), r_s is the leaf resistance to water vapour loss, assumed to be dominated by the stomatal resistance component (s m^{-1}), γ is the psychrometric constant (J kg^{-1}), R_{ni} is the net isothermal radiation (the net radiation for a leaf at air temperature) (Wm^{-2}), ρ is the density of the air (kg m^{-3}), c_p is the specific heat capacity of air (J kg^{-1} K^{-1}), D is the air vapour pressure deficit (Pa), and s is the slope of the curve relating saturating water vapour pressure to temperature (Pa K^{-1}).

One of the limitations of estimating g_s from temperature measurements is the fact that leaf/plant temperature is affected not only by g_s and related transpiration rates, but also by environmental variables like air temperature, solar radiation, and wind speed (Jones 1999a). This requires knowledge of these environmental inputs in addition to the leaf temperature values obtained by thermal imaging. Moreover, leaf/canopy temperature fluctuates with changing environmental conditions, which can mask the interesting patterns of g_s.

One of the solutions to overcome the problem is to use indices relating the temperature of the leaf to that of references (thermal indices) (Table 10.2) which partially alleviates the influence of meteorological factors (Jones 2004, Grant et al. 2006, 2007, Möller et al. 2007). Indices such as the crop water stress index (CWSI) or I_G are calculated from leaf/canopy temperatures relative to dry and wet references (Fig. 10.3). They provide an indication of relative g_s (Jones 1999a, b, Jones et al. 2002, Grant et al. 2006). I_G is theoretically proportional to stomatal conductance to water vapour (Table 10.2).

Table 10.2 Thermal indices derived from IR measurements

Stress index	Formula &/or background	References
Stress degree day	Uses the difference in temperature of a canopy and the air, measured soon after midday; a much elevated canopy temperature compared to air temperature indicates crop stress	Jackson et al. (1977)
Crop water stress index (CWSI)	$CWSI = (T_{plant} - T_{wet})/(T_{dry} - T_{wet})$ Relates the actual temperature of the canopy of interest (T_{plant}) to that of a very stressed canopy (with closed stomata) (T_{dry}) and of a well-irrigated canopy (with maximum conductance, non-stressed) (T_{wet}).	Idso (1982), De Lorenzi et al. (1993), Cohen et al. (2005), Grant et al. (2006), Möller et al. (2007)
I_G	$I_G = (T_{dry} - T_{leaf})/(T_{leaf} - T_{wet})$ Relates the temperature of a leaf with closed stomata (T_{dry}) with that of a leaf with fully open stomata (T_{wet}); T_{leaf} is the temperature of the leaf or canopy of interest. I_G is theoretically proportional to stomatal conductance (g_s)	Jones (1999a, b), Jones et al. (2002), Grant et al. (2007)

Fig. 10.3 Visible (*left*) and false coloured thermal image (IR Snapshot 525, 8–12 μm detector, 120 × 120 pixels) (*right*) showing a dry reference (*greased, upper leaf*), and a wet reference (*wet, bottom leaf*). The temperature of the dry reference (T_{dry}, 28.7± 1.9°C) and the temperature of the wet reference (T_{wet}, 22.1± 0.9°C) are used to estimate the thermal indices CSWI or I_G

Estimation of g_s from leaf temperature may integrate these reference temperatures. Alternatively, it can be calculated from meteorological data and leaf temperature alone (Leinonen et al. 2006).

10.2.7 Thermal Imaging and Stress Detection

Since abiotic or biotic stresses often result in a decrease in rates of photosynthesis and transpiration (e.g. Nilsson 1995, Chaerle and Van Der Straeten 2000) remote sensing of leaf temperature by thermal imaging can be a reliable way to detect changes in the physiological status of plants in response to different biotic and/or abiotic stress.

One of the major advantages of thermal imaging (as opposed to the use of infrared thermometry or porometry or leaf chambers) is the possibility of detecting variability within large areas of a crop, and hence the shorter time required to obtain large sets of data (Price et al. 2002). This allows for rapid measurement of the relative temperatures of the canopy of different plants. This avoids disturbances due to rapid changes of g_s brought about by changes in microclimate such as wind speed or radiation which can occur when using other techniques. Another advantage of thermal imaging is that leaf temperature can be measured continuously and non-destructively, which provides a tool to monitor transpiration of individual plants in response to a particular stress (Chaerle and Van Der Straeten 2000, Merlot et al. 2002, Jones 2004, Jones et al. 2009, Sirault et al. 2009). Besides, and contrary to techniques like porometry, thermal imaging can reveal spatial heterogeneity which has been shown to occur at the whole leaf level by different experimental approaches (Jones 1999b). Most methods for studying stomatal heterogencity are either not capable of giving quantitative information on variation in g_s or their resolution is rather low. The increasing availability of sensitive infrared imaging systems opened up the possibility of high resolution studies of stomatal variation over leaf surfaces and of stomatal dynamics (Jones 1999b).

10.3 Thermal Imaging Use in Grapevine

10.3.1 Stress and Stomatal Regulation in Grapevine

Stomata tightly regulate the influx of CO_2 needed for photosynthesis, and the loss of water vapour by transpiration. Stomata therefore allow plants to optimize the compromise between photosynthetic carbon gain and water loss under different environmental conditions (Schroeder et al. 2001, Nilson and Assmann 2007). Stomatal pores are flanked by two guard cells which integrate multiple signals, either environmental (light, humidity, CO_2) or physiological (hormones like abscisic acid, auxin; ions like K^+ and Ca^{2+}) and are able to regulate leaf gas-exchange (Schroeder et al. 2001). Stomatal regulation of leaf gas-exchange is critical for plant survival and adaptation to unfavourable growth conditions (drought, high radiation, diseases).

Vitis vinifera L. is considered well-adapted to dry and hot areas and it has been classified as a "drought avoiding" species due to the good control of water loss by stomata (Schultz 2003). However, the large genetic variation of this species results in different responses to stress (abiotic or biotic) depending on the cultivars (Chaves and Rodrigues 1987, Schultz 1996, Escalona et al. 1999, Medrano et al. 2002, Flexas et al. 2009). This variability in the response to stress is positive for breeding purposes, but makes it more difficult for growers to optimize crop management (irrigation, crop protection, plant nutrition), particularly because several cultivars are commonly grown in the same vineyard. Therefore, techniques that enable quick and reliable monitoring of grapevine physiological status could significantly improve crop management.

10.3.2 Stress Detection in Grapevine

10.3.2.1 Detection of Abiotic Stress: Monitoring of Plant Water Status and Irrigation

Irrigation has emerged as an important practice to guarantee wine quality or in more extreme cases, plant survival in regions affected by seasonal drought (Chaves et al. 2007, Costa et al. 2007). However, high doses of irrigation can result in excessive vegetative growth and yield and lower berry quality. In turn, severe drought results in stomatal closure and low photosynthesis and diminished shoot growth and berry sugar content (Chaves et al. 2007, Möller et al. 2007). Deficit irrigation (irrigation which does not completely replace evapotranspirational losses of water) emerged as a novel strategy to impose controlled and moderate water stress and optimize quality without reducing production (Dry et al. 2001, Chaves et al. 2007). In order

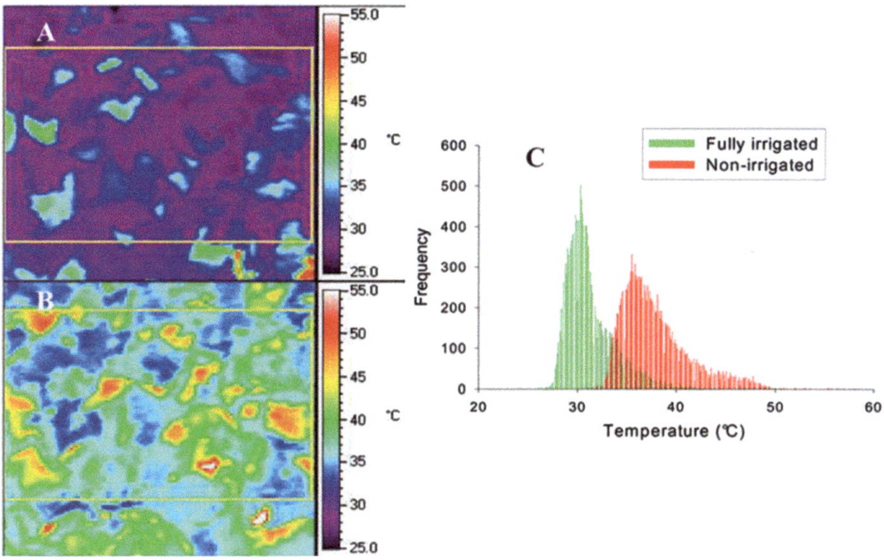

Fig. 10.4 False-coloured thermal images (IR Snapshot 525, 8–12 μm detector, 120 × 120 pixels) of the sunlit side of the canopy (50 cm × 50 cm) from field-grown grapevine plants subjected to (**a**) full irrigation and (**b**) no irrigation. (**c**) Frequency of pixel temperatures measured within the outlined areas of the thermal images showing that fully irrigated (FI) plants were cooler than non-irrigated (NI) plants

to guarantee the balance between plant vigour and quality of production, the wine industry needs methods that enable assessment of plant water status in the vineyard. The usefulness of canopy temperature as a measure of "crop water stress" has been recognized since the sixties (Tanner 1963) and the use of plant temperature measurements obtained by thermal imaging as an indicator of water stress has been recently established for grapevine (Jones et al. 2002, Fuentes et al. 2005, Grant et al. 2006, Leinonen et al. 2006, Grant et al. 2007, Möller et al. 2007) (Fig. 10.4).

10.3.2.2 Detection of Biotic Stress

Detection of biotic stress due to the attack of pathogens or insects and the optimization of related crop protection strategies are potentially relevant applications for thermal imaging in field conditions (Nilsson 1995). This could be particularly useful for grapevine which requires intensive use of pesticides, and grape growers face increasing pressures to reduce phytochemical treatments (Jacobson et al. 2005, Komárek et al. 2010). Late detection of fungal attack or poor spray coverage are major factors contributing to inefficient disease control in vineyards (Stoll et al. 2008a, b), with consequent losses of quality and yield.

Stomatal regulation and related transpiration can be affected by pests and diseases. The suitability of thermal imaging as a means to monitor the onset and progress of diseases in grapevine has recently been demonstrated in the case of

Plasmopara viticola (downy mildew), which is a major disease causing severe losses in yield and quality (Allègre et al. 2007). Alterations in leaf temperature were detected at an early stage after pathogen attack, and at least three days before symptoms were visible (Stoll et al. 2008a). The authors demonstrated that the pattern of variation in leaf temperature across grapevine leaves infected with downy mildew allowed a clear distinction between the impact of water stress and of mildew infection. The same authors showed that thermal imaging had the potential to assess the homogeneity of spray coverage within a canopy and consequently could optimize the efficiency of biocide application (Stoll et al. 2008b). Since diseases usually assume patchy patterns, unnecessary spraying occurs in disease-free areas. The remote identification of infected areas by thermal imaging would be useful to achieve more localised applications.

10.3.2.3 Breeding

The use of thermal imaging may also extend to genetic breeding programmes for grapevine. The potential for selecting genotypes with desirable stomatal function has been demonstrated with *Arabidopsis thaliana* (Merlot et al. 2002) and various crops (Raskin and Ladyman 1988, Kashiwagi et al. 2008). Preliminary studies have also shown that thermal imaging can help to characterize grapevine genotypes in terms of their stomatal characteristics and responses to mild drought stress (Costa et al. 2008) (Table 10.3) and in terms of frost hardiness (Fuller and Telli 2008).

Table 10.3 Pre-dawn leaf water potential (Ψ_{pd}, MPa), canopy temperature based on thermal imaging (T_{leaf}, °C) and stomatal conductance to water vapour (g_s, mol H_2O m^{-2} s^{-1}), net assimilation (A_n, μmol CO_2 m^{-2} s^{-1}) and intrinsic water use efficiency ($WUE_i = A_n/g_s$, μmol CO_2 mol^{-1} H_2O) measured at saturating light (1,200 μmol $m^{-2}s^{-1}$), 360 ppm CO_2, and 25°C, for field grown plants of five grapevine cultivars: Aragonez (ARA), Trincadeira (TRI), Syrah (SYR), Cabernet Sauvignon (CAB) and Touriga Nacional (TOU). Measurements were carried out in early August 2007, during the early afternoon. Values are means ± SE (n = 3–6)

Cultivar	Ψ_{pd}	T_{leaf}	g_s	A_n	WUE_i
ARA	−0.25 ± 0.01	30.5 ± 0.2	0.076 ± 0.006	15.2 ± 0.8	58.7 ± 4.6
TRI	−0.10 ± 0.02	30.7 ± 1.6	0.074 ± 0.005	14.1 ± 0.5	53.5 ± 3.6
SYR	−0.19 ± 0.01	34.4 ± 1.5	0.049 ± 0.008	12.1 ± 0.7	92.6 ± 11.7
CAB	−0.21 ± 0.02	31.4 ± 1.6	0.085 ± 0.008	12.4 ± 0.5	44.7 ± 4.4
TOU	−0.10 ± 0.01	29.5 ± 1.5	0.115 ± 0.007	15.6 ± 0.7	68.6 ± 14.1

10.4 Potential, Limitations and Future Developments of Thermal Imaging

The future of modern horticulture and wine production depends on the search for and use of novel crop management strategies that optimize the use of inputs like water, nutrients and biocides and minimise losses and negative impact on the environment. The use of remote sensing, including thermal imaging, can contribute to achieve this aim. However, limitations still exist that need to be minimised and solved (Costa and Grant 2005, Loveys 2008). One is the cost of the instruments and another is the diffusion of the know-how required to handle them properly and to analyze the results. Nevertheless, prices of thermal cameras have been decreasing and information on IR imaging is now more easily accessed, as a result of the support provided either by IR camera manufactures or consultancy companies in thermography and precision agriculture. Future developments in the use of thermal imaging in viticulture will depend on testing the robustness of the technique in different geographical and meteorological conditions and for different grapevine cultivars. The establishment of specific thermal signatures for specific types of stress and/or genotype together with the automatic processing of stress indices would assist real-time, practical applications under field conditions.

Precision viticulture is being progressively adopted by modern viticulture (e.g. in Australia) (Taylor 2004). Therefore, incorporation of data collected by imaging into crop management remains an important topic of research for viticulture.

Acknowledgments Research described here has been funded by the European Union, and the Fundação para a Ciência e a Tecnologia, Portugal (refs. SFRH/BPD/34429/2006, POCI/AGR/59079/2004, PPCDT/AGR/59079/2004 and WATERWEB).

References

Allègre M, Daire X, Heloir MC, Trouvelot S, Mercier L, Adrian M, Pugin A (2007) Stomatal deregulation in *Plasmopara viticola*-infected grapevine leaves. New Phytol 173:832–840

Bulanon DM, Burks TF, Alchanatis V (2008) Study on temporal variation in citrus canopy using thermal imaging for citrus fruit detection. Biosyst Engineer 101:161–171

Chaerle L., Van Der Straeten D (2000) Imaging techniques and the early detection of plant stress. Trends Plant Sci 5:495–501

Chaves MM, Rodrigues ML (1987) Photosynthesis and water relations of grapevines growing in Portugal – response to environmental factors. In: Tenhunen JD, Catarino FM, Lange OL, Oechel WC (eds) Plant response to stress – functional analysis in mediterranean ecosystems. Springer, Berlin, pp 379–390

Chaves MM, Santos TP, Souza CR, Ortuño MF, Rodrigues ML, Lopes CM, Maroco JP, Pereira JS (2007) Deficit irrigation in grapevine improves water-use efficiency while controlling vigour and production quality. Ann Appl Biol 150:237–252

Cifre J, Bota J, Escalona JM, Medrano H, Flexas J (2005) Physiological tools for irrigation scheduling in grapevine (*Vitis vinifera* L.) An open gate to improve water-use efficiency? Agric Eco Syst Environ 106:159–170

Cohen Y, Alchanatis V, Meron M, Saranga Y, Tsipiris J (2005) Estimation of leaf water potential by thermal imagery and spatial analysis. J Exp Bot 56:1843–1852.

Condon AG, Richards RA, Rebetzke GJ, Farquhar GD (2004) Breeding for high water-use efficiency. J Exp Bot 407:2447–2460

Costa JM, Grant OM (2005) Crop imaging can help to optimise production. Fruit Veg Tech 5.2: 30–32

Costa JM, Ortuño MF, Chaves MM (2007) Deficit irrigation as a strategy to save water: physiology and potential application to horticulture. J Int Plant Biol 49:1–14

Costa JM, Ortuño MF, Santos T, Lopes C, Chaves MM (2008) Caracterização eco-fisiológica de cinco cultivares de videira baseada na medição de trocas gasosas e na imagem térmica. Actas do IX Simposium Hispano-Português de Relaciones Hídricas en Plantas, pp 20–24

De Lorenzi F, Staghellini C, Pitacco A (1993) Water shortage sensing through infrared canopy temperature: timely detection is imperative. Acta Hortic 335:373–380

Dry PR, Loveys BR, McCarthy MG, Stoll M (2001) Strategic irrigation management in Australian vineyards. J Int Vigne Vin 35:129–139

Ehleringer JR (1989) Temperature and energy budgets. In: Pearcy RW, Ehleringer J, Mooney HA, Rundel PW (eds) Plant physiological ecology: field methods and instrumentation. Chapman and Hall, New York

Escalona JM, Flexas J, Medrano H (1999) Stomatal and non-stomatal limitations of photosynthesis under water stress in field-grown grapevines. Aust J Plant Physiol 26: 421–433

Flexas J, Galmés J, Gallé A, Gulias J, Pou A, Ribas-Carbo A, Tomás A, Medrano H (2009) Improving water use efficiency in grapevines: potential physiological targets for biotechnological improvement. Aust J Grape Wine Res 16:106–121

Fuentes S, Conroy JP, Kelley G, Rogers G, Collins M (2005). Use of infrared thermography to assess spatial and temporal variability of stomatal conductance of grapevines under partial rootzone drying: an irrigation scheduling application. Acta Hortic 689:309–316

Fuller MP, Telli G (2008). An investigation of the frost hardiness of grapevine (*Vitis vinifera*) during bud break. Ann Applied Biol 135:589–595

Gaussorgues G (1999) La thermographie infrarouge. Principes, Technologies, Applications. Editions TEC&DOC 4th edn. Paris, France, 587 pp

Grant OM, Chaves MM, Jones HG (2006) Optimizing thermal imaging as a technique for detecting stomatal closure induced by drought stress under greenhouse conditions. Physiol Plantarum 127:507–518

Grant OM, Tronina Ł, Jones HG, Chaves MM (2007) Exploring thermal imaging variables for the detection of stress responses in grapevine under different irrigation regimes. J Exp Bot 58:815–825

Hellebrand HJ, Beuche H, Linke M (2001) Determination of thermal emissivity and surface temperature distribution of horticultural products. http://www.atb-potsdam.de/hauptseite-deutsch/Institut/Abteilungen/Abt2/Mitarbeiter/jhellebrand/jhellebrand/Publikat/6th_symp_FNVE2001.pdf. Assessed on 1 December 2009

Idso SB (1982) Non-water stressed baselines: A key to measuring and interpreting plant water stress. Agric Meteor 27:59–70.

IPPC (2007) Climate change 2007. Impacts, adaptation, vulnerability. In: Parry ML, Canziani OF, Palutikof JP, Van der Linden CE, Hanson PJ (eds) Contribution of working group II to the 4th assessment report of the intergovernmental panel on climate change. Cambridge University Press, Cambridge, UK

Jackson RD, Reginator RJ, Idso SB (1977) Wheat canopy temperature: A practical tool for evaluating water requirements. Water Resou Res 13:651–656

Jacobson AR, Dousset S, Guichard N, Baveye P, Andreux F (2005) Diuron mobility through vineyard soils contaminated with copper. Environ Pollut 138:250–259

Jones HG (1992) Plants and microclimate, 2nd edn. Cambridge University Press, Cambridge

Jones HG (1999a) Use of infrared thermometry for estimation of stomatal conductance as a possible aid to irrigation scheduling. Agric For Meteorol 95:139–149

Jones HG (1999b) Use of thermography for quantitative studies of spatial and temporal variation of stomatal conductance over leaf surfaces. Plant Cell Environ 22:1043–1055

Jones HG (2004) Application of thermal imaging and infrared sensing in plant physiology and ecophysiology. Adv Bot Res 41:107–163

Jones HG, Schofield P (2008) Thermal and other remote sensing of plant stress. Gen Applied Plant Physiol 34:19–32

Jones HG, Stoll M, Santos T, de Sousa C, Chaves MM, Grant OM (2002) Use of infrared thermography for monitoring stomatal closure in the field: applications to grapevine. J Exp Bot 53:2249–2260

Jones HG, Serraj, R., Loveys, B.R., Xiong, L., Wheaton, A., Price, A.H. (2009). Thermal infrared imaging of crop canopies for the remote diagnosis and quiantification of plant responses to water stress in the field. Func Plant Biol 36:978–989

Kaplan H (2007) Practical applications of infrared thermal sensing and imaging equipment. Tutorial Texts in Optical Engineering Volume TT75, 3rd edn. SPIE Press, Bellingham, Washington, USA, 166p

Kashiwagi J, Krishnamurthy L, Upadhyaya HD, Gaur PM (2008) Rapid screening technique for canopy temperature status and its relevance to drought tolerance improvement in chickpea. J Sat Agric Res 6:1–4

Komárek M, Cadková E, Chrastný V, Bordas F, Bollinger J (2010). Contamination of vineyard soils with fungicides: a review of environmental and toxicological aspects. Environ Int 36: 138–151

Lambers H, Stuart Chapin FS, Pons TL (1998) Plant Physiological Ecology. Springer, New York, 540 pp

Leinonen I, Grant OM, Tagliavia CPP, Chaves MM, Jones HG (2006) Estimating stomatal conductance with thermal imagery. Plant Cell Environ 29:1508–1518

Liew OW, Chong PC, Li B, Asundi AK (2008) Signature optical cues: emerging technologies for monitoring plant health. Sensors 8:3205–3239

Loveys B (2008) Performance of vines under reduced irrigation: development of cost-effective irrigation scheduling tools. Final report. CSIRO Plant Industry. 11 pp. http://www.gwrdc.com.au/downloads/ResearchTopics/GWT%2007-02.pdf, Assessed 1 December 2009

Medrano H, Escalona JM, Bota J, Gulías J, Flexas J (2002) Regulation of photosynthesis of C_3 plants in response to progressive drought: Stomatal conductance as a reference parameter. Ann Bot 89:895–905

Merlot S, Mustilli AC, Genty B, North H, Lefebvre V, Sotta B, Vavasseur A, Giraudat J (2002) Use of infrared thermal imaging to isolate Arabidopsis mutants defective in stomatal regulation. Plant J 30:601–609

Möller M, Alchanatis V, Cohen Y, Meron M, Tsipris J, Naor A, Ostrovsky V, Sprintsin M, Cohen S (2007) Use of thermal and visible imagery for estimating crop water status of irrigated grapevine. J Exp Bot 58:827–838

Nilson SE, Assmann SM (2007) The control of transpiration. Insights from arabidopsis. Plant Physiol 143:19–27

Nilsson HE (1995) Remote sensing and image analysis in plant pathology. Annu Rev Phytopatol 15:489–527

Price AH, Cairns JE, Horton P, Jones HG, Griffiths H (2002) Linking drought resistance mechanisms to drought avoidance in upland rice using a QTL approach: progress and new opportunities to integrate stomatal and mesophyll responses. J Exp Bot 53:989–1004.

Raskin I, Ladyman JAR (1988) Isolation and characterization of a barley mutant with abscisic acid-insensitive stomata. Planta 173:73–78

Schroeder JI, Allen GJ, Hugovieux V, Kwak JM, Waner D (2001) Guard cell signal transduction. Ann Rev Plant Physiol Plant Mol Biol 52:627–658

Schultz HR (1996) Water relations and photosynthesis responses of two grapevine cultivars of different geographical origin during water stress. Acta Hortic 427:251–266

Schultz HR (2003) Differences in hydraulic architecture account for near isohydric and anisohydric behaviour of two field-grown *Vitis vinifera* L. cultivars during drought. Plant Cell Environ 26:1393–1405

Sirault XRR, James RA, Furbank RT (2009) A new screening method for osmotic component of salinity tolerance in cereals using infrared thermography. Funct Plant Biol 36:970–977.

Stoll M, Jones HG (2007) Thermal imaging as a viable tool for monitoring plant stress. J Int Sci Vigne Vin 41:77–84

Stoll M, Schultz HR, Berkelmann-Loehnertz B (2008a) Exploring the sensitivity of thermal imaging for *Plasmopara viticola* pathogen detection in grapevines under different water status. Funct Plant Biol 35:281–288

Stoll M, Schultz HR, Baecker G, Berkelmann-Loehnertz B (2008b) Early pathogen detection under different water status and the assessment of spray application in vineyards through the use of thermal imagery: Precision Agric 9:407–417

Tanner CB (1963) Plant temperatures. Agron J 55:210–211

Taylor JA (2004). Digital terroirs and precision viticulture: investigations into the application of information technology in Australian vineyards. Ph.D: Thesis, University of Sydney

Chapter 11
Grapevine Fruiting Cuttings: An Experimental System to Study Grapevine Physiology Under Water Deficit Conditions

M. Carmen Antolín, Héctor Santesteban, Marouen Ayari, Jone Aguirreolea, and Manuel Sánchez-Díaz

Abstract Partial rootzone drying (PRD) is an irrigation system that may improve the water use efficiency of crop production by means of better control of vegetative growth without reducing fruit yield (Kang and Zhang 2004). PRD uses biochemical response of plants to water stress to achieve a balance between vegetative and reproductive development (Davies and Hartung 2004). However, the underlying nature of how PRD affects berry ripening is only poorly defined, as indicated by the considerable controversy in the literature (Du Toit et al. 2003, Santos et al. 2003, Gu et al. 2004). This may be due to either poor implementation of PRD, or the interaction of many factors on grapevine growth and reproductive development. Given such ambiguities, it becomes necessary to design a PRD system that permits a fine control of most environmental factors with the aim to know the nature of PRD response of plants. Therefore, we have developed a method which uses fruiting cuttings subjected to PRD irrigation under controlled conditions. Emphasis is placed on physiological aspects including berry hormonal balance related to long-distance chemical signalling induced by soil drying.

Contents

M.C. Antolín (✉)
Facultades de Ciencias y Farmacia, Departamento Biología Vegetal, Sección Biología Vegetal Unidad Asociada al CSIC (EEAD, Zaragoza; ICVV, Logroño), Universidad de Navarra, Pamplona, 31008, Spain
e-mail: cantolin@unav.es

S. Delrot et al. (eds.), *Methodologies and Results in Grapevine Research*, 151
DOI 10.1007/978-90-481-9283-0_11, © Springer Science+Business Media B.V. 2010

11.1 Description of the Technique

11.1.1 Plant Material

Dormant *Vitis vinifera* (L.) cuttings of 400–500 mm long and 15–20 mm in diameter were obtained from 1-year-old cane-pruned Tempranillo and Superior Seedless, collected in winter from different experimental vineyards in Spain.

11.1.2 Production of Fruiting Cuttings

The cuttings were propagated by a technique which ensured that the formation of adventitious roots preceded bud burst using steps originally outlined in Mullins (1966) with some modifications described in Ollat et al. (1998) and Santa María (2004). Each cutting was sawn longitudinally 100 mm from the base towards the tip with a band saw and rooted in a heat-bed (25°C) inside a cool-room (5°C) for 30 days (Fig. 11.1).

After rooting, cuttings were transferred to a greenhouse with a 25/20°C and 70/80% RH (day/night) regime. They were illuminated for 15 h with natural daylight supplemented with high-pressure sodium lamps (SON-T Agro Phillips, Eindhoven, the Netherlands), providing a minimum photosynthetic photon flux density (PPFD) of 350 μmol m^{-2} s^{-1} at the level of inflorescence. Budbreak took place after one week under these conditions. Careful control of stem and leaf growth before flowering improves the partitioning of stored carbon towards the roots and the reproductive structures (Fig. 11.2). Thus, only a single flowering stem was allowed to develop on each plant during growth (Fig. 11.3). After berry set, growth conditions in the greenhouse were changed to a 25/15°C and 60/80% RH (day/night) regime with a PPFD

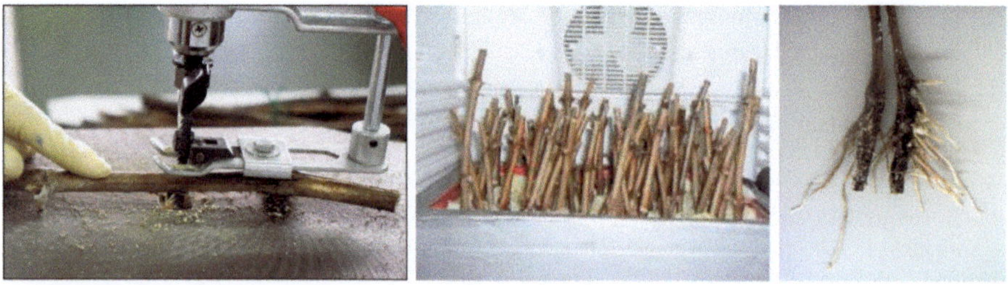

Fig. 11.1 Main steps to adapt fruiting cuttings to PRD studies

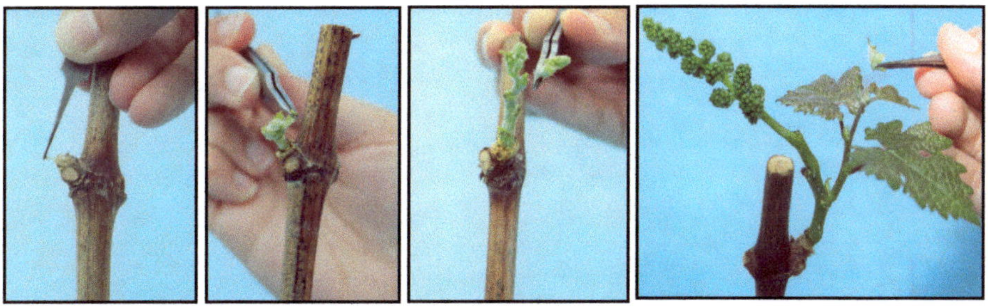

Fig. 11.2 Main steps to produce fruiting cuttings

Fig. 11.3 Main developmental stages of fruiting cuttings

of 600 μmol m^{-2} s^{-1} at the level of inflorescence. A proper nutrient solution provides a mineral nutrition in accordance with viticultural requirements (Ollat et al. 1998).

11.1.3 Technical Adaptation for PRD Studies

Split-rooted cuttings were planted such that each half of the cutting base was in a different 5 L plastic pot containing a potting mix comprising soil-peat (1:1, v/v) (Fig. 11.5). Utilization of split-rooted cuttings permit us to distinguish whether grapevine responses to PRD with respect to shoot growth, cropping and berry ripening are due "simply" to deficit irrigation per se, or whether the alternating wet/ dry cycles that are imposed on spatially separated portions of a grapevine's root system under a PRD regime have a physiologically-distinctive effect.

11.1.4 Water Treatments

Three water management treatments were applied after fruit set (E–L stage 27) (Coombe 1995). (1) Both pots maintained at pot capacity (control). (2) 50% of water given to controls, equally distributed between each side of the root system (sustained

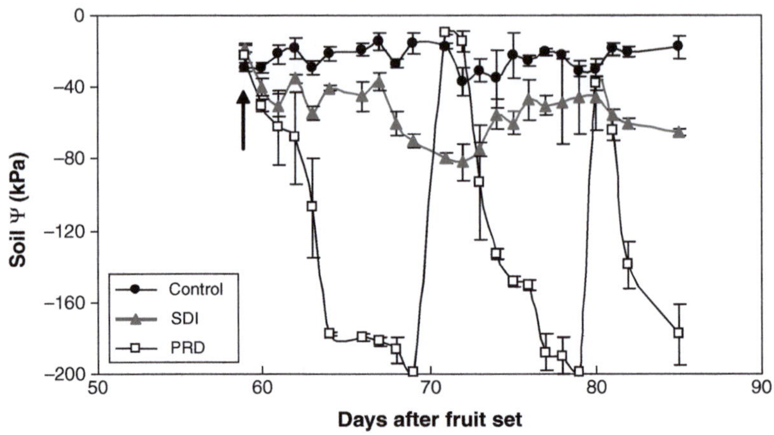

Fig. 11.4 Soil water potential (Ψ) measured from onset of veraison to maturity in pots subjected to different irrigation treatments: full irrigation (Control), deficit irrigation (SDI) or partial root drying (PRD). In PRD, data correspond to drying pattern in one pot. Values represent means (n = 5). Arrow indicates onset of veraison

deficit irrigated, SDI). (3) 50% of water given to controls alternatively supplied to only one side of the root system (partial root drying, PRD). Thus, under SDI, test vines were provided the same amount of water as the PRD-treated test vines, but the irrigation water was applied uniformly and simultaneously to both sides of the split-root system, rather than to alternate sides, as with PRD. All plants (including

Fig. 11.5 Measurement of soil Ψ with a Watermark sensor placed into each pot

controls) had split roots. The watering regime in PRD was reversed between the two pots when soil water potential (Ψ) of the dry pot reached levels about 10-fold less than in watered pots (approximately every 10 days) (Lovisolo et al. 2002). By this time, soil Ψ reached –0.2 MPa (Fig. 11.4). Soil Ψ was monitored with a granular matrix sensor (Watermark Soil Moisture Sensor, Irrometer Co, Riverside, CA, USA) placed within each pot (Fig. 11.5).

Pot capacity was previously assessed by determining water retained after free drainage. The surface of the plant containers was covered with quartz stones during the experiments to avoid water loss due to evaporation. Watering was performed with nutrient solution or deionised water in order to supply the different treatments with the same amount of nutrients during water deficit.

To evaluate the influence of irrigation system on berry growth and ripening, samples were obtained from berries collected at distinct stages of berry ripening: green berries beginning colour (in red cultivars) or to enlarge (in white cultivars), corresponding to Eichhorn and Lorenz (E–L) growth stage 35 (Coombe 1995) (onset of veraison); berries with intermediate Brix values (10–14°Brix depending on the cultivar utilized) (E–L 36 stage, middle veraison); red berries not quite ripe (approximately 16°Brix) (E–L 37 stage, end of veraison); and red, commercially ripe berries (20–22°Brix depending on the cultivar utilized) (E–L 37 stage, maturity).

11.2 Results

11.2.1 Main Vegetative Characteristics

The main developmental features of fruiting cuttings are presented in Fig. 11.6. Before flowering, a careful control of stem and leaf growth improved the partitioning of stored carbon towards the roots and the cluster. During ripening, vine vigour calculated as the leaf to fruit ratio was between 20 and 30 cm^2 of leaf area per gram of fruit, which is according to optimal values reported by Smart and Robinson (2006).

Vegetative growth occurs mainly between the end of veraison and harvest. Leaf appearance rates show that greater vegetative growth took place after veraison, during last phases of berry ripening (Fig. 11.7).

11.2.2 Grape Development

The clusters reached maturity 5 months after planting and the period of cluster development was not shortened. Fruit set was good and berry growth followed a typical double sigmoid curve (Ollat et al. 2002). After flowering, growth is by cell division and cell expansion, being the former process the main factor of berry growth during the first fortnight after anthesis. Afterwards, berry growth is mainly related to cell enlargement. Before veraison, growth slows down. It resumes rapidly at the onset of ripening and then, is only supported by cell expansion. Growth is

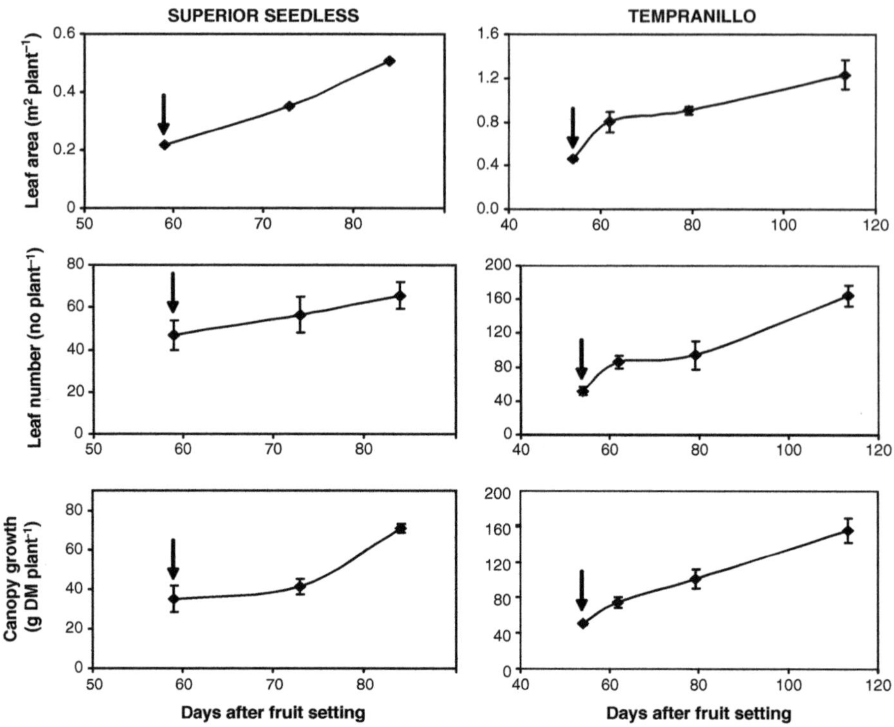

Fig. 11.6 Canopy development throughout berry ripening in a white table grape (Superior Seedless) and red grape wine (Tempranillo) fruiting cuttings under controlled conditions. Values represent means (n = 5). Arrows indicate onset of veraison

intense at the beginning of ripening and slows close to maturity. This pattern clearly appeared when berry growth rates were registered on clusters developing on fruiting cuttings of different cultivars (Fig. 11.7).

The main features of berries during ripening are presented in Fig. 11.8. Final size was similar than usual and the evolution of organics acids, sugars and phenolic compounds in white and red cultivars was also classical.

11.2.3 Grapevine Responses to Irrigation Regime

As expected, partial rootzone drying (PRD) irrigation system improved water use efficiency (WUE) because a better control of vegetative growth without reducing fruit yield was achieved (Fig. 11.9). Compared to well-watered control vines, yield per plant, and weight per bunch were both reduced significantly under SDI, but were sustained close to control values under the PRD regime in Tempranillo (Table 11.1) and Superior Seedless (Table 11.2). SDI and PRD thus differed significantly in their respective impacts on vine performance, and comparing just these two forms of deficit irrigation, PRD resulted in greater yield and higher bunch weight compared

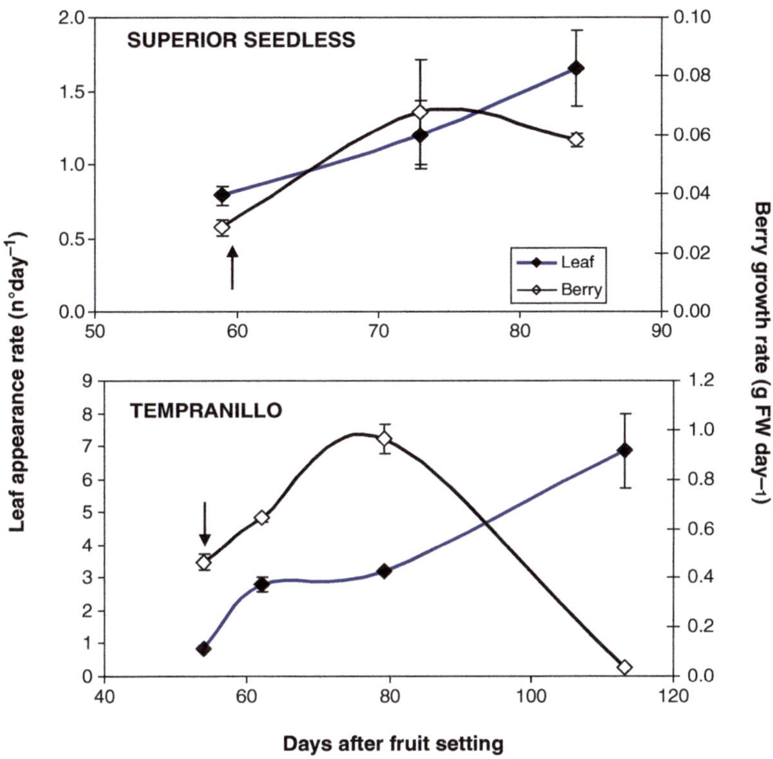

Fig. 11.7 Berry growth rate and leaf appearance rate registered throughout berry development in a white table grape (Superior Seedless) and red wine grape (Tempranillo) fruiting cuttings in a greenhouse. Values represent means (n=5). Arrows indicate onset of veraison

with the SDI treatment. Vegetative growth was severely decreased under both PRD and SDI treatments. In red cultivar, PRD also induced greater accumulation of skin anthocyanins at harvest, compared to SDI.

Little is known about how grapevine reproductive development is regulated when irrigated under PRD. The use of grapevine fruiting cuttings for physiological studies on irrigation responses permitted us to investigate the relationship between some endogenous growth regulators as abscisic acid (ABA) and polyamine (PA) levels during berry ripening, and how hormonal balance can be affected under an irrigation regime of PRD. Some authors (Gu et al. 2004) have questioned the implementation of a PRD design, arguing that the total amount of water available, rather than the irrigation method, seems to be the determining factor for constraining vegetative growth and enhancing yield. Our results showed that berry ABA concentration increased continuously throughout veraison in all treatments and achieved a maximum at the end of this period (Fig. 11.10). The increase in berry ABA concentration was earlier and faster in control and PRD berries. SDI treatments only showed a slight increase in berry ABA concentration at the end of veraison. These

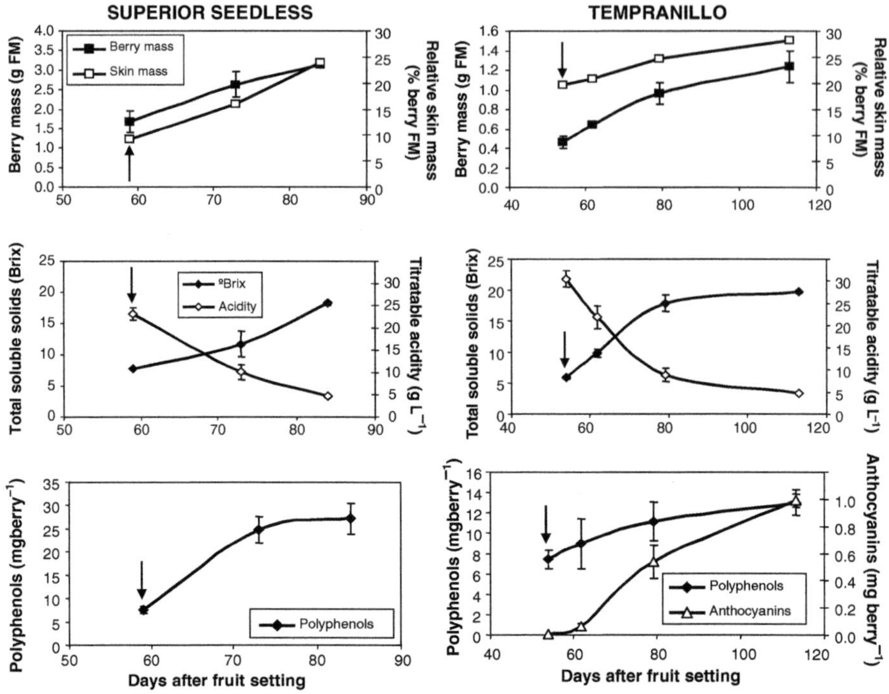

Fig. 11.8 Berry growth and fruit quality throughout berry development in a white table grape (Superior Seedless) and red wine grape (Tempranillo) fruiting cuttings under controlled conditions. Values represent means (n = 5). Arrows indicate onset of veraison

distinctive responses to PRD compared to SDI imply that the alternating wet-dry cycles of PRD, rather than simply a deficit irrigation, as in SDI, had a different effect on growth, ripening and berry composition. The data imply that these differential responses might be related to ABA physiology and especially to ABA levels in berries during ripening (Antolín et al. 2006).

Polyamines (PA) are involved in a broad range of physiological processes in plant growth and development (Kaur-Sawhney et al. 2003). In grapevines, conjugated to

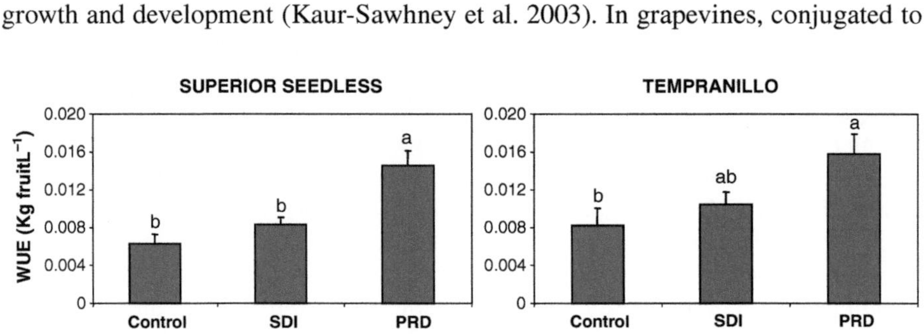

Fig. 11.9 Water use efficiency (WUE) in a white table grape (Superior Seedless) and red wine grape (Tempranillo) fruiting cuttings under controlled conditions. Values represent means (n = 5)

Table 11.1 Yield and fruit composition at harvest for Tempranillo subjected to different irrigation treatments: partial root drying (PRD), sustained deficit irrigation (SDI) or full irrigation (Control). Values represent means (n = 5)

Measurements	Control	SDI	PRD
Yield components			
Yield (g plant^{-1})	410.0a	261.5b	394.7a
Bunch weight (g)	440.1a	280.7b	428.5a
Vigour			
Leaf area (m^2)	1.234a	0.473c	0.603b
Fruit composition			
Total soluble solids (°Brix)	19.6b	22.6a	20.7b
Juice pH	4.3a	4.3a	4.2a
Titratable acidity (g L^{-1})	4.8a	4.2a	4.6a
Total polyphenols (mg g^{-1} FM skin)	45.9a	49.3a	48.3a
Anthocyanins (mg g^{-1} FM skin)	3.3b	3.7b	4.1a

More details in Antolín et al. (2006).

low molecular mass compounds have been considered as valuable markers of the onset of flowering and fruit set (Gény et al. 1999). Flowering and fruit set in the grape berry has been characterized by high PA levels while during ripening (after veraison) these concentrations decreased dramatically (Gény et al. 1997, Baigorri et al. 2001). Our results showed that at the beginning of veraison, PRD berries manifested higher free PA and free to bound PA ratio than other treatments (Table 11.3, Fig. 11.11). Since free PA could influence cell division and cell expansion and free to bound ratio was high during active growth stages, we suggest that berry growth rate

Table 11.2 Yield and fruit composition at harvest for superior seedless subjected to different irrigation treatments: full irrigation (Control), sustained deficit irrigation (SDI) or partial root drying (PRD). Values represent means (n = 5)

Measurement	Control	SDI	PRD
Yield components			
Yield (g plant^{-1})	226.4a	131.4b	244.5a
Bunch weight (g)	233.0a	135.6b	251.0a
Vine vigour			
Leaf area (m^2 plant^{-1})	0.51a	0.24b	0.29b
Fruit composition			
Total soluble solids (°Brix)	18.2b	20.5a	20.5a
Juice pH	3.6a	3.6a	3.7a
Titratable acidity (g L^{-1})	4.6a	4.4a	4.5a
Polyphenols (mg g^{-1} FM skin)	88.8a	63.9a	78.7a

More details in Antolín et al. (2008).

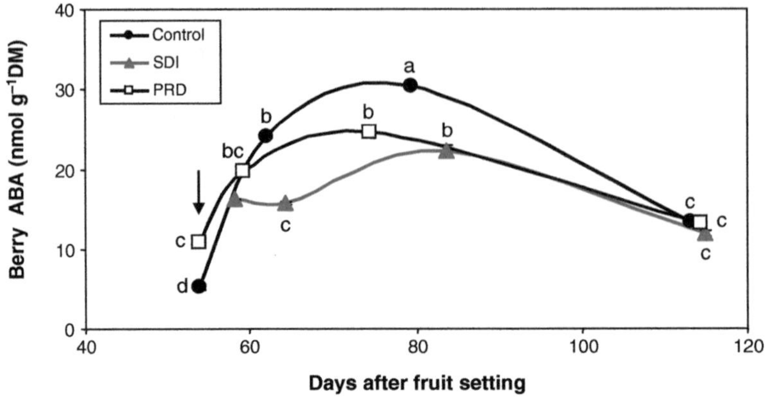

Fig. 11.10 Berry abscisic acid (ABA) evolution throughout fruit development in fruiting cuttings of Tempranillo grapevines subjected to different irrigation treatments. Arrow indicates onset of veraison. (Redrawn from Antolín et al. 2006)

in PRD vines could be faster than in SDI so that berry size and yield are maintained (Antolín et al. 2008).

We also found a significant correlation between free to bound PA ratio and ABA (Fig. 11.12), which evidenced that free PA increased as accumulation of ABA took place. As a consequence, results indicate that changes induced in berry ABA during PRD conditions could be counterbalanced by changes in PA metabolism.

Table 11.3 Polyamines (PA) free, conjugated to low-molecular mass compounds (SH-PA), or bound to different macromolecules (PH-PA) at various stages in berries from fruiting cuttings of superior seedless subjected to different irrigation treatments during berry ripening: full irrigation (Control), sustained deficit irrigation (SDI) or partial root drying (PRD). Values represent means (n = 5)

Developmental stage	Irrigation	Free PA (nmol·g^{-1} DM)	SH-PA (nmol·g^{-1} DM)	PH–PA (nmol·g^{-1} DM)
Onset of veraison	Control	106.50c	67.49a	81.54a
	SDI	203.52b	65.82a	60.72ab
	PRD	261.11a	35.71b	43.03ab
Middle of veraison	Control	71.46d	27.73b	22.51b
	SDI	39.52de	7.78b	47.67ab
	PRD	58.60de	6.87b	29.90b
Harvest	Control	28.75e	11.45b	33.96b
	SDI	46.24de	15.84b	42.10ab
	PRD	30.07e	9.92b	46.13ab

More details in Antolín et al. (2008).

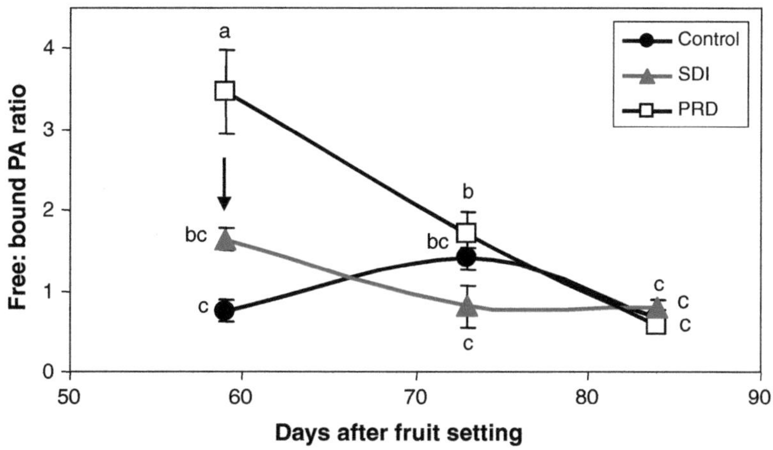

Fig. 11.11 Ratio of free to bound polyamines (PA) in berries from fruiting cuttings of Superior Seedless subjected to different irrigation treatments: full irrigation (Control), deficit irrigation (SDI) or partial root drying (PRD). Values represent means (n = 5)

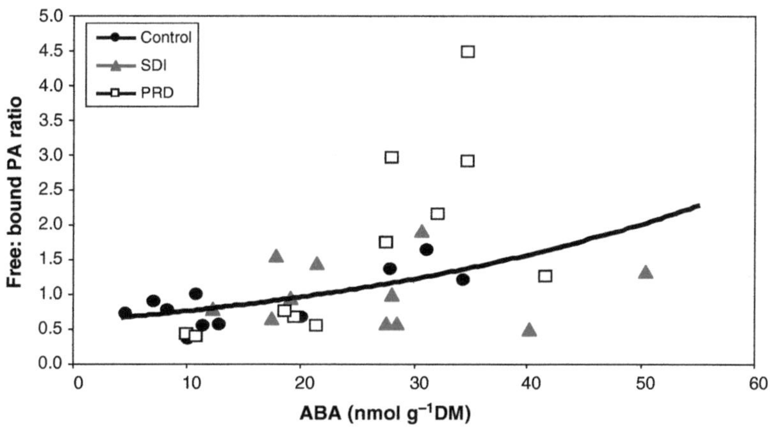

Fig. 11.12 Relationship between abscisic acid (ABA) and the free to bound polyamine ratio in berries from fruiting cuttings of Superior Seedless subjected to different irrigation treatments during fruit ripening. (See more details in Antolín et al. 2008)

11.3 Interest, Limitations and Progress for Our Understanding of Grapevine

The present data showed that fruiting cuttings behave mostly as conventional vines and could be a useful tool for grapevine physiology studies under water deficit conditions. The technical adaptation to PRD studies permitted a more exhaustive control of soil water content and grapevine irrigation during berry development, and improved our understanding about this irrigation system. Thus, our investigations have concluded that although the restriction of vegetative growth in PRD might be partly a response to the volume of water applied and therefore similar to a SDI response, specific hormonal factors may explain differential effects on the reproductive growth and yield.

Acknowledgements The authors thank J. Carreño (Instituto Murciano de Investigación y Desarrollo Agrario y Alimentario, IMIDA, Spain) and F. Aguirrezábal (Estación de Viticultura y Enología de Navarra, EVENA, Spain) for providing grapevine cuttings. This research was supported by Fundación Universitaria de Navarra (2001, 2005) and Genoma España (GrapeGen Project 2004–2007). M. Ayari was the recipient of grants from Agencia Española de Cooperación Internacional (Ministerio of Asuntos Exteriores, Spain) and the Instituto Agronómico Mediterráneo de Zaragoza (Spain). H. Santesteban was the recipient of a grant from Ministerio de Educación y Ciencia of Spain (Plan de Formación del Profesorado Universitario). The authors wish to thank the help of E. Santa María and the technical cooperation of A. Urdiáin and M. Oyarzun.

References

Antolín MC, Ayari M Sánchez-Díaz M (2006) Effects of partial rootzone drying on yield, ripening and berry ABA in potted Tempranillo grapevines with split roots. Aust J Grape Wine Res 12:13–20

Antolín MC, Santesteban H, Santa María E, Aguirreolea J, Sánchez-Díaz M (2008) Involvement of abscisic acid and polyamines in berry ripening of *Vitis vinifera* (L.) subjected to water deficit irrigation. Aust J Grape Wine Res 14:123–133

Baigorri H, Antolín MC, De Luis I, Gény L, Broquedis M, Aguirrezábal F, Sánchez-Díaz M (2001) Influence of training system on the reproductive development and hormonal levels of *Vitis vinifera* cv. Tempranillo. Am J Enol Vitic 52:357–363

Coombe BG (1995) Adoption of a system for identifying grapevine growth stages. Aust J Grape Wine Res 1:104–110

Davies WJ, Hartung W (2004) Has extrapolation from biochemistry to crop functioning worked to sustain plant production under water scarcity? In: Proceedings of the 4th international crop science congress, Brisbane Australia

Du Toit PG, Dry PR, Loveys BR (2003) A preliminary investigation on partial rootzone drying (PRD) effects on grapevine performance, nitrogen assimilation and berry composition. South Afr J Enol Vitic 24:43–54

Gény L, Broquedis M, Martin-Tanguy J Bouard J (1997) Free, conjugated, and wall-bound polyamines in various organs of fruiting cuttings of *Vitis vinifera* L. cv. Cabernet Sauvignon. Am J Enol Vitic 48:80–84

Gény L, Colin L, Brazillon I, Broquedis M (1999) Analyse des polyamines conjuguées aux acides hydroxycinnamiques dans les fleurs de vigne et les baies de raisin. Vitis 38:157–160

Gu S, Du G, Zoldoske D, Hakim A, Cochran R, Fugelsang K, Jorgensen G (2004) Effects of irrigation amount on water relations, vegetative growth, yield and fruit composition of Sauvignon blanc grapevines under partial rootzone drying and conventional irrigation in the San Joaquin Valley of California, USA. J Hort Sci Biotechnol 79:26–33

Kang S, Zhang J (2004) Controlled alternate partial root-zone irrigation: its physiological consequences and impacton water use efficiency. J Exp Bot 55:2437–2446

Kaur-Sawhney R, Tiburcio AF, Altabella T, Galston AW (2003) Polyamines in plants: an overview. J Cell Mol Biol 2:1–12

Lovisolo C, Hartung W, Schubert A (2002) Whole-plant hydraulic conductance and root-to-shoot flow of abscisic acid are independently affected by water stress in grapevines. Func Plant Biol 29:1349–1356

Mullins MG (1966) Test-plants for investigations of the physiology of fruiting in *Vitis vinifera* L. Nature 209:419–420

Ollat N, Gény L, Soyer JP (1998) Les boutures fructifères de vigne: validation d'un modèle d'étude de la physiologie de la vigne. I. Principales caractéristiques de l'appareil végétatif. J Int Sci Vigne Vin 32:1–9

Ollat N, Diakou-Verdin P, Carde JP, Barrieu F, Gaudillère JP, Moing A (2002) Grape berry development: a review. J Int Sci Vigne Vin 36:109–131

Santa María E (2004) Incidencia de *Botrytis cinerea* en relación con diferentes aspectos fisiológicos de la vid. Ph. D. Thesis. Universidad de Navarra, Pamplona (Spain)

Santos TP, Lopes CM, Rodrigues ML, de Souza CR, Maroco JP, Pereira JS, Silva JR, Chaves MM (2003) Partial rootzone drying: effects on growth and fruit quality of field-grown grapevines (*Vitis vinifera*). Func Plant Biol 30:663–671

Smart R, Robinson M (2006) Sunlight into wine: A hand-book for winegrape canopy management. Winetittles Phy. Ltd, Adelaide

Chapter 12
Nutritional Deficiencies

Luigi Bavaresco, Matteo Gatti, and Mario Fregoni

Abstract Fourteen mineral elements are considered essential for higher plant growth; these elements include 6 macronutrients (N, P, K, Ca, Mg, S) and 8 trace elements (Fe, Mn, Zn, Cu, Cl, B, Mo, Ni). The physiological role of each element is described, as well as the deficiency symptoms and the methods to prevent or cure the nutritional disorders. Leaf blade and/or petiole chemical analysis is an important tool to assess the nutritional status of the plant and the methodology to perform the leaf sampling is described as well as the optimum concentrations of each element. The effect of the grape variety, rootstock and environmental conditions on mineral nutrition is described, as well as the amount of element uptake in the vineyard, which is crucial for fertilizer recommendation.

Contents

L. Bavaresco (✉)
CRA-Research Centre for Viticulture, Viale XXVIII Aprile, 26, 31015 Conegliano (TV), Italy; Istituto di Frutti-Viticoltura, Università Cattolica S. Cuore, Piacenza, I-29100, Italy
e-mail: luigi.bavaresco@unicatt.it; luigi.bavaresco@entecra.it

S. Delrot et al. (eds.), *Methodologies and Results in Grapevine Research*,
DOI 10.1007/978-90-481-9283-0_12,

Abbreviations

EDDHA	Ethylenediamine-N,N′-bis(o-hydroxyphenylacetic) acid
EDDHMA	Ethylenediaminedi(o-hydroxy-p-methylphenylacetic) acid
EDDHSA	Ethylenediaminedi(2-hydroxy-5-sulfophenylacetic) acid
DTPA	Diethylenetriaminepentaacetic acid

12.1 Background

Grapevine needs fertilizer supply in order to keep a constant fruit load along time, together with a proper grape composition to give a high quality wine, because the natural soil reservoir of mineral elements is actually not sufficient to support the vine nutritional requirements throughout its entire life. There are 14 elements that are considered to be essential for higher plants; these are macronutrients (N, P, K, Ca, Mg, S) and trace elements (Fe, Mn, Zn, Cu, Cl, B, Mo, Ni). The average concentration in shoot dry matter (that is sufficient for adequate growth) for macronutrients is higher than 0.1%, while for trace elements it is lower than 100 ppm (Marschner 1995). Grapevine is a plant needing low levels of fertilizer supply, concerning especially nitrogen and phosphorus, unless it produces too heavy crop loads. Chemical analysis of the biomass produced along the annual growth cycle is utilized to calculate the nutritional needs of the vine, corresponding to the annual mineral element uptake by the whole plant (Fregoni 1980). The method of nutritional maps was provided to recommend fertilizer supply for the vineyards, taking in account soil composition, foliar or petiole analysis, nutrient uptake and losses (Fregoni 1984, 2009, Vercesi and Gatti 2007). According to the Table 12.1, nitrogen, phosphorus and potassium uptake by field growing vines increases as a consequence of increasing grape production, and their ratios change depending on the grape yield.

During the annual cycle, the vine can undergo mineral deficiencies, impairing the physiology of the plant, due to different factors such as high vegetative vigour, soil composition, meteorological conditions. Nutritional deficiencies affect the shape,

Table 12.1 N, P and K uptake (Kg/ha/year) of field growing vines depending on the crop load

t grapes /ha	Nitrogen (N) (kg/ha)	Phosphorus (P) (kg/ha)	Potassium (K) (kg/ha)
5.0	27	5	31
7.5	31	5	36
10.0	35	6	40
12.5	39	6	45
15.0	43	7	49
17.5	47	7	54
20.0	51	8	58
25.0	59	9	67

Average data obtained from 38 nutritional maps developed in representative Italian viticultural areas, including 41 varieties, 10 training systems, with grape yield ranging from 6.0 and 27.0 t/ha (Bavaresco 2005).

colour, chemical composition, performance and attainable age of individual organs of the whole vine, and they are the consequence of deficiency of one of more nutrients (Pearson and Goheen 1988). The visible symptoms provide clues about their cause, but soil and leaf (blade or petiole) analysis can confirm nutrient imbalance.

12.2 Mineral Element Function

The phloem mobility and the role of the elements in the vine physiology is reported in Table 12.2.

Table 12.2 Phloem mobility and main role of the mineral elements

Element	Phloem mobility	Main role
Nitrogen (N)	High	Constituent of chlorophyll, proteins, hormones, nucleic acids, lecithins, vitamins, alkaloids. It enhances vegetative growth (vigour), leading to bud burst delay, bigger shoot and leaf growth, higher drought and disease susceptibility, lower resveratrol synthesis, longer growing season, K and Fe deficiency. It enhances grape yield, delay grape ripening and reduce quality
Phosphorus (P)	High	Constituent of cell membrane, nucleic acids, vitamins, lecithins, proteins, ATP. It enhances the growth of shoot and root apex and leaves; it improves grape aromas
Potassium (K)	High	It enhances shoot growth, disease and winter frost resistance, sugar and starch accumulation (respectively in the berries and in the woody organs), it controls water status and berry acidity level, by activating many enzymes. It is required for protein synthesis
Calcium (Ca)	Low	Constituent of cell wall. It is involved in protein and carbohydrate synthesis and in transpiration
Magnesium (Mg)	High	Constituent of the chlorophyll and enzyme activator. It is involved in carbohydrate partitioning
Sulphur (S)	High	Constituent of some aminoacids, vitamins, CoA
Iron (Fe)	Intermediate	Constituent of cytochromes and other enzymes involved in photosynthesis, respiration, nitrate reduction
Boron (B)	Intermediate	It enhances root growth and lignin synthesis, membrane integrity and functioning, pollen germination and pollen tube growth, sugar accumulation
Manganese (Mn)	Low	It is involved in protein and oxidative metabolism, in photosynthesis and in hormone regulation. It improves bud fertility, fruit set, lignifications and wine bouquet
Zinc (Zn)	Intermediate	It improves bud fruitfulness, fruit set, drought and frost resistance, membrane integrity, wine bouquet
Copper (Cu)	Intermediate	It is involved in pollen formation and fertilization, in lignin synthesis and disease resistance
Molybdenum (Mo)	Intermediate	Constituent of nitrate reductase, it is involved in nitrogen metabolism
Chlorine (Cl)	High	It is involved in photosynthetic oxygen evolution and in stomatal regulation, and it stimulates tonoplast proton-pumping ATPase

12.3 Analytical Methods

Nutritional disorders can be avoided by monitoring the nutritional status of the vine, through leaf (blade) or petiole analysis, which has to be done periodically (every 2–3 years) even though no deficiency symptoms are visible in the vineyard (Christensen et al. 1978, Christensen 1984). The mineral composition of the leaf (blade or petiole) is actually representative of the whole plant nutritional status (Champagnol 1990, Delas 1990, Loué 1990, Vercesi et al. 1993). When the leaf levels are approaching the deficiency range, it is the time to act in order to avoid symptoms occurrence.

The methodology is reported in Table 12.3.

When the symptom is there, no chemical analysis of the blade/petiole is convenient, because, according to the literature, every mineral deficiency is corresponding to specific symptoms (Fregoni 1982).

The dried blades/petioles are analyzed to detect the total amount of each macronutrient and trace element, and the results are compared with normal (optimum) values (% and ppm on the dry weight basis), which are reported in Tables 12.4 and 12.5.

Table 12.3 Method for leaf blade and petiole sampling

	Blade	Petiole
Sampling time	Fruit set and veraison	Veraison
Number of vines	50–100 vines	100 vines
Leaf position in the shoot	Opposite the basal cluster	Opposite the clusters
Shoot position in the vine (cane pruning)	Middle of each cane	Middle of each cane
Shoot position in the vine (spur pruning)	Apical of a middle spur	Apical of a middle spur
Minimum leaf number	50–100	100
Leaf treatments	Water rinse[a]	Water rinse[a]
	Drying (3 d at 70°C)	Drying (3 d at 70°C)

[a]Only if the sampling is done after foliar spray, and for Cu, Zn, Mn analysis.

Table 12.4 Concentration ranges of mineral nutrients in leaf blade dry matter that are optimal for adequate growth (Fregoni 2005)

Element in the blade	Fruit set	Veraison
N (%)	2.08–2.95	1.41–2.20
P (%)	0.14–0.26	0.11–0.17
K (%)	0.78–1.40	0.62–1.24
Ca (%)	1.43–2.55	1.77–2.99
Mg (%)	0.19–0.37	0.20–0.43
K/Mg	2.17–6.21	1.62–5.56
K/Ca+Mg	0.28–0.64	0.19–0.55
Fe (ppm)	65–300	80–300
B (ppm)	20–70	15–60
Mn (ppm)	50–500	55–400
Zn (ppm)	20–250	14–160
Cu (ppm)	10–20	20–30

Table 12.5 Concentration ranges of mineral nutrients in petiole dry matter that are optimal for adequate growth (Fregoni 2005)

Element in the petiole	Veraison
N (%)	0.60–0.90
P (%)	0.15–0.60
K (%)	2.50–3.50
Ca (%)	1.20–1.80
Mg (%)	0.50–1.00
K/Mg	3–8
Fe (ppm)	25–60
B (ppm)	25–70
Mn (ppm)	20–150
Zn (ppm)	15–25
Cu (ppm)	3–6

12.4 Mineral Elements

12.4.1 Nitrogen

12.4.1.1 Symptoms

First appearance time: Spring.
Whole vine: Reduced vigour.
Leaves: Smaller, pale green and then yellow. Reddening of petiole (Figs. 12.1 and 12.2).
Shoots: Pink or red, reduced growth.
Fruit: Berries may be small (Fig. 12.3).

Fig. 12.1 N deficiency: adult leaves are pale green, turning *yellow*

Fig. 12.2 N deficiency: increasing symptoms from *left* to *right*

Fig. 12.3 N deficiency: the cluster is loose with small berries

12.4.1.2 Causes and Therapy

Causes

Very poor soils (low organic matter); soil green covering.

Therapy

Soil applications: 50–150 kg/ha N, during spring.

Foliar sprays: urea (0.3–0.5%), 1–4 times along the growing season, depending on the deficiency severity.

12.4.2 Phosphorus

12.4.2.1 Symptoms

First appearance time: Late spring, before flowering.
Whole vine: Reduced vigour.
Leaves: Reddening of young leaves. Red dots near the edges of adult leaves. Red dots may coalesce later into red bars at right angles to the vein; leaves can drop (Figs. 12.4 and 12.5).
Shoots: Reduced growth; lignifications is impaired.
Fruit: Small and loose clusters, due to reduced fruit set (Fig. 12.6).

12.4.2.2 Causes and Therapy

Causes
The deficiency is very rare in commercial vineyards, and symptoms are obtained in nutrient culture.
Therapy
Soil applications: 50–100 kg/ha P_2O_5, during autumn.

12.4.3 Potassium

12.4.3.1 Symptoms

First appearance time: Late spring (June).
Leaves: Reddening or yellowing along the margin of young leaves; then leaf margins dry and roll up. Late symptoms (after veraison) occur in adult leaves when blades become brown ("brunissure") or yellow (Figs. 12.7, 12.8, 12.9 and 12.10).
Shoots: Reduced growth; lignifications is impaired.
Fruit: Small and unevenly ripened berries.

Fig. 12.4 P deficiency: reddening near the teeth of a young leaf

Fig. 12.5 P deficiency: *red dots* that coalesce into *red bars*

Fig. 12.6 P deficiency: the cluster is short with small berries

Fig. 12.7 K deficiency: yellowing and rolling up along the margin of the leaf

Fig. 12.8 K deficiency: drying and rolling up of leaf margin

Fig. 12.9 K deficiency: "brunissure" on adult leaf

Fig. 12.10 K deficiency: severe leaf drop

12.4.3.2 Causes and Therapy

Causes

Clayey or sandy soil. Temporary soil dryness. Heavy nitrogen supply and fruit load. Varietal (ex Ugni blanc) and rootstock (ex. 1103 P) sensitivity.

Therapy

Soil applications: 500–800 kg K_2O /ha, along 2–3 years (clayey soil) or 250 kg K_2O/ha (sandy soil).

Foliar sprays: Potassium nitrate (0.4–1%), 3–4 times every 15 d before and after flowering. Other commercial fertilizers containing K can be used.

12.4.4 Calcium

12.4.4.1 Symptoms

First appearance time: Late spring (June).
Whole vine: Stunted growth.
Leaves: Necrosis at the margin of young leaves; necrotic dots, rolling up, leaden and yellow adult leaves (Fig. 12.11).
Shoots: Dark brown pimples up to 1 mm in diameter on the primary bark of the internodes (Fig. 12.12).
Fruit: Cluster stem necrosis and berry withering (after veraison).

Fig. 12.11 Ca deficiency: leaf symptoms

Fig. 12.12 Ca deficiency: stunted growth

12.4.4.2 Causes and Therapy

Causes
Acidic soils.
Therapy
Soil applications: quicklime, slaked lime, dolomite, calcium carbonate (0.5–1.5 t/ha).
Foliar sprays: calcium chloride (1%).

12.4.5 Magnesium

12.4.5.1 Symptoms

First appearance time: Late spring (after fruit set).
Leaves:Yellowing or reddening of basal leaves, beginning near the edge and moving inward between the primary and secondary veins (Figs. 12.13 and 12.14).
Fruit: Cluster stem necrosis and berry withering (after veraison) (Figs. 12.15, 12.16 and 12.17).

12.4.5.2 Causes and Therapy

Causes
Acidic, sandy, calcareous-clayey soils. Rainy meteorological conditions. High potassium supply. Heavy fruit load. Varietal (ex. Croatina, Rhein Riesling, Schiava, Cabernet Sauvignon, Sangiovese, etc.) and rootstock (ex. SO4, Fercal, Kober 5BB, etc.) sensitivity.

Fig. 12.13 Mg deficiency: leaf yellowing between the primary and secondary veins

Fig. 12.14 Mg deficiency: leaf reddening between the primary and secondary veins

Fig. 12.15 Mg and Ca deficiency: cluster stem necrosis

Therapy

Soil applications (only in non-calcareous soils): 50–100 kg MgO/ha

Foliar sprays: magnesium sulphate (2% if MgO = 32%, or 4% if MgO = 16%) or magnesium chloride / nitrate (1–1.5%) 3–6 times every week, beginning from fruit set.

Cluster sprays against stem necrosis: magnesium sulphate (as above), 3 times, at berry touch, veraison and 10 days later (Boselli et al. 1985).

12.4.6 Iron

12.4.6.1 Symptoms

First appearance time: Spring (few weeks after bud burst).

Whole vine: Stunted growth

Fig. 12.16 Mg and Ca deficiency: cluster stem necrosis

Fig. 12.17 Mg and Ca deficiency: berry withering

Leaves: Yellowing between the veins of young leaves and, under severe deficiency, leaf (young and adult) necrosis and drop (Fig. 12.18).

Shoots: Shortened internodes, bushy aspect (Figs. 12.19 and 12.20).

Fruit: Loose clusters, shot berries (Bavaresco et al. 2005) (Figs. 12.21 and 12.22).

12.4.6.2 Causes and Therapy

Causes

Fig. 12.18 Fe deficiency: young leaf yellowing with some necrotic spots

Fig. 12.19 Fe deficiency: symptoms on the shoot

Calcareous soils. Other contributing factors are: rainy spring; nitric fertilizer supply; soil tillage; high plant vigour; varietal (ex. Pinot, Cabernet Sauvignon, Cinsaut, Carmenère, etc.) and rootstock (3309 C, 101–114, etc.) sensitivity (Fig. 12.23).

Fig. 12.20 Fe deficiency: severe symptoms with the leaves of the whole vine completely chlorotic

Fig. 12.21 Fe deficiency: severe flower drop ("coulure"), causing yield loss

Therapy

Soil applications: ferrous sulphate (10% water dilution), 10 L/plant, applied in a furrow around the vine and covered right away; ferrous sulphate (powder or pellet), 3–4 t/ha, covered right away; ferrous sulphate + manure (pellet) 2–3 t/ha, covered right away; iron chelates (EDDHA, EDDHMA, EDDHSA, DTPA) 30–60 g/plant diluted in 4–10 L water, in a furrow or by an injector; pellet iron chelates (as above) 25–30 kg/ha, covered right away; organic fertilizers.

Foliar applications: ferrous sulphate (0.7%) + citric acid (0.1%); iron chelates (0.10–0.15%), 3–5 times every week, beginning with shoots 10 cm long, other organic compounds.

Fig. 12.22 Fe deficiency: loose clusters with shot berries

Fig. 12.23 Fe deficiency: chlorosis rating and corresponding SPAD values

12.4.7 Boron

12.4.7.1 Symptoms

First appearance time: Spring (few weeks after bud burst).

Whole vine: Stunted growth

Leaves: Yellowing or reddening dots, rolling down, corrugation and odd shapes in young leaves (Figs. 12.24, 12.25, 12.26 and 12.27).

Shoots: Bushy aspect, shortened internodes, many laterals. Tendril tips dry. Younger internodes swell and the pith becomes necrotic. Soot tip dries.

Fruit: Loose and odd clusters, with few seeded berries and small seedless berries (hen and chicken bunches); browning of hypodermal cells, berry break (Scienza et al. 1981) (Figs. 12.28, 12.29 12.30 and 12.31).

12.4.7.2 Causes and Therapy

Causes

Calcareous or sandy and acidic soils, especially under drought conditions.

Therapy

Soil applications (only in non-calcareous soils): borace (sodium tetraborate), 30–80 kg/ha.

Foliar sprays: solubor (20.5% B), 0.25%, 3 times every 15 d, beginning from 5th–6th leaf development stage (Bavaresco et al. 1989).

Fig. 12.24 B deficiency: odd shapes and rolling down of young leaves

Fig. 12.25 B deficiency: odd shapes of young leaves

Fig. 12.26 B deficiency: leaf yellowing, rolling down and corrugation

Fig. 12.27 B deficiency: leaf reddening, rolling down and corrugation

Fig. 12.28 B deficiency: loose cluster due to poor fruit set

Fig. 12.29 B deficiency: hen and chicken bunch

Fig. 12.30 B deficiency: berry browning (from Scienza et al. 1981)

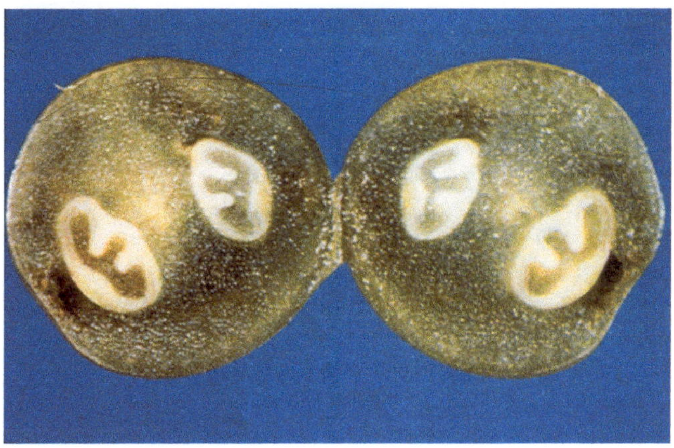

Fig. 12.31 B deficiency: browning of hypodermal berry cells (from Scienza et al. 1981)

12.4.8 Zinc

12.4.8.1 Symptoms

First appearance time: Spring (few weeks after bud burst).

Whole vine: Stunted growth

Leaves: Small asymmetric young leaf blades with opened petiolar sinuses and sharp teeth. The interveinal areas turn pale green to yellow or red in a mosaic pattern. The leaf veins also become clear (Figs. 12.32 and 12.33).

Shoots: Reduced internodes lengths; stunted lateral shoots (Fig. 12.34).

Fruit: "Hen and chicken" bunches of shot berries, which vary in size and stage of ripening (Fig. 12.35).

Fig. 12.32 Zn deficiency: small asymmetric leaf with interveinal yellowings

Fig. 12.33 Zn deficiency: small leaves with opened petiolar sinuses

Fig. 12.34 Zn deficiency: stunted growth of the shoot and the laterals

Fig. 12.35 Zn deficiency: hen and chicken bunch

12.4.8.2 Causes and Therapy

Causes
Calcareous or sandy soils.
Therapy
Soil applications (only in non-calcareous soils): borace (sodium tetraborate), 30–80 kg/ha.
Foliar sprays: zinc sulphate (0.5–1.0%) neutralized with calcium carbonate, 3 weeks before flowering; zinc chelates.

12.4.9 Manganese

12.4.9.1 Symptoms

First appearance time: Early summer (before veraison).
Whole vine: Stunted growth
Leaves: Interveinal chlorosis or reddening (yellow or red spots) of adult leaves. Mosaiclike arrangement of yellow/red spots, bordered by the smallest green veins (Figs. 12.36 and 12.37).
Shoots: Stunted growth.
Fruit: Delay of berry ripening.

12.4.9.2 Causes and Therapy

Causes
Calcareous or sandy soils.
Therapy
Foliar sprays: manganese sulphate (0.2–0.5%), 3 times, one before and two after flowering; manganese chelates.

Fig. 12.36 Mn deficiency: Mosaiclike arrangement of yellow interveinal spots

Fig. 12.37 Mn deficiency: symptom on white fruit variety (top). Symptoms on red fruit variety (bottom)

12.5 Interest, Limitation, Progress

Nutritional deficiencies (alone or more then one in the same vine) might occur in the vineyards due to the above indicated conditions. Nevertheless, in some cases, it is possible to prevent the deficiencies, by knowing the soil physical and chemical composition, before the vineyard is established. Low soil concentrations of some nutrients or situations impairing their uptake (i.e. the presence of antagonists, or particular pH values) affect the availability of the nutrient in the plant tissues. Soil supply of deficient nutrients before establish a new vineyard, or the choice of a proper rootstock are the most efficient methods to prevent some deficiencies. K, Mg and Fe are the most common deficiencies that can be prevented.

The soil availability ranges of K and Mg are reported in the Table 12.6 (Bavaresco 2005).

Under low K and Mg levels, soil applications are as follows:

- Potassium (K_2O): 300 kg/ha (sandy-clay soils); 400 kg/ha (loamy soils); 800–1,000 kg/ha (compact soils);
- Magnesium (MgO): 200 kg/ha (sandy-clay soils); 300 kg/ha (loamy soils); 400 kg/ha (compact soils).

Suitable rootstocks for low K soils are SO4, 44–53, 196-17, 110 R.

Table 12.6 Classification of K_2O and MgO soil concentration (ppm) depending on CEC

	CEC (meq/100 g soil)			
	<10 (low)	10–20 (medium)	>20 (high)	Classification
K_2O[a]	<18	<60	>90	Very low
	18–35	60–120	90–180	Low
	35–70	120–210	180–360	Normal
	70–120	210–360	360–600	High
	>120	>360	>600	Very high
MgO[b]	<8	<33	<50	Very low
	8–17	33–66	50–100	Low
	17–42	66–133	100–200	Normal
	42–100	133–300	200–498	High
	>100	>300	>498	Very high

[a] $K_2O = K \times 1.2$.
[b] $MgO = Mg \times 1.7$.

Table 12.7 Maximum tolerance of active lime soil concentration for grapevine rootstocks (Fregoni 1980)

Active lime (%)	Rootstocks
6	Riparia
9	101-14
11	3306C, 3309C
13	Ganzin 1
14	Rupestris du Lot
17	99R, 110R, 225Ru, 775P, SO4
20	Kober 5BB, 420A, 34EM, 779P, 1103P
22	157-11
25	161-49
40	41B, 333EM, 140Ru

Suitable rootstocks for low Mg soils are 1103 P, Rupestris du Lot.

As concerning iron deficiency, the most important soil parameter to consider is the active lime level, which impairs Fe uptake (Bavaresco et al. 1993). The rootstock choice, depending on the maximum lime tolerance, is reported in the Table 12.7.

References

Bavaresco L (2005) La fertilizzazione (della vite). In: Marenghi M (ed) Manuale di viticoltura. Edagricole, Bologna

Bavaresco L, Corazzina E, Ruini S (1989) Effetti del boro per via fogliare su vitigni in terreni carenti. Vignevini XVI (6):45–50

Bavaresco L, Fraschini P, Perino A (1993) Effect of the rootstock on the occurrence of lime-induced chlorosis of potted *Vitis vinifera* L. cv. “Pinot blanc”. Plant Soil 157:305–311.

Bavaresco L, Civardi S., Pezzutto S., Vezzulli S., Ferrari F (2005) Grape production, technological parameters, and stilbenic compounds as affected by lime-induced chlorosis. Vitis 44:63–65.

Boselli M, Bavaresco L, Fregoni M (1985) Possibilità di controllo del disseccamento del rachide della vite mediante applicazioni fogliari. Vignevini XII(5):39–45

Champagnol F (1990) Rajeunir le diagnostic foliaire. Progrès Agric Vitic 107 :343–351

Christensen LP (1984) Nutrient level comparison on leaf petioles and blades in twenty-six grape cultivar over three years, 1979–1981. Am J Enol Vitic 35:124–133

Christensen LP, Kasimatis AN, Jensen FL (1978) Grapevine Nutrition and Fertilization in the San Joaquin Valley. University of California, Berkeley, CA, USA

Delas J (1990) Diagnostic foliaire: aspect historique, pratique actuelle. Progrès Agric Vitic 107 :399–402

Fregoni M (1980) Nutrizione e fertilizzazione della vite. Edagricole, Bologna

Fregoni M (1982) Vademècum sulle carenze e tossicità degli elementi meso e micro nutritivi della vite. Vignevini IX (3):19–25

Fregoni M (1984) Nutrient needs in wine production. 18th Coll. Int. Potash Institute Bern:319–332

Fregoni M (2005) Viticoltura di qualità. Tecniche Nuove, Milano

Fregoni M (2009) Atlante nutrizionale della vite. Tecniche Nuove, Milano

Loué A (1990) Le diagnostic foliaire (ou pétiolaire) dans les enquêtes de nutrition minérale des vignes. Progrès Agric Vitic 107 :439–453

Marschner H (1995) Mineral Nutrition of Higher Plants. Academic Press, London

Pearson RC, Goheen AC (1988) Compendium of Grape Diseases. APS, St. Paul, MN, USA.

Scienza A, Miravalle R, Boselli M, Dorotea G (1981) Effets de la carence en B sur le dévelopment et sur la composition chimique des baies de *Vitis vinifera* cv. Barbera. Vignevini VIII (11) : 37–42

Vercesi A, Bozzalla L, Fregoni M (1993) Diagnostica fogliare in viticoltura: 20 anni di esperienze. Vignevini 3:30–33

Vercesi A, Gatti M (2007) La concimazione del vigneto con il metodo delle carte nutrizionali. VQ 1:30–35

Chapter 13
Polyamines in Grapevine Research

Panagiotis N. Moschou and Kalliopi A. Roubelakis-Angelakis

Abstract Polyamines have been correlated with numerous cellular functions, including growth, development and responses to stresses. During the past years polyamine metabolism has gained new interest following the isolation and characterization of the biosynthetic and catabolic genes/enzymes from model plants. In grapevine the corresponding polyamine biosynthetic/catabolic enzymes/genes have not yet been completely characterized. The present Chapter presents the recent advances in polyamine research focusing on relevant experimental techniques in grapevine.

Contents

K.A. Roubelakis-Angelakis (✉)
Department of Biology, University of Crete, Crete, Greece
e-mail: poproube@biology.uoc.gr

S. Delrot et al. (eds.), *Methodologies and Results in Grapevine Research*,
DOI 10.1007/978-90-481-9283-0_13,

Abbreviations

ADC	Arginine decarboxylase
Arg	Arginine
BSA	Bovine serum albumine
DAO	Diamine oxidase
EDTA	Ethylene diamine tetraacetic acid
HPLC	High performance liquid chromatography
ODC	Ornithine decarboxylase
Orn	Ornithine
PA	Polyamine
PAO	Polyamine oxidase
Put	Putrescine
PVPP	Polyvinylpyrrolidone
RT	Room temperature
SAMDC	S-Adenosyl-L-methionine decarboxylase
Spd	Spermidine
SPDS	Spermidine synthase
Spm	Spermine
TCA	Trichloracetic acid
TLC	Thin layer chromatography

13.1 Introduction

The polyamines (PAs) spermidine (Spd), spermine (Spm) and their diamine precursor putrescine (Put) are small aliphatic polycationic amines, ubiquitous in all plant cells. Their initial discovery was made as early as in 1678, when van Leeuwenhoek described them as crystals in human semen. Because the described polyamine occurs at high concentration in sperm, the name Spm [*N*,*N*′-bis (3-aminopropyl)butane-1,4-diamine] was designated by Ladenburg and Abel in 1888. Ever since, it took long time for its structure and chemical synthesis to be documented (Wrede 1925, Dudley et al. 1926, Dudley et al. 1927). Two other naturally occurring PAs, Put (butane-1,4-diamine) and cadaverine (pentane-1,5-diamine, Cad), which are fragrant volatile compounds (Brieger 1885) result from bacterial decomposition. Both contribute to the foul odour of the putrefying flesh of cadaver, which gives them their names. Their structures were established by comparison with the already synthesized molecules (Ladenburg 1886, von Udransky and Baumann 1888). Cad is less abundant than Put, Spd and Spm.

PAs have been proposed to participate in various molecular and physiological functions. Put is known to activate the transcription of the genes involved in PA uptake and utilization. Also, PAs can bind to various cellular macromolecules, including DNA, RNA, chromatin and proteins by electrostatic bonds. Furthermore, covalent linkages can lead to cross-link formation of proteins forming cytotoxic derivatives, or to their conjugation with micro- or macromolecular structures, such as hydroxycinnamic derivatives or pectins, respectively.

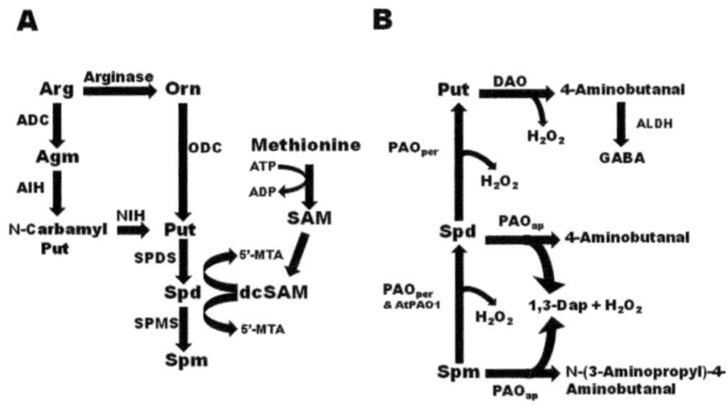

Fig. 13.1 The PA anabolic and catabolic pathways in plants. **a**, The PA biosynthetic pathway. **b**, The PA catabolic pathway

Thus, PAs have been implicated in a myriad of fundamental cellular processes, including regulation of gene expression, translation, cell proliferation, modulation of cell signalling, and membrane stabilization (Tabor and Tabor 1984, Cohen 1998, Igarashi and Kashiwagi 2000). Interestingly, thermophilic bacteria contain two additional categories of unique PAs, proposed to be involved in high temperature tolerance (Hamana et al. 1998, Oshima 2007). These include PAs with longer chains such as caldopentamine and caldohexamine, and branched PAs, such as tris(3-aminopropyl-amine) (so-called mitsubisine) and tetrakis (3-aminopropyl)ammonium (Fig. 13.1, Igarashi and Kashiwagi 2000). PAs can also regulate cell death, particularly apoptosis (Thomas and Thomas 2001, Seiler and Raul 2005, Moschou et al. 2008a, b, 2009).

Intensive research on PA function(s) in mammals, fungi and bacteria has revealed that they are essential regulators of gene transcription, and translation as well as growth (Thomas and Thomas 2003, Childs et al. 2003, Umekage and Ueda 2006, Wortham et al. 2007, Chattopadhyay et al. 2008). PAs and their metabolism are of medical and pharmacological importance. They are present at relatively high concentrations in the mammalian brain and are believed to be involved in the pathophysiological processes underlying brain ischemia (Li et al. 2007, Kim et al. 2009). Owing to the high turnover of the intestinal mucosal cells, they have a high requirement for PAs. They contribute to the maintenance of normal gut function, the maturation of the intestinal mucosa and its repair after injury (Seiler and Raul 2007). In addition, the PAs metabolic pathway is a recognized drug target for cancer prevention, as there is a strong positive correlation between PAs content and cancer cell growth (Casero and Marton 2007, Saunder and Wallace 2007). In plants, PA homeostasis within meristems, leaves, petioles, internodes and roots was documented as a part of the control of cell cycle progression, cell division/expansion and differentiation in the developing tobacco and grapevine plant (Paschalidis and Roubelakis-Angelakis 2005a, b, Paschalidis et al. 2009a, b).

As mentioned previously, PAs exist as soluble (S-), as conjugated to micro- (SH-) or to macro- (PH-) molecules. The exact role of each of these fractions remains unknown. PA contents, primarily the S-form, have been analyzed in various plant organs and tissues (Kakkar and Sawhney 2002). Recently, PA titers and conjugation were analyzed in a tissue-specific manner along the tobacco and grapevine plant axis and were correlated with detailed anabolic and catabolic gene expression patterns (Paschalidis and Roubelakis-Angelakis 2005a, b, Paschalidis et al. 2009a). Plant PAs have been suggested to play important roles in growth and development, such as organogenesis, morphogenesis, embryogenesis, leaf senescence, programmed cell death, as well as abiotic and biotic stress responses (Kumar et al. 1997, Bouchereau et al. 1999, Alcázar et al. 2006, Yoda et al. 2006, Groppa and Benavides 2007, Kusano et al. 2007, Moschou et al. 2008a, b, 2009, Paschalidis et al. 2009b). Plant PAs are also responsible for characteristics of agro-economical importance, including phytonutrient content, fruit quality and vine life (Mehta et al. 2002, Mattoo et al. 2006).

The pathway of PA biosynthesis is ubiquitous in living organisms and rather short (Fig. 13.1). Most of the participating enzymes have been characterized and the corresponding genes/cDNAs cloned from different plant sources (Kaur-Sawhney et al. 2003, Kusano et al. 2007). Orn decarboxylase (ODC, EC 4.1.1.19) and Arg decarboxylase (ADC, EC 4.1.1.19) catalyze the removal of the carboxyl group from Orn and Arg, respectively, to yield the precursor of higher PAs Put. Primikirios and Roubelakis-Angelakis (1999, 2001) characterized and studied the expression pattern of *ADC* gene in grapevine. S-adenosyl-L-methionine decarboxylase (SAMDC, EC 4.1.1.50) decarboxylates S-adenosyl-L-methionine (SAM) producing the corresponding decarboxylated SAM molecule, which acts as an aminopropyl donor in the conversion of Put to Spd and Spm. The transfer of the aminopropyl moiety is catalyzed by the enzymes Spd synthase (SPDS, EC 2.5.1.16) and Spm synthase (EC 2.5.1.22) (Hashimoto et al. 1998, Panicot et al. 2002).

The catabolic pathway in plants involves the final deamination of Put, Spd and Spm by the action of amine oxidases or the backconversion of Spm to Put (Moschou et al. 2008c). More specifically, both pathways depend on the family consisting of diamine oxidases (DAO, EC 1.4.3.6) and the flavoproteins polyamines oxidases (PAO, EC 1.5.3.3), which oxidize Spd and Spm at their secondary amino groups. DAOs produce Δ^1-Pyrroline, 1,3-diamino-propane (Dap), ammonia and H_2O_2. The apoplastic PAOs yield Δ^1-pyrroline and 1,5-diazabicyclononane from Spd and Spm, respectively, along with Dap and H_2O_2. Recently, Tavladoraki et al. (2006) and Moschou et al. (2008c) showed that in *Arabidopsis* PAOs are able to convert Spm to Spd and Spd to Put, with the concomitant production of H_2O_2. Interestingly, Moschou et al. (2008c) showed that these PAOs are localized to plant peroxisomes, thus implicating peroxisomes with amine oxidation.

In the tobacco plant, PA titers and anabolism decrease with age (Paschalidis and Roubelakis-Angelakis 2005a), but generally, concentration gradients of PAs decrease towards the base or the apex of the stem or the root depending on the tissue, the organ or the plant species (Dumortier et al. 1983, Federico and Angelini 1988, Altamura et al. 1993, Torrigiani and Scoccianti 1995, Paschalidis et al. 2001, Della Mea et al. 2007, Paschalidis et al. 2009a). More details on PAs homeostasis in grapevine are presented in Paschalidis et al. (2009b).

Greater insight into PA homeostasis would increase our understanding of the role of PAs. For this reason, in this Chapter we give a detailed analysis of the most relevant techniques for studying PA metabolism.

13.2 Determination of Key Enzymatic Activities Associated with Polyamine Metabolism in Grapevine

13.2.1 Protein Extraction

Although protein extraction from plant tissues for enzymatic assays is based on the efficient disruption of cells walls, the homogenization conditions should not be too harsh to avoid any concomitant denaturation of the enzymes. Among the variables that affect the solubility of proteins are the ionic strength and pH of the extraction buffer, the concentration and type of the detergent used, and the presence of cofactors. Extraction of membrane-bound and hydrophobic proteins is less affected by the ionic strength of the lysis buffer but often requires a mixture of ionic and non-ionic detergents. In addition, inhibitors of proteases are commonly included in extraction buffers (Sambrook et al. 1989).

For the extraction of proteins from grapevine tissues, total proteins are extracted according to Papadakis et al. (2001). More specifically, the tissue is pulverized using a prechilled mortar and pestle. Homogenization is performed in a buffer consisting of 100 mM Tris–HCl, pH 8.0, 1 mM EDTA, 50 μM pyridoxal phosphate, 5 mM dithiothreitol, 0.5 mM phenylmethylsulfonyl fluoride, 10 μM leupeptin, 10% (v/v) glycerol, 0.2% Triton X-100, and also 20% PVPP (polyvinylpolypyrrolidone). Addition of PVPP is necessary in order to decrease the content of soluble phenolics, that are abundant in grape, which could inhibit enzymatic assays. Moreover, polyphenols are responsible for the interaction with proteins that causes the haze and precipitates. For each g of fresh weight, 3–5 mL of buffer are usually added. The homogenates are centrifuged for at least 20 min at 16,000*g* at 4°C and the supernatants (soluble fractions) are desalted in Bio-gel P-6 (Bio-Rad, Hercules, CA), after being filtered through miracloth. The previous step can be avoided if tissue purity is high, but is highly recommended.

The pellets (cell walls, nucleus, plastids, and mitochondria; de Marco and Roubelakis-Angelakis 1996) represent the particulate fraction and can be further dissolved in half-strength buffer, by increasing the ionic strength with 1 M NaCl in the extraction buffer. After a centrifugation step at 16,000*g* at 4°C, they give the particulate fractions.

13.2.2 Protein Content Determination

Enzymatic activities are more accurately expressed as units per mg protein rather than units per mg fresh weight. Thus, a method for total protein content determination should be employed. Protein content is determined according to Lowry et al.

(1951) and by gel visualization. The latter is the optimum choice since it gives more accurate results.

The principle of the Lowry method in determining protein content lies in the reactivity of the peptide nitrogen with the Cu^{2+} ions under alkaline conditions and the subsequent reduction of the Folin-Ciocalteau phosphomolybdic-phosphotungstic acid to heteropolymolybdenum blue by the Cu-catalyzed oxidation of aromatic acids (Folin and Ciocalteau 1927, Dunn 1992). The Lowry method is sensitive to pH changes and therefore the pH should be maintained at 10–10.5.

The Lowry method is sensitive to low content of protein (0.10–2 mg of protein/mL; Dunn 1992), while Price (1996) suggests content of 0.005–0.10 mg/mL. The solutions used in this method are the A (2% Na_2CO_3 in 0.1 N NaOH, 1% NaK tartrate in H_2O), and B (0.5% $CuSO_4$.5 H_2O in H_2O) and the phenol reagent [1 part Folin-Phenol (2x) dissolved in 1 part water]. Generally, 20 μL of the protein extract include sufficient protein amount for precise determination, regardless of the tissue used. Thus, in 20 μL of protein extract 20 μL of 20% TCA are added and the proteins are precipitated, following 30 min incubation at 4°C, and centrifugation at 10,000*g* for 20 min at RT. The supernatant is aspirated off and the protein pellet is re-dissolved in 100 μL solution A. A 30 min incubation with periodic vortexing will increase the solubility of the pellet, thus the accuracy of the protein content determination. Solutions A and B are mixed [10:0.2 (v/v), respectively], and from the mixture 1 mL is added to the samples. Samples are incubated for 15 min and 100 μL of Folin-Phenol (diluted with 1 part water) are added, and samples are incubated for 30 min at RT. Absorbance is red at 625 nm, after preparation of a standard curve using BSA.

This method is 20 times more sensitive than the measurement of the UV absorption at 280 nm (which cannot be used in plant material due to the high amount of phenolics within tissues) and is several fold more sensitive than other methods, like ninhydrin reaction. Moreover, it is simple and easy to adapt for small scale analyses. Free amino acids give less colour than proteins. There are three major disadvantages: (1) The amount of colour varies with different proteins and is not strictly proportional to concentration. (2) The narrow pH range within which it is accurate. However, when using very small volumes of sample, changes in the pH are negligible. (3) Several compounds interfere with the Lowry procedure. These include some amino acid derivatives, certain buffers, drugs, lipids, sugars, salts, nucleic acids and sulfudryl reagents (Dunn 1992). Price (1996) notes that ammonium ions, zwitterionic buffers, nonionic buffers and thiol compounds may also interfere with the Lowry reaction. These substances should be removed or diluted before running Lowry assays.

13.2.3 Determination of Arginase

Arginase is responsible for Orn formation, the substrate for ODC, in the urea cycle. Arginase activity can be colorimetrically measured as the rate of Orn formation, as described by Roubelakis and Kliewer (1978).

Total proteins for this assay are extracted as described above and the reaction mixture contains 100 mM Tris–HCl, pH 9.7, 25 mM L-Arg (adjusted to pH 9.7). Total protein extract of 0.1 mL, and 1.5 mM $MnCl_2$ (10 min of preincubation at RT) are additionally added, in a total volume of 1 mL. The reaction is terminated after 1 h by addition of 0.5 mL of 15% perchloric acid. The arginase activity is a linear function of incubation time, under these conditions, at least for 1 h, using substrate concentrations up to 20 mM and of the amount of protein extract under these conditions. Boiled enzyme preparations and reactions terminated with perchloric acid before incubation should be used as negative controls, whereas 200 μM N^{ω}-hydroxy-L-Arg inhibits arginase activity in grapevine tissues by 96%.

Enzyme activity is determined as the Orn present in the reaction mixture. Orn is determined spectrophotometrically by the method of Chinard modified as follows (Roubelakis and Kliewer 1978). In screw cap tubes, 500 μL of the reaction mixture are mixed with 1.5 mL of glacial acetic acid and 500 μL of ninhydrin [25 mg/ml ninhydrin in glacial acetic acid/phosphoric acid/dH_2O, 3:1:1 (v/v/v)] solution is added, mixed and boiled in a water bath for 1 h. The ninhydrin is dissolved in acetic acid by continuous agitation in a water bath at 60°C. Then, the two other components are added and the solution is chilled before use. After the samples are let in a water bath to cool down, absorbance is red at 515 nm. At the same time, Orn standards are prepared and the absorbance is linear between 0 and 60 μM of Orn.

13.2.4 Determination of PA Metabolism-Involved Decarboxylases

The determination of ADC, ODC and SAMDC in grapevine extracts is achieved by the isotopic method of $^{14}CO_2$ determination, previously described by Kaur-Sawhney et al. (1982). Special precautions should be taken, given that the method is based on radioactive volatile material. A detailed analysis of the spatial and temporal distribution of these enzymatic activities in grapevine is given in Paschalidis et al. (2009a, b, see also Fig. 13.2). Total protein extract is used and extraction is performed as described above. Please note that speeding up the procedure is an important factor for the precision of the analysis.

More specifically, the substrates L-[1-^{14}C]Arg, L-[1-^{14}C]Orn, and adenosyl-L-Met S-[carboxyl-^{14}C] are used as radioactive substrates, respectively (obtained by the ARC, St. Louis). Usually, 50–100 μL of the protein extract are incubated with 25 μL (0.078 μCi) L-[1-^{14}C] Arg (0.626 mCi/ mmol, 3.13 μCi/ mL, 1 mM final concentration of Arg), 25 μL (0.078 μCi) L-[1-^{14}C]Orn (0.626 mCi/ mmol, 3.13 μCi/ mL, 1 mM final Orn concentration), and 25 μL (0.0156 μCi) L-[1-^{14}C]adenosyl-L-Met (0.1252 mCi/ mmol, 0.626 μCi/ mL, 1 mM final adenosyl-L-Met concentration) in polypropylene tubes (Fig. 13.3). Immediately, the tubes are sealed with a plastic cap penetrated by a 22 gauge needle, with its external end sealed with an air-tight sealing, and inside the tube the needle is holding a paper disc of 8 mm diameter. The discs are impregnated in 40 μL of 2 N KOH. The tubes are

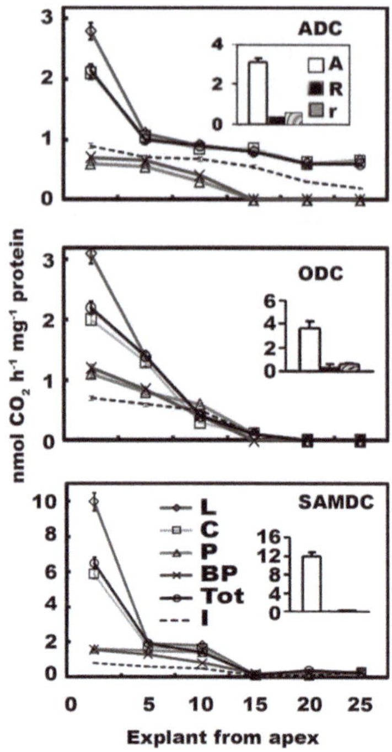

Fig. 13.2 Specific activities of decarboxylases along the developmental axis of grapevine plant. Specific activities of ADC, ODC, SAMDC in the soluble fractions from grapevine explants of different developmental stages. L, apical and marginal leaf lamina; P, acropetal petiole; Tot, total leaf with petiole; I, internode. Numbering started from the apical organ being designated as 1st. Inserted graphs depict specific activities in: A, shoot apex; R, primary root; *r*, secondary root (modified from Paschalidis et al. 2009a)

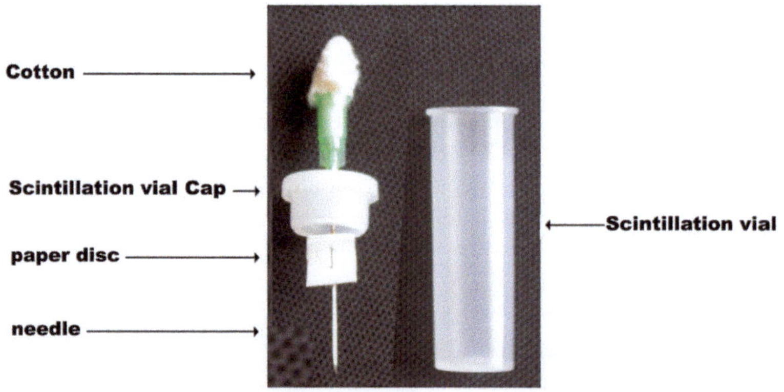

Fig. 13.3 The vials used for the assay of decarboxylases. A hypodermic needle is used air sealed with a wet cotton, penetrating a plastic cap and a paper disc impregnated with KOH which traps CO_2. A scintillation vial is used for the assay and the cap is placed on the top of it immediately after the reaction starts by placing the protein extract and the substrate on the bottom of the vial

subsequently incubated for 1 h under constant agitation in a water bath at 37°C. The reaction is terminated by infiltrating 150 μL 10% (w/v) TCA through the needle without removing the cap.

The mixture is further incubated at 49°C for at least 45 min to trap on the disc any remaining $^{14}CO_2$. The discs are removed and left to dry and then they are placed within 5 mL scintillation vials, containing 4 mL of scintillation liquid 5% [w/v] 2,5-diphenyloxazole and 0.05% [w/v] 1,4-bis[5-phenyloxazoyl] benzene in toluene. The radiolabelled CO_2 that was trapped on the discs, is estimated in a scintillation counter LS6000 SE (Beckman Instruments Inc., Palo Alto, CA).

For more precise determination of ADC, 200 μM N^w-hydroxy-L-Arg (a competitive inhibitor of arginase; Iniesta et al. 2001) is also added in the reaction mixture. Labelled CO_2 is counted in an LS 6000SE (Beckman, Fullerton, CA) scintillation counter or another counter.

Boiled enzymes are also used as negative controls. Note that liberation of CO_2 is not achieved when boiling the enzymes. Enzymatic activity can be expressed as nM of $^{14}CO_2$ liberated/mg protein or /gr FW after calibrating the counter with the corresponding standards.

13.2.5 Determination of SPDS and SPMS

The determination of SPDS and SPMS in grapevine is achieved by measuring the formation of Spd and Spm, respectively. The method is a modification of the method described by Tabor (1962a), and is based on the principle that a propylamine group (aminopropyl moiety) of 5'-deoxyadenosyl-(5'),3-aminopropyl-(1), methylsulfonium salt, prepared as described in Tabor (1962b) by using recombinant SAMDC, is added to Put, or Spd, resulting in the formation of Spd and Spm, respectively (Fig. 13.4).

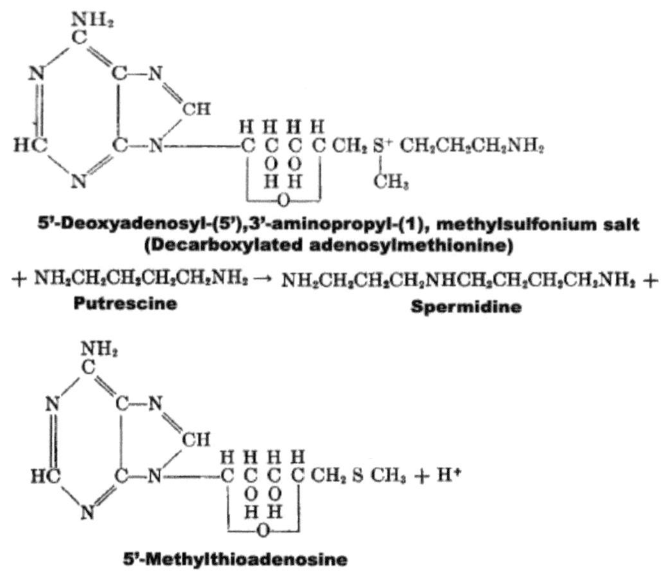

Fig. 13.4 Reaction catalyzed by SPDS. Put is converted to Spd, by the transfer of an aminopropyl moiety from dcSAM to Put, converting dcSAM to MTA (methylthioadenosine)

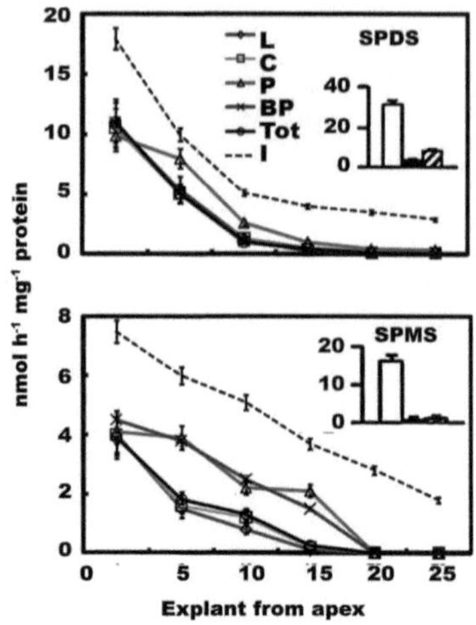

Fig. 13.5 Specific activities of higher PA synthases along the developmental axis of grapevine plant. Specific activities of SPDS and SPMS, in the soluble fractions from grapevine explants of different developmental stages. Numbering started from the apical organ being designated as 1st. Inserted graphs depict specific activities. For more information, see legend to Fig. 13.2 (modified from Paschalidis et al. 2009a)

The products of this reaction can be separated and quantified using HPLC as described later. A detailed analysis of the spatial and temporal distribution of these enzymatic activities in grapevine is given in Paschalidis et al. (2009a, b, Fig. 13.5).

SPDS and SPMS are assayed by measuring the formation of Spd and Spm, respectively (Fig. 13.3). The assay mixture for SPDS consists of 100 mM Tris–HCI buffer, pH 9.0, 3 mM Put, 0.2 mM dSAM, and the protein extract in a total volume of 200 μL (the volume can be scaled down). The reaction is performed at 37°C for 1 h and is terminated by the addition of 200 μL of 65 mM borate-KOH buffer, pH 10.5, followed with 1 mL of 2 N NaOH and 10 μL of benzoylchloride. The products are separated with an HP 1100 HPLC system as described later. SPMS is assayed similarly to SPDS by replacing Put with Spd in the assay mixture.

13.2.6 Determination of DAO and PAO

Determination of DAO and PAO in grapevine tissues is achieved either by a radiometric or a colorimetric assay. A detailed analysis of the spatial and temporal distribution of these enzymatic activities in grapevine is given in Paschalidis et al. (2009a, Fig. 13.6).

DAO and PAO enzyme activities are estimated by a modification of the radiometric method of Biondi et al. (2001), using [1,4-^{14}C]Put and [1,4-^{14}C]Spd (Amersham, Buckinghamshire, UK; specific activities 4.37 GBq $mmol^{-1}$; Santanen and Simola 1994) as labelled substrates, using total protein extract. The assay mixture contains 0.5 mL of the extract, 1 mM unlabelled Put or Spd, and 3.7 KBq of [1,4-^{14}C]Put or [1,4-^{14}C]Spd. After incubation on a shaker at 37°C for 60 min, the reaction is terminated by adding 150 μL of saturated sodium carbonate (Bhatnagar et al. 2002). Δ^1-[^{14}C]pyrroline from [1,4–^{14}C]putrescine (4.03 TBq mol^{-1}, Amersham Phamacia Biotech Italia) during a 30 min incubation at 37 °C (Scaramagli et al.

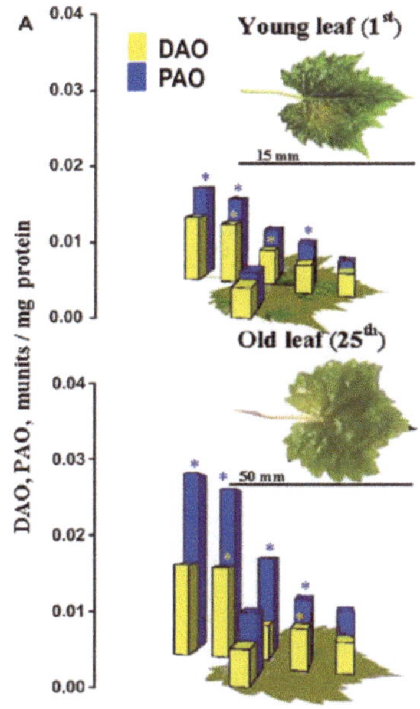

Fig. 13.6 Specific activities of PA catabolic enzymes along the developmental axis of grapevine plant DAO and PAO activities in the soluble fractions of the youngest (1st) and the oldest leaf (25th) (modified from Paschalidis et al. 2009a)

1999). Labelled Δ^1-pyrroline is extracted immediately in 1 mL toluene. The 0.5-mL aliquots are placed in scintillation liquid [0.5% (w/v) 2,5-diphenyloxazole and 0.05% (w/v) 1,4-bis(5-phenyloxa-zoyl)benzene in toluene] and counted in an LS 6000SE (Beckman, Fullerton, CA) scintillation counter. One unit of enzyme (U) represents the amount of enzyme catalyzing the formation of 1 μmol of Δ^1-[^{14}C]pyrroline/min.

A spectrophotometric assay can be also used for DAO and PAO enzyme activities, according to Holmsted et al. (1961) and Federico et al. (1985, 1988). The principle of the method is that Δ^1-pyrroline formed by the enzymatic oxidation of Spd, can react with o-aminobenzaldehyde to produce a yellowish-colored dihydroquinazolinium derivative (Larson and Bronquist 1962, Fig. 13.7).

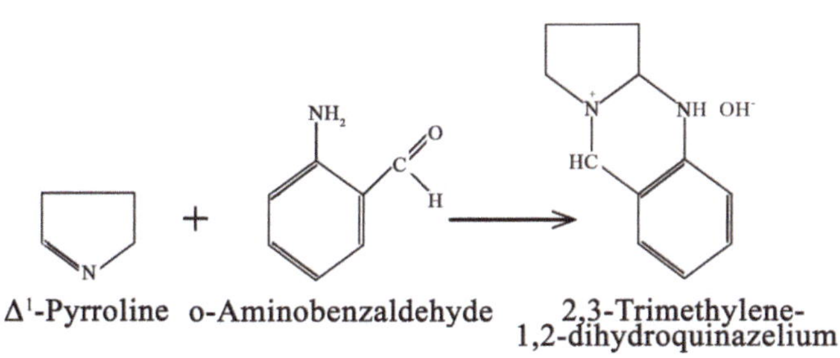

Fig. 13.7 Formation of the yellowish complex from the reaction of Δ1-pyrroline with o-aminobenzaldehyde

The extraction buffer for DAO and PAO determination is different from the one described above. More specifically, the extraction buffer for DAO and PAO consists of 0.1 M K-phosphate, pH 6.5, 10 μ pyridoxal phosphate and 2 mM dithiothreitol (DTT). The reaction mixture contains the extract, 12 μL Put or Spd 0.1 M, for DAO and PAO respectively, and up to 1 mL 0.1 M K-phosphate, pH 6.5. Reactions are carried out for 1 h at 37°C and the reaction is terminated by the addition of 0.125 μL of 10% (w/v) TCA followed by 12.5 μL o-aminobenzaldehyde (10 mg mL^{-1}) dissolved in ethanol. Under these conditions, Δ^1-pyrroline forms the yellowish complex (Fig. 13.7). After a centrifugation step at 12,000*g* at 4°C to remove proteins, the complex absorbance is red at 430 nm ($\varepsilon = 1.86\,103\,\text{mol}^{-1}\text{cm}^{-1}$). One Unit represents the amount of enzyme catalyzing the formation of 1 μmol of Δ^1-pyrroline/min.

The colorimetric assay described by Tavladoraki et al. (2006) and Moschou et al. (2008a), did not work efficiently as the o-aminobenzaldehyde assay, under our experimental conditions when using grapevine tissues.

13.3 Determination of Polyamine Titers

13.3.1 Polyamine Extraction

In order to extract PAs, plant tissue is pulverized using a mortar and pestle chilled with liquid N_2 and homogenized in 5% (v/v) perchloric acid. For every gram of fresh weight, 5 ml of 5% (v/v) perchloric acid are added. Homogenates are incubated on ice for 1 h and centrifuged subsequently at 27,000*g* for 20 min at 4°C. Aliquots of the supernatant and pellet (the latter is re-dissolved in 1 N NaOH) are hydrolyzed by mixing with equal volume of 12 N HCl (200 μL for supernatant and 400 μL for the pellet), in screw cap tubes and placed at 110°C for 18 h. The hydrolyzed product is centrifuged at 3,000*g* to remove the carbonized material and evaporated at 70°C in a thermoblock. The dried pellet is re-dissolved in 200 μL PCA. PA standards are treated the same way.

13.3.2 Polyamine Determination

Long ago, PAs were separated and quantified by using thin-layer chromatographic separation (TLC), following pre-chromatic derivatization with dansyl chloride (1-dimethylaminonapthalene-5-sulfonyl chloride). Although dansylated PAs are highly fluorescent and detectable in small quantities, dansyl chloride is non-specific since it exhibits high cross-reactivity with phenols, alcohols and carbohydrates. Moreover, it tends to give side products by self- or cross-reaction. On the other hand, using benzoyl chloride instead is advantageous since this agent produces a small number of by-products thus shortening the elution programmes.

The HPLC separation is much more efficient than the TLC one, allowing better separation and determination of PAs. Flores and Galston (1982) initially developed an analytical HPLC method for the separation and determination of PAs in plant extracts. Thus, benzylpolyamines are separated at room temperature through a C-18 column. A modified HPLC method was described by Kotzabasis et al. (1993). In this method, the column used by Flores and Galston (1982) was replaced by a C-18 narrow-bore 2.1×200 mm column with 5 μM particle size. This method is about 5 times more sensitive than the initial method and saves about 80% solvent. Moreover, the method has been widely used by our laboratory producing satisfactory results for both tobacco and grapevine (see our most recent publications Papadakis et al. 2005, Paschalidis and Roubelakis-Angelakis 2005a, b, Moschou et al. 2008a, b, c, Paschalidis et al. 2009a, b).

In order to benzoylate PAs, 1 ml 2 N NaOH and 10 μL of benzoylchloride (Sigma) are added to 200 μL of each sample and vortexed for 30 sec. Samples are incubated for 20 min at 25°C, and subsequently 2 mL of saturated NaCl are added to stop the reaction. Benzoyl-PAs are extracted by mixing with 2 mL of diethyl-ether (stabilized with about 7 ppm 2,6-di-tert-butyl-4-methylphenol). Samples are centrifuged at 3,000*g* for 5 min and the ether phase is evaporated in a water bath at 60°C, after being transferred to a new tube. The benzoyl-PAs are re-dissolved in 200 μL of 63% methanol and 20–50 μl of the extract are injected into the HPLC preferably using an autosampler. Analysis can be performed using a Hewlett-Packard 1100 HPLC, equipped with a C-18 narrow bore column (2.1×200 mm, 5 μm particle size, Hypersyl; Hewlett-Packard). Separation temperature is kept at 25°C. Flow rate is 200 μL min^{-1}. Regression curves are used for quantification of PAs. The lower detection limit is 5 pmol, equal to that of the dansylated derivatization.

To ensure appropriate separation of Put from Cad, a gradient elution should be used. The most appropriate separation is achieved when using 55% methanol at the onset to 84% methanol at 23 min and 84% from 23 to 26 min. Residual benzoyl-chloride appears at a retention time of 8.3 min, between Cad and Spd. The HPLC standards are Put, 6.6 min, Cad, 7.5 min; Spd 10.5 min; Spm 13.9 min; Agm 23.9 min. In addition, in each sample a known amount of internal standard should be added, such as Cad, after the extraction of the PAs. A typical HPL chromatogram of PAs is shown in Fig. 13.8.

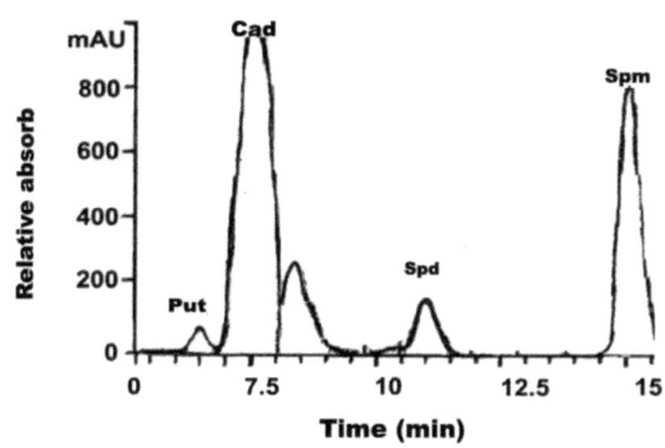

Fig. 13.8 A typical HPLC chromatograph of polyamines (modified from Moschou et al. 2008c)

References

Alcázar R, Marco F, Cuevas JC, Patron M, Ferrando A, Carrasco P, Tiburcio AF, Altabella T (2006) Involvement of polyamines in plant response to abiotic stress. Biotechnol Lett 28: 1867–1876

Altamura MM, Torrigiani P, Falasca G, Rossini P, Bagni N (1993) Morpho-functional gradients in superficial and deep tissues along tobacco stem: polyamine levels, biosynthesis and oxidation, and organogenesis *in vitro*. J Plant Physiol 142:543–551

Bachrach U, Oser IS (1962) Enzymatic assay for spermidine. J Biol Chem 258

Bhatnagar P, Minocha R, Minocha S (2002) Genetic manipulation of the metabolism of polyamines in poplar cells: the regulation of putrescine catabolism. Plant Physiol 128:1455–1469

Biondi S, Scaramagli S, Capitani F, Maddalena M, Patrizia A, Torrigiani P (2001) Methyl jasmonate upregulates biosynthetic gene expression, oxidation and conjugation of polyamines, and inhibits shoot formation in tobacco thin layers. J Exp Bot 52:231–242

Bouchereau A, Aziz A, Larher F, Martin-Tanguy J (1999) Polyamines and environmental challenges: recent development. Plant Sci 140:103–125

Brieger L (1885) Ueber Spaltungsprodukte der Bacterien, Zweite Mittheilung. Zeitschr Physiol Chem 9:1–7

Casero RA Jr, Marton LJ (2007) Targeting polyamine metabolism and function in cancer and other hyperproliferative diseases. Nat Rev Drug Discov 6:373–390

Chattopadhyay MK, Myung HP, Tabor H (2008) Hypusine modification for growth is the major function of spermidine in *Saccharomyces cerevisiae* polyamine auxotrophs grown in limiting spermidine. Proc Natl Acad Sci USA 105:6554–6559

Childs AC, Metha DJ, Gerner EW (2003) Polyamine-dependent gene expression. Cell Mol Life Sci 60:1394–1406

Cohen SS (1998) A guide to the polyamines. Oxford University Press, New York, pp 1–595

Della Mea M, De Filippis F, Genovesi V, Serafini Fracassini D, Del Duca S (2007) The acropetal wave of developmental cell death of tobacco corolla is preceded by activation of transglutaminase in different cell compartments. Plant Physiol 144:1211–1222

de Marco A, Roubelakis-Angelakis KA (1996) The complexity of enzymic control of hydrogen peroxide concentration may affect the regeneration potential in plant protoplasts. Plant Physiol 110:137–145

Dudley HW, Rosenheim O, Starling WW (1926) The chemical constitution of spermine. III. Structure and synthesis. Biochem J 20:1082–1094

Dudley HW, Rosenheim O, Starling WW (1927) The constitution and synthesis of spermidine, a newly discovered base isolated from animal tissues. Biochem J 21:97–103

Dumortier, F.M., Flores, H.E., Shekhawat, N.S., Galston, A.W. (1983) Gradients of polyamines and their biosynthetic enzymes in coleoptiles and roots of corn. Plant Physiol 72: 915–918

Dunn M J, (1992) Protein determination of total protein concentration. In: Harris ELV, Angal S (eds) Protein purification methods. IRL Press, Oxford

Federico R, Angelini R, Cesta A, Pini C (1985) Determination of diamine oxidase in lentil seedlings by enzymic activity and immunoactivity. Plant Physiol 79:62–64

Federico R, Angelini R (1988) Distribution of polyamines and their related catabolic enzymes in etiolated and light-grown leguminosae seedlings. Planta 173:317–321

Flores HE, Galston AW (1982) Analysis of polyamines in higher plants by high performance liquid chromatography. Plant Physiol 69:701–706

Folin O, Ciocalteau V (1927) On tyrosine and tryptophane determinations in proteins. J Biol Chem 73:627–650

Groppa MD, Benavides MP (2007) Polyamines and abiotic stress: recent advances. Amino Acids 34:35–45

Hamana K, Niitsu M, Samejima K, Itoh T, Hamana H, Shinozawa T (1998) Polyamines of the thermophilic eubacteria belonging to the genera *Thermotoga, Thermodesulfovibrio,*

Thermoleophilum, Thermus, Rhodothermus and *Meiothermus*, and the thermophilc archaebacteria belonging to the genera *Aeropyrum, Picrophilus, Metanobacterium* and *Methanococcus*. Microbios 94:7–21

Hashimoto T, Tamaki K, Suzuki K, Yamada Y (1998) Molecular cloning of plant spermidine synthases. Plant Cell Physiol 39:73–79

Holmsted B, Larsson L, Tham R (1961) Further studies on spectophotometric method for the determination of amine oxidase activity. Biochim Bioph Acta 48:182–186

Igarashi K, Kashiwagi K (2000) Polyamines: mysterious modulators of cellular functions. Biochem Biophys Res Commun 271:559–564

Kakkar RK, Sawhney VK (2002) Polyamine research in plants-a changing perspective. Physiol Plant 116:281–292

Kaur-Sawhney R, Shih LM, Flores HE, Galston AW (1982) Relation of polyamine synthesis and titer to ageing and senescence in oat leaves. Plant Physiol 69:405–410

Kaur-Sawhney R, Tiburcio AF, Atabella T, Galston AW (2003) Polyamines in plants: an overview. J Cell Mol Biol 2:1–12

Kim GH, Komotar RJ, McCullough-Hicks ME, Otten ML, Starke RM, Kellner CP, Garrett MC, Merkow MB, Rynkowski M, Dash KA, Connolly ES (2009) The role of polyamine metabolism in neuronal injury following cerebral ischemia. Can J Neurol Sci 36:14–19

Kotzabasis K, Christakis-Hampsas MD, Roubelakis-Angelakis KA (1993) A Narrow-Bore HPLC method for the identification and quantification of free, conjugated, and bound polyamines. Anal Biochem 214:484–489

Kumar A, Altabella T, Taylor M, Tiburcio AF (1997) Recent advances in polyamine research. Trends Plant Sci 2:124–130

Kusano T, Yamaguchi K, Berberich T, Takahashi Y (2007) Advances in polyamine research in 2007. J Plant Res 120:345–350

Ladenburg A (1886) Über die Identität des Cadaverin mit dem Pentamethyldiamin. Ber Dtsch Chem Ges 19:2585–2586

Ladenburg A, Abel J (1888) Über das Aethylenimin (Spermin?). Ber Dtsch Chem Ges 21: 758–766

Larson RL, Broquist HP (1963) A color reaction between reduced pyridine nucleotides and o-aminobenzaldehyde of potential analytical use. Anal Biochem 5:255–269

Li J, Doyle KM, Tatlisumak T (2007) Polyamines in the brain: distribution, biological interactions, and their potential therapeutic role in brain ischaemia. Curr Med Chem 14: 1804–1813

Lowry O H, Rosebrough N J, Farr AL, Randall RJ (1951) Protein measurement with the Folin-Phenol reagents. J Biol Chem 193:265–275

Mattoo AK, Sobolev AP, Neelam A, Goyal RK, Handa AK, Segre AL (2006) Nuclear magnetic resonance spectroscopy-based metabolite profiling of transgenic tomato fruit engineered to accumulate spermidine and spermine reveals enhanced anabolic and nitrogen–carbon interactions. Plant Physiol 142:1759–1770

Mehta RA, Cassol T, Li N, Ali N Handa AK, Matto AK (2002) Engineered polyamine accumulation in tomato enhances phytonutrient content, juice quality, and vine life. Nat Biotechnol 20:613–618

Moschou PN, Delis ID, Paschalidis KA, Roubelakis-Angelakis KA (2008a) Transgenic tobacco plants over-expressing polyamine oxidase are not able to cope with oxidative burst generated by abiotic factors. Physiol Plant 133:140–156

Moschou PN, Paschalidis KA, Delis ID, Andriopoulou AH, Lagiotis GD, Yakoumakis DI, Roubelakis-Angelakis KA (2008b) Spermidine exodus and oxidation in the apoplast induced by abiotic stress is responsible for H_2O_2 signatures that depict tolerance responses in tobacco. Plant Cell 20:1708–1724

Moschou PN, Sanmartin M, Andriopoulou AH, Rojo E, Sanchez-Serrano JJ, Roubelakis-Angelakis KA (2008c) Bridging the gap between plant and mammalian polyamine catabolism: A novel peroxisomal polyamine oxidase responsible for a full back-conversion pathway in *Arabidopsis*. Plant Physiol 147:1845–1857

Moschou PN, Sarris PN, Scandalis N, Andriopoulou AH, Paschalidis KA, Panopoulos NJ, Roubelakis-Angelakis KA (2009) Engineered polyamine catabolism pre-induces SA-independent immunity and enhances tolerance to bacteria and fungi in tobacco. Plant Physiol 149:1970–1981

Oshima T (2007) Unique polyamines produced by an extreme thermophile, *Thermus thermophilus*. Amino Acids 33:367–372

Panicot M, Minguet EG, Ferrando A, Alcazar R, Blazquez MA, Carbonell J, Altabella T, Koncz C, Tiburcio AF (2002) A polyamine metabolon involving aminopropyl transferase complexes in *Arabidopsis*. Plant Cell 14:2539–2551

Papadakis AI, Siminis CI, Roubelakis-Angelakis KA (2001) Reduced activity of antioxidant machinery is correlated with supression of totipotency in plant protoplasts. Plant Physiol 126:434–444

Papadakis AK, Roubelakis-Angelakis KA (2005) Polyamines inhibit NADPH oxidase-mediated superoxides generation and putrescine prevents programmed cell death induced by polyamine oxidase-generated hydrogen peroxide. Planta 220:826–837

Paschalidis KG, Aziz A, Geny L, Primikirios NI, Roubelakis-Angelakis KA (2001) Polyamines in *Vitis vinifera L.* In: Roubelakis-Angelakis KA (ed) Molecular Biology and Biotechnology of the Grapevine. Kluwer Academic, The Netherlands, pp 109–152

Paschalidis KA, Roubelakis-Angelakis KA (2005a) Spatial and temporal distribution of polyamine levels and polyamine anabolism in different organs/tissues of the tobacco plant. Correlations with age, cell division/expansion, and differentiation. Plant Physiol 138:142–152

Paschalidis KA, Roubelakis-Angelakis KA (2005b) Sites and regulation of polyamine catabolism in the tobacco plant. Correlations with cell division/expansion, cell cycle progression, and vascular development. Plant Physiol 138:2174–2184

Paschalidis KA., Moschou PN, Toumi I, Roubelakis-Angelakis KA (2009a) Polyamine anabolic/catabolic regulation along the woody grapevine plant axis. J Plant Physiol 166: 1508–1519

Paschalidis KA, Moschou PN, Aziz A, Toumi I, Roubelakis-Angelakis KA (2009b) New insights on polyamines in the grapevine. In: Roubelakis-Angelakis KA (ed) Grapevine Molecular Physiology & Biotechnology. Springer, ISBN 978-90-481-2304-9 e-ISBN 978-90-481-2305-6, DOI 10.1007/978-90-481-2305-6, pp 207–228

Price NC (1996) Proteins. Academic Press, Labfax, Oxford

Primikirios NI, Roubelakis-Angelakis KA (1999) Cloning and expression of an arginine decarboxylase cDNA from *Vitis vinifera* L. cell-suspension cultures. Planta 208:574–582

Primikirios NI Roubelakis-Angelakis KA (2001) Indications for post-translational regulation of *Vitis vinifera* L. arginine decarboxylase. Plant Mol. Biol 45:669–678

Roubelakis KA, Kliewer WM (1978) Enzymes of Krebs-Henseleit Cycle in *Vitis vinifera* L III *In vivo* and *in vitro* studies of arginase. Plant Physiol 62:344–347

Saunder FR, Wallace HM (2007) Polyamine metabolism and cancer prevention. Biochem Soc Trans 35:364–368

Sambrook J, Fritsch EF, Maniatis T (1989) Molecular Cloning, 2nd edn. Cold Spring Harbor Laboratory Press

Santanen A, Simola LK (1994) Catabolism of putrescine and spermidine in embryogenic and non-embryogenic callus lines of *Picea abies*. Physiol Plant 90:125–129

Scaramagli S, Biondi S, Torrigiani P (1999) Methylglyoxal (bis-guanylhydrazone) inhibition of organogenesis is not due to S-adenosylmethionine decarboxylase inhibition/polyamine depletion in tobacco thin layers. Physiol Plant 107:353–360

Seiler N, Raul F (2005) Polyamines and apoptosis. J Cell Mol Med 9:623–642

Seiler N, Raul F (2007) Polyamines and the intestinal tract. Crit Rev Clin Lab Sci 44: 365–411

Tabor CW (1962a) In: Colowick SP, Kaplan NO (eds) Methods in Enzymology, vol 5. Academic Press, New York, p 761

Tabor CW (1962b) In: Colowick SP, Kaplan NO (eds) Methods in Enzymology, vol 5. Academic Press, New York, p 756

Tabor CW, Tabor H (1984) Polyamines. Annu Rev Biochem 53:749–790

Tavladoraki P, Rossi MN, Saccuti G Perez-Amador MA, Polticelli F, Angelini R, Federico R (2006) Heterologous expression and biochemical characterization of a polyamine oxidase from *Arabidopsis* involved in polyamine back conversion. Plant Physiol 141:1519–1532

Thomas T, Thomas T (2001) Polyamines in cell growth and cell death: molecular mechanisms and therapeutic applications. Cell Mol Life Sci 58:244–258

Thomas T, Thomas T (2003) Polyamine metabolism and cancer. J Cell Mol Med 7: 113–126

Torrigiani P, Scoccianti V (1995) Regulation of cadaverine and putrescine levels in different organs of chickpea seed and seedlings during germination. Physiol Plant 93:512–518

Umekage S, Ueda T (2006) Spermidine inhibits transient and stable ribosome subunit dissociation. FEBS Lett 580:1222–1226

von Udransky L, Baumann E (1888) Über die Identität des Putrescins und des Tetramethylendiamins. Ber Dtsch Chem Ges 21:2938–2941

Wortham BW, Patel CN, Oliveira MA (2007) Polyamines in bacteria: pleiotropic effects yet specific mechanisms. Adv Exp Med Biol 603:106–115

Wrede F (1925) Über die aus menschlichen Sperma isolierte Base Spermin. Disch Med Wochenschr 51:24

Yoda H, Hiroi Y, Sano H (2006) Polyamine oxidase is one of the key elements for oxidative burst to induce programmed cell death in tobacco cultured cells. Plant Physiol 142:193–206

Chapter 14
Field Assessment and Diagnostic Methods for Detection of Grapevine Viruses

Giorgio Gambino, Elisa Angelini, and Ivana Gribaudo

Abstract The diagnosis of grapevine viral diseases and their associated viruses is usually performed by visual observation of symptoms or by using laboratory tests. The observation of symptoms can be carried out directly in the field, in case of manifest pathologies, or by means of biological indexing with susceptible indicators, in case of latent diseases. The laboratory assays include various serological and biomolecular tests, based on the detection of viral capsid and nucleic acid, respectively. Serological assays are largely used, also thanks to the availability of commercial kits. Biomolecular assays underwent a dramatic evolution in the last years, from dot-blot hybridisation to various PCR-based techniques and finally to microarrays, improving their sensitivity and reliability.

Contents

Abbreviations

GFLV Grapevine fanleaf virus
GLRaV Grapevine leafroll-associated viruses

I. Gribaudo (✉)
Istituto Virologia Vegetale CNR, U.O.S. Grugliasco, Via L. da Vinci 44, Grugliasco, TO 10095, Italy
e-mail: i.gribaudo@ivv.cnr.it

S. Delrot et al. (eds.), *Methodologies and Results in Grapevine Research*,
DOI 10.1007/978-90-481-9283-0_14,

GRSPaV	Grapevine rupestris stem pitting-associated virus
GVA	Grapevine virus A
GVB	Grapevine virus B
PCR	Polymerase chain reaction
RT-PCR	reverse transcription-polymerase chain reaction

14.1 Viruses of Grapevine

More than 50 viruses have been described in grapevine, most of them being phloem-limited (Martelli and Boudon-Padieu 2006). The most important viral diseases, in terms of spreading and economical impact, are the following:

- Infectious degeneration, caused by *Grapevine fanleaf virus* (GFLV), which belongs to the *Nepovirus* genus. It is widespread in all Europe.
- Grapevine leafroll, caused by at least 9 serologically distinct viruses, denoted *Grapevine leafroll-associated viruses 1* to *9* (GLRaV-1 to 9); eight of them belong to the *Ampelovirus* genus, GLRaV-2 to the *Closterovirus* genus. The most widespread leafroll viruses are GLRaV-1 and 3, followed by GLRaV-2; the other leafroll-associated viruses are found only sporadically in Europe.
- Rugose wood complex, which consists of four distinct diseases, associated with different etiological agents, as follows: rupestris stem pitting, caused by several strains of *Grapevine rupestris stem pitting-associated virus* (GRSPaV); Kober stem grooving, caused by *Grapevine virus A* (GVA); corky bark, caused by *Grapevine virus B* (GVB); LN33 stem grooving, whose etiological agent is still unknown. *Grapevine virus C* and *Grapevine virus D* have also been associated with the rugose wood complex, but their role is still not clear. The most widespread viral diseases of the rugose wood complex are rupestris stem pitting, which is almost ubiquitous, and Kober stem grooving.
- Fleck complex, which consists of several diseases and viruses, the most widespread being the *Grapevine fleck virus* (GFkV).
- Vein necrosis, which has been recently associated to the presence of a specific strain of GRSPaV. It is very widespread in grapevine.

The diagnosis of the grapevine viral diseases and the associated viruses is usually performed by means of visual observation of symptoms or using laboratory tests. The observation of symptoms can be performed directly in the field, in case of evident pathologies, or by means of biological indexing on susceptible indicators, in case of latent diseases. The laboratory assays include various serological and biomolecular tests, based on the detection of viral capsid and nucleic acid, respectively. The laboratory tests are necessary in order to identify and confirm the etiological agents of the apparent and latent viral diseases.

14.2 Visual Symptoms and Biological Tests (indexing)

Symptoms vary in type and severity according to the type of virus, the susceptibility of varieties and the interactions with the environment, soil and climate. Some grapevine viruses show clear symptoms on *Vitis vinifera* and/or most American *Vitis* species and rootstock hybrids; symptoms can be directly observed in the field on leaves, bunches, canes and trunk. However, some other diseases are evident only on susceptible varieties and latent in most species, especially in *V. vinifera* varieties. In these cases, visual field observation does not help in identifying the disease and the use of biological indexing trials with susceptible indicators is needed for the observation of the specific symptoms. Biological indexing trials consist of grafting between an indicator and the plant to be tested, with subsequent observations on grafted materials. Suitable grapevine indicators of the different grapevine diseases are available for biological indexing and must be grafted as a scion or as a rootstock, according to the viral disease (Table 14.1). Wood and green grafting can be performed; the green grafting technique allows a significant shortening of the period of observation of symptoms, from 2–3 years to 3–6 months (Pathirana and McKenzie 2005).

It is important to be able to recognize the typical manifestation of symptoms of all the viruses, because each viral disease has its own characteristics and a specific period for symptoms breakout. Moreover, the intensity of the symptoms and the virulence also depend on the susceptibility of the cultivars.

Table 14.1 List of the most suitable grapevine indicators for the diagnosis of the most important grapevine viral diseases using biological indexing

Viral disease	Associated virus	Grapevine indicator	Use of indicator
Fanleaf	GFLV	*V. rupestris* du Lot, *V. rupestris* St. George	Scion
Leafroll	GLRaV-1 to 9	Cabernet franc, C. Sauvignon, Carmenere, Merlot, Pinot noir	Scion
Rugose wood: rupestris stem pitting	GRSPaV	*V. rupestris* du Lot	Rootstock
Rugose wood: kober stem grooving	GVA	*V. Berlandieri* x *V. riparia* Kober 5BB	Rootstock
Rugose wood: corky bark	GVB	Hybrid LN33	Scion
Rugose wood: LN33 stem grooving	Unknown	Hybrid LN33	Scion
Fleck	GFkV	*V. rupestris* St. George, *V. rupestris* du Lot	Scion
Vein necrosis	GRSPaV-1	*V. rupestris* x *V. Berlandieri* cv. 110 Richter	Scion

Fig. 14.1 Fanleaf symptoms: (**a**) malformation of leaves; (**b**) malformation of canes; (**c**) irregularly ripening of berries

Grapevines infected with fanleaf virus show typical deformations of leaves and canes. Leaves are asymmetrical with acute denticulation, may show enlarged petiolar sinuses and chlorotic mottles or bright yellow discoloration of veins, depending on the viral strain (Fig. 14.1a). Foliar symptoms develop in the early vegetative season, but become less evident in late summer. Canes show abnormal branching, double nodes and fasciations (Fig. 14.1b). Moreover, the infected plants show dropping off of flowers and berries, which are smaller and ripen irregularly (Fig. 14.1c). Symptoms are evident on most *V. vinifera* varieties.

Typical symptoms of leafroll are rolling and reddening (red varieties) or yellowing (white varieties) of leaves, which become thick and brittle; usually the primary and secondary veins remain green (Fig. 14.2a, b). Foliar symptoms start during the summer and progress throughout the vegetative season, becoming more evident in autumn and on the older leaves. Bunches from infected plants are often smaller in number and size and ripen later (Fig. 14.2c). The intensity of the foliar symptoms depends on the virus type and viral strain. Symptoms are easier to identify on red *V. vinifera* varieties, more difficult on white varieties.

The four distinct diseases that constitute the rugose wood complex show typical symptoms mostly in the indicator varieties and are latent in many European varieties. Typical symptoms of rupestris stem pitting are visible on the rootstock indicator *V. rupestris* when the bark is removed: the wood below the graft union appears wrinkled and longitudinal pits are clearly visible on the surface (Fig. 14.3a); symptoms are not present in Kober 5BB and LN33. Also Kober stem grooving shows symptoms in the wood surface of the rootstock indicator, which is Kober 5BB, when the bark is removed, but the symptoms are different from stem pitting and consist of marked stem grooving (Fig. 14.3b); symptoms do not occur on *V. rupestris* and LN33. Canes of susceptible grapevines affected by corky bark show internodal swelling and longitudinal cracks and develop abnormal corky overgrow of bark; moreover, leaves from red cultivar turn red and roll downwards in summer. LN33 stem grooving shows the same symptomatology of Kober stem grooving, but only when LN33 hybrid is used as an indicator; no internodal swelling of the shoots nor foliar discoloration are present.

Fig. 14.2 Leafroll symptoms: on leaves of white (**a**) and black (**b**) varieties; (**c**) irregularly ripening of berries

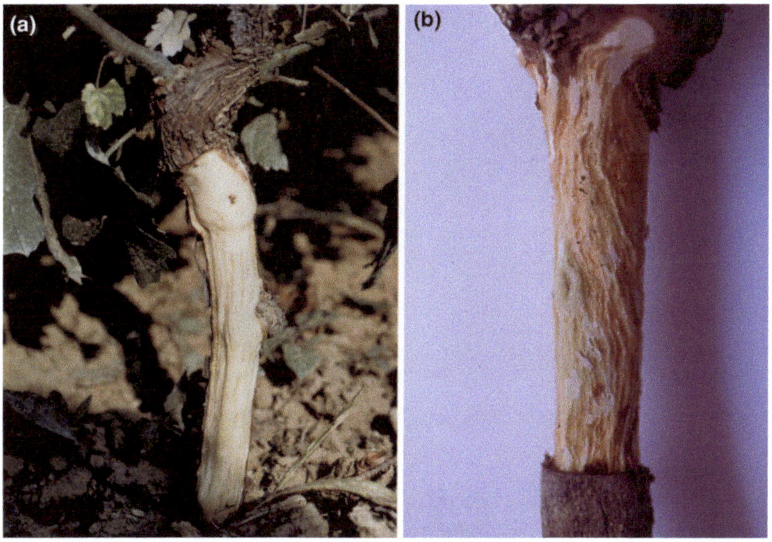

Fig. 14.3 Rugose wood on indicators: (**a**) rupestris stem pitting on *Vitis rupestris*; (**b**) Kober stem grooving on Kober 5BB

Fleck occurs as a latent infection in European varieties and most grapevine species. On *V. rupestris* the virus shows a typical clearing of the veins of third and fourth order, producing localized translucent spots, which are evident especially in spring (Fig. 14.4). Clear spots tend to disappear with the warm temperature.

Vein necrosis is latent in many cultivars, but shows particularly clear symptoms on leaves of 110 Richter. The typical symptom is the blackish necrosis on leaf veins, evident especially on the abaxial surface of the basal leaves, which extends to

Fig. 14.4 Fleck symptoms on leaves of the indicator *Vitis rupestris*

Fig. 14.5 Vein necrosis on leaves of the indicator 110 Richter

the other leaves as the season advances (Fig. 14.5). This disease is very serious in nursery, because it causes strong decrease of the graft take.

For a complete photographic review of the symptoms of viral diseases in grapevine, see Bovey et al. (1980).

14.3 Serological Methods

Serological methods, targeting the coat protein of the viruses, are useful for the diagnosis of the viral diseases whose etiological agent is known. Detection of grapevine viruses is possible using ELISA (enzyme-linked immunosorbent assay) or dot immunobinding on nylon membranes, with polyclonal antisera or monoclonal antibodies when available (Torrance and Jones 1981). Reliable antisera are available in the form of ELISA commercial kits for many grapevine viruses; however, no consistent ELISA method has been yet developed for the detection of GRSPaV (Minafra et al. 2000). The ELISA technique in general is reliable, sensitive, fast and easy to perform (Clark and Adams 1977, Crowther 2001); therefore, it is the serological method of choice for routine diagnosis of grapevine viruses. The ELISA kits available on the market for grapevine viruses are based on direct or indirect detection. Moreover, several variants of the ELISA have been established: DAS (double antibody sandwich), TAS (triple antibody sandwich) and PTA (plate trapped antibody) ELISA, all of them being reliable (Koenig and Paul 1982). However, the consistency of the results obtained using the ELISA strongly depends on the type of virus, the material to be tested (leaves or wood) and the season of sampling. In general, the best antigen sources for serological diagnosis of all viruses are cortical phloem shavings from mature dormant canes.

Many ELISA kits for the detection of GFLV, agent of fanleaf, are available on the market (Huss et al. 1986, Rowhani 1992). The detection of GFLV by ELISA is very reliable and the test does not present particular difficulty. Infected leaves collected from May to the end of the vegetative season and dormant canes always give positive results; however, positive responses are more evident from mature leaves and dormant canes. Apical leaves can sometimes give false negative response. Results are consistent in both European varieties and rootstocks.

Reliable ELISA kits are available only for GLRaV-1, 2, 3, 4 and 6 (Hu et al. 1990, Zimmerman et al. 1990a,b, Rowhani 1992, Monis and Bestwick 1997, Ling et al. 2000, Ling et al. 2007). Antibodies for the other leafroll viruses are not working properly or have not yet been developed. The ELISA detection of GLRaV-1 and 2 in leaves of infected plants can sometimes give false negative results caused by the low titre of the two viruses; the most suitable material to be tested is dormant canes, where the assays reach a good reliability. It is worth to note that infected dormant canes analysed later than 2–3 months from the collection can show negative results to the ELISA, especially in the case of GLRaV-2. Several ELISA kits for the detection of GLRaV-1 and 2 are available on the market, but not all of them show the same reliability and broad spectrum of detection (Bertazzon et al. 2002). The detection of both the viruses is more reliable on European varieties than on American rootstocks. The diagnosis of GLRaV-3 by ELISA is very reliable on leaves and mature canes from both *V. vinifera* and other grapevine species. The responses are always very good, except for infected young leaves collected at the beginning of the vegetative season, which sometimes can provide false negative results. The quality of the different ELISA kits available on the market for GLRaV-3 is similar.

Only two of the viruses associated with the rugose wood complex can be detected by serological methods: GVA and GVB (Boscia et al. 1992, Rubinson et al. 1997, Saldarelli et al. 2005). The diagnosis of GVA by ELISA test can be difficult to establish, due to the characteristics of the ELISA kits available on the market; however, once well established, it works properly on dormant canes from both European and American grapevines, if the conservation of the wood is shorter than 3 months. The detection on the leaves is not reliable, especially on young leaves, where the false negative results can reach the 100% of the responses. The ELISA kits specific for GVA available on the market can yield different results, as most of them are able to react only against a limited number of viral strains; therefore, the use of more than one kit or of other detection methods is suggested. It is also inadvisable to store the commercial GVA kits for many months, as it has been observed that their sensibility can decrease with time. The requirements for the serological detection of GVB are the same than for GVA; however, the reliability of the test is lower. Moreover, the availability of ELISA kits on the market is limited.

Several good ELISA kits are available on the market for the detection of fleck virus (Walter and Cornuet 1993, Boscia et al. 1995, Schieber et al. 1997). The reliability of the test is very good on every kind of tissue and grapevine species and the results are consistent in every season; however, positive responses are more evident on dormant canes. Apical leaves can sometimes give negative response in the late vegetative season.

14.4 Molecular Methods

Many new techniques have been developed for the detection of plant viruses, with the potential for increased sensitivity, specific or broad-spectrum detection. The most frequently used molecular methods have been molecular hybridization and, more recently, polymerase chain reaction (PCR). The new techniques are in some cases complex and usually require expensive equipment and reagents: that could hamper their routine use, especially when the financial resources are limited. However, the molecular technologies provide many advantages, and modifications that reduce complexity and cost will result in an increase in their application for large-scale diagnosis of plant viruses.

14.4.1 Molecular Hybridization

The technique involves the direct application of a nucleic acid solution extracted from samples to a solid support, as nylon membrane, and the subsequent detection with appropriate specific probes. The most common molecular hybridization format for the detection of viruses is non-isotopic dot-blot using digoxigenin-labelled probes. The use of non-radioactive precursors to label the probes has made this techniques more accessible for routine virus diagnosis. Greater preference is given to RNA probes than to DNA probes, since the majority of plant viruses have RNA genomes and RNA-RNA hybrids are more stable: this increases specificity and lowers non-specific hybridization signals. Concerning grapevine, several authors have developed probes for the detection of the major viruses affecting the species. For example a probe for GFLV was developed by Fuchs et al. (1991); probes for GVA by Kominek et al. (2008); for GLRaV-3 by Saldarelli et al. (1994) and Habili et al. (1995); for GFkV by Sabanadzovic et al. (1996) and Elbeaino et al. (2001), and probes for GLRaV-1 detection were used by Kominek and Bryxiova (2005).

The tissue-printing technique was developed for the detection and the localization of viruses inside the explants without the RNA extraction, thus simplifying the sample preparation. Further reductions of time, labour and costs are allowed by multiplex versions of molecular hybridization: mixtures of up to six riboprobes or synthesis of polyprobes (where partial nucleic acid sequences of different viruses are cloned in tandem to obtain the synthesis of a unique riboprobe; Herranz et al. 2005), were used to detect several viruses simultaneously. However, problems like the relatively low sensitivity and the complexity have hampered the practical applications of the hybridization techniques.

14.4.2 RT-PCR

PCR was developed over 30 years ago, and its use in diagnosis of plant viruses (and plant diseases in general) has become very common in recent years, thanks to its higher sensitivity over bioassays and ELISA. It is particularly useful if antisera for ELISA are not available or not suitable. As the majority of viruses are RNA viruses, an initial step of reverse transcription that converts RNA to cDNA is necessary for PCR amplification. However, the reliability of RT-PCR is limited by the presence, in plant extracts, of inhibitors of the reverse transcriptase and/or polymerase, resulting generally in decreased sensitivity. Extraction of high-quality RNA from the tissues of woody and perennial plants is particularly challenging because of high concentrations of polysaccharides, polyphenols, and other secondary metabolites. Although several protocols have been developed for the extraction of total RNA of grapevine, most of them are time consuming and technically complex. From our experience two protocols allow rapid isolation of high quality RNA extraction from different grapevine tissues: RNeasy (RNeasy Plant Mini Kit, Qiagen) commercial kit modified (MacKenzie et al. 1997) and the rapid CTAB-based protocol described by Gambino et al. (2008). With the aim of reducing the time and cost of the extraction without reducing quality and yield of RNA, we set up a rapid CTAB protocol by introducing several changes to the original procedure of Chang et al. (1993). The rapid CTAB method gave high-quality RNA from several tissues of grapevine, as well as of other woody species, at low cost in just 3 h. The commercially RNeasy method performed following the manufacturer's instructions (Qiagen) gave low yields of poor-quality RNA for some grapevine tissues (Nassuth et al. 2000, Gambino et al. 2008). An additional extraction step with chloroform:isoamyl alcohol before using the RNeasy kit allowed removal of some contaminants and increased the binding capacity of RNeasy columns thus permitting efficient RNA purification from all grapevine tissues (Gambino and Gribaudo 2006). With these modifications, the protocol proved to be suitable thanks to its simplicity, reproducibility and rapidity, but it was more costly and the RNA quality was slightly lower in comparison with the rapid CTAB method (Gambino et al. 2008).

Other sample preparation methods were suggested specifically for virus detection in grapevine. Examples include direct release of different viruses from excised tissue (Thompson and Dietzgen 1995), binding of virion from crude extract onto

ELISA microplates (Rowhani et al. 1995) or onto nylon membranes (La Notte et al. 1997, Dovas and Katis 2003), direct spotting of crude sap derived from infected leaf, petiole or cambial tissue onto positively charged membranes (Osman and Rowhani 2006). Although these methods have been used for virus detection, little information was reported about the quality and quantity of RNA extracted and in few cases an RNA internal control was used in the reaction. Besides the RNA quality, the RT-PCR efficiency is controlled by many parameters such as primer selection, polymerase type, buffer composition and stability, concentrations of dNTPs and cycling conditions. The primers are generally designed on conserved sequence regions of each virus to allow amplification of all or several variants of the virus. In some cases a few mutations could occur within the primers-binding sites and the failure of the primer to bind to its recognition site can lead to false negatives. The accurate detection of viruses by RT-PCR is influenced also by sampling protocols, since molecular methods generally prescribe very small volume of samples. Considering the possible uneven distribution of viruses within a given plant (Dovas and Katis 2003), it is advisable to mix samples from several shoots or canes of the same plant.

14.4.2.1 Nested RT-PCR

Sensitivity and specificity problems associated with conventional RT-PCR can be reduced by using nested RT-PCR methods, based on two consecutive rounds of amplification. The products of the first amplification are subjected to a second amplification carried out using one or two internal primers. Sensitivity is increased, but two rounds of amplification also increase the risk of contamination.

Over the past 10 years many RT-PCR protocols have been developed by several authors for the diagnosis of the main grapevine viruses, so that it would be difficult to list them in detail in this context. Many references on grapevine viruses and their diagnosis can be found on the website of the International Council for the Study of Virus and Virus-like Diseases of the Grapevine (http://www.icvg.ch/).

14.4.2.2 Polyvalent RT-PCR

PCR with polyvalent primers that hybridize in a region conserved between different targets is a strategy developed for the simultaneous detection of several viruses in a single reaction. The simplest approach is to synthesize a pool of degenerate primers containing most or all of the possible sequences present in the target. One limitation is that as the level of degenerancy of the pool of primers increases, the concentration of each individual primer decreases. Consequently, the primers that are actually able to prime synthesis in a reaction may be quite few, reducing amplification efficiency and the overall sensitivity of the amplification process. Another limitation is the low-stringency amplification conditions needed to allow hybridization of the primer to the target sequence: some primers in the pool may hybridize to non-target molecules. Nevertheless, this strategy has been applied for the amplification of groups of grapevine viruses. *Vitivirus* genus was amplified by Saldarelli et al. (1998), *Nepovirus* by Wetzel et al. (2002) and Digiaro et al. (2007). Dovas and

Katis (2003) used degenerate deoxyinosine (dI)-substituted primers (dI is a modified nucleotide able to base pair with all four normal nucleotides) for the generic detection of *Vitivirus*, *Foveavirus* and *Closterovirus* in the same reaction tube. Maliogka et al. (2008) used degenerate primers for the generic detection of *Closterovirus* using HSP70h as target followed by a nested PCR, which detects and differentiates all virus-members. Bertazzon and Angelini (2004) designed primers that allowed the simultaneous detection of all isolates of GLRaV-2, a highly polymorphic virus.

Since primers involved in polyvalent PCR target highly conserved regions, the visualization of a PCR band by gel electrophoresis may not provide information on the viruses present in the sampled material. RFLP analysis of the amplified band, direct sequencing or cloning the PCR products, may provide such information even if interpretation of results may be complicated in the case of mixed infection.

14.4.2.3 Multiplex RT-PCR (mRT-PCR)

Another strategy for simultaneous detection of several viruses is the mRT-PCR, a variant of PCR in which two or more loci are amplified simultaneously, thus providing quick, reliable and cost-effective routine diagnosis. Development of a multiplex assay is often complex: each individual amplification has to function under identical conditions, and interference and competition may occur between the individual reactions. In the literature there are many examples of simultaneous amplification by mRT-PCR of RNAs from two to five plant viruses, but few cases have been reported in which more than five plant viruses were amplified in a single mRT-PCR. The design of a mRT-PCR is based on the use of compatible primers specific to different targets evaluated theoretically *in silico* and tested empirically *in vitro*.

Simultaneous detection of grapevine viruses was reported by Minafra and Hadidi (1994), La Notte et al. (1997) and Nassuth et al. (2000), which used a multiplex procedure to detect two or three grapevine viruses. More recently, Faggioli and La Starza (2006) developed one-step mRT-PCR for the detection of eight viruses in a sanitary selection program. Gambino and Gribaudo (2006) reported the simultaneous detection of nine grapevine-infecting viruses ArMV, GFLV, GVA, GVB, GRSPaV, GFkV, GLRaV-1, to 3 and RNA internal control (18S rRNA) used as indicator of RNA quality and RT-PCR effectiveness (Fig. 14.6). This mRT-PCR protocol allowed reliable detection of these viruses within 1 day. This may help to save time and reduce costs, and could replace presently used techniques like ELISA or bioassays for the indexing of these viruses. The mRT-PCR was optimized by varying reaction components and cycling conditions one at a time. The primers used did not show primer-primer interactions both *in silico* and *in vitro*. Primer concentrations for the viruses and the internal control were optimized to determine which combination of primer concentrations gave best amplification of the expected specific fragments in several different extracts. Although individual viruses and the internal control were specifically amplified at 56 –58°C, lowering the annealing temperature to 50°C was required initially for co-amplification in the multiplex reaction, as well as increasing extension time, PCR buffer, dNTPs, $MgCl_2$ and Taq DNA polymerase concentrations above those used in a single RT-PCR. In the years following

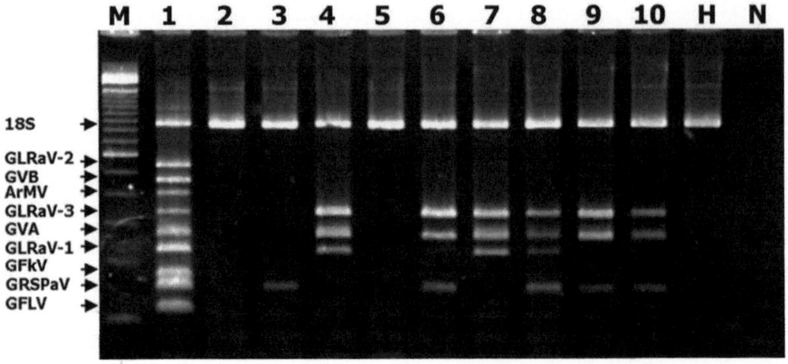

Fig. 14.6 Agarose gel electro-phoretic analysis of DNA fragments amplified by mRT-PCR. Lane 1: positive control; lanes 2–10: naturally infected grapevines. H: healthy control, N: water, M: 100 bp DNA ladder

the publication of the original work (Gambino and Gribaudo 2006), the mRT-PCR protocol was further optimized by rising the annealing temperature to 55°C, the $MgCl_2$ to 3.2 mM and decreasing the Taq polymerase to 1.25 unit. This allowed to avoid the extra bands that were sometimes produced in samples with low yields of poor-quality RNA. The inclusion of an internal control for the RT-PCR assay is essential for detecting false negatives due to RNA degradation or the presence of inhibitors of the reaction. The detection limits of mRT-PCR were lower than those of single RT-PCR for seven viruses, because in the multiplex assay the cocktail of primers compete for all the templates rather than for one. However, the level of sensitivity could be regarded as adequate for testing routine samples. Specific bands on the gel could be detected for all viruses up to 1:1,000 dilution, corresponding to 0.1 mg of infected tissue in 100 mg in total. This allows detection of infection of one positive sample in a bulk extract of several plants, which could be useful for large scale indexing. Up to now more than 700 samples of grapevines from different geographical areas of Italy have been analyzed by mRT-PCR. The reaction was able to reliably identify all combinations of the infecting viruses regardless of the plant tissue, with the same efficiency observed in a single RT-PCR, and allowed the detection of some viruses that were not detectable by ELISA.

14.4.2.4 Real-Time RT-PCR

Real-time RT-PCR is a powerful diagnostic tool that allows rapid and reliable detection and quantification of viruses. With conventional PCR, amplicons are detected after the reaction by electrophoresis, while real-time PCR provides a continuous monitoring along the entire amplification process by fluorescence-based detection chemistries. Assays are more rapid than end-point RT-PCR, since electrophoretic analysis is not required and the amplification time is usually shorter (real-time PCR works better with small amplicons: 50–200 bp). Fluorescent detection chemistries may be specific (probe or primer-based chemistry: TaqMan, Scorpion, Molecular

Beacons) or nonspecific (intercalating dye-based chemistry: SYBR green), and the mechanisms for generating fluorescent signals depend on the type of detection chemistry.

Most real-time PCR assays for plant viruses relied on TaqMan probes (oligonucleotides of 20–30 bases that contain a fluorescent dye and a quencher) that are designed to recognize specific targets. In grapevine, real-time TaqMan RT-PCR assays were developed to detect GLRaV-2 variants (Beuve et al. 2007), GLRaV-1 to 5 and 9 (Osman et al. 2007) and the viruses associated with Rugose wood complex: GRSPaV, GVA, GVB and GVD (Osman and Rowhani 2008). Comparisons between the conventional one step RT-PCR and TaqMan RT-PCR for the detection of these viruses showed that real time PCR was more sensitive and could detect viruses at 32- and 256-fold higher dilutions for purified RNA and crude extract respectively, compared to standard RT-PCR (Osman and Rowhani 2008). Stewart et al. (2007) used a real-time assays based on non-specific chemistry (SYBR green) coupled with melt-curve analyses for the detection of *Tomato ring spot virus* (ToRSV) in grapevine. The utility of real-time assays based on SYBR green has been demonstrated for other viruses and the use of this chemistry reduced cost as well as the risk that a specific probe-based assay may fail to recognize the target. In the TaqMan system, any mutation (strains and subgroups of viruses) in the probe-binding site can inhibit binding and lead to false negative results.

Now that many fluorophores are available, probe-based assays such as TaqMan may be easily adapted in multiplex format for the detection of several viruses. However, in practice it is difficult to optimize multiplex reaction: probe design for multiplex assays is critical, and the reagents to be used are also costly. Currently, there are no examples of multiplex real-time RT-PCR for detection of grapevine viruses.

14.4.3 Microarray Technology

Since the development of microarray technology for gene expression studies (Schena et al. 1995), new approaches are extending their application to the detection of pathogens. DNA microarrays or biochips are made of a solid surface including microplates, blotting membranes, and solid nonporous surfaces such as glass which are linked to multiple capture probes. Solid nonporous surfaces allow the precise deposition of small amounts of biochemical reagents, facilitating the development of microarrays or biochips with the potential for over 100,000 probes per chip. Each probe is complementary to a specific DNA or RNA sequence and hybridisation with the labelled complementary sequence provides a signal that can be detected and analysed. Probe design and quality determine the effectiveness of an array, and array probes are available in two forms: oligonucleotides (20–70 nt in length) and cDNA (100–500 bases in length). Both cDNA and oligonucleotide probes (Boonham et al. 2003) have been employed in microarray detection of plant viruses. The main advantage of cDNA probes is that long probes may be synthesized cheaply, but there are drawbacks including purification requirements and lack of flexibility. Oligonucleotide probes allow the development of microarrays with greater specificity and flexibility (Boonham et al. 2003). The power of microarray technology in the detection and diagnosis of plant diseases and other quarantine pathogens is very high, due to the multiplex capabilities of the system.

In grapevine, at the moment a few works have reported the use of the array technologies for virus detection. Engel et al. (2006) produced a microarray chip containing oligonucleotides complementary to highly conserved as well as to variable domains of all viruses known to infect grapevine. This chip allows the detection of already known viruses and new unknown viruses if their genomes have partial homology with the printed probes. Osman et al. (2008) designed a low-density arrays (LDA), based on the real-time TaqMan RT-PCR assays, for the specific detection of 13 viruses that infect grapevines in addition to the housekeeping gene 18S rRNA. Primers and TaqMan probe mixes for the viruses and the 18s rRNA were dried onto the plastic surface of 384-well plates. The plates were used for the screening of several grapevine samples with an efficiency greater than other diagnostic methods. The authors showed that the LDA technology is a time-saving tool for detecting plant viruses and allows simultaneous analyses of different pathogens in the same sample in a single reaction set up. Abdullahi and Rott (2009) developed an antibody microarray procedure for the detection of several grapevine and tree fruit viruses. In antibody arrays immobilized capture antibodies are exposed to samples containing the target antigens, which can then be detected by fluorescence chemiluminescence or chromogenic substrates. The authors showed that the antibody microarray system was similar to ELISA with respect to sensitivity and specificity.

Although there is great potential for microarray technology in the diagnosis of plant diseases, the practical development of this application is still in progress. Disadvantages of array analysis include the need for complex skills, the difficulty of adapting the technology for routine screening of large numbers of samples, and the high costs associated with equipment required to create the microarrays and to analyse the results.

14.5 Conclusions

The ability of performing a correct diagnosis on the viral infections in grapevine is based on a good knowledge of the specific symptomatology and of the laboratory tests available. None of the diagnostic methods listed here can be completely exhaustive, as each of them possesses its own limits (Rowhani et al. 2005). The visual diagnosis of symptoms directly in the field requires long field training, it depends upon the period of observation and it is not achievable for the latent diseases. The biological indexing requires long time (1–3 years), but it is the only diagnostic methods able to detect all the known viral diseases. Moreover, new viruses, not yet isolated or characterized, associated with known diseases are detectable only by indexing. However, laboratory assays are necessary in order to identify and characterize the exact virus type, for example in the case of leafroll-associated viruses. The ELISA test is easy and fast and allows a high throughput of results; moreover, it shows a good sensitivity and is a broad-spectrum assay, generally good in identifying also new divergent strains of a known virus. The sensitivity is good, though some problems can arise from the analysis of rootstocks. The conventional and real time PCR tests are the most sensitive techniques, though the processing of the samples is longer than for the ELISA; however, the establishment of PCR tests directly from sap and of multiplex PCR assays can reduce significantly the time required for the analyses. The major limit of the molecular techniques is the knowledge of nucleic acid sequences of all known viruses. Indeed, many viruses and divergent viral variants do exist which are not yet amplifiable by PCR. For all these reasons, the integration of the different diagnostic techniques is desirable for a correct diagnosis, especially in the case of the viruses whose detection is more difficult. In any case, the use of one or another diagnostic technique depends on the aim of the work.

Acknowledgments Figures 14.1, 14.2, 14.3, 14.4 and 14.5 are courtesy of Dr. Michele Borgo, CRA-VIT, Conegliano (TV, Italy), whose help is thankfully acknowledged.

References

Abdullahi I, Rott M (2009) Microarray immunoassay for the detection of grapevine and tree fruit viruses. J Virol Methods 160:90–100

Bertazzon N, Angelini E (2004) Advances in the detection of Grapevine leafroll-associated virus 2 variants. J Plant Pathol 86:283–290

Bertazzon N, Angelini E, Borgo M (2002) Detection of grapevine leafroll-associated virus-2 (GLRaV-2) by ELISA and RT-PCR. J Plant Pathol 84:175

Beuve M, Sempe L, Lemaire O (2007) A sensitive one-step real-time RT-PCR method for detecting Grapevine leafroll-associated virus 2 variants in grapevine. J Virol Methods 141: 117–124

Boonham N, Walsh K, Smith P, Madagan K, Graham I, Barker I (2003) Detection of potato viruses using microarray technology: towards a generic method for plant viral disease diagnosis. J Virol Methods 108:181–187

Boscia D, Aslouj E, Elicio V, Savino V, Castellano MA, Martelli GP (1992) Production, characterization and use of monoclonal antibodies to grapevine virus A. Arch Virol 127: 185–194

Boscia D, Elicio V, Savino V, Martelli GP (1995) Production of monoclonal antibodies to grapevine fleck virus. Plant Pathol 44:160–163

Bovey R, Gartel W, Hewitt WB, Martelli GP, Vuittenez A (1980) Virus and virus-like diseases of grapevine, colour atlas of symptoms. La Maison Rustique, Paris

Chang S, Puryear J, Cairney J (1993) A simple and efficient method for isolating RNA from pine trees. Plan Mol Biol Rep 11:113–116

Clark MF, Adams AN (1977) Characteristics of the microplate method of enzyme-linked immunosorbent assay for the detection of plant viruses. J Gen Virology 34:475–483

Crowther JR (2001) The ELISA guidebook. Humana, Totowa, NJ

Digiaro M, Elbeaino T, Martelli GP (2007) Development of degenerate and species-specific primers for the differential and simultaneous RT-PCR detection of grapevine-infecting Nepoviruses of subgroups A, B and C. J Virol Methods 141:34–40

Dovas CI, Katis NI (2003) A spot multiplex nested RT-PCR for the simultaneous and generic detection of viruses involved in the aetiology of grapevine leafroll and rugose wood of grapevine. J Virol Methods 109:217–226

Elbeaino T, Sabanadzovic S, Digiaro M, Abou Ghanem-Sabanadzovic N, Rowhani A, Kyriakopoulou PE, Martelli GP (2001) Molecular detection of Grapevine fleck virus-like viruses. Vitis 40:65–68

Engel E, Arrendono V, Girardi C, Gonzalez A, Escobar P, Fiore N, Arce P, Valenzuela PDT (2006) Grapevine viruses in Chile: genomics and detection based on immunocapture and microarray technologies. In: Proceedings of the 15th ICVG Meeting, Ext. Abstr., Stellenbosch (South Africa), pp 118–119

Faggioli F, La Starza S (2006) One-step multiplex RT-PCR for simultaneous detection of eight grapevine viruses and its application in a sanitary selection program. In: Proceedings of the 15th ICVG Meeting, Ext. Abstr., Stellenbosch (South Africa), pp :120–121

Fuchs M, Pinck M, Etienne L, Pinck L, Walter B (1991) Characterization and detection of Grapevine fanleaf virus by using cDNA probes. Phytopathology 81:559–565

Gambino G, Gribaudo I (2006) Simultaneous detection of nine grapevine viruses by multiplex RT-PCR with coamplification of a plant RNA internal control. Phytopathology 96:1223–1229

Gambino G, Perrone I, Gribaudo I (2008) A rapid and effective method for RNA extraction from different tissues of grapevine and other woody plants. Phytochem Anal 19:520–525

Habili N, Fazeli Cf, Ewart A, Hamilton R, Cirami R, Saldarelli P, Minafra A, Rezaian MA (1995) Natural spread and molecular analysis of Grapevine leafroll-associated virus-3 in Australia. Phytopathology 85:1418–1422

Herranz MC, Sanchez-Navarro JA, Aparicio F, Pallás V (2005) Simultaneous detection of six stone fruit viruses by non-isotopic molecular hybridization using a unique riboprobe or 'polyprobe'. J Virol Methods 124:49–55

Hu JS, Gonsalves D, Boscia D, Namba S (1990) Use of monoclonal antibodies to characterize grapevine leafroll associated closteroviruses. Phytopathology 80:920–925

Huss B, Walter B, Etienne L, Van Regenmortel MHV (1986) Grapevine fanleaf virus detection in various organs using polyclonal and monoclonal. Vitis 25:178–188

Koenig R, Paul HL (1982) Variants of ELISA in plant virus diagnosis. J Virol Methods 5:113–125

Kominek P, Bryxiova A (2005) Comparison of three techniques for detection of Grapevine leafroll-associated virus 1. Acta Virol 49:37–43

Kominek P, Kominkova-Bryxiova M, Jandurova OM, Holleinova V (2008) An RNA Probe for the Detection of *Grapevine virus A*. J. Phytopathol 156:762–764

La Notte P, Minafra A, Saldarelli P (1997) A spot-PCR technique for the detection of phloem-limited grapevine viruses. J Virol Methods 66:103–108

Ling KS, Zhu HY, Jiang ZY, Gonsalves D (2000) Effective application of DAS-ELISA for detection of grapevine leafroll associated closterovirus-3 using a polyclonal antiserum developed from recombinant protein. Eur J Plant Pathol 106:301–309

Ling KS, Zhu HY, Petrovic N, Gonsalves D (2007) Serological detection of grapevine leafroll virus 2 using an antiserum developed against the recombinant coat protein. J Phytopathol 155:65–69

MacKenzie DJ, Mclean MA, Mukerji S, Grenn M (1997) Improved RNA extraction from woody plants for the detection of viral pathogens by reverse transcription-polymerase chain reaction. Plant Dis 81:222–226.

Maliogka VI, Dovas CI, Katis NI (2008) Generic and species-specific detection of viruses belonging to an evolutionary distinct lineage within the Ampelovirus genus. J Virol Methods 154:41–47

Martelli G, Boudon-Padieu E (2006) Directory of infectious diseases of grapevines and viroses and virus-like diseases of the grapevine: bibliographic report 1998–2004. Options méditerranéennes Series B: 55, CIHEAM, Bari, Italy

Minafra A, Hadidi A (1994) Sensitive detection of grapevine virus A, B, or leafroll-associated III from viruliferous mealybugs and infected tissue by cDNA amplification. J Virol Methods 47:175–188.

Minafra A, Casati P, Elicio V, Rowhani A, Saldarelli P, Savino V, Martelli GP (2000) Serological detection of grapevine rupestris stem pitting-associated virus (GRSPaV) by a polyclonal antiserum to recombinant virus coat protein. Vitis 39:115–118

Monis J, Bestwick RK (1997) Serological detection of grapevine associated closteroviruses in infected grapevine cultivars. Plant Dis 81:802–808

Nassuth A, Pollari E, Helmeczy K, Stewart S, Kofalvi SA (2000) Improved RNA extraction and one-tube RT-PCr assay for simultaneous detection of control plant RNA plus several viruses in plant extracts. J Virol Methods 90:37–49

Osman F, Rowhani A (2006) Application of a spotting sample preparation technique for the detection of pathogens in woody plants by RT-PCR and real-time PCR (TaqMan). J Virol Methods 133:130–136

Osman F, Rowhani A (2008) Real-time RT-PCR (TaqMan ®) assays for the detection of viruses associated with Rugose wood complex of grapevine. J Virol Methods 154:69–75

Osman F, Leutenegger C, Golino D, Rowhani A (2007) Real-time RT-PCR (TaqMan ®) assays for the detection of Grapevine Leafroll associated viruses 1–5 and 9. J Virol Methods 141:22–29

Osman F, Leutenegger C, Golino D, Rowhani A (2008) Comparison of low-density arrays, RT-PCR and real-time TaqMan® RT-PCR in detection of grapevine viruses. J Virol Methods 149: 292–299

Pathirana R, McKenzie MJ (2005) A modified green-grafting technique for large-scale virus indexing of grapevine. Scientia Hortic 107:97–102

Rowhani A (1992) Use of $F(ab')_2$ antibody fragment in ELISA for detection of grapevine viruses Am J Enol Vitic 43: 38–40

Rowhani A, Maningas MA, Lile LS, Daubert SD, Golino DA (1995) Development of a detection system for viruses of woody-plants based on PCR analysis of immobilized virions. Phytopathology 85:47–352

Rowhani A, Uyemoto JK, Golino DA, Martelli GP (2005) Pathogen testing and certification of *Vitis* and *Prunus* species. Annu Rev Phytopathol 43:61–278

Rubinson E, Galiakparov N, Radian S, Sela I, Tanne E, Gafny R (1997) Serological detection of grapevine virus A using antiserum to a nonstructural protein, the putative movement protein. Phytopathology 87:041–1045

Sabanadzovic S, Saldarelli P, Savino V (1996) Molecular diagnosis of Grapevine fleck virus. Vitis 35: 137–140

Saldarelli P, Minafra A, Martelli GP, Walter B (1994) Detection of Grapevine leafroll-associated closterovirus-III by molecular hybridization. Plant Pathol 43:1–96

Saldarelli P, Rowhani A, Routh G, Minafra A, Digiaro M (1998) Use of degenerate primers in a RT-PCR assay for the identification and analysis of some filamentous viruses, with special reference to clostero- and vitiviruses of the grapevine. Eur J Plant Pathol 104: 945–950

Saldarelli P, Keller H, Dell'Orco M, Schots A, Elicio V, Minafra A (2005). Isolation of recombinant antibodies (scFvs) to grapevine virus B. J Virol Methods 124:191–195

Schena M, Shalon D, Davis RW, Brown PO (1995) Quantitative monitoring of gene-expression patterns with a complementary-dna microarray. Science 270:467–470

Schieber O, Seddas A, Belin C, Walter B (1997) Monoclonal antibodies for detection, serological characterization and immunopurification of grapevine fleck virus. Eur J Plant Pathol 103: 767–774

Stewart EL, Qu XS, Overton BE, Gildow FE, Wenner NG, Grove DS (2007) Development of a real-time RT-PCR SYBR green assay for Tomato ring spot virus in grape. Plant Dis 91: 1083–1088.

Thompson D, Dietzgen RG (1995) Detection of DNA and RNA plant-viruses by PCR and RT-PCR using a rapid virus release protocol without tissue homogenization. J Virol Methods 54:85–95

Torrance L, Jones RAC (1981) Recent developments in serological methods suited for use in routine testing for plant viruses. Plant Pathol 30:1–24

Walter B, Cornuet P (1993) ELISA detection of grapevine fleck virus (GFkV). Agronomie 13: 651–657

Wetzel T, Fuchs M, Bobko M, Krczal G (2002) Size and sequence variability of the Arabis mosaic virus protein 2A. Arch Virol 147:643–1653

Zimmermann D, Bass P, Legin R, Walter B (1990a) Characterization and serological detection of four closterovirus-like particles associated with leafroll disease on grapevine. J Phytopathol 130:205–218

Zimmermann D, Sommermeyer G, Walter B, Van Regenmortel MHV (1990b) Production and characterization of monoclonal antibodies specific to closterovirus-like particles associated with grapevine leafroll disease. J Phytopathol 130:277–288

Chapter 15
Real-Time PCR Detection Methods for Economically Important Grapevine Related Bacteria

Matjaž Hren, Tanja Dreo, Jana Erjavec, Petra Nikolić, Jana Boben, Kristina Gruden, Marina Dermastia, Marjana Camloh, and Maja Ravnikar

Abstract Preventive measures are extremely important for control of diseases that are caused by plant pathogenic bacteria on economically important plants. For this reason, fast and reliable detection methods are required. Along with the time-consuming conventional detection methods such as isolation and culturing of bacteria on media, PCR-based methods have been introduced as supplementary tests for better diagnosis of plant pathogenic bacteria. In the case of non-culturable bacteria, such as Phytoplasmas, PCR-based tests even became indispensable. Real-time PCR has become a widely used platform in nucleic acid detection and quantification in diagnostics. Despite the general guidelines for some of the steps, there are no common guidelines or standard operating procedures for introduction of real-time PCR-based detection systems for pathogens for routine laboratory use. Four cases of detection systems for grapevine pathogenic bacteria that were developed at the National Institute of Biology are presented and discussed here, providing a practical overview of the whole process from the initial assay design to the implementation of the assay into the laboratory.

Contents

M. Hren (✉)
National Institute of Biology, Department of Biotechnology and Systems Biology, Ljubljana 1000, Slovenia
e-mail: matjaz.hren@nib.si

S. Delrot et al. (eds.), *Methodologies and Results in Grapevine Research*,
DOI 10.1007/978-90-481-9283-0_15,

Abbreviations

AY	Aster yellows
BN	Bois noir
DNA	Deoxyribonucleic acid
ELISA	Enzyme-linked immunosorbent assay
EPPO	European and Mediterranean Plant Protection Organisation
FD	Flavescence dorée
GYs	Grapevine yellows
OEPP	Organisation Européenne et Méditerranéenne pour la Protection des Plantes
PCR	Polymerase chain reaction
qPCR	Real-time PCR

15.1 Introduction

Plant pathogenic bacteria cause a variety of symptoms on grapevine which can result in damage leading to significant yield losses, especially if the severe outbreaks occur. Preventive and control measures for phytopathogens rely on fast and reliable detection methods which allow timely removal of diseased plants or their parts. Therefore, the methods have to be quick and preferably sensitive enough to detect also latent infections, where the titer of pathogen can be very low (López et al. 2006).

Fast and sensitive methods are even more important in case of high-risk pathogens such as grapevine bacterial pathogens *Xylella fastidiosa*, *Xylophilus ampelinus* and Phytoplasma associated with a grapevine yellows disease Flavescence dorée (Council Directive 2000/29/EC; EPPO/CABI 2003). Of these only *X. fastidiosa* that causes Pierce's disease is not yet present in the European area.

Detection techniques depend on the biology of the specific bacterium. A detection of bacteria that readily grow on artificial media is commonly performed through their isolation in pure culture and further identification (Lelliott and Stead 1987); however this is not an easy task with grapevine bacterial pathogens. *X. fastidiosa* and slow growing *X. ampelinus* can be easily overgrown by other bacteria and are thus difficult to isolate even when present in high concentrations. On the other hand, Phytoplasmas still resist attempts to be cultured independently from the plants. Apart from that, methods based on culturing are usually time consuming, giving the results within one or several weeks; thus, they are not appropriate as routine detection methods on large number of samples. In addition, they tend not to detect stressed, injured or viable but non-culturable bacteria (VBNC) that do not divide sufficiently in growing cultures and therefore do not produce visible colonies on solid media (López et al. 2006). For overcoming the mentioned detection limitations, serological techniques (e.g. indirect immunofluorescence assay-IIF and enzyme-linked immunosorbent assay-ELISA) and the PCR-based methods have been applied. While serological methods often lack the required sensitivity (typically around 10^5 cells/ml), PCR has technically advanced into the real-time PCR (from now on referred to as qPCR) in the past 10 years. The latter has proven to be robust, reliable, specific, sensitive, time and cost-saving, especially when coupled with automatic sample processing and nucleic acid isolation.

The PCR-based methods have been widely accepted to provide evidence for the presence/absence of pathogens and are an important supplementary tool for better diagnosis. In order to introduce them into different testing schemes, they have to be compared and validated against traditional methods. Exceptions are non-culturable phytopathogic bacteria, such as Phytoplasmas, where PCR-based methods are indispensible (López et al. 2006).

15.2 Real-Time PCR

In the qPCR, the amount of amplified product is measured at each cycle throughout the PCR reaction. There are several qPCR chemistries available. Each of them based on a slightly different technological solution ensuring the same common principle: measurement of the level of fluorescence, emitted by fluorescent reporter molecules, which is directly related to the amount of amplified target (compared in Buh Gašparič et al. 2008). The most commonly used chemistries are SYBR Green® and TaqMan® (Bustin 2005).

Main reasons why qPCR has become a popular technique in nucleic acid detection and quantification both for research and diagnostic purposes are: (i) the combination of DNA amplification and detection in a homogeneous assay obviates the need for post-PCR processing (e.g. electrophoresis); (ii) a wide dynamic range allows straightforward comparison between very low and very high abundance of target; and (iii) real-time PCR assays can be used in both quantitative and qualitative ways (Bustin and Mueller 2005). Another important application of qPCR is reverse transcription (RT) coupled real-time PCR (qRT-PCR) that enables detection and quantification of RNA molecules and can be used for detection of RNA viruses and disease specific markers (i.e. gene expression markers) (Bustin et al. 2005). In recent years the number of phytopathogen detection tests based on qPCR has been increasing very fast. In our laboratory alone, we have introduced several qPCR-based assays for detection of various plant pathogens, in addition to the ones presented here (Boben et al. 2007, Gutiérrez-Aguirre et al. 2009, Kogovšek et al. 2008).

15.3 General Guidelines for Introduction of a New Real-Time PCR Diagnostic Test

The process of introducing a new qPCR-based detection system into laboratory for routine detection of plant pathogens consists of several stages (Fig. 15.1).

Steps following an assay design (Fig. 15.1) that are chosen according to the main standard used for testing and calibration laboratories (ISO 17025), are already part of the validation procedure. Such a method is considered to be "in house validated". If desired or otherwise required, the assay validation can be extended to inter-laboratory comparisons (e.g. ring tests).

The whole process requires expertise on target pathogen biology, qPCR methodology, detection/diagnostics procedures and quality control. Currently, there are no general guidelines or standard operating procedures (SOPs) for introducing a new qPCR-based detection system for any pathogen into routine laboratory detection. However, there are general recommendations for some steps (López et al. 2006), guidelines for publication of newly developed qPCR assays (Bustin et al. 2009) and efforts of European and Mediterranean Plant Protection Organisation (EPPO)

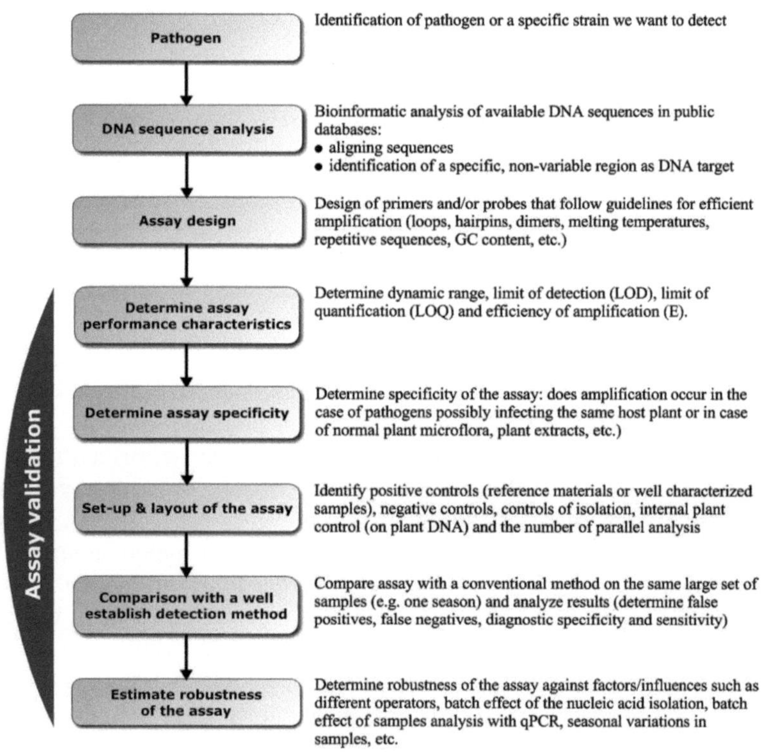

Fig. 15.1 qPCR assay implementation workflow

to extend general guidelines on validation of methods to qPCR assays in order to formulate a common strategy of development of new qPCR assays (Camloh et al. 2008, Žel et al. 2008).

When the decision is made that qPCR will be introduced as the method of choice for the detection of a certain pathogen, a few possible scenarios exist:

- qPCR detection system needs to be set-up from the very beginning (no adequate existing qPCR detection systems)
- qPCR detection system is already published but has to be complemented and implemented into the laboratory
- a known/published qPCR detection system has to be implemented into the laboratory.

15.4 Development and Implementation of New Real-Time PCR Assays for Plant Pathogenic Bacteria

Four qPCR-based detection systems for plant pathogenic bacteria that were developed in the laboratory of the National Institute of Biology, Slovenia will be demonstrated and discussed together with results from the implementation of the assays. Each system is different in terms of the level of implementation into routine diagnostics (from fully to partially implemented systems) and in terms of the target organism (i.e. non-culturable vs. culturable and slow growing bacteria).

15.4.1 Grapevine Yellows Phytoplasma and X. ampelinus: Designing a New Detection System

15.4.1.1 Background

Phytoplasmas are plant pathogenic bacteria belonging to the class Mollicutes (Bertaccini 2007, Hogenhout and Loria 2008). They are cell wall-free, and both their cell size (0.1–0.8 μm in diameter) and genome size (0.5–1.3 Mbp) are the smallest among bacteria. They are transmitted by propagation of infected plant material or from plant to plant by sap-feeding insect vectors and they multiply within the cytoplasm of both insects and plants. In plants, they are exclusively found in nutrient-rich phloem tissues, where they have been documented by electron microscopy in sieve elements, companion cells and phloem parenchyma cells (Christensen et al. 2005).

Grapevine yellows (GYs) are Phytoplasma-associated diseases which were identified in the majority of grapevine growing countries worldwide. Several molecularly distinct groups of Phytoplasmas which cause the GY have been identified, among them Phytoplasmas associated with Bois noir (BN Phytoplasma, stolbur group, 16SrXII-A) and Flavescence dorée (FD Phytoplasma, "*Candidatus* Phytoplasma *vitis*", elm yellows group, 16SrV) being the main cause of GY in Europe. Typical symptoms of the GYs are leaf curling and discoloration of leaf veins and laminas, interveinal yellowing or reddening, uneven or total lack of cane lignification, flower abortion and berry withering. Leaves on the affected shoots often have a hard, brittle texture. Eventually, a severe decline in plant vitality and the potential death of sensitive grapevine varieties occur, resulting in a serious decrease of crop quality and yield with great economic impact. FD is recognized by European Union (Council Directive 2000/29/EC) and the European and Mediterranean Plant Protection Organization (EPPO) as a quarantine harmful organism listed in A2 list (EPPO/CABI 2003).

Bacterial blight of grapevine is caused by a slow-growing bacterium *X. ampelinus* (Panagopoulos 1969) comb nov. (Willems et al. 1987). It can cause extensive damage and severe economic loss through the decay of newly developed vine shoots or complete plants.

15.4.1.2 Existing Detection Methods

Several methods for detection of non-culturable FD or BN Phytoplasmas have been developed: (i) early serological methods including ELISA and PCR-ELISA (Seddas et al. 1996, EPPO/CABI 2003). (ii) PCR-based methods including nested and multiplex-PCR that amplify either ribosomal or non-ribosomal phytoplasmic DNA (Daire et al. 1997, Clair et al. 2003, Lee et al. 1998). (iii) RFLP methods using different restriction enzymes on ribosomal PCR products (Lee et al. 1998, Marzachi et al. 2001). (iv) More recently, real-time PCR based assays for either universal Phytoplasma detection or detection of specific Phytoplasma groups have been developed (Christensen et al. 2004, Baric and Dalla-Via 2004, Torres et al. 2005, Bianco et al. 2004, Galetto et al. 2005, Angelini et al. 2007, Margaria et al. 2009, Hodgetts et al. 2009, Pelletier et al. 2009).

In contrast to Phytoplasma, *X. ampelinus* can be grown on artificial media but can be quickly overgrown by other bacteria due to its slow growth. Other methods for detection in plant extracts include serological tests that have not been widely accepted in routine testing due to their low sensitivity (Manceau et al. 2000) and PCR tests (Botha et al. 2001, Manceau et al. 2005).

15.4.1.3 Design and Implementation of New Detection System

Among available qPCR detection systems for FD and BN Phytoplasmas developed for implementation into routine diagnostics, a system developed by Hren et al. (2007) will be discussed and compared with classical nested PCR-based detection system. A new qPCR detection system based on highly specific TaqMan® chemistry with TaqMan® MGB™ probes was designed and validated according to the scheme described in Fig. 15.1. The detection system was designed as an assay triplet, consisting of three individual qPCR assays: FDgen for detection of FD Phytoplasma (genomic region), BNgen for detection of BN Phytoplasma (genomic region) and UniRNA for universal Phytoplasma detection (16S ribosomal DNA). Further details on the assay design and the evaluation of amplification parameters of individual assays can be found in Hren et al. (2007).

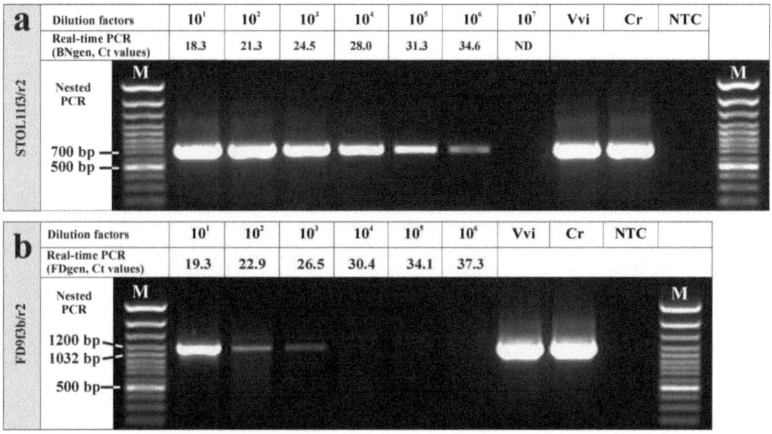

Fig. 15.2 Comparison of classical method (nested PCR) and newly developed real-time PCR on the same artificially prepared reference material (DNA from healthy grapevine was spiked with DNA from Phytoplasma infected grapevine and dilution series in range from 10 to 10^7-fold was prepared). (a.) shows comparison of detection of BN: qPCR results are presented as Ct values. Nested PCR results (primer pair STOL11f3/r2) are shown as photograph of gel-electrophoresis. (b.) shows comparison of detection of FD: q PCR results are presented as Ct values, nested PCR results (primer pair FD9f3b/r2) are shown as photograph of gel-electrophoresis. Vvi and Cr are positive controls for the PCR: Phytoplasma infected grapevine sample and *Catharanthus roseus* infected with Phytoplasma, respectively. NTC (no-template-control) is a negative real-time PCR control. ND = not detected. The figure was originally published in Hren et al. (2007) with kind permission from the Journal and British Society of Plant Pathology and Blackwell Publishing

The comparison of the newly developed qPCR detection system with classical PCR (Daire et al. 1997, Clair et al. 2003) used in routine diagnostics in the laboratory was done on two levels. At the first level, a detailed comparison on the same reference material gave an estimation of the limits of detection of both methods and direct visual comparison of both methods (Fig. 15.2). At the second, the detection systems were compared on a larger number of samples spanning throughout the sampling season and geographically covering all Slovenian wine growing regions (altogether 153 samples). This step allowed a comparison of the end results of both detection systems followed by analysis of parameters such as diagnostic specificity and sensitivity (Fig. 15.3). Such a comparison produces the most relevant data for a diagnostic laboratory.

The new qPCR detection system for FD and BN Phytoplasmas has proven to be more sensitive and specific than classical PCR-based detection method. In addition, qPCR results from the screening of a full growing season of samples together with amplification characteristics of each separate assay carried out on artificially prepared reference samples enabled us to set-up cut-off Ct values (cycle threshold values) for each assay, i.e. true limits of detection were set-up avoiding false amplification being considered as positive result.

By using new qPCR detection system, each grapevine sample is tested with three assays for the presence of Phytoplasmas (two Phytoplasma group specific

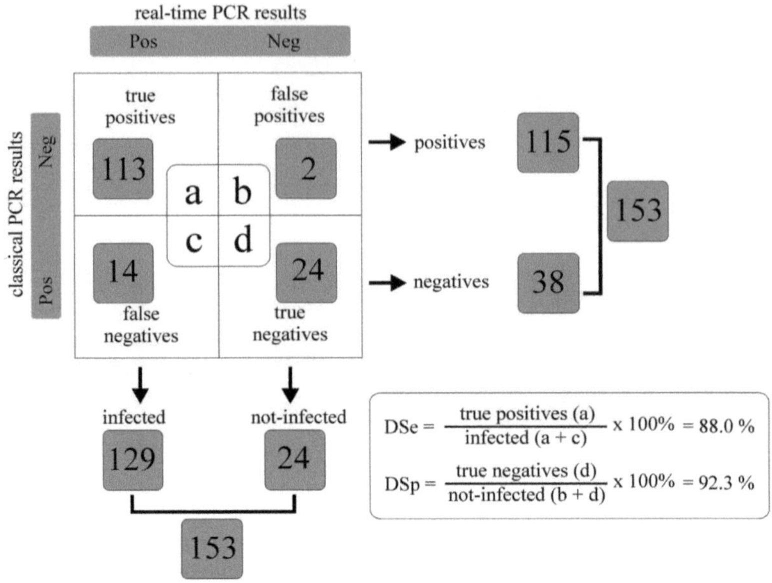

Fig. 15.3 Comparison of end-results of new qPCR detection system and a classical nested PCR detection system for detection of FD and BN on 153 field collected grapevine samples. Data in gray fields were used to calculate diagnostic sensitivity (DSe) and diagnostic specificity (DSp) of classical PCR detection method compared to qPCR detection method (qPCR was taken as the reference method). Conclusion can be made that the classical PCR method is less sensitive and specific compared to the new qPCR detection method

assays and a universal one) and one assay for detection of plant DNA (detection of cytochrome oxydase – COX or 18S rRNA gene). The latter one is co-isolated with phytoplasmic DNA during the DNA isolation process and serves as an endogenous quality control of DNA extraction. In case of insect samples, COX needs to be replaced by amplicon for detection of 18S rDNA. A sample is positive for the presence of FD/BN Phytoplasma, when the Ct values are in the safe range, i.e. above limit of detection for FDgen/BNgen amplicon and UniRNA and when we are able to detect plant DNA. Assay triplet therefore ensured a high level of reliability of results.

A qPCR system was developed for *X. ampelinus* using TaqMan® MGB™ probe targeting a previously reported sequence that was obtained through subtractive hybridization (Dreo et al. 2007). Tests of spiked grapevine samples of leaves, vines and field collected samples from an infected vineyard from three seasons showed that the assay was fast, reliable and *X. ampelinus* specific. On a low number of naturally infected samples, a correlation between successful isolation on media and results of qPCR was observed with bacterial isolation only successful from samples with Ct values below 26. The method is suitable as a sensitive and reliable screening test for this bacterium.

Apart from superior specificity compared to classical PCR, qPCR has several other advantages, such as reduction in duration of analysis, much better possibilities of automatisation and higher throughput, i.e. larger amount of samples can be processed at the same time. Because of its high sensitivity, faster nucleic acid extraction procedures can also be used.

15.4.2 Aster Yellows Phytoplasma

15.4.2.1 Background

Apart from the Phytoplasmas associated with Flavescence dorée and Bois noir, strains belonging to aster yellows Phytoplasma (16SrI, "*Candidatus* Phytoplasma *asteris*") have been sporadically found to infect grapevine in Europe. They also occur in mixed infections with Phytoplasmas from elm yellows group 16SrV (Alma et al. 1996). Aster yellows group is among the most diverse and geographically most widespread groups of Phytoplasmas, affecting mostly plants from family *Asteraceae*.

15.4.2.2 Existing Detection Methods

A qPCR method based on TaqMan® chemistry for detection of Phytoplasmas from aster yellows group was published by Angelini et al. (2007).

15.4.2.3 Implementation of the Detection System

To increase the sensitivity of the assay, a novel qPCR was designed to target the same gene for 16S rRNA as in Angelini et al. (2007). TaqMan® MGB™ probe was used and the two assays compared.

The guidelines summarized in Fig. 15.1 were followed. In order to determine the most appropriate DNA region where an amplicon (primers and MGB™ probe) for specific detection of aster yellows group Phytoplasma could be designed, the nucleotide sequences of 16S rDNA gene from public accessible GeneBank database for "*Ca.* Phytoplasma *asteris*", BN Phytoplasma, and FD Phytoplasma were obtained and aligned (Table 15.1). The specificity of amplicon was analyzed *in silico* by BLAST algorithm.

Table 15.1 Nucleotide sequence of the designed primers and probe of newly developed qPCR assay (AYnib)

Name		Nucleotide sequence (5′ – 3′)
AYnibF	Forward primer	GGGTTAAGTCCCGCAACGA
AYnibR	Reverse primer	TCTTGCTAAAGTCCCCACCATTAC
AYnibS	Probe	FAM-CAACCCTTATTGTTAGTTRCCAG-MGB[a]

[a]On the 5′ of the probe is a reporter FAM (6-carboxifluoresceine), on the 3′ is a non-fluorescent quencher MGB (minor groove binder). Note that AYnibS probe contains a degenerated nucleotide in order to amplify broader range of Phytoplasma strains from AY group.

For further specificity tests of the designed detection system, several different DNA isolates of Phytoplasmas, samples from healthy field collected grapevine, and isolates of bacteria that are present as normal microflora on grapevine were tested (Table 15.2).

In parallel to specificity tests, comparisons of the newly developed qPCR assay (from now on referred to as AYnib) with the assay introduced by Angelini et al. (2007) (from now on referred to as AYan) were carried out on the same samples.

Both detection systems distinguished Aster yellows Phytoplasma from those belonging to the groups 16SrV and 16SrXII that infect grapevine. They also have similar dynamic range. However, there was a difference in specificities of both detection systems (see Table 15.2). Cross-reactivity of AYan was observed with several epiphytic bacterial isolates from grapevine, which can give rise to false positive results in diagnostics of grapevine samples. On the other hand AYnib amplified DNA isolates from the 16SrII, 16SrX, and 16SrVI groups of Phytoplasma. Since these are not associated with grapevine they would not likely affect the detection of Phytoplasma in grapevine samples. However, for detection on related hosts AYan might be preferential.

In conclusion, it seems that combination of both systems can give most reliable results in determination of aster yellows Phytoplasma in grapevine.

15.4.3 Implementing an Existing Real-Time PCR Detection System: Case X. fastidiosa

15.4.3.1 Background

X. fastidiosa (Wells et al. 1987) is a small, Gram negative bacterium that causes Pierce's disease (PD) on grape. In many areas in the United States, the disease is endemic and a limiting factor for grapevine growth (Hopkins and Purcell 2002). *X. fastidiosa* is limited to the xylem of the infected host and is spread by several xylem sap feeding insect vectors (Purcell and Hopkins 1996). The most characteristic symptoms of primary infection on grapevine are leaf scorch and sudden dying of a part of a green leaf. As disease progresses, the desiccation spreads and the whole leaf may shrivel and drop, leaving only the petiole attached. Infected plants rarely survive a year or two (Hopkins 1981, Hopkins and Purcell 2002, EPPO/OEPP 2004).

15.4.3.2 Existing Detection Methods

X. fastidiosa is a fastidious bacterium and colonies can be observed on the media only after 10–14 days (Campanharo et al. 2003). The isolation of the pathogen is difficult even from symptomatic samples.

X. fastidiosa can also be observed in vessels, in cross-sections of petioles and by light and electron microscopy (French et al. 1978, Carbajal et al. 2004). Due to lower

Table 15.2 Specificity tests for newly developed aster yellows Phytoplasma qPCR assay and comparison with assay published in Angelini et al. (2007)

Specificity tests		Origin of DNA No. of samples	Amplification (No. of positive samples/No. of tested samples) AYnib[a]	AYan[b]
DNA of Phytoplasmas that infect grapevine	Aster yellows (16SrI)	*C. roseus* (4)	4/4	4/4
		E. purpurea (1)[c]	1/1	1/1
	Elm yellows (16SrV)	*V. vinifera* (3)	0/3	0/3
		C. roseus (1)	0/1	0/1
	Stolbur (16SrXII)	*V. vinifera* (22)	0/22	0/22
		C. roseus (1)	0/1	0/1
DNA of Phytoplasmas that do not infect grapevine	Peanut WB (16SrII)	*C. roseus* (6)[d]	4/6	0/6
	Coconut lethal yellows (16SrIV)	*C. roseus* (1)[d]	0/1	0/1
	Clover proliferation (16SrVI)	*C. roseus* (1)[d]	1/1	0/1
	Apple proliferation (16SrX):			
	–AP	Apples (3)	1/4	0/4
		C. rosesus (2)	0/2	0/2
	–ESFY	Stone fruits (3)	0/4	0/4
		C. roseus (1)	0/1	0/1
	–PD	pears (3)	2/4	0/4
		C. roseus (1)[d]	0/1	0/1
	Rice yellow dwarf (16SrXI)			
DNA isolates of unknown Phytoplasma type		*E. purpurea* (1)	0/1	0/1
Bacterial isolates from plant extracts of various grapevine cultivars		Various bacteria (41)	0/41	3/41
DNA isolates of healthy grapevine samples		*V. vinifera* (19)	0/19	0/19

[a]AYnib – newly developed assay for AY Phytoplasma.
[b]AYan – assay published by Angelini et al. (2007).
[c]Sample provided by Sebastjan Radišek, PhD, Slovenia.
[d]Samples provided by Matt Dickinson, PhD, UK.

sensitivity of both methods low concentrations of bacteria in latent infected samples cannot be detected. While ELISA has been useful for detecting the pathogen in symptomatic samples late in the season (Smart et al. 1998), it also lacks the sensitivity needed to detect latent infections early in the season (Schaad et al. 2002).

With the development of classical PCR-based techniques (Minsavage et al. 1994, Smart et al. 1998), sensitivity has improved significantly. Also, the time needed for diagnosis is shorter compared to isolation and pathogenicity test, but tests are still time-consuming (Schaad et al. 2002). The need for sensitive and rapid detection method for diagnosis of Pierce's disease has led to development of qPCR procedure for on-site diagnosis of this bacterium. The system enables rapid detection of *X. fastidiosa* in asymptomatic plant material and has been proven highly sensitive in detecting latent infections (Schaad et al. 2002, Francis et al. 2006). qPCR protocols for detection of *X. fastidiosa* have been described by Schaad et al. (2002) and Francis et al. (2006), both on the portable qPCR cycler apparatus (Smart Cycler TD system, Cepheid).

15.4.3.3 Implementation of the Detection System

Since neither *X. fastidiosa* nor insect vectors are known to be present in European area, they are listed on I.A.I quarantine list of harmful organisms (Council Directive 2000/29/EC). To prevent potential spread of the disease it is very important to have a system of diagnostic tests available for rapid and reliable detection of this pathogen. Therefore, we implemented a set of previously described diagnostic tests (Schaad et al. 2002) in our laboratory to be able to monitor plant samples for *X. fastidiosa* presence. qPCR described by Schaad et al. (2002) was selected for implementation based on preliminary tests, which showed slightly higher sensitivity compared to the one developed by Francis et al. (2006). Results were compared to ELISA test (Agdia, USA).

X. fastidiosa strains causing Pierce's disease and periwinkle wilt were cultured in liquid and on solid medium and served as positive control for further tests. Grapevine samples were collected from five Slovenian vineyards. Extracts were prepared from leaf nodes and petioles. In addition xylem fluid was collected from shoots by applying negative pressure to one end of the shoot. DNA was isolated using DNA-extraction method based on magnetic beads (QuickPick^{TM}) (Pirc et al. 2009). qPCR was carried out in a portable system Smart Cycler with ITS primers and probe and run-program as published by Schaad et al. (2002).

Sensitivities of the qPCR assay and ELISA assay (DAS ELISA protocol, Agdia) in terms of limit of detection (LOD) were estimated on the same set of 10-fold serial dilutions of *X. fastidiosa* pure culture cells, ranging from 10^8 to 10^1 cells per ml (Fig. 15.4, Table 15.3). In the case of qPCR DNA was isolated from each serial dilution sample prior to quantification. The same set of dilutions was used to determine performance characteristics of qPCR assay (e.g. dynamic range and efficiency of amplification).

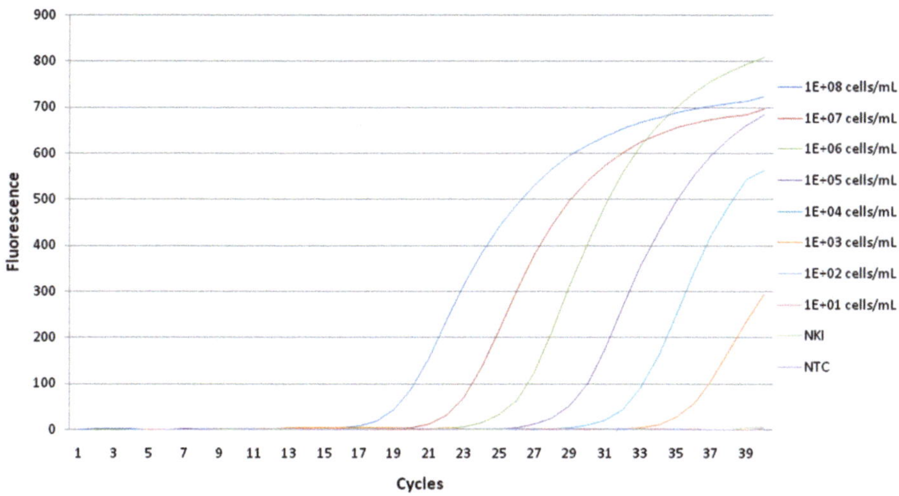

Fig. 15.4 Amplification of *Xylella fastidiosa* DNA from pure culture diluted 10-fold serially from 10^8 to 10 cells/mL. The limit of detection was estimated to 10^3 cell/mL or 2 cells per reaction. The left axis is the change in fluorescence and the bottom axis is the cycle number

Table 15.3 Comparison of ELISA and real-time PCR sensitivity

X. fastidiosa concentration (cells/ml)	qPCR result	ELISA result
10^8	+	NA
10^7	+	NA
10^6	+	+
10^5	+	+
10^4	+	–
10^3	+	–
10^2	–	–
10^1	–	–

LOD of qPCR assay was determined to be 10^3 cell per ml or 2 cells per reaction (Fig. 15.4) compared to LOD of 10^5 cells per ml in ELISA (Table 15.3). The determined sensitivities of ELISA and qPCR were similar to those reported by Schaad et al. (2002).

In the next step specificity of qPCR assay was estimated by analyzing extracts from healthy grapevine, bacterial isolates of normal grapevine microflora (bacteria isolated from healthy grapevine samples), DNA from Phytoplasma infected *C. roseus* plants and pure cultures of bacteria that can be potentially present in the grapevine samples. The same samples were also tested with ELISA. No cross reactivity was observed with either test (Table 15.4). In addition, plants from five Slovenian vineyards were analyzed using ITS primers and probe and the results were negative as expected (Table 15.4).

Table 15.4 Specificity of real-time PCR analysis of vine samples from five Slovenian vineyards (right) and cross reactivity testing of ELISA and ITS primers and probe (left)

Cross-reactivity sample ID	qPCR result	ELISA result	field sample ID	qPCR result
V. vinifera extract	–	–	vineyard 1 (5 samples)	–
C. roseus extract	–	–	cv. Tokaj	
N. tabaccum extract	–	–		
A.vitis	–	–	vineyard 2 (3 samples)	–
A. rhizogenes	–	–	cv. Tokaj	
A. tumefacines	–	–		
Xylophilus ampelinus	–	–	vineyard 3 (1 samples)	–
Flavescence dorée (FD)	–	–	cv. Chardonnay	
Aster yellows (AY)	–	–		
Bois noir (BN)	–	–	vineyard 4 (1 samples)	–
vine microflora 1	–	–	cv. Šentlovrenka	
vine microflora 2	–	–		
vine microflora 3	–	–	vineyard 5 (2 samples)	–
vine microflora 4	–	–	cv. Kerner	
vine microflora 5	–	–		

We have successfully implemented a qPCR assay for detecting *X. fastidiosa* in grapevine samples in our laboratory. As expected, we did not detect *X. fastidiosa* in samples from Slovenian vineyards with real-time PCR, but the monitoring of disease symptoms and insect vectors is still of great importance.

15.5 Conclusions

Developing new qPCR detection methods requires a systematic approach from the very beginning. Even when an assay is already available from scientific papers it might not yet be ready for direct use for detection purposes because (i.) most of the published methods were not evaluated in ring trials and therefore they might not perform similarly in different laboratories; (ii.) the method might not be assessed entirely for the specificity and cross reactivity with similar strains of bacteria, with plant material or nonpathogenic plant microflora from the geographical region where the test is going to be used; (iii.) it was not applied for screening of larger number of samples; (iv.) its diagnostic specificity and sensitivity was not compared to a conventional ("gold standard") detection method on a larger number of diagnostic samples. The last two points are closely linked to the first step of assay design – search and analysis of nucleic acid sequences for the specific pathogen and related organisms (e.g. from the same genus). In case where many sequences of a target genomic fragment are available (or maybe even complete genomes), the *in silico* analysis alone will result in the most appropriate target region for primer and probe design. In this way, high specificity of the assay is expected on real samples. However, in case of poorly characterized taxons (which is mostly the case for plant pathogens and their relatives) in terms of availability of genome sequences, only thorough screening of a large number of "real" diagnostic samples and specificity tests on as many isolates as possible (e.g. bacteria that might be present in the natural microflora) will ensure the specificity of the assay.

In contrast to conventional detection methods such as conventional PCR or PCR followed by RFLP (restriction fragment length polymorphism), where we have only two possibilities of result: positive (specific band or pattern on the gel) or negative (absence of specific band or incorrect pattern on the gel), qPCR produces numerical data (Ct values) which can be misleading and need to be thoroughly evaluated in the process of assay implementation before the numbers can be taken for granted. With qPCR, we can quickly step into the zone of false positive or false negative results. For example, false negative results can be obtained because of the inhibition of amplification due to the presence of various inhibitors in samples. qPCR amplicons are much shorter than classical PCR amplicons (as short as 70 base pairs compared to a few hounded base pairs in case of PCR) that is why they are theoretically less sensitive to inhibition of amplification and to DNA/RNA degradation that might occur in sample material. Different amplicons may also have different performance characteristics, such as sensitivity to inhibitors (Cankar et al. 2006). For this reason it is wise not to rely on a single target for detection of a specific pathogen, but to include at least two assays that detect different regions of DNA of the same pathogen, e.g. an amplicon specific for that pathogen and an amplicon that detects a broader group of bacteria (e.g. a combination of FD specific and universal Phytoplasma amplicons, presented in Section 15.4.1).

It is very important that detailed analysis of results is assessed in few seasons from the assay implementation on, so that the robustness of the method and any seasonal variation that might occur due to different eco-physiological conditions is analyzed.

When a qPCR assay is thoroughly assessed, laboratory can thus increase the throughput of samples and reduce the time to produce reliable results.

Acknowledgments This work was partially supported by the Phytosanitary Administration and Phytosanitary Inspectorate of the Republic of Slovenia, Slovenian Ministry of Education, Science and Sport and Ministry of Agriculture, Forestry and Food. We thank Norman W. Schaad (USDA) and Igor Zidarič (Agricultural Institute of Slovenia) for bacterial strains, Gabrijel Seljak, Msc. for field work and Magda Tušek Žnidarič and Nataša Mehle for help with ELISA.

References

Alma A, Davis RE, Vibio M, Danielli A, Bosco D, Arzone A, Bertaccini A (1996) Mixed infection of grapevines in northern Italy by Phytoplasmas including 16S rRNA RFLP subgroup 16SrI-B strains previously unreported in this host. Plant Dis 80:418–421

Angelini E, Bianchi GL, Filippin L, Morassutti C, Borgo M (2007) A new TaqMan method for the identification of Phytoplasmas associated with grapevine yellows by real-time PCR assay. J Microbiol Methods 68:613–622

Baric S, Dalla-Via J (2004). A new approach to apple proliferation detection: a highly sensitive real-time PCR assay. J Microbiol Methods 57:135–145

Bertaccini A, 2007. Phytoplasmas: diversity, taxonomy, and epidemiology. Front Biosci 12: 673–689

Bianco PA, Casati P, Marziliano N (2004) Detection of Phytoplasmas associated with grapevine flavescence dorée disease using real-time PCR. J Plant Pathol 86, 257–61

Boben J, Kramberger P, Petrovič N, Cankar K, Peterka M, Štrancar A. Ravnikar M (2007) Detection and quantification of Tomato mosaic virus in irrigation waters. Eur J Plant Pathol 118:59–71

Botha WJ, Serfontein S, Greyling MM, Berger DK (2001) Detection of *Xylophilus ampelinus* in grapevine cuttings using a nested polymerase chain reaction. Plant Pathol 50, 515–526

Buh Gašparič M, Cankar K, Žel J, Gruden K, 2008. Comparison of different real-time PCR chemistries and their suitability for detection and quantification of genetically modified organisms. BMC Biotechnol. doi:10.1186/1472-6750-8-26

Bustin SA (2005) Real-time, fluorescence-based quantitative PCR: a snapshot of current procedures and preferences. Expert Rev Mol Diagn 5:493–498

Bustin SA, Benes V, Garson JA, Hellemans J, Huggett J, Kubista M, Mueller R, Nolan T, Pfaffl MW, Shipley GL, Vandesompele J, Wittwer CT (2009) The MIQE guidelines: minimum information for publication of quantitative real-time PCR experiments. Clin Chem 55:611–622

Bustin SA, Benes V, Nolan T, Pfaffl MW (2005) Quantitative real-time RT-PCR-a perspective. J Mol Endocrinol 34:597–601

Bustin SA, Mueller R, 2005. Real-time reverse transcription PCR (qRT-PCR) and its potential use in clinical diagnosis. Clin Sci 109:365–379

Camloh M, Dreo T, Žel J, Ravnikar M (2008) The flexible scope of accreditation in GMO testing and its applicability to plant pathogen diagnostics. EPPO Bulletin 38:178–184

Campanharo JC, Lemos MVF, Lemos E (2003). Growth optimization procedures for the phytopathogen *Xylella fastidiosa*. Cur Microbiol 46:99–102

Cankar K, Štebih D, Dreo T, Žel J, Gruden K (2006). Critical points of DNA quantification by real-time PCR – effects of DNA extraction method and sample matrix on quantification of genetically modified organisms. BMC Biotechnol. doi:10.1186/1472-6750-6-37

Carbajal D, Morano KA, Morano LD (2004). Indirect immunofluorescence microscopy for direct detection of *Xylella fastidiosa* in xylem sap. Cur Microbiol 49:372–375

Christensen NM, Axelsen KB, Nicolaisen M, Schulz A (2005). Phytoplasmas and their interactions with hosts. Trends Plant Sci 10:526–535

Christensen NM, Nicolaisen M, Hansen M, Schulz A (2004). Distribution of Phytoplasmas in infected plants as revealed by real-time PCR and bioimaging. Mol Plant Microbe Interact 17:1175–1184

Clair D, Larrue J, Aubert G, Gillet J, Cloquemin G, Boudon-Padieu E (2003). A multiplex nested-PCR assay for sensitive and simultaneous detection and direct identification of Phytoplasma in the Elm yellows group and Stolbur group and its use in survey of grapevine yellows in France. Vitis 42:151–57

Council Directive 2000/29/EC of 8 May 2000 on protective measures against the introduction into the Community of organisms harmful to plants or plant products and against their spread within the Community (L169/1,10/07/2000)

Daire X, Clair D, Reinert W, Boudon-Padieu E (1997) Detection and differentiation of grapevine yellows Phytoplasmas belonging to the elm yellows group and to the stolbur subgroup by PCR amplification of non-ribosomal DNA. Eur J Plant Pathol 103:507–514

Dinnes J, Deeks J, Kirby J, Roderick P (2005) A methodological review of how heterogeneity has been examined in systematic reviews of diagnostic test accuracy. Health Technol Assess 9:1–128

Dreo T, Gruden K, Manceau C, Janse JD, Ravnikar M (2007) Development of a real-time PCR-based method for detection of *Xylophilus ampelinus*. Plant Pathol 56:9–16

EPPO/CABI (2003) Data sheets on quarantine pests – grapevine Flavescence dorée Phytoplasma. In: Smith IM, McNamara DG, Scott PR, Holderness M (eds) Quarantine Pests for Europe. CABI International, Wallingford, UK

EPPO Standards/Normes OEPP (2004) Diagnostic protocols for regulated pests, *Xylella fastidiosa*. Bulletin OEPP/EPPO: Bulletin 34:187–119

Francis M, Lin H, Rosa JC-L, Doddapaneni H, Civerolo EL (2006) Genome-based PCR primers for specific and sensitive detection and quantification of *Xylella fastidiosa*. Eur J Plant Pathol 115:203–213

French WJ, Stassi DL, Schaad NW (1978) The use of immunofluorescence for the identification of peach phony bacterium. Phtopathol 68:1106–1108

Galetto L, Bosco D, Marzachı C (2005) Universal and groups pecific real-time PCR diagnosis of flavescence doreé (16Sr-V), bois noir (16Sr-XII) and apple proliferation (16Sr-X) Phytoplasmas from field-collected plant hosts and insect vectors. Ann Appl Biol 147: 191–201

Gutiérrez-Aguirre I, Mehle N, Delić D, Gruden K, Mumford R, Ravnikar M, 2009. Real-time quantitative PCR based sensitive detection and genotype discrimination of Pepino mosaic virus. J Virol Methods 162:46–55

Hodgetts J, Boonham N, Mumford R, Dickinson M (2009) Panel of 23S rRNA gene-based real-time PCR assays for improved universal and group-specific detection of Phytoplasmas. Appl Environ Microbiol 75:2945–2950

Hogenhout SA, Loria R (2008) Virulence mechanisms of Gram-positive plant pathogenic bacteria. Cur Opin Plant Biol 11:449–456

Hopkins DL (1981) Seasonal concentration of Pierce′s disease bacterium in grapevine stems, petioles and leaf veins. Phytopathol 71:415–418

Hopkins DL, Purcell AH (2002) *Xylella fastidiosa*: cause of Pierce′s disease of grapevine and other emergent diseases. Plant Dis 86:1056–1066

Hren M, Boben J, Rotter A, Kralj P, Gruden K, Ravnikar M (2007) Real-time PCR detection systems for Flavescence dorée and Bois noir Phytoplasmas in grapevine: comparison with conventional PCR detection and application in diagnostics. Plant Pathol 56:785–796

Kogovšek P, Gow L, Pompe-Novak M, Gruden K, Foster GD, Boonham N, Ravnikar M (2008) Single-step RT real-time PCR for sensitive detection and discrimination of Potato virus Y isolates. J Virol Methods 149:1–11

Lee I–M, Gundersen–Rindal DE, Davis RE, Bartoszyk IM (1998). Revised classification scheme of Phytoplasmas based on RFLP analyses of 16S rRNA and ribosomal protein gene sequences. Int Journal Syst Evol Microbiol 48:1153–1169

Lelliott RA, Stead DE (1987) Diagnostic procedures for bacterial plant diseases. Methods for the diagnosis of bacterial diseases of plants. Blackwell Scientific Publications, Oxford, UK

López MM, Bertolini E, Marco-Noales E, Llop P,Cambra M (2006) Update on molecular tools for detection of plant pathogenic bacteria and viruses. In: Rao JR, Fleming CC, Moore JE, (eds) Molecular diagnostics: current technology and applications. Horizon Bioscience, Norfolk, UK

Marzachi C, Palermo S, Boarino A, Veratti F, D′Aquilio M, Loria A, Boccardo G (2001) Optimization of a one–step PCR assay for the diagnosis of Flavescence dorée–related Phytoplasmas in field–grown grapevines and vector populations. Vitis 40:213–217

Margaria P, Turina M, Palmano S (2009) Detection of Flavescence doreée and Bois noir Phytoplasmas, Grapevine leafroll associated virus-1 and -3 and Grapevine virus A from the same crude extract by reverse transcription-RealTime TaqMan assays. Plant Pathol 58: 838–845

Manceau C, Coutaud MG, Guyon R (2000) Assessment of subtractive hybridization to select pecies and subspecies specific DNA fragments for the identification of *Xylophilus ampelinus* by polymerase chain reaction (PCR). Eur J Plant Pathol 106: 243–254

Manceau C, Grall S, Brin C, Guillaumes J (2005) Bacterial extraction from grapevine and detection of *Xylophilus ampelinus* by a PCR and microwell plate detection system. Bulletin OEPP 34: 55–60

Minsavage GV, Thompson CM, Hopkins DL, Leite RMVBC, Stall RE (1994) Development of a polymerase chain reactionprotocol for detection of *Xylella fastidiosa* in plant tissue. Phytopathol 84:456–461

Pelletier C, Salar P, Gillet J, Cloquemin G, Very P, Foissac X, Malembic-Maher S (2009) Triplex real-time PCR assay for sensitive and simultaneous detection of grapevine Phytoplasmas of the 16SrV and 16SrXII-A groups with an endogenous analytical control. Vitis 48:87–95

Pirc M, Ravnikar M, Tomlinson J, Dreo T (2009) Improved fireblight diagnostics using quantitative real-time PCR detection of *Erwinia amylovora* chromosomal DNA. Plant Pathol 58:872–881

Purcell AH, Hopkins DL (1996) Fastidious xylem-limited bacterial plant pathogens. Ann Rev Phytopathol 34:131–151

Schaad NW, Opgenorth D, Gaush P (2002) Real-time polymerase chain reaction for one-hour on-site diagnosis of Pierce's disease of grape in early season asymptomatic vines. Phtopathol 92:721–728

Seddas A, Meignoz R, Daire X, Boudon-Padieu E (1996) Generation and characterization of monoclonal antibodies to Flavescence dorée Phytoplasma: serological relationships and differences in electroblot immunoassay profiles of Flavescence dorée and elm yellows Phytoplasmas. European Journal of Plant Pathology 102:757–764

Smart CD, Hendson M, Guilhaert MR, Saunders S, Friebertshouser G, Purcell A, Kirkpatrick BC (1998) Seasonal detection of *Xylella fastidiosa* in grapevines with culture, ELISA and PCR. Phytopathol 88, S83(suppl.)

Torres E, Bertolini E, Cambra M, Monton C, Martin MP (2005) Real-time PCR for simultaneous and quantitative detection of quarantine Phytoplasmas from apple proliferation (16SrX) group. Mol Cell Probes 19:334–340

Wells JM, Raju BC, Hung HY, Weisburg WG, Mandeico-Paul L, Brenner DJ (1987) *Xylella fastidiosa* gen. Nov., sp. Nov.: Gram-negative, xylem-limited, fastidious plant bacteria related to Xanthomonas spp. Int J Syst Evol Bacteriol 37:136–143

Willems A, Gillis M, Kersters K, Van Den Broecke L, de Ley J (1987) Transfer of *Xanthomonas ampelina* Panagopoulos 1969 to a new genus, *Xylophilus* gen. nov., as *Xylophilus ampelinus* (Panagopoulos 1969) comb. nov. Int J Syst Bacteriol 37: 422–430

Žel J, Mazzara M, Savini C, Cordeil S, Camloh M, Štebih D, Cankar K, Gruden K, Morisset D, Van den Eede G (2008) Method Validation and Quality Management in the Flexible Scope of Accreditation: An Example of Laboratories Testing for Genetically Modified Organisms. Food Anal Method 1:61–72

Chapter 16
Field Assessment and Diagnostic Methods for Detection of Grapevine Phytoplasmas

Elisa Angelini

Abstract Grapevine phytoplasma occurring in Europe belong to two different species and are associated with two grapevine yellows diseases: *Flavescence dorée* and *Bois noir*. The diagnosis of the phytoplasma infections is usually carried out by both visual observation of the symptoms and molecular assay. The symptom observation directly in the field is the prerequisite for a correct diagnosis; however, only the molecular test allows the exact identification of the phytoplasma species, which is fundamental for disease monitoring. Presently, in routine mass screening of grapevine samples it is advisable to use the real time PCR approach, which is fast, robust and sensitive, while nested PCR/RFLP or sequencing of the amplicons are required for a finer differentiation of the phytoplasma strains.

Contents

Abbreviations

GY Grapevine yellows
FD *Flavescence dorée*
BN *Bois noir*

E. Angelini (✉)
C.R.A. VIT Centro di Ricerca per la Viticoltura Conegliano (TV) 31015, Italy
e-mail: elisa.angelini@entecra.it

S. Delrot et al. (eds.), *Methodologies and Results in Grapevine Research*,
DOI 10.1007/978-90-481-9283-0_16,

16.1 Phytoplasmas of Grapevine

Phytoplasmas are plant pathogenic wall-less gram-positive bacteria living in the phloem sieve tubes of numerous plants. In grapevine, they are associated with grapevine yellows (GY) disease, which cause very serious damage in viticulture and wine industry, ranging from a lower yield of berries and wine to the death of vines. Phytoplasmas belong to the newly-established genus *Candidatus* Phytoplasma (*Ca.* Phytoplasma), which includes at least 27 different species (IRPCM 2004). Only some of them occur in grapevine: *Ca.* P. asteris, present in USA, Europe, Israel, Chile, Australia and South Africa; *Ca.* P. aurantifolia, which occurs only in Australia; *Ca.* P. pruni, which has been found in USA and Israel; *Ca.* P. vitis, present only in Europe; *Ca.* P. solani, which occurs in Europe and the Mediterranean basin; and *Ca.* P. australiense, present only in Australia. The diseases occurring in Europe are *Flavescence dorée* (FD) and *Bois noir* (BN), caused by the species *Ca.* P. vitis and solani, respectively. FD is strongly epidemic and a quarantine pest in the European Community; on the contrary, BN shows an endemic behaviour in Europe and usually infects only a low percentage of the plants in vineyard.

FD is caused by several isolates that belong to the 16SrV-C and -D phytoplasma phylogenetic subgroups, both included in the species *Ca.* P. vitis. The FD-D isolate (also called FD92 or FD88), the only FD isolate included in the 16SrV-D subgroup, is the most widespread; it has been detected in grapevine in Northern and Central Italy, in Southern and Central France, in the North-Eastern areas of Spain and in an area of Southern Switzerland (Daire et al. 1997, Martini et al. 1999, Angelini et al. 2001, Martini et al. 2002, Arnaud et al. 2007). FD70, FD2000 and the related isolates that belong to the 16SrV-C subgroup have only been described in France (Caudwell et al. 1970, Boudon-Padieu 2002, Arnaud et al. 2007). Another cluster in the 16SrV-C subgroup included the remaining isolates: FD-C and the closely related isolates, identified only in Italy and the Balkans (Martini et al. 1999, Credi et al. 2002, Martini et al. 2002, Maixner 2006, Botti and Bertaccini 2007, Kuzmanović et al. 2008).

The phytoplasma associated to BN in grapevine, which is also called stolbur phytoplasma, occurs in all Europe and the Mediterranean basin (Maixner 2006). It is less polymorphic than the FD phytoplasma. Indeed, only recently three different stolbur isolates have been found in grapevine: VKI, VKII and VKIII (Langer and Maixner 2004). Molecular studies on the stolbur genome are currently ongoing and have shown a much greater variability of this phytoplasma in other less conserved genes (Cimerman et al. 2009, Pacifico et al. 2009).

Diagnosis of GY is possible by means of several methods, but not all of them have the same features and reliability. The direct visual observation of symptoms is the basis for the diagnosis of GY and is possible only during the vegetative season. Transmission by grafting to indicators and observation at the electron microscopy are possible, but unsuitable for disease monitoring. In vitro culture is not possible, as phytoplasmas are mandatory parasites and can live only inside a host. Serological procedures have been raised to BN and FD associated phytoplasmas, but their sensitivity is quite low and the antisera are not available on the market (Seddas et al.

1996). DNA and RNA-based probes have been used in dot hybridisation for the detection of GY phytoplasmas (Daire et al. 1992, Davis et al. 1993). Finally, several biomolecular assays based on PCR have been established and have greatly improved GY diagnosis and phytoplasma detection. Indeed, currently the most useful tools for diagnosis of grapevine yellows are the direct observation of symptoms and the PCR-based molecular assays.

16.2 Visual Symptoms

All GY diseases show similar symptoms in *Vitis vinifera*, though different species of phytoplasmas can be associated with the disease. The main symptoms associated to GY are evident on leaves, bunches and canes. The symptoms on the leaves start to appear from June and become more evident as the vegetative season advances. The leaves infected by GY are crispy, brittle, downwards rolling and show reddening (orange to purple) in red varieties and yellowing (golden to chlorotic) in white varieties (Fig. 16.1a, b). Discoloration involves always also the main veins; sectorial discoloration of the blades occurs in some varieties (Fig. 16.1c). Moreover, when leaves fall down, the petioles can stand on the canes longer. Symptoms on canes involve the presence of short internodes, necrosis of terminal buds and black/brown pustules on the basis, evident since June; canes appear rubbery and weeping (Fig. 16.2a, b). A partial or total lack of lignification in canes of infected grapevines occurs in autumn. When the symptoms appear early in the season, the flowers wither and falling down; when the manifestation of the GY symptomatology occurs later in the season, the bunches wither and shrivel (Fig. 16.3a–c). The presence of characteristic symptoms on leaves, canes and flowers or bunches is strong evidence of phytoplasma infection, while the presence of symptoms only in one organ is easy to confuse with other biotic or abiotic stresses. In particular, the constriction of grapevine canes, caused by wires or insects (for instance, the buffalo treehopper *Stictocephala bisonia*) leads to symptoms identical to those caused by GY.

The identification of the phytoplasma species occurring in a symptomatic vine is not possible by visual observation. However, as far as Italy is concerned, it was

Fig. 16.1 GY symptoms on leaves on infected plants from different cultivars: (**a**) white variety; (**b**) red variety; (**c**) sectorial discoloration of the blades

Fig. 16.2 GY symptoms on canes: (**a**) weeping appearance of whole plant; (**b**) pustules on the basis of the cane

Fig. 16.3 GY symptoms on flowers and bunches in different phenological periods

observed that generally the grapevines with the earliest symptoms were mostly infected with FD phytoplasma; as the season advanced, a clear increase in BN infection was evidenced (Fig. 16.4, Angelini et al. 2006).

Environmental conditions being equal, symptom expression and sensitivity vary according to the virulence of the phytoplasma strain and the variety. Indeed, a differential sensitivity could be influenced by the genomic characteristics of the grapevine variety/biotype or by the feeding behaviour of the vectors. In general, however, the number of infected plants in vineyard and the gravity of symptoms depend on the pressure of the disease. American and other rootstock varieties usually are quite tolerant. It is possible to observe GY symptoms on several rootstocks (partial lack of lignification, downwards rolling of leaves), though they do not show all the typical GY symptoms (Fig. 16.5a, b) (Moutous 1977, Caudwell et al. 1994). Infection of rootstocks has been detected only for FD so far; however, it cannot be excluded for other phytoplasmas.

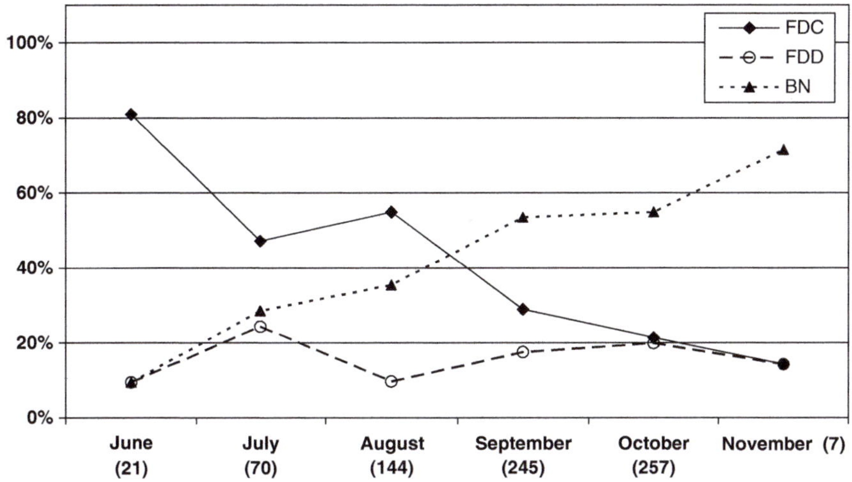

Fig. 16.4 Trends of occurrence, in percentage, of BN, FD-C and FD-D phytoplasmas from June to November. The results report data obtained from 1999 to 2004 in different Italian vineyards. The graph shows the number of samples tested per month (from Angelini et al. 2006)

Fig. 16.5 GY symptoms on rootstocks: (**a**) 420 A; (**b**) 140 Ruggeri

16.3 Molecular Methods

Molecular methods for the diagnosis of GY disease include DNA extraction and amplification and strain identification by RFLP. Movement of phytoplasmas is slow, their distribution is erratic and the phytoplasma titre in grapevine is very low. Therefore, all organs of the plants may be infected, however only symptomatic leaves and petioles are useful organs for reliable diagnosis. No consistent diagnosis is possible from canes and propagation wood material, even though they are infected and show GY symptoms. Symptomatic leaves to be analysed must be in good conditions and without evident necrosis or rot, which can contain compounds inhibiting molecular analyses. Detection is usually done from main veins and/or petioles of the symptomatic leaves.

Different methods for the extraction of phytoplasma DNA have been described. They essentially followed either of three different strategies. One group of techniques used an enrichment procedure to prepare a phytoplasma-rich pellet subsequently submitted to DNA extraction (Kirkpatrick et al. 1987, Ahrens and Seemüller 1992, Daire et al. 1992, Prince et al. 1993, Bertaccini et al. 1995). A second group of techniques used fast extraction of total DNA after grinding tissues in the presence of a strong detergent (Daire et al. 1997, Angelini et al. 2001). Further protocols are based on the use of commercial affinity columns for DNA extraction (Green et al. 1999, Bianco et al. 2004); the advantage of these protocols is the quality of the extracted DNA, but they are more expensive. Comparison between some of the listed extraction methods has been performed for phytoplasma DNA extracted from periwinkle and grapevine (Boudon Padieu et al. 2003).

Phytoplasma detection and characterization in grapevine are generally carried out by means of DNA amplification with polymerase chain reaction (PCR), followed by restriction fragments length polymorphism (RFLP) analysis of amplicons. However, the concentration of phytoplasmas can be very low in grapevine and the presence of compounds inhibiting PCR can be high (Daire et al. 1992, Gibb et al. 1999, Constable et al. 2003), therefore nested-PCR or real-time PCR are the methods of choice.

Numerous PCR-based procedures have been developed for detection of grapevine phytoplasmas, using a variety of primers. Universal and group-specific primers have been developed, which target conserved ribosomal phytoplasma DNA fragment or more variable non ribosomal genes. Generally, universal primers are used in direct PCR, while the nested PCR is performed using specific primers. However, several protocols involved both pairs of specific primers, while other protocols are based on two couples of universal primers; in the last case the identification of the specific phytoplasma is achieved using RFLP.

Many universal primers have been developed in the past, targeting the well conserved ribosomal RNA operon. Most of them are currently used for a generic detection of grapevine phytoplasmas and coupled in different manners: P1 (forward, Deng and Hiruki 1991), P7 (reverse, Smart et al. 1996), U1/U4 (Ahrens and Seemüller 1992), U5/U3 (Lorenz et al. 1995), 16R758F/16R1232R (Gibb et al. 1995), 16S723f/m23SR (Padovan et al. 1995), R16F1, R16F2 (Davis and Lee 1993), R16R0, R16R2 (Lee et al. 1993), R16Fn2 (Gundersen and Lee 1996), R16mF2, R16mR1 (Lee et al. 1998). Most of these primers target also some ribosomal bacterial sequences, which can sometimes interfere with the diagnosis from field collected grapevines.

Other primers have been designed on the polymorphic sites of the ribosomal genes and allow the specific amplification of one or more phytoplasma species. In particular, primer pairs R16(V)F1/R1 (Lee et al. 1994) and fAY/rEY (Marcone et al. 1996, Marzachì et al. 2001) target specifically *Ca.* P. vitis and related species belonging to the same phylogenetic group. Primer pairs R16(I)F1/R1 (Lee et al. 1994) and G35p/m (Davis et al. 1992) target only *Ca.* P. solani, *Ca.* P. australiense and *Ca.* P. asteris and related species belonging to the same phylogenetic groups; primer pair fStol/rStol target specifically *Ca.* P. solani, the agent of BN (Maixner et al. 1995).

Therefore, FD and BN associated phytoplasmas can be detected separately by using these specific primer pairs.

Some sets of primers have been designed on the ribosomal protein operon of phytoplasmas, containing the 3' end of *rpl22* gene and the *rps3* gene (Lee et al. 1998). In particular, primer pair rpF1/rpR1 allows the amplification of all phytoplasma DNA species, while primers rpVF1 and rpVF2 target specifically FD phytoplasma and related species.

Alternatively, other group-specific primer pairs have been designed to amplify PCR fragments of non ribosomal DNA of phytoplasmas. Primers specific for amplification of FD and related phytoplasmas target *secY* and *map* genes (Daire et al. 1997, Angelini et al. 2001, Clair et al. 2003, Arnaud et al. 2007). Primers specific for the detection of BN and related phytoplasmas have been designed in several genomic fragments of the genome, not all of them being yet assigned as a specific gene: the STOL4 fragment (Daire et al. 1997); the STOL11 fragment (Daire et al. 1997, Clair et al. 2003); a genomic fragment delimited by the primer pair M1/P8 (Marzachì et al. 2000); and the *tuf* gene (Schneider et al. 1997). A multiplex nested PCR assay for the simultaneous detection and identification of FD and BN associated phytoplasmas from grapevine, which revealed to be very useful for routine analyses, have been developed in France using four primer pairs targeting non-ribosomal regions (Clair et al. 2003).

All the primers listed above are reliable for the detection of phytoplasmas from grapevine. However, nested protocols have not been developed in all the phytoplasma genomic regions, especially the non-ribosomal ones; therefore, in some non-ribosomal fragments only direct PCR is currently possible. In these cases, it is worth to remember that the sensitivity of direct PCR is much lower than the nested approach, thus leading to false negative results. Comparison between some of the listed nested PCR protocols for detection of FD and BN associated phytoplasmas has been performed in periwinkle and grapevine (Boudon Padieu et al. 2003).

PCR protocols and reactions conditions are numerous and vary according to the primers. One of the most used PCR thermal protocols was developed by Schaff et al. (1992); although it is quite time-consuming, it proved to increase the sensitivity of the PCR assay.

RFLP digestion is required for the identification of PCR amplicons obtained using universal primers, but it is also advisable for confirmation of results obtained using group-specific primers. The RFLP patterns derived from the ribosomal gene region in all known phytoplasma groups are exhaustively described by Lee et al. (1998) and Wei et al. (2007). However, one of the most used PCR/RFLP protocol for the identification of phytoplasmas associated with FD and BN involves the use of 16R758F/m23SR ribosomal primer pair, followed by enzymatic digestion with Taq*I*. The RFLP profiles obtained with the last protocol allow an easy distinction of FD and BN phytoplasmas from each other and from phytoplasmas belonging to other phylogenetic groups; moreover, the two different phylogenetic subgroups of FD (FD-C and FD-D) show clear distinct RFLP profiles (Martini et al. 1999, Angelini et al. 2001, Martini et al. 2002). A finer distinction between FD strains can be reached with the digestion of the amplicons obtained from the *secY*, *map* and *rp* genes, using different restriction enzymes (Arnaud et al. 2007, Botti and Bertaccini 2007, Filippin et al. 2009). The three subgroups of phytoplasmas associated with BN, named VKI, VKII and VKIII, can be differentiated by Hpa*II* digestion of *tuf* gene amplicons (Langer and Maixner 2004).

A few real-time PCR protocols for diagnosis of grapevine phytoplasma have been developed so far. Two TaqMan systems using universal primers and probe targeting all plant phytoplasmas have been developed (Christensen et al. 2004, Hren et al. 2007). A very specific TaqMan assay for detecting FD-infected vines has been described, which involved a preliminary conventional PCR step before the real-time assay (Bianco et al. 2004). Two real-time PCR assays with SYBR® Green detection proved to work on BN- and FD-infected grapevines and on insect samples (Galetto et al. 2005). Two research groups designed different TaqMan primer/probe systems that are specific for FD and BN associated phytoplasmas (Angelini et al. 2007, Hren et al. 2007). A TaqMan assay targeting *Ca.* P asteris and another one targeting a grapevine gene, to be used as an internal control in real time amplification, have been also reported (Angelini et al. 2007). The most recent real time protocol published allows the distinction between the two BN phytoplasma subtypes VKI and VKII (Berger et al. 2009).

Both methods, nested-PCR coupled with RFLP and real time PCR, are very sensitive, specific and reliable for the diagnosis of GY phytoplasmas. The advantage of nested-PCR/RFLP is the possibility to distinguish between different strains of FD and BN phytoplasmas; however, it is a time-consuming analysis, involving many labour-intensive steps. Moreover, cross contamination and false positive results are a concrete risk, due to the long manipulation of the samples. The real time PCR procedure is faster, involves less manipulation of samples and is compatible with high throughput testing, as required by mass diagnostic screenings and quarantine disease monitoring. Therefore, it solves most of the disadvantages of conventional nested-PCR and can be a convenient alternative.

Preliminary promising results have been obtained using high throughput methods for the detection of phytoplasma DNA, though often these methods do not have the same consistence and sensitivity of real time PCR. A rapid protocol for extracting sap from symptomatic GY leaves and successive one-tube amplification by reverse-transcription PCR has demonstrated to be as sensitive as the conventional nested-PCR protocols (Margaria et al. 2007). LDR (Ligase Detection Reaction) technique associated with micro-hybridisation on universal arrays allowed to detect and distinguish between FD and BN phytoplasmas in grapevine (Frosini et al. 2002). A method using universal PCR followed by oligonucleotide microarray hybridization has been established for the diagnosis and identification of several species of phytoplasmas, included those infecting grapevine; however they have not been tried directly on grapevine so far (Nicolaisen and Bertaccini 2007). Finally, a nanobiotransducer carrying an oligonucleotide probe specific for FD phytoplasma produced encouraging results in the detection of FD-infected grapevine samples (Firrao et al. 2005).

16.4 Conclusions

The ability to perform a correct diagnosis of the phytoplasma infections in grapevine relies upon a good knowledge of the specific symptomatology in the field and a good experience on PCR detection. The visual observation of symptoms directly in the field is the prerequisite for a correct diagnosis; however, only the molecular test allows the exact identification of the phytoplasma species. The identification of FD or BN phytoplasma by means of biomolecular assay is fundamental for disease monitoring in order to direct agronomic and control strategies in vineyards, as the two diseases show different epidemiology. Both the diagnostic methods, visual observation and molecular assay, are therefore needed and must be integrated. In order to obtain reliable data, it is important to reach a good agreement between visual observation of symptoms and results of the laboratory assay (at least 90%) and to increase the personal experience with a continuous comparison between field observations and laboratory analysis results. At present, it is advisable to use the real time PCR approach, which is fast, robust and sensitive, for routine mass screening of grapevine samples, while nested PCR/RFLP or sequencing of the amplicons is required for a finer differentiation of the phytoplasma strains.

Acknowledgments All photos are courtesy of Dr. Michele Borgo, CRA-VIT, Conegliano (TV, Italy).

References

Ahrens U, Seemüller E (1992) Detection of DNA of plant pathogenic mycoplasmalike organisms by a polymerase chain reaction that amplifies a sequence of the 16S rRNA gene. Phytopathology 82:828–832

Angelini E, Clair D, Borgo M, Bertaccini A, Boudon-Padieu E (2001) *Flavescence dorée* in France and Italy – Occurrence of closely related phytoplasma isolates and their near relationships to Palatinate grapevine yellows and an alder yellows phytoplasma. Vitis 40:79–86

Angelini E, Filippin L, Michielini C, Bellotto D, Borgo M (2006) High occurrence of *Flavescence dorée* phytoplasma early in the season on grapevines infected with grapevine yellows. Vitis 45:151–152

Angelini E, Bianchi GL, Filippin L, Morassutti C, Borgo M (2007) A new TaqMan method for the identification of phytoplasmas associated with grapevine yellows by real-time PCR assay. J Microbiol Methods 68:613–622

Arnaud G, Malembic-Maher S, Salar P, Bonnet P, Maixner M, Marcone C, Boudon-Padieu E, Foissac X (2007) Multilocus sequence typing confirms the close genetic interrelatedness of three distinct *Flavescence dorée* phytoplasma strain clusters and group 16SrV phytoplasmas infecting grapevine and alder in Europe. Appl Environ Microbiol 73:4001–4010

Berger J, Dalla Via J, Baric S (2009) Development of a TaqMan allelic discrimination assay for the distinction of two major subtypes of the grapevine yellows phytoplasma *Bois noir*. Eur J Plant Pathol 124:521–526

Bertaccini A, Vibio M, Stefani E (1995) Detection and molecular characterization of phytoplasmas infecting grapevine in Liguria (Italy). Phytopath Medit 34:137–141

Bianco PA, Casati P, Marziliano N (2004) Detection of phytoplasmas associated with grapevine *Flavescence dorée* disease using real-time PCR. J Plant Pathol 86:259–264

Botti S, Bertaccini A (2007) Grapevine yellows in Northern Italy: molecular identification *of Flavescence dorée* phytoplasma strains and of *Bois noir* phytoplasmas. J Appl Microbiol 103:2325–2330

Boudon-Padieu E 2002. *Flavescence dorée* of the grapevine: knowledge and new developments in epidemiology, etiology and diagnosis. In: Atti Giornate Fitopatologiche. Baselga di Pinè, Italy, 1:15–34

Boudon-Padieu E, Béjat A, Clair D, Larrue J, Borgo M, Bertotto L, Angelini E (2003) Grapevine yellows: comparison of different procedures for DNA extraction and amplification with PCR for routine diagnosis of phytoplasmas in grapevine. Vitis 42:141–149

Caudwell A, Kuszala C, Bachelier JC, Larrue J (1970) Transmission de la Flavescence dorée de la vigne aux plantes herbacées par l'allongement du temps d'utilisation de la cicadelle *Scaphoideus littoralis* Ball et l'étude de sa survie sur un grand nombre d'espèces végétales. Ann Phytopathol 2:415–428

Caudwell A, Larrue J, Tassart V (1994). Caractère « porteur de la flavescence dorée » chez les vignes porte-greffes, en particulier le 3309 Couderc et le Fercal. Agronomie 2:83–94

Christensen NM, Nicolaisen M, Hansen M, Schulz A (2004) Distribution of phytoplasmas in infected plants as revealed by real-time PCR and bioimaging. Mol Plant Microbe Interact 17:1175–1184

Cimerman A, Pacifico D, Salar P, Marzachì C, Foissac X (2009) Striking diversity of *vmp1*, a variable gene encoding a putative membrane protein of the stolbur phytoplasma. Appl Environ Microbiol 75:2951–2957

Clair D, Larrue J, Aubert G, Gillet J, Cloquemin G, Boudon-Padieu E (2003) A multiplex nested-PCR assay for sensitive and simultaneous detection and direct identification of phytoplasma in the elm yellows group and stolbur group and its use in survey of grapevine yellows in France. Vitis 42:151–157

Constable FE, Gibb KS, Symons RH (2003) Seasonal distribution of phytoplasmas in Australian grapevines. Plant Pathol 52:267–276

Credi R, Terlizzi F, Stimilli G, Nardi S, Lagnese R (2002) Flavescenza dorata della vite nelle Marche. L'Inf Agrario 22:61–63

Daire X, Boudon-Padieu E, Berville A, Schneider B, Caudwell A (1992) Cloned DNA probes for detection of grapevine *Flavescence dorée* mycoplasma-like organism (MLO). Ann Appl Biol 121:95–103

Daire X, Clair D, Reinert W, Boudon-Padieu E (1997) Detection and differentiation of grapevine yellows phytoplasmas belonging to the elm yellows group and to the stolbur subgroup by PCR amplification of non-ribosomal DNA. Eur J Plant Pathol 103:507–514

Davis RE, Lee IM (1993) Cluster-specific polymerase chain reaction amplification of 16S rDNA sequences for detection and identification of mycoplasmalike organisms. Phytopathology 83:1008–1011

Davis RE, Dally EL, Bertaccini A, Credi R, Lee IM, Osler R, Carraro L, Barba M (1992) Cloned DNA probes for specific detection of Italian periwinkle virescence mycoplasmalike organism (MLO) and investigation of genetic relatedness with other MLOs. Phytopath Medit 31:5–12

Davis RE, Dally EL, Bertaccini A, Lee IM, Credi L, Osler R, Savino V, Carraro L, Di Terlizzi B, Barba M (1993) Restriction fragment length polymorphism analyses and dot hybridizations distinguish mycoplasmalike organisms associated with *Flavescence dorée* and southern European grapevine yellows disease in Italy. Phytopathology 83:772–776

Deng S, Hiruki C (1991) Amplification of 16S rRNA genes from culturable and nonculturable *Mollicutes*. J Microbiol Methods 14:53–61

Filippin L, Jović J, Cvrković T, Forte V, Clair D, Toševski I, Boudon-Padieu E, Borgo M, Angelini E (2009) Molecular characteristics of phytoplasmas associated with *Flavescence dorée* in clematis and grapevine and preliminary results on the role of *Dictyophara europaea* as a vector. Plant Pathol 58:826–837

Firrao G, Moretti M, Rosquete MR, Gobbi E, Locci R (2005) Nanobiotransducer for detecting *flavescence dorée* phytoplasma. J Plant Pathol 87:101–107

Frosini A, Casati P, Bianco PA, Bordoni R, Consolandi C, Castiglioni B, Mezzelani A, Rizzi E, Battaglia C, Belli G, Rossi Bernardi L, De Bellis G (2002). Ligase detection reaction and universal array as a tool to detect grapevine infecting phytoplasmas. Minerva Biotec 14: 265–267

Galetto L, Bosco D, Marzachí C (2005) Universal and group-specific real-time PCR diagnosis of *Flavescence dorée* (16Sr-V), *Bois noir* (16Sr-XII) and apple proliferation (16Sr-X) phytoplasmas from field-collected plant hosts and insect vectors. Ann Appl Biol 147:191–201

Gibb K, Padovan AC, Mogen BD (1995) Studies on sweet potato little-leaf phytoplasma detected in sweet potato and other plant species growing in northern Australia. Phytopathology 85:169–174

Gibb KS, Constable FE, Moran JR, Padovan AC (1999) Phytoplasmas in Australian grapevines – detection, differentiation and associated diseases. Vitis 38:107–114

Green MJ, Thompson DA, MacKenzie DJ (1999) Easy and efficient DNA extraction from woody plants for the detection of phytoplasmas by polymerase chain reaction. Plant Disease 83: 482–485

Gundersen DE, Lee IM (1996) Ultrasensitive detection of phytoplasmas by nested-PCR assays using two universal primer pairs. Phytopathol Medit 35:144–151

Hren M, Boben J, Rotter A, Kralj P, Gruden K, Ravnikar M (2007) Real-time PCR detection systems for *Flavescence dorée* and *Bois noir* phytoplasmas in grapevine: comparison with conventional PCR detection and application in diagnostics. Plant Pathol 56:785–796

IRPCM (2004). "*Candidatus* Phytoplasma", a taxon for the wall-less, non-helical prokaryotes that colonize plant phloem and insects. Int J System Evol Microbiol 54:1243–1255

Kirkpatrick BC, Stenger DC, Morris TJ, Purcell AH (1987) Cloning and detection of DNA from a nonculturable plant pathogenic mycoplasma-like organism. Science 238:197–200

Kuzmanović S, Martini M, Ermacora P, Ferrini F, Starović M, Tosić M, Carraro L, Osler R (2008) Incidence and molecular characterization of *Flavescence dorée* and stolbur phytoplasmas in grapevine cultivars from different viticultural areas of Serbia. Vitis 47:105–111

Langer M, Maixner M (2004) Molecular characterisation of grapevine yellows associated phytoplasmas of the stolbur-group based on RFLP-analysis of non-ribosomal DNA. Vitis 43:191–199

Lee IM, Hammond RW, Davis RE, Gundersen DE (1993) Universal amplification and analysis of pathogen 16S rDNA for classification and identification of mycoplasmalike organisms. Phytopathology 83:834–842

Lee IM, Gundersen DE, Hammond RW, Davis RE (1994) Use of mycoplasmalike organism (MLO) group-specific oligonucleotide primers for nested-PCR assays to detect mixed-MLO infections in a single host plant. Phytopathology 84:559–566

Lee IM, Gundersen–Rindal DE, Davis RE, Bartoszyk IM (1998) Revised classification scheme of phytoplasmas based on RFLP analyses of 16S rRNA and ribosomal protein gene sequences. Int J System Evol Microbiol 48:1153–1169

Lorenz KH, Schneider B, Ahrens U, Seemüller E (1995) Detection of the apple proliferation and pear decline phytoplasmas by PCR amplification of ribosomal and nonribosomal DNA. Phytopathology 85:771–776

Maixner M (2006) Grapevine yellows – current developments and unsolved questions. In: Extended Abstracts 15th Meeting ICVG. Stellenbosch, South Africa, addendum. http://www.icvg.ch/data/maixner.pdf

Maixner M, Ahrens U, Seemüller E (1995) Detection of the German grapevine yellows (Vergilbungskrankheit) MLO in grapevine, alternative hosts and a vector by a specific PCR procedure. Eur J Plant Pathol 101:241–250

Marcone C, Ragozzino A, Schneider B, Lauer U, Smart CD, Seemüller E (1996) Genetic characterization and classification of two phytoplasmas associated with spartium witches'-broom disease. Plant Dis 80:365–371

Margaria P, Rosa C, Marzachì C, Turina M, Palmano S (2007) Detection of *flavescence dorée* phytoplasma in grapevine by reverse-transcription PCR. Plant Dis 91:1496–1501

Martini M, Murari E, Mori N, Bertaccini A (1999) Identification and epidemic distribution of two *Flavescence dorée*-related phytoplasmas in Veneto (Italy). Plant Dis 83:925–930

Martini M, Botti S, Marcone C, Marzachì C, Casati P, Bianco PA, Benedetti R, Bertaccini A (2002) Genetic variability among *flavescence dorée* phytoplasmas from different origins in Italy and France. Mol Cell Probes 16:197–208

Marzachì C, Veratti F, d'Aquilio M, Vischi A, Conti M, Boccardo G (2000) Molecular hybridization and PCR amplification of non-ribosomal DNA to detect and differentiate stolbur phytoplasma isolates from Italy. J Plant Pathol 82:201–212

Marzachì C, Palermo S, Boarino A, Veratti F, d'Aquilio M, Loria A, Boccardo G (2001) Optimisation of a one-step PCR assay for the diagnosis of *Flavescence dorée*-related phytoplasmas in field-grown grapevines and vector populations. Vitis 40:213–217

Moutous G (1977) Définitions des symptômes de la "flavescence dorée" sur les variétés porte-greffes. Rev Zool Agr Pathol Végétale 76:90–98

Nicolaisen M, Bertaccini A (2007) An oligonucleotide microarray-based assay for identification of phtytoplasma 16S ribosomal groups. Plant Pathol 56:332–336

Pacifico D, Alma A, Bagnoli B, Foissac X, Pasquini G, Tessitori M, Marzachi C (2009) Characterization of *Bois noir* isolates by restriction fragment length polymorphism of a stolbur-specific putative membrane protein gene. Phytopathology 99:711–715

Padovan AC, Gibb KS, Bertaccini A, Vibio M, Bonfiglioli RE, Magarey PA, Sears BB (1995) Molecular detection of the Australian grapevine yellows phytoplasma and comparison with grapevine yellows phytoplasmas from Italy. Aust J Grape Wine Res 1:25–31

Prince JP, Davis RE, Wolf TK, Lee IM, Mogen BD, Dally EL, Bertaccini A, Credi R, Barba M (1993) Molecular detection of diverse mycoplasmalike organisms (MLOs) associated with grapevine yellows and their classification with aster yellows, X-disease, and elm yellows MLOs. Phytopathology 83:1130–1137

Schaff D, Lee IM, Davis RE (1992) Sensitive detection and identification of mycoplasma-like organisms in plants by polymerase chain reactions. Biochem Biophys Res Comm 186: 1503–1509

Schneider B, Gibb KS, Seemüller E (1997) Sequence and RFLP analysis of the elongation factor Tu gene used in differentiation and classification of phytoplasmas. Microbiology 143: 3381–3389

Seddas A, Meignoz R, Daire X, Boudon-Padieu E (1996) Generation and characterization of monoclonal antibodies to *Flavescence dorée* phytoplasma: serological relationships and differences in electroblot immunoassay profiles of *Flavescence dorée* and elm yellows phytoplasmas. Eur J Plant Pathol 102:757–764

Smart CD, Schneider B, Blomquist CL, Guerra LJ, Harrison NA, Ahrens U, Lorenz KH, Seemüller E, Kirkpatrick BC (1996) Phytoplasma-specific PCR primers based on sequences of the 16S–23S rRNA spacer region. Appl Environ Microbiol 62:2988–2993

Wei W, Davis RE, Lee IM, Zhao Y (2007) Computer-simulated RFLP analysis of 16S rRNA genes: identification of ten new phytoplasma groups. Int J System Evol Microbiol 57:1855–1867

Chapter 17
NICT: New Tools to Control Phytochemical Treatments and Traceability

Vincent de Rudnicki, Bernadette Ruelle, and Laurent Scheyer

Abstract In France the concept of precision viticulture is a topical subject in relation with the "Grenelle of environment" interministerial meeting and recommendation. Crop productivity must be increased, but in the same time a strong reduction of the environmental impact is needed. New technologies of Information and Communication (NTIC) provide potential responses to this challenge of quality. These technologies give solutions to manage pesticide applications thanks to field work assistance tools. They also allow continuous monitoring and recording of the field operations to implement traceability and automatic field logbooks. This paper presents the results of LIFE AWARE & TICSAD projects based on NTIC. It will deal with the possible tracks allowing a better management of news methods of treatment.

Keywords Winegrower precision viticulture · NICT · Phytochemical products · spraying · Embedded equipment · Traceability

Contents

V. de Rudnicki (✉)
Cemagref, UMR ITAP, Montpellier 34196, France
e-mail: vincent.derudnicki@cemagref.fr

New Information and communication technologies.

S. Delrot et al. (eds.), *Methodologies and Results in Grapevine Research*,
DOI 10.1007/978-90-481-9283-0_17,

Abbreviations

AWARE	European project: A Water Assessment to Respect the Environment
LIFE	European tool projects
MHYDAS	Modélisation Hydrologique Distribuée en AgroSystèmes
NDVI	Normalized Difference Vegetation Index
NICT	New technologies of communication and information
OPTIDOSE	Project aiming to reduce the quantity of pesticide for better practices
POD MILDIUM	Project aiming to reduce the number of treatment for better practices
TERRAVITIS	Rules and tools for sustainable winegrowing
TICSAD	French project: "Technologies de l'Information et de la Communication au Service d'une Agriculture Durable".

17.1 Introduction

Water pollution caused by pesticides is one of the biggest problems faced by aquatic ecosystems. To improve water quality, it is necessary to assess the impact of agricultural practices on the environment.

In France, studies conducted by "l'Institut Français de l'Environnement" revealed that both surface and ground waters exhibit significant contamination. In particular, 49% of surface water sampling locations were of average to poor quality and 27% of ground water sampling locations would need pesticide elimination treatment in order to provide acceptable drinking water.

The Water Framework Directive commits Member States to achieve "good" ecological water status for their water bodies by 2015. Pesticides are one of the causes of water deterioration.

In this context, Cemagref has been carrying out studies since 2000 to show how environmental pollution could be mitigated through (De Rudnicki et al. 2008): the optimisation of pesticide application techniques using NICT technologies, the implementation of inside plot of land operation traceability and through educational campaigns that aim to increase awareness of stakeholders in viticulture. The two last studies aiming to these objectives are the Life AWARE and the TICSAD projects.

17.1.1 The Project Life AWARE

One of the objectives of the Life AWARE (Sinfort 2007) project is to link the quantities of pesticide product spread during treatments with those found in surface water. Consequently an assessment of product distribution on the plants, soil and air during and after application is essential in order to establish the link between pesticide use and pesticide contamination of surface waters.

Vine sprayers are commonly operated with air assistance that improves pesticide penetration within the canopy, but on the other hand transports droplets to the air or onto the ground. Field measurements measuring ground and air spray losses for different spraying systems showed sizeable losses that varied according to the machine used (soil losses: 7–15%, air losses: 14–45% depending on vegetation growth stages).

Sprayers commonly used are often hydraulic or pneumatic machines that work along 2–4 rows. Different variables come into play during treatment: nozzle orientation, cleaning and maintenance of sprayers between treatments, product flow rate settings (outlet rates according to volume/Ha sought and dosage/Ha) to limit treatment to targeted vegetation (Fig. 17.1).

Monitoring of farmers' activities showed that with the type of existing machinery, it is very hard and tedious to adjust sprayers for each treatment and even more so

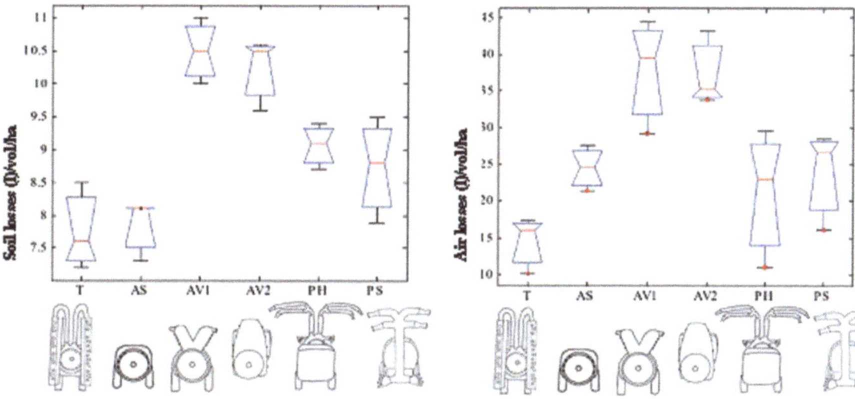

Fig. 17.1 Losses recorded for each type of machine (after Sinfort 2007)

when moving between 2 fields (treatment passage from 2 to 3 rows or change of row span). Most cases of plant-health, product over-and-under dosages during treatments are due to incorrect settings or poor maintenance of sprayers.

The problem to solve is: how to help the farmers *to optimize the settings of the sprayer* and to have a back-up of their work in order to improve their practices and reduce environmental impact?

Within the framework of the European project "Life AWARE", in the last 3 years (2006–2009) novel on-board system technology was developed to closely follow phytochemical application located on a catchments' basin. 15 vine growers were equipped with these embedded systems and were followed step by step to improve the use of the system.

17.1.2 The Project TICSAD

This project, that is a continuation of AWARE, is running to this day, geographically encompassing land areas from Bourgogne to Languedoc, with 20 winegrowers equipped with the new TICSAD system. This new system takes into account the improvements enacted by the users of AWARE. By studying how the farmers use the system, we optimise their practices. The goal is to improve the system and its robustness. A second aim is to also teach the users the new NICT tools and methods to improve operating practices; different levels of students analyse the formation within the system and the software. The embedded system is going to be robustly tested and finalized during this project before it can be transferred to an industrial firm.

17.1.3 Inventory of On-Board NICT Systems

There are different devices such as calculators and/or regulators (Land Manager from Dickey-John, Spraymat from Muller Elektronik. . .....), which enable measurement, and in certain cases, recording of treatment related data for cereal crop sprayers. However, they are generally cost prohibitive or unsuitable for vine and fruit tree sprayers (WTK-Elektronik). When compared to the cost of the sprayers used in viticulture and arboriculture, they are overly expensive. In addition, they do not satisfy the vine or fruit growers requirements in terms of continuous machinery monitoring based on simple parameters (right, left or partial flow rates, tank level, nozzle clogging detection, plot surface covered etc.) nor to adjust settings on a daily basis: pressure flow rate settings, nor to check nozzle performance, nor to transmit recorded date for simple traceability.

The systems, which we describe in this study, meet the need for pesticide reduction adapted to this form of agriculture by facilitating proper sprayer settings and the collection of objective data required to optimize plant health product treatments with the result of analyses realised by the added software.

17.2 Material and Methods

17.2.1 The AWARE and TICSAD Embedded Equipment

The onboard system measures and records, second by second, product application parameters (flow rates, volume, meteorological information), which are geo-referenced using GPS technology. The system architecture is comprised of treatment software and embedded equipment linked with the software by USB flash drives or WIFI network (Life AWARE only).

The embedded equipment includes an embedded electronic computer unit fitted on the tractor (MPU) (Fig. 17.2) and one more on the sprayer (APU) as well as measurement sensors.

The electronic MPU integrates acquired data management, Media data transfer (WIFI or USB), GPS-based geo-referencing, the weather station, data display and the Human Machine Interface as a very user friendly design providing pull-down menus similar to those used on mobile phones to guarantee easy usage and to counteract potential dangerous situations.

The electronic « **APU** » manages the acquisition of treatment-related data of the sprayer.

Consequently, the parameters measured are: Left and right flow rates, tank level, supply pressure, meteorological information (wind speed and directions, temperature & humidity) and GPS positioning. This Ultrasound static « marine » type measurement station is free of all mechanical problems. The electro magnetic flow-rate meters tested are not affected by clogging due to sticky or powder products and do not depend on pressure-based measurements of tank residues (circuit purging and rinsing). An ultrasound sensor, the signal of which is especially corrected, measures the tank level (Figs. 17.3, 17.4, 17.5 and 17.6).

The system is designed so that the farmer has no extra constraints, or operations to conduct during treatments, which means that if a fault occurs, spraying operations do not have to be halted.

Fig. 17.2 Tractor – sprayer equipment, the tractor MPU (photo Cemagref)

Fig. 17.3 *Electro magnetic flow-rate meters*

Fig. 17.4 *Tank level sensor*

Fig. 17.5 *On-board IHM system*

Fig. 17.6 *Weather station*

17.2.2 The Onboard Systems have 3 Roles

- *Prior to application*: Assist the winegrower during sprayer tank filling (direct and accurate reading of tank level), by allowing him to easily set flow rates based on plot parameters (row spacing, volume/ha etc.). Weather information (wind speed <19 km/h and humidity levels) allow for better product applications. Verify product from among those listed and set or change quantity levels.
- *During application*: Continuously visualize machine operation parameters during plot treatment i.e. monitoring. This allows the winegrower to detect malfunctions (clogging, unbalanced spraying, tank residue levels, etc). Plant-health product traceability is facilitated by GPS referencing.
- *Post treatment*: Transfer and then automatic generation of treatment logs. Inter plot traceability is rendered possible as the GPS module can differentiate vine rows of plots.

Before beginning the work, the winegrower has initialized the work with the programmed USB key, (using the software TICSAD) so he can verify the list of tasks, the settings of the sprayer and each product used. He chooses the task he has to do and, once the tractor is running, the GPS module of the onboard system synchronises itself so that when the winegrower arrives at the plot of land he starts manually to record the data. When he arrives at the end of the designed plot, he stops manually to record and can then choose the next task.

The default mode of the onboard system displays flow rates and tank levels for application monitoring but the grower can choose other options if required, e.g. Volume per hectare with ground speed or weather measurement (wind speed, temperature and humidity). The farmer can program phytochemical treatment tasks and also every other task required, like land measurement, tillage, topping and others with geo-referenced ways.

17.2.3 The Traceability Software

Traceability data are collected (low rates, tank levels, weather information), analyzed and then compared with winegrowers declared data, to provide farmers with strategies designed to improve their practices. The treatment software aims to prepare the fieldwork and to treat the measurement's data issued from the embedded system to generate an automatic field logbook report.

The software permits to configure the list of plots of land, the tools (sprayer and others), the users, the products for each treatment and the different tasks. Thus, the farmer prepares the work, which will be transferred with a configuration file using the USB key.

After work, the data are collected from the USB key to the database which collects and processes received data (trajectory computations and merging of machine data) and generate:

- Treatment analysis results related to plots treated and application dates.
- The traceability of realised work and geo-referenced quantities product used.

17.2.4 Processing of Treatment Product Data

For the Life AWARE project, a geographical information system (GIS) was deployed to create, collect and then represent the entire set of data: define catchments' basin border, wine-growers plots, hydrographical network, relief, plant health products applied etc. so that maps and spatial analyses of the catchments' basin could be derived. For the TICSAD project, the information system was designed for only one farm and so contains the geographical records of all the plots of the wine grower.

The information recorded, second-by-second, includes: GPS (metric) sprayer positioning, left/right flow rates, volume of mixture tank residues and weather conditions (temperature, hygrometry, wind and speed direction). Each on-board system has an ID allowing it to reference the farm and thereby these plots of land. Information related the each parcel is extracted from each treatment event using a Geographical Information System (GIS). Cross-referenced analysis of the data is performed to derive essential criteria to assist decision-making.

The totality of available data means that in-depth scientific study of the application can be conducted.

17.2.5 Generation of Plot Log

In relation to its role in application quality assessment, the system generates, in graphical form, plot related results where measured parameters, or those issued from computations, are displayed (Vol./ha, wind speed /direction). A written report, providing information on different parameters, is also generated: commonly used

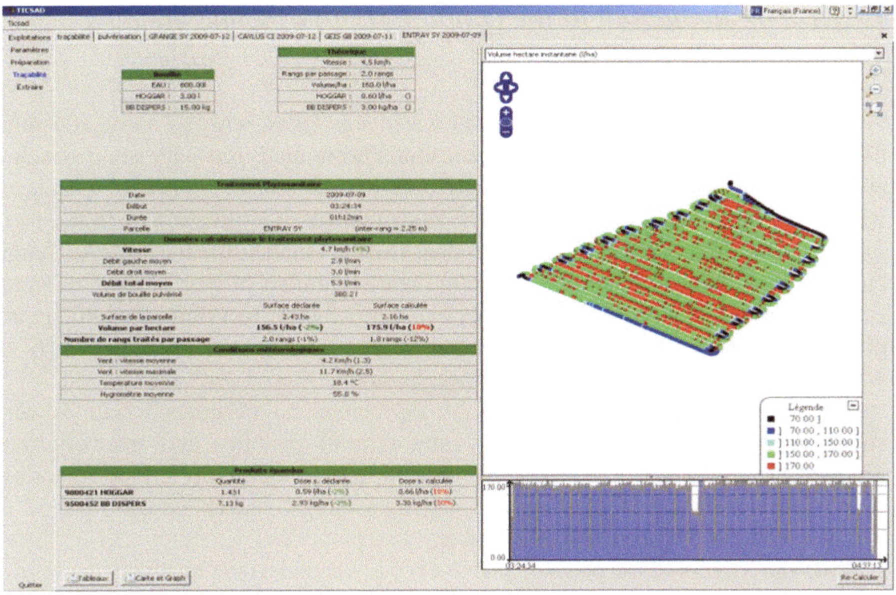

Fig. 17.7 Plot log and automatic field logbook

sprayer settings, measurements, surface area treated, number of rows, dosage levels applied etc.

Once treated, the data summarize results in treatment sheets in which information provided by the winegrowers can be compared with those objectively recorded by the system. Data processing automatically generates a plot log and an automatic field logbook (Fig. 17.7). That can be modified to satisfy related representations of this kind of information as well as sustainable agriculture specifications like those of TERRAVITIS.

The aim is to provide the winegrower easy to use information that can be instantly understood.

Other parameters can also be presented in graphic formats like wind force or humidity to enhance analysis.

Once all the data have been gathered, they are communicated to farmers via a data-exchange intranet site in an educative format designed for easy interpretation. The data can be visualized, interpreted and changed to improve the practices.

17.3 Results

One of the results of these projects concerns the reduction of pesticides found at the outlet of the catchments' basin. A second result concerns the use of the NICT embedded system.

17.3.1 The Reduction of the Quantities of Pesticides Related to Better Practices

For the Life AWARE project, analyzed water was taken with an automatic sampler to the catchments basin outlet. 22 samples were analyzed with multi residues analysis (more than 350 active substances and metabolites). 33 active substances were detected either with the top of the threshold of quantification or in the form of "presence". Weed-killers (58%) (Glyphosate or its metabolite the AMPA), fungicides (36%), insecticides (6%) were found with a number of weed-killers of larger concentrations. This finding is possibly explained by the fact that weed killers are directly applied to the ground and therefore are immediately absorbed by superficial waters generated by heavy rainfalls or precipitation.

The annual quantitative variation of phytochemical products found, is very strongly correlated with pluviometry and rainfall quantity and intensity. The Mediterranean climate of the zone is characterized by one rainy period in spring, followed by a dry one (July–August) which in turn is succeeded by a period of heavy rains from September to November.

Over the 3 years of the project, year 2006 was characterized by a very dry spring (10 mm). Consequently, there was no pressure from mildew and the adventitious ones did not develop. The wine growers thus performed less treatment in 2006 as compared to 2007–2008. Moreover, as it did not rain to a significant degree, there was no transfer of pesticides towards water. On the other hand, years 2007 and 2008 exhibited important rains during spring and autumn, therefore supporting the transfer of pesticides. During those 2 years the wine growers had to carry out more treatments, as the parasitic pressure was stronger. These two combined mechanisms therefore explain the elevated presence, in terms of quantity and composition, of plant health products during 2007 and 2008 in VAILLELE River waters.

It appears difficult on a time span of 3 years to generate conclusions on the reduction of the quantities of pesticides related to better practices. Modelling could make it possible to test various scenarios according to the ground observations collected during the 3 years of the project.

17.3.2 Model of Plant Protection Product Transfer

Model MHYDAS (Moussa et al. 2008) used in the project is a model rain-flow developed by UMR LISAH with the aim of studying the effects of agricultural installations on the hydrological behaviour of catchments' basins during the rainy events (Fig. 17.8).

It models the hydrological surface-underground exchanges, assessments, the transport of pollutants, and the transport of suspended matter. The functions of simulation of the transfer of pollutants make it possible to simulate the principal organic-physico-chemical processes governing to become it and the transfer of the plant health products in the environment.

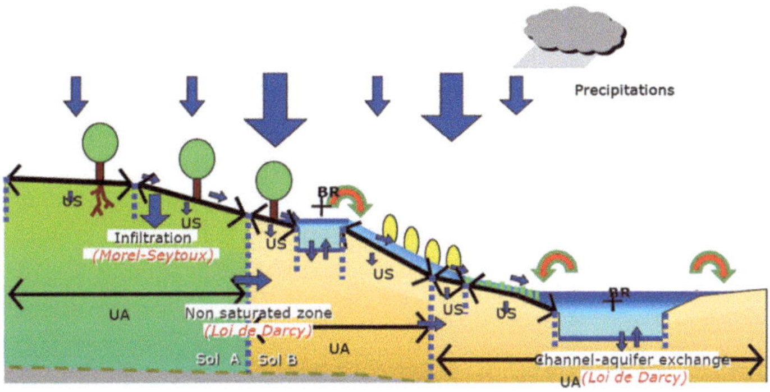

Fig 17.8 Main hydrological processes in MHYDAS (US : Surface Unit; UA : Aquifer Unit; BR :Channel network; from Ghesquière 2008)

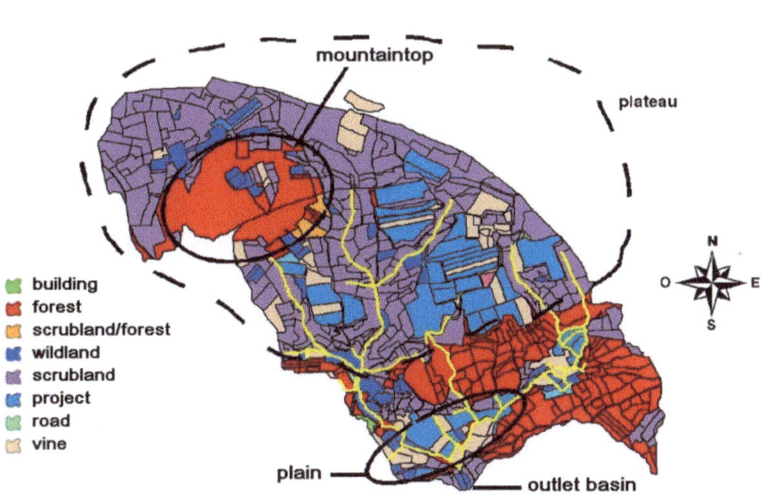

Fig. 17.9 Segmentation of the catchments' basin for MHYDAS

The model has been calibrated (Fig. 17.9). The whole area has been parted into homogeneous hydrological zones. We can apply rainfall and pesticides quantities on each plot and let the model to calculate the river flow and the concentration of pesticides molecules in the river.

17.3.3 The Use of the NICT Embedded System

17.3.3.1 Role of the NICT for sprayer tuning

The Embedded system sensing devices have been used at two steps by the farmer.

First, during filling up, the tank level sensor is very useful as it allows to stop filling when necessary. This operation was much less comfortable when using the sprayer level indicators; specific pollution is limited.

Second, during spraying, the farmer can monitor the spraying and external parameters and therefore adapt its speed or detect any malfunction (ex: stuck nozzle. . .). The displayed parameters are presently: the right and left flows (in l/min) and the weather parameters (then the farmer can stop if the wind is strengthening).

Finally, thanks to the embedded system and the adjustments, the wine grower takes particular care to modify his apparatus, not only at the hydraulics but also on the spray nozzles direction and the rinsing of the tank.

17.3.3.2 Role of NICT for Farmer Practices Improvements

After data processing, useful information is presented in the form of graphs and maps, which can be used by the farmers and advisors for evaluation and improvement of farming practices. Using these types of maps, the farmer can get involved into a self-teaching scheme.

17.3.3.3 Role of NICT for Traceability Improvement

Data can also be organised in order to fill a "traceability book", similar to the one that is obligatory for the farmer to fill in. Outputs like the one presented in Fig. 17.9, can be automatically generated in order to help the farmer to have a guaranteed "traceability book".

First comparisons made with the manual traceability books show several discrepancies. These are always due to farmer errors (due to a delay in filling up the book, to writing errors. . .). Therefore, the AWARE & TICSAD systems offer a secure way for elaborating the "traceability book".

17.4 Prospects

17.4.1 The Future of TICSAD with Scientific Approach

The objective, which was assigned at the end of the "GRENELLE" meeting on the environment is to reduce by half the use of pesticides within 10 years. A pesticide-free viticulture seems utopian. The treatments will be always necessary whatever the product. On the other hand, the reduction of the quantities of pesticides used and lost in the environment can be reduced by evaluating/controlling the adjustments on the sprayers and by using innovative solutions developed by the technical institutes and the research centers. Research Programs like POD MILDIUM (INRA – CEMAGREF)(Léger et al. 2007)] and OPTIDOSE (IFV) (Davy 2007) show that the fight against parasitic attacks can be won by using the right amount of pesticide at the right time, which avoids excess applications and decreasing further contamination of water sources.

To better manage these treatment methods, the TICSAD embedded system provides an obvious help.

17.4.2 The POD "Mildium" Approach

Evaluation of a decision rule designed to reduce the number of fungicide applications against grapevine powdery mildew.

This Operational process of decision (POD), prepared by a collection of expert knowledge, presents several originalities in relation to current practices of wine growers by defining the objectives of protection with the intent not to alleviate all symptoms but to avoid any loss of harvest, by the simultaneous management of the populations of two bio aggressors, Mildew and Oidium, in the same process and the reduction of the number of observations required for decision-making. Its overall design is centred on a limited number of mandatory treatments aimed at controlling small scale epidemics, supplemented by optional treatments to counteract against the eventuality of a severe epidemic attack.

The aim is also to adapt this process, developed at the plot, at the level of an exploitation that is relevant to decision making. The application of the principles of this process must allow, eventually, a significant reduction in the use of phytochemical inputs.

The traceability AWARE & TICSAD system would be able in this context to dictate the right settings for the sprayers, to obtain the traceability of phytochemical treatments but also to reference geographically, places of infestation in the plot thanks to the integrated management of the tasks and events. A fine and objective analysis at the vine stock scale could be obtained. This approach is currently in evaluation at the Cemagref of Montpellier (Fig. 17.10).

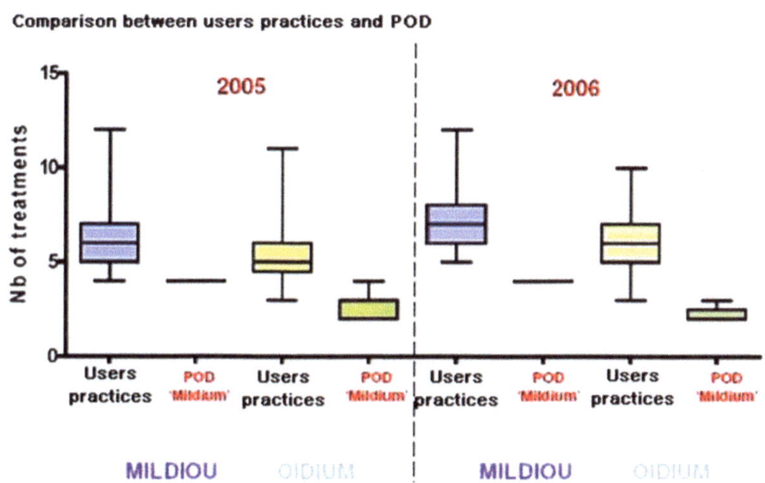

Fig 17.10 Comparison between users practices and POD

17.4.3 The "OPTIDOSE" Approach

Adaptation of the amount according to the vegetative development and the parasitic pressure

This project, developed by the IFV institute, aims to optimize the amount of products used in the vineyard while ensuring an optimal efficacy of the treatment. Whereas the homologation of the products remains based on the ratio of the amount by hectare irrespective of the conditions during treatment, the OPTIDOSE project uses an approach that adapts the amount of product to the conditions met at the time of the treatment.

Indeed, the approved amount is calculated to remain effective when all factors (climatology, vegetative development, compartmental sensitivity...) are favourable to the development of mushrooms, which is not always the case in practice.

The principle is based on different evaluations and measures that consist of:

- The Observation of the symptoms at the plot to define the parasitic pressure to which the models of the agricultural warnings are added (EPI Potential state of infection & FTA Theoretical frequency of attack)
- The measurement of the Leaf area Index at key stages in order to determine the vegetative expression at the time of the treatment
- The evaluation of the quantity of active matter necessary for optimal protection

A table with triple entries indicates the percentage of the amount approved for application according to the plant surface (LAI Leaf Area Index), the parasitic pressure and the seasonal phenology (Fig. 17.11). This table makes it possible to adapt the method and to determine its limits of application. This grid is tested since

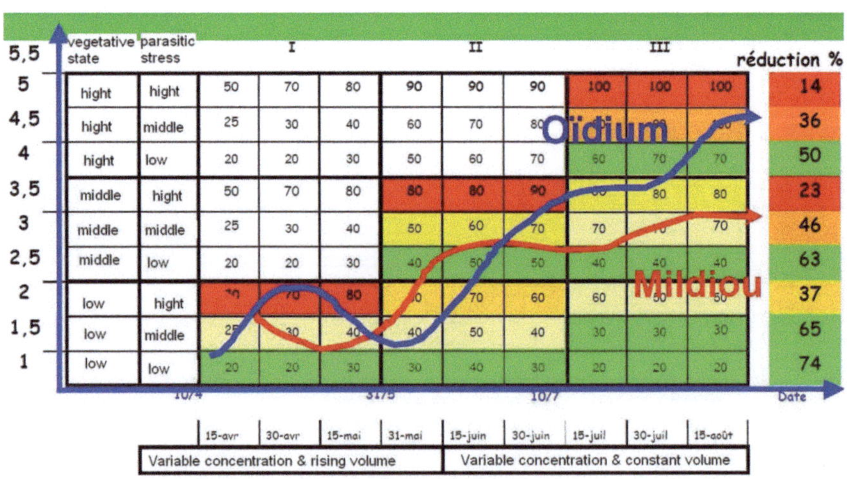

Fig 17.11 OPTIDOSE table : image © Alexandre Davy – IFV

2007 on mildew and OïDIUM with very distinct parasitic profiles of pressure. The results show that the plots treated with lower amounts (average reduction of about 20 to 50%) have more small scale attacks than those treated with the full amount. However, these differences are small and the protection generated by the application of optimised amounts does not affect the quantity and quality of the final product.

The Life AWARE & TICSAD systems could, in the same manner as described for POD MILDIUM paragraph, make it possible to obtain the traceability of geographically referenced phytochemical treatments but also to place the infestation in the plot thanks to the integrated management of the tasks and events with a fine and objective analysis.

Since the adjustments of the sprayers must be adapted according to the parasitic pressure and the vegetative stages, the embedded system is in addition able to correctly set the sprayers daily.

17.4.4 Volume Modulation According to the Vigour Variability of the Vine

Another way of reducing the quantity of pesticides is to modulate in real time the volume spread according to the variability inside the plot of the vigour of the vine and its vegetative expression.

In order to achieve this, it is necessary to measure foliar volume (i.e. TRV tree row volume [Gil 2008]) with ultrasound sensors (Gil 2008) or NDVI sensors like Green-seeker (Goutouly et al. 2006) for example. It is possible to modulate the spray according to the mapping of the NDVI[1] by air or satellite imagery of a heterogeneous plot of land. However, if satellite NDVI measurements are more relevant, those by embedded measurement are going to be to validated (Goutouly et al. 2006, Driss et al. 2009). These measurements are adapted better for the differentiation of classes that indicate quality during harvest (Tisseyre et al. 2006). The real time modulation of volumes is not yet implemented to date on the wine sprayers.

A number of studies (Gil 2008) have already been carried out or are in hand. It is necessary to compare the quality of the protection of the vine between the various processes OPTIDOSE & POD-MILDIUM with embedded real time modulation.

17.5 Conclusions

The NICT systems offer farmers new technology, enabling the automatic generation of reliable data, automatic feedback on current practices and real traceability on spraying and pesticide product usage. The analysis of these readings, which

[1] Normalized Difference Vegetation Index: NDVI=(NIR-RED/NIR+RED).

involves the different stakeholders, enables detection of different malfunctions and provides avenues to improve practices and equipment, thereby reducing quantities of pesticide products used and mitigating environmental contamination. It provides objective data automatically. The most important hydraulic parameters involved in sprayer functioning can be visualized and, if needed, easily corrected.

The quality of treatment is undoubtedly the first means by which the vine grower can reduce the use of pesticides. A good adjustment of the sprayer makes it possible to limit air and soil losses and drifts (they can represent from 7 to 50% according to the vegetative stage) and to avoid the over and under proportioning.

The main objective of these projects is to enable farmers to completely control the operations of their every day working tools. These tools enable the farmer to check and correct sprayer parameters on a daily basis. Geo-referenced inter-plot traceability of treatments provides an objective assessment of application quality, which allows the winegrower to analyze and correct operations. Thanks to the assistance provided by this tool, users pay greater attention to both hydraulic and mechanical (nozzle direction) settings, and to the general maintenance of their sprayers, with the aim of obtaining a specific result that can be immediately assessed after treatment.

NICT tools give the operators the means to control pesticide application and to meet the objective of significantly reducing soil and air losses, thereby mitigating surface and groundwater contamination and atmospheric pollution. Associated with new methods of treatments like POD MILDIUM and OPTIDOSE, the objective of reducing the amount of pesticides used can be achieved in a short time frame, without compromising the quality of the protection necessary to achieve high yields of good quality products.

References

Bellon-Maurel V, Vallet A, Tinet C, De Rudnicki V, Ruelle B, Douchin M (2008) Pulvérisation modulée utilisant mesures et tic (Chili). Ageng 2008

Davy A (2007) Le programme Optidose : optimisation agronomique et environnementale de la pulvérisation. (France), Euroviti 157–162.

De Rudnicki V, Ruelle B, Douchin M, Bellon-Maurel V (2008) Reducing pesticide-related water pollution by improving crop protection practices : The use of embedded ICT technologies. (France), 13th IWRA World Water Congress 2008

Driss R et al (2009) Non-destructive measurement of grapevine leaf area by ground normalized difference vegetation index. Agron J 101(1):226–231

Gil E (2008) Tecnologías para la optimización de las aplicaciones de fitosanitarios, Spain, Life AWARE project meeting

Goutouly JP et al (2006) Characterization of vine vigour By ground based NDVI measurements. In: Procedings of the VIth International Terroir Congress, France, pp 237–241

Léger B, Cartolaro P et al (2007) An expert based crop protection decision strategy against grapevine's powdery and downy mildews epidemics: Part (1) formalization. (Sicil), IOBC/WPRS Working Group "Integrated Control in Viticulture"

Moussa R et al (2008) MHYDAS – Spatially distributed hydrological modelling of agrosystems: model development and application cases for water and pollutant transfer in various agro-hydro-meteorological conditions. In: Proceeding of the 13th IWRA World Water Congress, France

Sinfort C (2007) Les pertes de produits phytosanitaires dans l'environnement pendant les applications : le rôle du matériel , France, Euroviti

Tisseyre B, Ojeda H, Taylor J (2006). New technologies and methodologies for site-specific viticulture, Journal International des Sciences de la Vigne et du Vin, France, GESCO Meeting

Chapter 18
Isolation and Use of Protoplasts from Grapevine Tissues

Natacha Fontes, Hernâni Gerós, Anastasia K. Papadakis, Serge Delrot, and Kalliopi A. Roubelakis-Angelakis

Abstract Highly pure, intact and functional protoplasts can be obtained from plant tissues, which are readily amenable for challenging with exogenous sugars, acids, analogues, transport inhibitors and drugs. Thus, they may be used as models for both basic research and biotechnological approaches. Some of these studies require the regeneration of plants from protoplasts; however most agronomically important plant species, including grapevine, are recalcitrant to plant regeneration. Oxidative stress has been considered as a crucial factor accounting for the recalcitrance of grapevine protoplasts, as supported by the profiles of generated reactive oxygen species (ROS) and ROS-scavenging enzymes, the modified cell redox state, as well as the altered endogenous titers of polyamine levels. In the present work, methods for the purification of intact and functional protoplasts from grape berry mesocarp tissue and for the isolation and culture of mesophyll protoplasts are described. Methods for the detection of ROS in grapevine protoplasts, together with assays for antioxidant enzyme and antioxidant biomolecules are also detailed.

Keywords Biotechnology · Cell viability · Cellulose-degrading enzymes · Protoplast isolation · Membrane integrity · Totipotency · Reactive oxygen species · Antioxidant biomolecules · Antioxidant enzymes

Contents

H. Gerós (✉)
Departamento de Biologia, Centro de Investigação e de Tecnologias Agro-Ambientais e Biológicas (CITAB), Universidade do Minho, Braga 4710-057, Portugal
e-mail: geros@bio.uminho.pt

S. Delrot et al. (eds.), *Methodologies and Results in Grapevine Research*,
DOI 10.1007/978-90-481-9283-0_18, © Springer Science+Business Media B.V. 2010

Abbreviations

APO	Ascorbate peroxidase
BAP	6-Benzylaminopurine
CAT	Catalase
cpm	Counts per minute
DHAR	Dehydroascorbate reductase
DTT	Dithiothreitol
EDTA	Ethylenediaminetetraacetic acid
FDA	Fluorescein diacetate
GR	Glutathione reductase
GSH	Reduced glutathione
GSSG	Oxidized glutathione
H_2O_2	Hydrogen peroxide
MDHAR	Monodehydroascorbate reductase
MES	2-(*N*-morpholino)ethanesulfonic acid
NAA	1-Naphthaleneacetic acid
$O_2{}^{\cdot -}$	Superoxide radical
PAs	Polyamines
PCA	Perchloric acid
PMSF	Phenylmethanesulfonyl fluoride
ROS	Reactive oxygen species
SOD	Superoxide dismutase
TEMED	Tetramethylethylenediamine

18.1 Introduction

Plant cell walls are highly complex structures providing structural and mechanical support to the cells, maintaining cell shape, direction of cell growth and ultimately the architecture of the plant. They are mainly responsible for the integrity and texture of tissues and, therefore, determine fruit processing. In higher plants, about

90% of primary cell wall material consists of polysaccharides, such as cellulose, hemicellulose, and pectin, the rest (ca. 10%) being structural proteins (Barnavon et al. 2000). Due to the rigid cellulose wall, it has been relatively difficult to handle plant cells. In 1892, crude preparations of naked cells called protoplasts, were obtained following plasmolysis of leaf tissue cells of *Stratiotes aloides* (Klercker 1892). Only about 100 years later, the isolation of protoplasts in large scale was achieved when cell wall hydrolytic enzymes were employed (Cocking 1960, 1961), based on the observation that during fruit ripening natural enzymatic degradation of cell walls occurs. Since then, other methods based on mechanical removal of cell walls and on the use of solvents have been used, although the methods involving the use of hydrolytic enzymes have been the most popular (Fontes et al. 2010a). Also, approaches for grapevine protoplast culture and manipulation have gained enormous advances (reviewed by Papadakis et al. 2001a, 2009).

Grapevine protoplasts have been isolated from leaves (Nishimura et al. 1984, DeFilippis and Ziegler 1985, Wright 1985, Barbier and Bessis 1988, Lee and Wetzstein 1988, Barbier and Bessis 1990, Deswarte et al. 1994, Theodoropoulos and Roubelakis-Angelakis 1989, 1990, 1991, Katsirdakis and Roubelakis-Angelakis 1992a, b, Papadakis and Roubelakis-Angelakis 1999), stems (Reustle and Alleweldt 1991), roots (Reustle and Alleweldt 1990), callus (Skene 1975, Ui et al. 1990, Mii et al. 1991) and embryogenic tissue (Reustle et al. 1995, Zhu et al. 1997). Of special significance has been the use of axenic leaves from in vitro grown plants for the isolation of protoplasts (reviewed by Papadakis et al. 2001a, 2009). Recently, protoplasts from grape berry mesocarp have been successfully purified (Fontes et al. 2009, 2010a).

Plant protoplasts have provided a useful tool for physiological, biochemical and molecular studies (Davey et al. 2004). Also, they have been used for genetic transformation of plants. The present work describes methods for protoplast isolation from grapevine tissues. Highly pure, intact and functional protoplasts can be obtained with these methods readily amenable for challenging with exogenous sugars, acids, analogues, transport inhibitors and drugs, treatments that cannot always be performed in plant tissues. Thus, they may be used as models for both basic research and biotechnological approaches, such as solute uptake and compartmentation, toxicological assessments and grapevine breeding programs. Some of these studies require the regeneration of plants from protoplasts; this is not the case for most agronomically important plant species (perennial woody species, cereals and legumes), including grapevine (Roubelakis-Angelakis 1993, Papadakis et al. 2009). Oxidative stress has been considered as a potential crucial factor affecting recalcitrance of grapevine protoplasts, as supported by the profiles of generated reactive oxygen species (ROS), hydrogen peroxide (H_2O_2) and superoxide ($O_2^{\cdot -}$), and ROS-scavenging enzymes, the modified cell redox state, as well as the altered endogenous titers of polyamine levels (reviewed by Papadakis et al. 2009). The present work also describes methods for the isolation and culture of mesophyll protoplasts, and for the detection of ROS in grapevine protoplasts, together with assays for antioxidant enzyme and antioxidant biomolecules.

18.2 Methods for the Isolation of Protoplasts from Grapevine Tissues

18.2.1 Protoplast Isolation from Mesocarp Tissue

18.2.1.1 Digestion of Mesocarp Tissue

The present method of protoplast isolation and purification from grape berry mesocarp cells is based on the patent application P103851 (Fontes et al. 2009), that was recently explored and discussed in detail (Fontes et al. 2010a). Due to the low pH of berry tissue and its fragility, and to the wide range of secondary metabolites present, cell wall digestion and protoplast purification is performed in a very special environment to maintain their integrity and viability. Berries are collected 14 weeks after flowering (harvesting). After berries are deseeded and weighed, skins are peeled and flesh tissue is cut into discs of 3–6 mm thickness and washed with pre-incubation buffer (Gambor B5, 0.3 M mannitol, 0.5 mM $CaCl_2$, 0.25 mM $MgCl_2$, 10 mM MES, pH 5.6, 1 mM DTT, 1 mg/mL NAA and BAP) with constant shaking at 4°C. Flesh discs are then incubated with medium I (205 mM KCl, 65 mM $CaCl_2$, 1 mM DTT, 10 mM MES, pH 5.6) containing 0.03% cellulase Y-C and 0.003% pectolyase Y-23 (1 g flesh tissue/mL medium). A relatively long digestion period of 12 h is used, at 22°C under constant shaking (15–25 rpm). The resulting protoplasts are gently collected, filtered through 0.5 mm mesh pore size and then purified (Fig. 18.1).

18.2.1.2 Protoplast Purification

In a first step, protoplasts are separated by filtration through a 0.5 mm mesh pore size and subsequently separated by sedimentation, at 150 g for 8 min. The resulting pellet is washed with medium A (400 mM sucrose, 30 mM K-gluconate, 2 mM $MgCl_2$, 2 mM EDTA, 1 mM DTT, 10 mM Mops-Tris, pH 7.2), at a ratio 1:1 (v/v). The sedimented protoplasts are gently collected and diluted by adding 4 vol of medium A. A discontinuous gradient is prepared by overlaying 1/2 vol of medium B (500 mM sorbitol, 30 mM K-gluconate, 2 mM $MgCl_2$, 2 mM EDTA, 1 mM DTT, 10 mM Mops-Tris, pH 7.2) on the diluted protoplasts. After centrifuging at 500*g* for 8 min, protoplasts are recovered from the interface of the gradient (Fig. 18.1c, d), resuspended in 2 vol of medium I and sedimented for 8 min at 150 g. The pellet is resuspended in the same medium and stored at 4°C (Fontes et al. 2009, 2010a).

18.2.1.3 Protoplast Yield, Viability and Integrity

The isolation of a sufficient amount of high quality protoplasts is a prerequisite for using protoplasts either as an experimental model for basic research or for biotechnological applications (Papadakis et al. 2009, Fontes et al. 2010a). The protoplasts

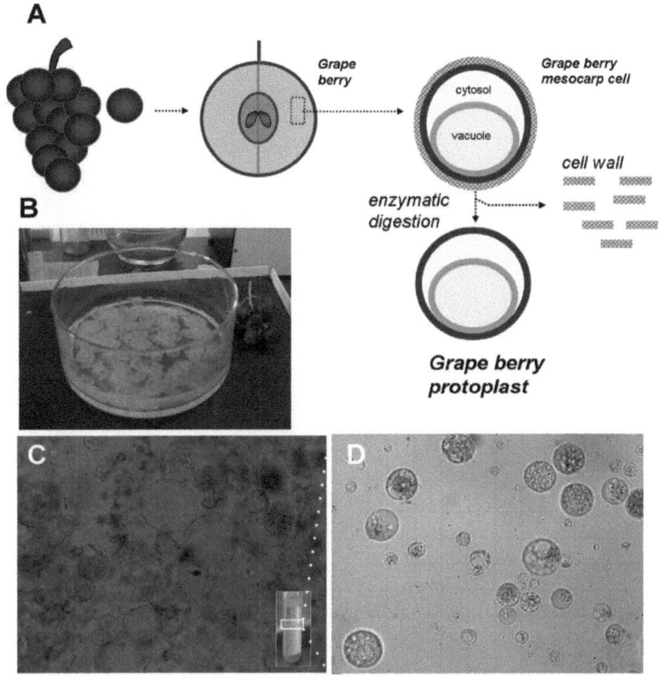

Fig. 18.1 (**a**) Isolation and purification of protoplasts from grape berry mesocarp and (**b–d**) specific steps of the protocol. (**b**) Grape berry mesocarp discs under overnight incubation with the enzyme mixture, (**c**) protoplast sample observed under the light microscope prior the purification step and (**d**) after purification by a density centrifugation step. *Inset*: Density gradient composed by an upper layer of sorbitol over a sucrose layer; protoplasts are collected at the interface (Fontes et al. 2009, 2010a)

are counted in a Malassez chamber under the light microscope (Fig. 18.1d). A protoplast yield of 10×10^6 protoplasts /20 g of flesh tissue is routinely obtained (Fontes et al. 2010a). Protoplast viability is assessed after incubating the protoplasts with Fluorescein diacetate (FDA). The intact plasma membrane is permeable to FDA, and FDA is converted to a green fluorescent dye, fluorescein, by internal esterases, displaying a green fluorescence in viable cells (Jones and Senft, 1985). Observations are performed under a Leica Microsystems DM-5000B epifluorescence microscope with appropriate filter settings. Images are acquired with a Leica DCF350FX digital camera and processed with LAS AF Leica Microsystems software. For the staining protocol, 1 ml of protoplast suspension is incubated with 10 μl of FDA stock solution in the dark for 10 min at room temperature and observed further. Figure 18.2a depicts a typical protoplast population labelled with FDA observed under UV light. Comparison of the epifluorescence light with visible light images shows that most protoplasts remained viable immediately after isolation, displaying an intense green fluorescence (Fontes et al. 2010a).

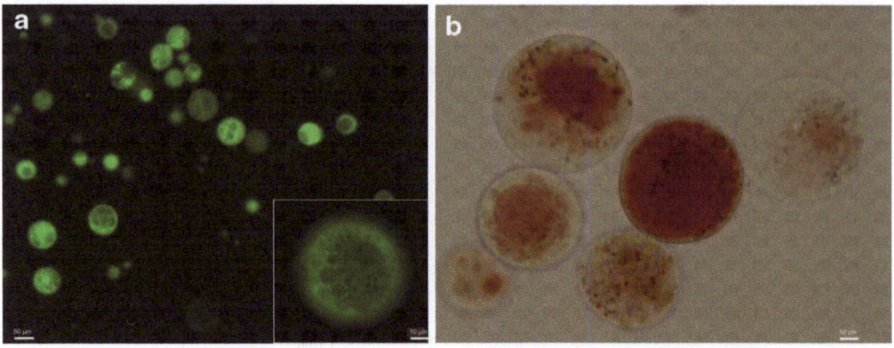

Fig. 18.2 (**a**) Isolated protoplasts observed under UV light (epifluorescence) after staining with fluorescein diacetate (FDA) to measure viability. Inset: a close-up view of an intact protoplast highlighting the complex vacuolar apparatus that is not labelled with FDA. (**b**) Intact protoplasts labelled with Neutral Red showing the acid pH of the vacuolar apparatus (Fontes et al. 2009, 2010a)

18.2.1.4 Visualization of the Vacuolar Apparatus with Neutral Red

In the mesocarp of fleshy fruits, the vacuoles play a prominent role in cell expansion, fruit size and fruit quality. This is the case for the vacuoles of the grape berry cells, which accumulate high concentrations of sugars, organic acids and secondary metabolites. These compounds, which all play a key role for the taste and flavour of the fruit are imported and/or compartmented in various cells (mesocarp, skin), by specific plasma membrane and tonoplast transporters (Conde et al. 2007). To visualize the vacuolar apparatus, Neutral Red staining is performed by incubating the protoplast sample with the lipophilic phenazine dye (Sigma-Aldrich) (prepared at 4 μM in 0.5 M mannitol pH 7.0 prior to its use) at a ratio 1:10 (v/v), and the preparation is observed under a light microscope (Fontes et al. 2010a). In contrast to grape cultured cells that possess a large central vacuole (Fontes et al. 2010b), mesocarp cells show a complex vacuolar apparatus (Fig. 18.2b; Fontes et al. 2010a), corroborating the idea of the vacuole being a complex organelle (Marty, 1999), in contrast to the initial ideas on vacuolar structure (Taiz, 1992). The possibility opened by this method to purify intact vacuoles from grape berry protoplasts, as performed before from protoplasts derived from grape cultured cells (Fontes et al. 2009a), will allow several basic and applied approaches, namely the detailed analysis of vacuole content along with ripening, proteomics and compartmentation studies.

18.2.2 Protoplast Isolation from Mesophyll Cells

18.2.2.1 Plant Material

Protoplasts are isolated from fully expanded, but not senescent, leaves of in vitro grown *Vitis vinifera* L. cv Sultanina (Fig. 18.3; Roubelakis-Angelakis and

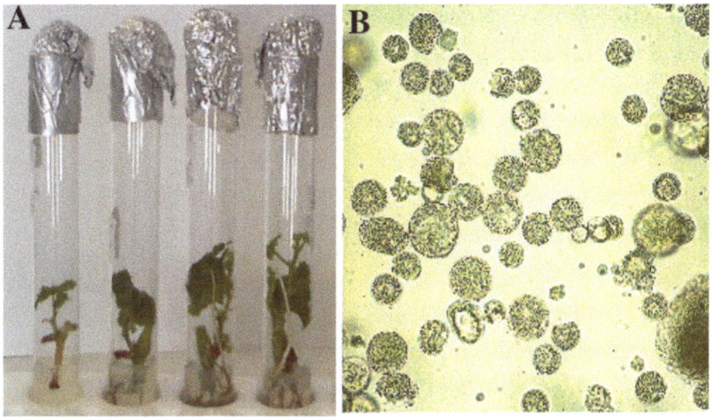

Fig. 18.3 (**a**) In vitro-grown grapevine plants and (**b**) freshly isolated protoplasts from leaf tissue

Zivanovitc, 1991). Virus-free plants are grown at 25 ± 0.5°C, under a 16/8 h photoperiod provided by cool white fluorescent lamps (50 μmol m^{-2} s^{-1} total energy).

18.2.2.2 Isolation and Culture of Mesophyll Protoplasts

Small leaf segments (2 mm) from the in vitro plants are punched aseptically and placed into petri dishes with isolation medium (IM) (30 mL g^{-1}FW) consisting of 25 mM MES, Murashige and Skoog (MS, Murashige and Skoog, 1962) salts and microelements, 0.02 mg L^{-1} biotin, 2 mg L^{-1} myo-inositol, 0.2 mg L^{-1} nicotinic acid, 0.2 mg L^{-1} pyridoxine-HC1, 0.1 mg L^{-1} thiamine-HCl, 0.2 mg L^{-1} pantothenic acid, and 43 mg L^{-1} EDFS, pH 5.7 ± 0.1. The medium is also supplemented with 0.7 M sucrose, 0.1 mg L^{-1} 6-BAP and 1 mg L^{-1} NAA. Osmolality of IM is determined with a Gonotek cryoscopic Osmomat 30 osmometer equal to 1050 mOsmol L^{-1}. The standard macerating enzyme solution is added to IM, containing 1% (w/v) Cellulase R-10 Onozuka (Yakult Honsha Co., Tokyo, Japan), known to contain xylanase (Fuchs et al. 1989), and 0.5% (w/v) Macerozyme R-10 Onozuka (Yakult Honsha Co., Tokyo, Japan). Cellulase Onozuka can be replaced by 0.25% (w/v) purified cellulase Worthington (Worthington Biochemical Corporation). The duration of the maceration period is 4 to 16 h, in the dark at 25°C (Theodoropoulos and Roubelakis-Angelakis 1990).

Then, the enzymatic solution is replaced by IM without enzymes. Petri dishes are shaken at 120 rpm for 10 min; the mixture is filtered through cheesecloth into centrifuge tubes and an adequate volume of culture medium (CM; with the same composition of IM, but 0.7 M sucrose is replaced with 0.7 M glucose) is added to produce two layers. Protoplasts are isolated in the interphase after centrifugation at 165*g* for 15 min.

The protoplasts are then washed twice with CM. The sedimented protoplasts of the second centrifugation are gently collected and resuspended in the same medium (Fig. 18.3b). Protoplast viability is tested by Evans blue staining technique and a hematocytometer is used to count isolated protoplasts; they are further cultured at a density of 10^5 protoplasts mL^{-1} CM, in the dark at 25°C.

18.2.2.3 Transport Experiments

Uptake of D-[U-^{14}C]glucose can be used to test the integrity and functioning of plasma membrane (Theodoropoulos and Roubelakis-Angelakis 1990). The solution consists of the CM containing protoplasts at a final density of 5×10^5 protoplasts per mL. The total assay volume is 3.5 mL; NaN_3 is added 30 s prior to adding 2 μCi of ^{14}C-compound. Final concentration of glucose is 2 μM. Uptake assaying is performed at 30°C and pH 5.7. At defined intervals, aliquots of 250 μL are withdrawn, and rapidly filtered through nitro-cellulose membranes (Millipore, 0.22 μm). Following rinsing of membranes with 2 mL of CM, they are dried and placed into scintillation vials filled with 5 mL of scintillant. Radioactivity is counted in a Packard scintillation spectrometer and results are expressed as pmoles of accumulated compound per 10^6 viable protoplasts.

Cultured grapevine protoplasts rapidly deplete glucose from the culture medium. Concentration dependent uptake of labelled glucose is linear for concentrations higher than 1.5 mM, at lower concentrations a saturating pattern is observed (Theodoropoulos and Roubelakis-Angelakis 1990). The transport system is hexose specific and the stereospecificity is closely related to carbon-1 of the glucose molecule. Glucose structural analogues are not metabolized beyond the stage of phosphorylation. The sugars enter the plasma membrane by a carrier, which is driven by a proton motive force, probably in an uncharged form (Theodoropoulos and Roubelakis-Angelakis 1991).

18.3 Detection of Reactive Oxygen Species in Grapevine Leaf Tissue and Protoplasts

In order to test if reactive oxygen species are implicated in protoplast recalcitrance, the elicitation of ROS production can be studied in both grapevine leaf tissue and protoplasts (Papadakis and Roubelakis-Angelakis, 1999). Leaf strips are intensively wounded with a razor blade and floated in the CM; at zero time, two different cellulase preparations with similar enzymatic activity are added to leaf segments; 1% (w/v) cellulase R-10 Onozuka or 0.25% (w/v) purified cellulase Worthington. ROS accumulation in the culture medium is monitored over a period of 16 h. Protoplasts that are isolated with purified cellulase Worthington and macerozyme Onozuka, are cultured and at zero time, 1% (w/v) cellulase R-10 Onozuka is added and ROS accumulation is monitored over a period of 4 h.

18.3.1 Luminol-Dependent Chemiluminescence Assay for H_2O_2

The production of H_2O_2 from leaf cells and protoplasts is determined by the chemiluminescence assay of luminol as described by Papadakis and Roubelakis-Angelakis (1999). Leaf segments (50 mg mL^{-1}) are floated in the CM and this medium is used for H_2O_2 assays. The production of H_2O_2 from protoplasts, corresponding to the same fresh weight of leaf tissue, is determined in the culture medium after centrifugation at 1000 rpm for 15 s. The assay is conducted in a total volume of 2 mL by placing 0.8 mL reaction buffer containing 10 mM Tris-MES, pH 7.0, 1 mM $CaCl_2$ and 0.1 mM KCl, 0.2 mL of 1 mM luminol solution, 0.1 unit of peroxidase in 20 mM potassium phosphate buffer (pH 7.4) and 1 mL of culture medium in a scintillation vial (Auh and Murphy 1995). The scintillation vial is immediately placed in a scintillation spectrometer (LS 8000, Beckman) and chemiluminescence is detected. Counts are reported every 15 s for 1 min and the last two values are averaged. A standard curve is produced correlating the chemiluminescence values with standard concentrations of H_2O_2.

18.3.2 Lucigenin-Dependent Chemiluminescence Assay for $O_2{}^{\cdot -}$

The accumulation of $O_2^{\cdot -}$ is measured by the chemiluminescence of lucigenin, (Papadakis and Roubelakis-Angelakis 1999). The assay is conducted in a total volume of 2 mL by placing 0.2 mL of 1 mM lucigenin in 0.1 M glycine-NaOH buffer (pH 9.0) containing 1 mM EDTA. All the other conditions described for H_2O_2 assays are also followed for the $O_2^{\cdot -}$ assays. Counts are reported every 6 s for 30 s and the last two values are averaged. The system xanthine/xanthine oxidase is used, as described by Murphy and Auh (1996), in order to convert chemiluminescence data to production rates. One million counts per minute (10^6 cpm in the chemiluminescence assay) are equivalent to 38.21 ± 2.16 pmol $O_2{}^{-}$ produced min^{-1} (mean $\pm$ SE, three experiments).

18.3.3 Assay for $O_2{}^{\cdot -}$ - Synthase

The assay is described in Papadakis and Roubelakis-Angelakis (1999); one mL of the reaction mixture contains 100 mM glycine-NaOH, pH 9, 1 mM EDTA, 200 μM NADH or 100 μM NADPH, 0.02% (v/v) Triton X100 and 0.4 mM lucigenin. At zero time, 5 μg of plasma membrane protein is added and the mixtures are counted for 1 min in a liquid scintillation counter.

Plasma membranes are isolated from untreated grapevine protoplasts (controls) and from protoplasts treated with 1% (w/v) cellulase Onozuka for 30 min, as previously described (Papadakis and Roubelakis-Angelakis 1999). Four volumes of the extraction buffer (50 mM Tris–HCl, pH 7.5, 20% (w/v) sorbitol, 1 mM ascorbate, 1 mM EDTA, 10 mM DDT, 10 μM leupeptin, 0.3% (v/v) Triton X-100) are added to protoplasts and after intense vortexing, the homogenate is centrifuged for 20 min at 13,000*g*. The resulting supernatant is further centrifuged for 50 min at 85,000*g* for separating the microsomal fraction, which is then resuspended in 50 mM Tris–HCl, pH 7.0, 250 mM sucrose, 0.5 M KCl, 10% (v/v) glycerol and re-centrifuged twice in order to remove any unspecifically bound enzyme. Plasma membranes are isolated using a two-phase partition system, as described by de Marco and Roubelakis-Angelakis (1996a).

18.4 Assays for Antioxidant Enzyme and Antioxidant Biomolecules

Reactive oxygen species are generated during isolation and culture of plant protoplasts (Papadakis and Roubelakis-Angelakis, 1999). Use of non-purified cellulase Onozuka during maceration induces a burst of the ROS, $O_2^{\cdot -}$ and H_2O_2 accumulation in grapevine leaf. When protoplasts isolated with purified cellulase are treated with non-purified cellulase, ROS are also generated (Papadakis and Roubelakis-Angelakis 1999). In grapevine protoplasts and plasma membrane vesicles only one ROS generating activity is detected corresponding to a NAD(P)H oxidase-peroxidase, which is responsible for the generation of both ROS (Papadakis and Roubelakis-Angelakis 1999).

Living organisms have developed a wide range of antioxidant strategies for protection from oxidative damage, based on direct radical-scavenging ability of certain chemical species such as ascorbate, glutathione, α-tocopherol and polyamines, and also enzymatic reactions (Apel and Hirt 2004). Superoxide dismutase (SOD, EC 1.15.1.1) is the first enzyme in the detoxifying process; it disproportionates $O_2^{\cdot -}$ to H_2O_2, which is further reduced to H_2O by catalase in peroxisomes and by ascorbate peroxidase (APO, EC 1.11.1.11) in chloroplasts and the cytosol. Oxidized ascorbate resulting from APO activity is reduced by monodehydroascorbate reductase (MDHAR, EC 1.6.5.4), and dehydroascorbate reductase (DHAR, EC 1.8.5.1) in reactions, which exploit NADPH and glutathione; glutathione reductase (GR, EC 1.6.4.2) completes the cycle maintaining a high ratio between reduced glutathione (GSH) and its oxidized form (GSSG) in the cell.

18.4.1 Antioxidant Enzymes

Total proteins are extracted from leaf tissue and protoplasts as already described (Papadakis et al. 2001b). In brief, extraction buffer consists of 0.2 M Tris-HCl, pH 8.0, 5 mM DTT, 0.5 mM PMSF, 10 μM leupeptin, 10% (w/v) glycerol, 0.25% (w/v) Triton X100 and 20% (w/v) insoluble polyvinylpolypyrrolidone. Ascorbate (1 mM) is included when the protein extract is to be used for APO assay. The samples are homogenized with extraction buffer using a Polytron (Ultra Turrax T25, probe S15 N 10G) at a speed of 20,000 rpm. The homogenates are centrifuged at 40,000*g* for 30 min and the supernatants divided into aliquots and frozen at −80°C. The

Table 18.1 Determination of antioxidant enzymes activity

Enzyme	Reaction buffer	nm	ε ($mM^{-1} \cdot cm^{-1}$)	References
APO	50 potassium phosphate, pH 7.0 0.5 mM ascorbate 0.2 mM H_2O_2	290	2.8	Nakano and Asada (1981)
MDHAR	50 mM Tris–HCl, pH 7.6 1 mM ascorbate 1 unit APO 0.2 mM NADH	340	6.22	Serrano et al. (1994)
DHAR	50 mM potassium phosphate, pH 6.5 0.1 mm EDTA 5 mM GSH 0.5 mM dehydroascorbate	265	14	Asada (1984)
GR	100 mM Tris–HCl, pH 7.5, 1 mM EDTA 1 mM GSSG 0.05 mM NADPH	340	6.22	Foyer and Halliwell (1976)
SOD	10 mM potassium phosphate, pH 7.5 0.222 mM o-dianisidine 0.014 mM riboflavin (under cool white)	460	Compared to 1 Unit Horseradish SOD, Sigma	Misra and Fridovich (1977)

entire extraction procedure is performed at 4°C. Protein determination is performed according to Lowry et al. (1951).

Total enzyme activity (APO, MDHAR, DHAR, GR, SOD) is determined in crude protein extracts using the methods that are summarized at Table 18.1. All enzyme assays are carried out at 25°C. Isoenzymic analysis of antioxidant enzymes is performed as described in Table 18.2 and is presented in Fig. 18.4.

For RNA blotting, extraction of RNA from grapevine protoplasts is performed as described by Papadakis et al. (2001b). Total RNA is quantified by spectroscopy and further confirmed by gel electrophoresis and ethidium bromide staining. For RNA blot analysis, 15 μg of total RNA is denaturated in formaldehyde, electrophoresed and transferred to GeneScreen membranes (NEN) by capillary blotting. RNA is fixed on the membranes using 30 sec UV exposure followed by 2 h incubation at 90°C. Membranes are hybridized at 60°C with a specific ^{32}P-labeled probe, prepared with random priming (Feinberg and Vogelstein 1983), using as template a cDNA fragment of 494 bp corresponding to cytoplasmic SOD of *Nicotiana plumbaginifolia* (Tsang et al. 1991) or to cytoplasmic APO (Chatfield and Dalton 1993).

18.4.2 Antioxidant Biomolecules

Ascorbate and glutathione have central and interrelated roles in oxidative stress, acting both chemically and as substrates of enzyme-catalyzed detoxification reactions (Foyer and Noctor, 2005). Total and reduced ascorbate are measured as described

Table 18.2 Separation of isoenzymes of antioxidant enzymes

Enzyme	Electrophoresis	Reaction buffer	References
APO	*Stacking gel*: 4% polyacrylamide* running at 15 mA *Resolving gel:* 10% polyacrylamide* running at 20 mA 10 mM ascorbate in electrophoresis buffer (pre-running for 30 min at 20 mA)	*Buffer 1* (30 min in dark): 50 mM potassium phosphate, pH 7.0 2 mM ascorbate *Buffer 2* (30 min): 50 mM potassium phosphate, pH 7.0 4 mM ascorbate 2 mM H_2O_2 *Buffer 3* (light): 50 mM potassium phosphate, pH 7.8 14 mM TEMED 1.2 mM 4-nitroblue tetrazolium chloride	Rao et al. (1995)
GR	*Stacking gel*: 4% polyacrylamide* running at 15 mA *Resolving gel:* 7.5% polyacrylamide* running at 25 mA	*Buffer 1* (5–10 min): 250 mM Tris–HCl, pH 8.4 4 mM GSSG 1.5 mM NADPH 2 mM 5,5′-dithiobis-2-nitrobenzoic acid	Foyer et al. (1991)
SOD	*Stacking gel*: 4% polyacrylamide* running at 15 mA *Resolving gel:* 7.5% polyacrylamide* running at 25 mA	*Buffer 1* (30 min in dark): 50 mM potassium phosphate, pH 7.0 42 mg mL^{-1} 4-nitroblue tetrazolium chloride *Buffer 2* (20 min in dark): 50 mM potassium phosphate, pH 7.4 0.1 mg mL^{-1} riboflavin 0.25% (v/v) TEMED Light induces the development of isoenzyme zones	Beauchamp and Fridovitch (1971)

by Wang et al. (1991); total and oxidized glutathione are measured according to Akerboom and Sies (1981).

The diamine putrescine (Put), the triamine spermidine (Spd) and the tetramine spermine (Spm) are the main polyamines (PAs) found in all living cells. PAs may occur in the free molecular form (S, soluble), or/and associated with small molecules like phenolic acids (SH, soluble hydrolyzed, soluble conjugated forms), or/and with various macromolecules like proteins (PH, pellet hydrolyzed, insoluble conjugated forms). PAs have been implicated in a number of cellular and molecular processes in eukaryotic cells; in higher plants, PAs act as modulators of several developmental processes, but they also have a significant impact in plant defence and adaptation to

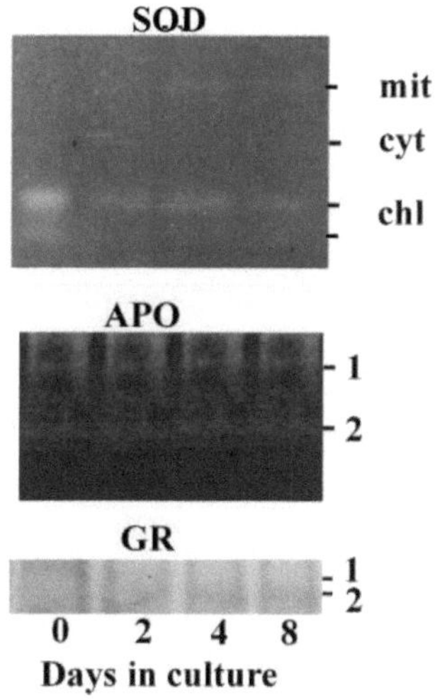

Fig. 18.4 Activity staining of SOD, APO and GR during grapevine protoplast culture

stress conditions (Groppa and Benavides 2008). Endogenous polyamines (PAs) are extracted and their quantities are determined according to Papadakis et al. (2005). Leaf tissue, already pulverized in liquid nitrogen, and protoplasts are homogenized in 1:5 (w/v) of 5% (v/v) ice-cold perchloric acid (PCA). The homogenates are centrifuged at 20,000*g* for 30 min (Sorvall, Dupont). Aliquots of the supernatant and the resuspended pellet (in 1 N NaOH) and their hydrolyzed products are evaporated and redissolved in 200 μL of PCA. [5% (v/v)]. Two hundreds μL of the original supernatant, that contained the free PAs (S), of the soluble hydrolyzed fraction, that contained the conjugated with micromolecules PAs (SH) and of the hydrolyzed pellet, that contained the conjugated with macromolecules PAs (PH), are benzoylated. The benzoyl-PAs are further extracted with diethyl ether; the ether phase is collected and evaporated to dryness and the remainder is redissolved in 200 μl 63% (v/v) methanol. The benzoyl-PAs are separated in a C-18 narrow-bore column with a Hewlett-Packard 1100 HPLC system (Hewlett-Packard, Wadbronn, Germany). Quantification of Put, Spd and Spm is achieved using standard curves. Further information about PAs can be found in Chapter 13 of this book and also in Paschalidis et al. (2009).

18.5 Final Remarks

Plant cells possess the unique property of totipotency, which consists in the ability of somatic cells or protoplasts from fully differentiated, non-dividing cells, to dedifferentiate, re-enter the cell cycle, and proliferate, eventually regenerating the whole plant (Papadakis et al. 2009). Therefore, when given the correct chemical and physical stimuli, each protoplast is theoretically capable of regenerating a new wall and of undergoing repeated mitotic division to produce daughter cells from which fertile plants may be regenerated via the tissue culture process. Protoplast-to-plant systems are available for many species, with an extensive literature relating to their exploitation upon somatic hybridization by protoplast fusion or genetic transformation by gene transfer mediated through *Agrobacterium* and biolistics (Davey et al. 2004, Papadakis et al. 2009). Due to lack of cell walls, the transfer of genes into the plant genome via protoplast transformation is not complicated and plants derived from protoplasts are generally clonal in origin which eliminates chimerism in transgenic plants (Visser et al. 2003).

However, protoplasts from some species (perennial woody species, including grapevine, cereals and legumes), are unable to express their totipotency, exhibiting recalcitrance to plant regeneration, which limits their utilization as tools for plant breading programs. Oxidative stress has been considered as a potential crucial factor accounting for recalcitrance, as supported by the profiles of generated ROS and ROS-scavenging enzymes, the modified cell redox state, as well as the altered endogenous titers of polyamine and phytoalexin levels. Nevertheless, efforts in developing techniques were initiated 10 years ago, with the first successful case in plant regeneration of grapevine protoplasts derived from embryogenic callus (Reustle et al. 1995, Zhu et al. 1997).

As single cell systems, protoplasts have also been widely used for the development of new scientific knowledge involving cell membrane functioning, synthesis of pharmaceutical products, and toxicological assessments (Davey et al. 2004). Grape berry protoplasts may also be used to purify the vacuole, which occupies a large part of the cell and accumulates the most important grape components such as sugars, acids, phenolics and ions like potassium, and even toxic compounds like pesticides.

References

Akerboom T, Sies H (1981) Assay of glutathione, glutathione disulfide, and glutathione mixed disulfides in biological samples. Methods Enzymol 77:373–382

Apel K, Hirt H (2004) Reactive oxygen species: metabolism, oxidative stress, and signal transduction. Annu Rev Plant Biol 55:373–399

Asada K (1984) Chloroplasts: formation of active oxygen and its scavenging. Methods Enzymol 105:422–429

Auh CK, Murphy TM (1995) Plasma membrane redox enzyme is involved in the synthesis of $O_2{}^{\cdot-}$ and H_2O_2 by *Phytophthora* elicitor-stimulated rose cells. Plant Physiol 107:1241–1247

Barbier M, Bessis R (1988) Effets de differents facteurs contributant a l'amelioration de l'isolement de protoplastes a partir de feuilles de vignes (*Vitis vinifera* L.). Bull Soc Bot Fr Lett Bot 135:251–261

Barbier M, Bessis R (1990) Isolation and culture of grapevine cultivar Chardonnay leaf protoplasts. Euphytica 47:39–44

Barnavon L, Doco T, Terrier N, Ageorges A, Romieu C, Pellerin P (2000) Analysis of cell wall neutral sugar composition, β-galactosidase activity and a related c-DNA clone throughout the development of *Vitis vinifera* grape berries. Plant Physiol. Biochem. 38(4):289–300

Beauchamp C, Fridovitch I (1971) Superoxide dismutase: improved assays and an assay applicable to acrylamide gels. Anal Biochem 44:276–287

Chatfield M, Dalton DA (1993) Ascorbate peroxidase from soybean root nodules. Plant Physiol 103:661–662

Cocking EC (1960) A method for the isolation of plant protoplasts and vacuoles. Nature 187: 927–929

Cocking EC (1961) Properties of isolated protoplasts. Nature 191:780–782.

Conde C, Silva P, Fontes N, Dias ACP, Tavares RM, Sousa MJ, Agasse A, Delrot S, Gerós H (2007) Biochemical Changes throughout grape berry development and fruit and wine quality. Food 1:1–22

Davey MR, Anthony P, Power JB, Lowe KC (2004) Plant protoplasts: status and biotechnological perspectives. Biotechnol Adv 23:131–171

DeFilippis LF, Ziegler H (1985) The physiology of grapevine (*Vitis vinifera*) protoplasts isolated from green and senescing leaves. Biochem Physiol Pflanzen 180:645–653

de Marco A, Roubelakis-Angelakis KA (1996a) The complexity of enzymic control of hydrogen peroxide concentration may affect the regeneration potential of plant protoplasts. Plant Physiol 110:137–145

de Marco A, Roubelakis-Angelakis KA (1996b) Hydrogen peroxide plays a bivalent role in the regeneration of protoplasts. J Plant Physiol 149:109–114

Deswarte C, Rouquier P, Roustan JP, Dargent R, Fallot J (1994) Ultrastructural changes produced in plantlet leaves and protoplasts of *Vitis vinifera* cv Cabernet Sauvignon by eutypine, a toxin from *Eutypa lata*. Vitis 33:185–188

Feinberg AP, Vogelstein B (1983) A technique for radiolabelling DNA restriction endonuclease fragments to high specificity. Anal Biochem 132:6–13

Fontes N, Delrot S, Gerós H (2009) Method to obtain intact, viable protoplasts from grape berry mesocarp cells and biotechnological applications. Portuguese patent application no 103851

Fontes N, Delrot S, Gerós H (2010a) A method for the isolation of protoplasts from grape berry mesocarp tissue. Recent Patents Biotechnol 4:125–129

Fontes N, Silva R, Vignault C, Lecourieux F, Gerós H, Delrot S (2010b) Purification and functional characterization of protoplasts and intact vacuoles from grape cells. BMC Res Notes (http://www.biomedcentral.com/1756-0500/3/19).

Foyer C, Lelandais M, Galap C, Kunert KJ (1991) Effects of elevated cytosolic glutathione reductase activity on the cellular glutathione pool and photosynthesis in leaves under normal and stress conditions. Plant Physiol 97:863–872

Foyer CH, Halliwell B (1976) The presence of glutathione and glutathione reductase in chloroplasts: a proposed role in ascorbic acid metabolism. Planta 133:21–25

Foyer CH, Noctor G (2005) Redox homeostasis and antioxidant signalling: a metabolic interfase between stress perception and physiological responses. Plant Cell 17:1866–1875

Fuchs Y, Saxena YA, Gamble HR, Anderson JD (1989) Ethylene biosynthesis-inducing protein from cellulisin is an endoxylanase. Plant Physiol 89:138–143

Groppa MD, Benavides MP (2008) Polyamines and abiotic stress: recent advances. Amino Acids 34:35–45

Jones KH, Senft JA (1985) An improved method to determine cell viability by simultaneous staining with fluorescein diacetate–propidium iodide. J Histochem Cytochem 33:77–79

Katsirdakis KC, Roubelakis-Angelakis KA (1992a) A modified culture medium and culture conditions increase viability and cell wall synthesis in grapevine (*Vitis vinifera* L. cv Sultanina) leaf protoplasts. Plant Cell Tiss Org Cult 28:255–260

Katsirdakis KC, Roubelakis-Angelakis KA (1992b) Ultrastructural and biochemical aspects of cell wall regeneration in recalcitrant and regenerating leaf protoplasts. *In Vitro* Cell Dev Biol 28:90–96

Klercker I (1892) Eine methode zur isolierung lebender protoplasten. Ofvers vet-Akad Forhdl. 9:463–474

Lee N, Wetzstein HY (1988) Protoplasts isolation and callus production from leaves of tissue-cultured *Vitis* spp. Plant Cell Rep 7:531–534

Lowry OH, Rosebrough NJ, Farr AL, Randall RJ (1951). Protein measurement with the Folin-Phenol reagents. J. Biol Chem 193:265–275

Marty F (1999) Plant vacuoles. Plant Cell 11:587–599

Mii M, Zou YM, Sugiyama T, Yanagihara S, Iizuka M (1991) Highfrequency callus formation from protoplasts of *Vitis labruscana* Bailey and *Vitis thunbergii* Sieb. et Zucc. by embedding in gellan gum. Scientia Hort 46:253–260

Misra HP, Fridovich I (1977) Superoxide dismutase: a photochemical augmentation assay. Arch Biochem Biophys 181:308–312

Paschalidis KA, Moschou PN, Aziz A, Toumi I, Roubelakis-Angelakis KA (2009) Polyamines in grapevine: an update. In: Roubelakis-Angelakis KA (ed) Grapevine molecular physiology and biotechnology, 2nd edn. DOI 10.1007/978-90-481-2305-6_16. Springer Academic Publishers, Netherlands

Murashige I, Skoog F (1962) A revised medium for rapid growth and bioassays with tobacco tissue cultures. Physiol Plant 15:473–497

Murphy TM, Auh CK (1996) The superoxide synthases of plasma membrane preparations from cultured rose cells. Plant Physiol 110:621–629

Nakano Y, Asada K (1981) Hydrogen peroxide is scavenged by ascorbate-specific peroxidase in spinach chloroplasts. Plant Cell Physiol 22:867–880

Nishimura M, Hara-Nishimura I, Robinson SP (1984) Isolation of metabolically competent protoplasts from grapevine leaves. Plant Sci Lett 37:171–175

Papadakis AK, Fontes N, Gerós H, Roubelakis-Angelakis KA (2009) Progress in grapevine protoplast technology. In: Roubelakis-Angelakis KA (ed) Grapevine molecular physiology and biotechnology, 2nd edn. DOI 10.1007/978-90-481-2305-6_16. Springer Academic Publishers, Netherlands

Papadakis AK, Reustle GM, Roubelakis-Angelakis KA (2001a) Protoplast technology in grapevine. In: Roubelakis-Angelakis KA (ed) Molecular biology and biotechnology of the grapevine. Kluwer Academic Publishers, The Netherlands

Papadakis AK, Paschalidis KA, Roubelakis-Angelakis KA (2005) Biosynthesis profile and endogenous titers of polyamines differ in totipotent and recalcitrant plant protoplasts. Physiol Plant 1250:10–20

Papadakis AK, Roubelakis-Angelakis KA (1999) The generation of active oxygen species differs in *Nicotiana* and *Vitis* plant protoplasts. Plant Physiol 121:197–245

Papadakis AK, Siminis CI, Roubelakis-Angelakis KA (2001b) Reduced activity of antioxidant machinery is correlated with suppression of totipotency in plant protoplasts. Plant Physiol 126:434–444

Rao M, Hale BA, Ormrod DP (1995) Amelioration of ozone-induced oxidative damage in wheat plants grown under high carbon dioxide. Plant Physiol 109:421–432

Reustle G, Alleweldt G (1990) Isolation and culture of grapevine protoplasts. 5th International Symposium on Grape Breeding (*Vitis* special issue), St. Martin, Germany

Reustle G, Alleweldt G (1991) Callus formation from protoplasts of grapevine (*Vitis* spp). Phys Plant 82:97

Reustle G, Harst M, Alleweldt G (1995) Plant regeneration of grapevine (*Vitis* sp.) protoplasts isolated from embryogenic tissue. Plant Cell Rpts 15:238–241.

Roubelakis-Angelakis KA (1993) An assessment of possible factors contributing to recalcitrance of plant protoplasts. In: Roubelakis-Angelakis KA, Tran Thanh Van K (eds) Morphogenesis in plants: molecular approaches. Plenum, New York

Roubelakis-Angelakis KA, Zivanovit SA (1991) A new culture medium for in vitro rhizogenesis of grapevine (*Vitis* spp.) genotypes. Hort. Sci 26:1551–1553

Serrano A, Cordoba F, Gonzales-Reyes JA, Navas P, Villalba JM (1994) Purification and characterization of two distinct NAD(P)H dehydrogenases from onion (*Allium cepa* L) root plasma membrane. Plant Physiol 106:87–96

Skene KGM (1975) Production of callus from protoplasts of cultured grape pericarp. Vitis 14: 177–180

Taiz L (1992) The Plant Vacuole. J Exp Biol 172:113–122.

Theodoropoulos PA, Roubelakis-Angelakis KA (1989) Mechanism of arginine transport in *Vitis vinifera* L. protoplasts. J Exp Bot 40:1223–1230

Theodoropoulos PA, Roubelakis-Angelakis KA (1990) Progress in leaf protoplast isolation and culture from virus-free axenic shoot cultures of *Vitis vinifera* L. Plant Cell Tiss Org Cult 20: 15–23

Theodoropoulos PA, Roubelakis-Angelakis KA (1991) Glucose transport in *Vitis vinifera* L. protoplasts. J Exp Bot 42:477–483

Tsang EWT, Bowler C, Herouart D, Van Camp W, Villarroel R, Genetello C, Van Montagu M, Inzé D (1991) Differential regulation of superoxide dismutases in plants exposed to environmental stress. Plant Cell 3:783–792

Ui S, Suzuki M, Kubota S, Masuda H, Muraki H, Yamakawa Y, Sato T (1990) Cooperative effect of activated charcoal and gellun gum on grape protoplast culture. Agric Biol Chem 54:207–209

Visser RGF, Raemakers CJJ, Bergervoet van Deelen JEM (2003) Methods for producing and transforming Cassave protoplasts. *United States patent* US 6551 827

Wang SY, Jiao HJ, Faust M (1991) Changes in ascorbate, glutathione, and related enzyme activities during thiadiazuron-induced bud break of apple. Physiol Plant 82:231–236

Wright DC (1985) Factors affecting isolation of protoplasts from leaves of grape (*Vitis vinifera*). Plant Cell Tiss Org Cult 4:95–100

Zhu YM, Hoshino Y, Nakano M, Takahashi E, Mii M (1997) Highly efficient system of plant regeneration from protoplasts of grapevine (*Vitis vinifera* L.) through somatic embryogenesis by using embryogenic callus cultures and activated charcoal. Plant Sci 123:151–157

Chapter 19
RNA Extraction from Grapevine Woody Canes for Gene Expression Analysis by Real-Time RT-PCR

Sophie Bordiec, Fanja Rabenoelina, Florence Mazeyrat-Gourbeyre, Christophe Clément, and Fabienne Baillieul

Abstract RNA extraction from grapevine, as from all woody plants, is known to be problematic. Several methods have been developed last years, among which some are available for extraction from wood, but they are usually time-consuming, or validated for northern hybridization. Here we present a rapid method for RNA extraction effective for studying grapevine wood gene expression by real-time RT-PCR.

Keywords Grapevine · RNA · Real-time PCR · Gene expression · Wood

Contents

Abbreviations

CTAB	cetyltrimethylammonium bromide
DEPC	diethylpyrocarbonate
RT-PCR	reverse-transcription polymerase chain reaction
TAE	Tris-Acetate-EDTA

F. Baillieul (✉)
Laboratoire de Stress, Défenses et Reproduction des Plantes, Unité de Recherche Vigne et Vins de Champagne, Stress et Environnement, UPRES EA 2069, UFR Sciences Exactes et Naturelles, Université de Reims, Champagne-Ardenne Reims 51687, France
e-mail: fabienne.baillieul@univ-reims.fr

S. Delrot et al. (eds.), *Methodologies and Results in Grapevine Research*,
DOI 10.1007/978-90-481-9283-0_19,

19.1 Introduction

RNA quality and integrity are the main limiting factors in the application of RT-PCR techniques. RNA extraction from woody plants, such as grapevine, is particularly difficult due to their high content in carbohydrates, acids, tannins and phenolics (Loulakakis et al. 1996) which may limit extraction yield, RNA quality and inhibit subsequent enzymatic reactions (Wang et al. 2000).

Many methods have been developed to extract RNA from woody plants such as forest trees (Allona et al. 1998, Wang et al. 2000, Le Provost et al. 2007) and fruit trees (Gambino et al. 2008). They involve the use of ultracentrifugation steps (Loulakakis et al. 1996, Gambino et al. 2008), CTAB (cetyltrimethylammonium bromide) (Gambino et al. 2008) or extraction/purification kits (Wang et al. 2000, Le Provost et al. 2007, Gambino et al. 2008). These protocols are usually not adapted for RNA extraction from wood or are time-consuming (2-day extraction, treatment of plastic/glasswares and solutions with DEPC (diethylpyrocarbonate)).

Several methods exist for RNA extraction from grapevine but they were developed for in vitro plantlets (Thomas and Schiefelbein 2002, Thomas et al. 2003), leaves and roots (Loulakakis et al. 1996, Moser et al. 2004, Gambino et al. 2008), cell cultures (Loulakakis et al. 1996) or for the detection of pathogens in wood (MacKenzie et al. 1997, Nassuth et al. 2000, Osman and Rowhani 2008). Recently, Gambino et al. (2008) proposed a rapid CTAB method for extracting RNA from mature canes suitable for northern blotting analysis.

Here we present another rapid and simple method for RNA extraction from wood, validated for gene expression analysis by real-time RT-PCR.

19.2 Material and Methods

19.2.1 Plant Material and Sampling

Woody canes were collected from vegetative or fruiting woody cuttings. Woody canes were cut in 1 cm long pieces after removal of epidermis. They were frozen in liquid nitrogen, then stored at −80°C until use. Samples were ground in liquid nitrogen to a fine powder with a laboratory ball mill with inox grinding balls of 30 mm diameter (Prolabo).

19.2.2 RNA Extraction Protocol

Solutions were not treated with DEPC and all experiments until PCR were realized with disposable RNase-free tips and microcentrifuge tubes.

The Concert^{TM} Plant RNA reagent method for small scale RNA isolation was used according to the manufacturer's instructions (Invitrogen, ref 12322-012) but with modifications. One hundred mg of powdered sample were extracted by adding 0.5 mL of cold "Plant RNA reagent". Following a strong vortexing for 90 s, the mixture was incubated for 5 min at room temperature. After centrifugation at 12,000*g* for 2 min, the supernatant was extracted twice with 300 μL of chloroform/isoamyl alcohol (24:1, v/v) and centrifugated at 12,000*g* for 10 min at 4°C. The upper phase was collected in a new tube and RNA was precipitated with an equal volume of isopropanol and 100 μL of 5 M NaCl. Following 10 min incubation at room temperature, RNA was pelleted by centrifuging at 12,000*g* for 10 min at 4°C. The supernatant was discarded and the pellet was washed with 75% ethanol, air-dried for approximately 20 min and resuspended in 10 μL of milliQ water. Genomic DNA was removed with RNase-free DNase treatment (Promega, ref 9PIM610). Five μL of total RNA were treated with 1 U of enzyme according to manufacturer's instructions.

19.2.3 RNA Analysis

RNA purity and concentration were assessed by determining the spectrophotometric absorbance of the samples at 260 and 280 nm and ratios A_{260}/A_{280}. RNA integrity was evaluated from the 28S and 18S rRNA bands on 1% agarose gel after electrophoresis in 0.5X TAE (Tris-Acetate-EDTA), staining with ethidium bromide and visualisation under UV light.

19.2.4 Real-Time RT-PCR

One hundred and fifty ng of total RNA were reverse-transcribed by using the Verso SYBR Green 2-Step QRT-PCR Rox kit (ThermoElectron, ref AB-4113/A). RT conditions were 42°C for 30 min, then 2 min at 95°C.

PCR reactions were carried out in 96-well plates (15 μL per well) in a reaction buffer containing 1X SYBR green I mix (Absolute Blue QPCR SYBR Green, ThermoElectron, ref AB-4162/B; including *Taq* polymerase, dNTPs, SYBR Green), 280 nM primers (forward and reverse) and 5 μL of cDNA diluted 10 times. PCR conditions were 95°C for 15 s (denaturation) and 60°C for 1 min (annealing/extension) for 40 cycles on a Chromo4 thermocycler (Bio-Rad). PCR efficiency of the primers sets was calculated by performing real-time PCR on several dilutions. We used the following *VvEF1α* primers: *VvEF1α*-F (5′-GAACTGGGTGCTTGATAGGC-3′) and *VvEF1α*-R (5′-AACCAAAATATCCGGAGTAAAAGA-3′).

19.3 Results and Discussion

Canes were ground in a fine powder with a ball grinder that allowed easy and fast processing of numerous samples. Epidermis was removed from the canes before grinding. Initially, extractions were done with whole wood cuttings, but RNA was difficult to resuspend in water and solutions were sticky, suggesting the presence of

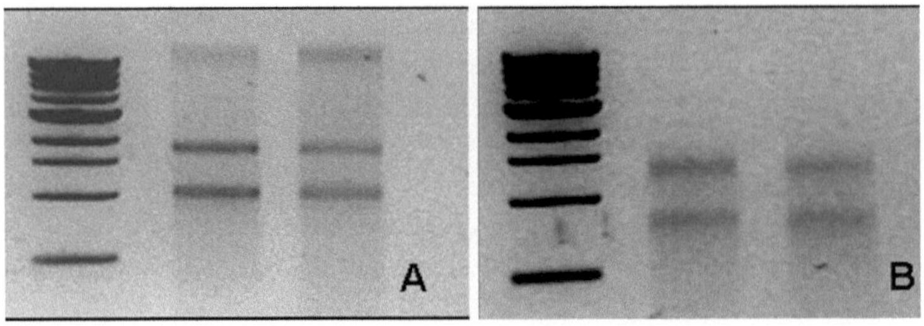

Fig.19.1 Total RNA extracted before (**a**) and after (**b**) DNase treatment for 2 wood samples. On each gel electrophoresis, the first lane corresponds to 1 Kb DNA ladder (Ozyme, ref N3232S). (**a**) One μL of extract was loaded. (**b**) Control of DNase efficiency for 150 ng of treated total RNA

undesirable compounds which may interfere for further analysis. Moreover, RNA gel electrophoresis revealed some RNA degradations (data not shown). So we tested extractions without the epidermis which is a dead tissue. However, the suber, which is a living tissue, was also eliminated because it was linked to epidermis. In this case, RNA resuspended easily, and for all RNA samples tested, distinct 28S and 18S rRNA bands (about 1.7 and 1 Kbp respectively) without degradation were observed after gel electrophoresis analysis (Fig. 19.1a), indicating that we obtained good integrity RNA. Unfortunately genomic DNA contamination was also observed (band above 10 Kbp).

DNA is particularly troublesome in gene expression studies using PCR because its presence contributes to false representation of gene expression within a given RNA sample. DNase treatment was then included at the end of the RNA extraction protocol and efficiently removed DNA without RNA degradation (Fig 19.1b). Lack of DNA contamination was also checked by performing a standard PCR with primers designed to amplify a fragment of the housekeeping gene actin (accession number AF369524), using directly DNase-treated RNA as template, omitting the prior RT step. No amplification was observed (data not shown).

Quality and quantification of extracted RNA (after DNase treatment) were estimated by spectrophotometry. The A_{260}/A_{280} ratio was comprised between 1.6 and 2, and the average yield was about 4 μg total RNA per 100 mg of starting plant material. A_{260}/A_{280} ratio lower than 1.8 suggests minor protein contamination, but this did not interfere with further molecular analysis (RT and real-time PCR). Usually, within a series of extractions, RNA yield and quality were close for all samples, but they varied between experiments, possibly depending on the physiological status of the plant or room temperature. In some cases, yield reached 10 μg RNA per 100 mg of starting material. Compared with methods established on stem from grapevine plantlets (Thomas and Schiefelbein 2002, Thomas et al. 2003) or mature canes (Gambino et al. 2008), this protocol provided RNA of good quality with a lesser or equivalent yield. In fact, except for the modified CTAB method from Gambino et al. (2008), "fast extraction" methods usually gave RNA of good quality associated with a low yield compared to time-consuming protocols. Nevertheless, yield of extracted RNA was over than necessary for subsequent real-time RT-PCR

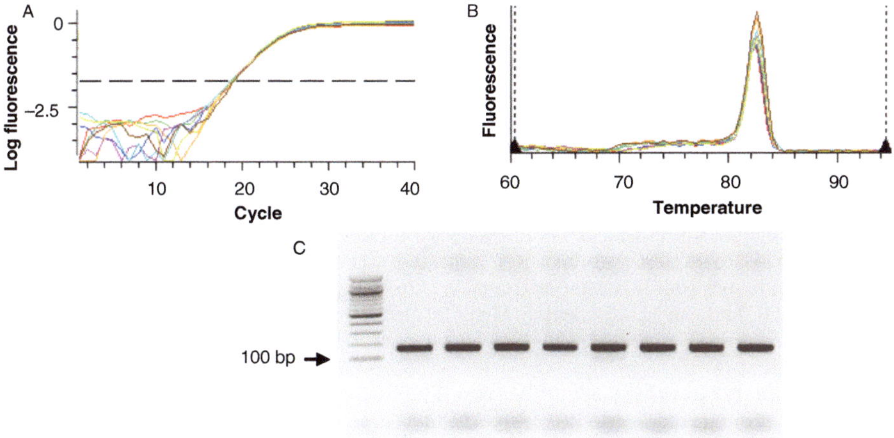

Fig.19.2 (**a**) Amplification plot of real-time RT-PCR targeting *VvEF1α* gene. (**b**) Melting curve for *VvEF1α* amplification products. (**c**) Agarose gel electrophoresis analysis of the real-time RT-PCR products. The first lane corresponds to 100 bp DNA ladder (Ozyme, ref N3231S). The presented data are a duplicate of 4 different samples

analysis because only 150 ng of total RNA were used for reverse-transcription and cDNA were diluted 30 times (final volume) for gene expression analysis.

The cDNAs were amplified by PCR and the housekeeping gene *VvEF1α* (Terrier et al. 2005) was used for normalization. Real-time PCR was successfully performed (Fig. 19.2a). Analysis of the melting curves as well as agarose gel electrophoresis of PCR *VvEF1α* products revealed specific amplification (Fig. 19.2b, c). Efficiency of *VvEF1α* primers was well-established with 96% efficiency validated over 4 log (data not shown). The average Ct for *VvEF1α* gene expression was around 20. Efficiency of primers specific to other genes, like those implicated in defense mechanisms, were also tested with good efficiencies (data not shown). For example, the Ct values, corresponding to basal expression, of *VvGluc* coding for a glucanase (a pathogenesis-related protein) and *VvSTS* coding for the stilbene synthase (implicated in the resveratrol biosynthesis pathway) were respectively around 26 and 23.

Our protocol is also available for RNA extraction from other grapevine tissues from plantlets (roots, stem, leaves) or cuttings (roots, leaves and flowers). On cuttings, RNA yields from roots and leaves were respectively 7.6 ± 0.9 and 22 ± 5 μg per 100 mg of plant material, our method giving a better yield for leaves than the modified CTAB method (Gambino et al. 2008). RNA extracted from cuttings leaves were directly free of contaminant genomic DNA.

In conclusion, the simple protocol described here allows rapid (less than 3 h) extraction of RNA from grapevine woody canes. This method results in RNA quality and yield which are consistent for gene expression analysis by real-time RT-PCR and semi-quantitative RT-PCR.

References

Allona I, Quinn M, Shoop E, Swope K, St Cyr S, Carlis J, Riedl J, Retzel E, Campbell MM, Sederoff R, Whetten RW (1998) Analysis of xylem formation in pine by cDNA sequencing. Proc Natl Acad Sci USA 95:9693–9698

Gambino G, Perrone I, Gribaudo I (2008) A rapid and effective method for RNA extraction from different tissues of grapevine and other woody plants. Phytochem Anal 19:520–525

Le Provost G, Herrera R, Paiva JAP, Chaumeil P, Salin F, Plomoin C (2007) A micromethod for high throughput RNA extraction in forest trees. Biol Res 40:291–297

Loulakakis KA, Roubelakis-Angelakis KA, Kanellis AK (1996) Isolation of functional RNA from grapevine tissues poor in nucleic acid content. Am J Enol Vitic 47:181–185

MacKenzie DJ, McLean MA, Mukerji S, Green M (1997) Improved RNA extraction from woody plants for detection of viral pathogens by reverse transcription-polymerase chain reaction. Plant Dis 81:222–226

Moser C, Gatto P, Moser M, Pindo M, Velasco R (2004) Isolation of functional RNA from small amounts of different grape and apple tissues. Mol Biotechnol 26:95–99

Nassuth A, Pollari E, Helmeczy K, Stewart S, Kofalvi SA (2000) Improved RNA extraction and one-tube RT-PCR assay for simultaneous detection of control plant RNA plus several viruses in plant extracts. J Virol Methods 90:37–49

Osman F, Rowhani A (2008) Real-time (TaqMan®) assays for the detection of viruses associated with rugose wood complex of grapevine. J Virol Methods 154:69–75

Terrier N, Glissant D, Grimplet J, Barrieu F, Abbal P, Couture C, Ageorges A, Atanassova R, Léon C, Renaudin JP, Dédaldéchamp F, Romieu C, Delrot S, Hamdi S (2005) Isogene specific oligo arrays reveal multifaceted changes in gene expression during grape berry (*Vitis vinifera* L.) development. Planta 222:832–847

Thomas P, Lee MM, Schiefelbein J (2003) Molecular identification of proline-rich protein genes induced during root formation in grape (*Vitis vinifera* L.) stem cuttings. Plant Cell Environ 26:1497–1504

Thomas P, Schiefelbein JW (2002) Improved method for purification of RNA from stem tissue of grapevine and its use in mRNA profiling. Am J Enol Vitic 53:231–234

Wang SX, Hunter W, Plant A (2000) Isolation and purification of functional total RNA from woody branches and needles of Sitka and white spruce. BioTechniques 28:292–296

Chapter 20
A Method for Isolating Total RNA from Mature Buds and Other Woody Tissues of *Vitis Vinifera*

Atiako Kwame Acheampong, Ariel Rotman, Chuanlin Zheng, Alexandra Keren, Tamar Halaly, Omer Crane, Aliza Ogrodovitch, and Etti Or

Abstract Toughness of plant materials and their high secondary plant metabolites, polysaccharides and proteins that bind to and/or co-precipitate with the RNA, are some of the major constraints when extracting total RNA from woody tissues such as mature buds and woody stems. Here, we detail an efficient method for isolating total RNA from woody tissues of grapes. RNA was extracted with high ionic strength buffer at 65°C. Proteins were denatured, and secondary metabolites removed by repeated phenol:chloroform:isoamyl alcohol extractions. The RNA was separated from the DNA by selective precipitation with lithium chloride (LiCl) solution. Though the procedure is laborious and time-consuming, the yield and quality of the RNA extracted were higher compared to other conventional extraction protocols. Yield and purity were spectrophotometrically monitored by UV absorbance (A_{260}/A_{280} and A_{260}/A_{230}). The yield was about 180 μg total RNA per gram of tissue, and the A_{260}/A_{280} and A_{260}/A_{230} absorbance ratios were greater than 2.0. Standard reverse transcription PCR (RT-PCR) yielded 3.5 kb products from RNA isolated by this protocol. Isolated RNA has also been applied in other molecular application such as real-time PCR, Northern blot, dot blot and gene expression studies using Affymetrix Chips.

Contents

A.K. Acheampong (✉)
Institute of Plant Sciences, A.R.O. Volcani Center, Bet Dagan 50250, Israel
e-mail: vhettior@agri.gov.il

S. Delrot et al. (eds.), *Methodologies and Results in Grapevine Research*,
DOI 10.1007/978-90-481-9283-0_20,

Abbreviations

DEPC	Diethyl pyrocarbonate
CTAB	Cetyltrimethyl ammonium bromide
PVP	Polyvinylpyrrolidone
EDTA	Ethylenediaminetetraacetic acid
SDS	Sodium dodecyl sulfate
CIA	Chloroform:Isoamyl alcohol
PCIA	Phenol:Chloroform:Isoamyl alcohol

20.1 Introduction

The importance of high quality RNA in post-genomic studies cannot be overemphasized. The establishment of EST libraries, studies of gene expression and functions, microRNA (miRNA) and small nucleolar RNA (snoRNA) studies largely depend on the quality and quantity of RNA extracted. The major challenges to the extraction of high-quality RNA from woody plant tissues include high concentrations of polysaccharides, polyphenols, and other secondary metabolites which bind to and co-precipitate with the RNA in the presence of alcohols, subsequently interfering with downstream applications (Salzman et al. 1999). The extremely high concentrations of these polysaccharides, proteins and secondary metabolites in *Vitis vinifera* (grapes) make RNA extraction and purification more daunting. Most conventional extraction protocols using modified cetyltrimethyl ammonium bromide (CTAB) protocol of Chang et al. (1993) are useful for extracting RNA from less woody tissues and organs such as leaves, shoot-tip, tendrils and rachis (Boss et al. 1996, Loulakakis et al. 1996, Thomas and Schiefelbein 2002, Iandolino et al. 2004, Moser et al. 2004, Tattersall et al. 2005). These protocols however, failed to recover either maximum quality or yield RNA from woody tissues such as dormant buds and old canes. Furthermore, none of the widely used conventional protocols for extracting RNA from other woody plants (Kolosova et al. 2004, Meisel et al. 2005, Asif et al. 2006, Le Provost et al. 2007) has been effectively employed to extract RNA from these woody grape tissues.

Guanidine-based protocols (Chomczynski and Sacchi 1987, Gambino et al. 2006, Li et al. 2006, Bilgin et al. 2009) and commercial RNA isolation kits such as TRIZOL^{TM} (Gibco-BRL Life Technologies, Gaithering, MD), and RNAeasy^{TM} (RNeasy Plant Mini Kit, Qiagen, Valencia, CA) (Gambino and Gribaudo 2006) have also been found to be unsuitable for extraction from these hard tissues. The fibrous cellulose-binding method of Geuna et al. (1998), sodium perchlorate method (Boss et al. 1996), and a modified silica capture method of Rott and Jelkmann (2001) were all successfully used for extraction from grape berry skins but yielded poor-quality RNA from dormant buds.

Using a modified CTAB-LiCl protocol of Chang et al. (1993), an efficient protocol was developed for extracting high-quality RNA from dormant buds and mature, woody stems. Furthermore, the amount of RNA isolated by this protocol may be 15–20 fold higher than that from other conventional methods. Finally, DNA contamination with this method is comparatively low.

20.2 Materials and Methods

20.2.1 Plant Material

Mature buds and dried stems were collected from field-grown *V. vinifera* cv. Perlette, snap-frozen in liquid nitrogen and stored at −80°C prior to extraction.

20.2.2 Reagents and Solutions

All solutions used in this extraction protocol were treated with 0.1% (V/V) diethyl pyrocarbonate (DEPC), stirred overnight, and autoclaved to inactivate RNases before use.

- CTAB Buffer: 0.1 M Tris HCl (pH 8.0), 25 mM ethylenediaminetetraacetic acid (EDTA; pH 8.0), 2.0 M NaCl, 2% Cetyltrimethyl ammonium bromide (CTAB), 2% polyvinylpyrrolidone (PVP; Mr. 40,000), 2% β-mercaptoethanol (added just before use). All component solutions were prepared from DEPC-treated water and autoclaved twice.
- SSTE: 1.0 M NaCl, 10 mM Tris HCl (pH 8.0), 0.5% Sodium dodecyl sulfate (SDS). All component solutions, except SDS, were prepared from DEPC-treated, double-distilled water, and autoclaved twice.
- 10 M LiCl(aq)
- Chloroform:Isoamyl alcohol (CIA) (24:1 V/V)
- Acid Phenol:Chloroform:Isoamyl alcohol (PCIA)(25:24:1 V/V)
- 70% ethanol (prepared with DEPC-treated water)

20.2.3 Glassware and Plasticware

All glassware and plasticware were sterilized by autoclaving.

20.2.4 Procedure

20.2.4.1 Day 1

- Plant material (2.3 g) was pulverized into finely-ground powder with pre-chilled mortar and pestle in liquid nitrogen. RNA yield depends on extent of grinding and fineness of extracting material. When grinding mature buds, hairy balls, produced after a preliminary stage of grinding, which interfere with the extraction, were removed.

To prevent RNase degradation, it is imperative to keep plant material frozen at all times. If necessary, slowly add extra liquid nitrogen to the mortar during grinding.

- The frozen powder was transferred into a pre-chilled 40 mL Sorvall centrifuge tube (Thermofisher Scientific, Waltham, MA). Pre-heated (65°C) CTAB buffer (20 mL) containing 2% β-mercaptoethanol was added to the frozen powder and mixed well by vortexing. The mixture was incubated at 65°C for 10 min in a water-bath, with intermittent vortexing.

- The mixture was allowed to cool to room temperature and 20 mL of CIA were added and mixed by gently inverting tube for 5 min.
- The mixture was centrifuged at 10,000 rpm (Rotor SS34, Sorvall RC6 centrifuge) for 10 min at room temperature.
- The supernatant was transferred to a new tube and purification repeated by adding 1 volume of CIA, and mixing by inverting tube repeatedly for 5 min.
- Mixture was centrifuged again at 10,000 rpm for 10 min at room temperature, and the upper aqueous phase transferred into a clean 40 mL centrifuge tube.
- Finally, 1/3 volume of 10 M LiCl(aq) was added to the supernatant, mixed briefly by vortexing twice (3 s each), and stored overnight at 4°C to precipitate RNA.

20.2.4.2 Day 2

- The RNA was pelleted by centrifuging at 10,000 rpm for 20 min at 4°C. The supernatant was discarded, and pellet washed with 1 mL ice-cold 70% ethanol.
- Mixture was centrifuged at 10,000 rpm for 3 min at 4°C, and ethanol discarded.
- The remaining ethanol was removed by drying pellets under vacuum at room temperature for 7 min, and again air-dried for 3 min on working bench.
- Pellet was suspended in 1 mL SSTE, and vortexed strongly.
- To the mixture was added 1 mL of PCIA (25:24:1), and mixed by vortexing strongly for 20 s.
- Mixture was transferred into a sterile 2 mL Eppendorf tube (Eppendorf, Barkhausenweg, Hamburg, Germany), centrifuge at 12,000 rpm (Eppendorf bench-top centrifuge) for 5 min at room temperature, and the upper aqueous phase carefully transferred into a sterile 2 mL Eppendorf tube.
- The mixture was further purified by adding 1 volume of CIA, mixed well by vortexing, and centrifuged at 13,000 rpm on a bench-top centrifuge (Eppendorf Centrifuge 5417C) for 5 min.
- The upper aqueous phase was carefully divided into two sterile 2 mL Eppendorf tubes.
- To precipitate RNA, 2 volumes of 100% ethanol were added, mixed by gently inverting tube several times, and stored at −20°C overnight.

20.2.4.3 Day 3

- RNA was pelleted by centrifuging mixture at 13,000 rpm for 15 min.
- The supernatant was discarded and pellet was washed with 100 μL ice-cold ethanol (70%).
- RNA pellet was recovered by centrifuging at 13,000 rpm for 2 min, and discarding ethanol. To remove residual ethanol, tubes were pulsed and ethanol removed

with sharp tips. Pellets were air-dried by keeping on ice for 10 min on working bench.

- RNA pellet was finally resuspended in 30 μL of DEPC-treated, sterile double distilled water, and stored at −80°C until use.

If RNA is to be used for sensitive applications such as Microarray analysis, it must be further cleaned with the RNAeasy™ Plant Mini Kit columns according to manufacturer's instructions.

20.3 Results and Discussion

This protocol is suitable for isolating RNA from dormant buds and dried, woody tissues of grapes. Though laborious and time-consuming compared to commercial kits, it is relatively cheaper and produces large amounts of high-quality RNA.

The tissues were mechanically disrupted by grinding in liquid nitrogen, making most of the nuclear and cytosolic RNA available for extraction. We chose a high-temperature (65°C) extraction protocol because, unlike extraction at room temperature, the yield is about 2 folds higher (Liu et al. 1998). CTAB, a major component of the extraction buffer, chelates most cations required for DNase and RNase activity. In the absence of these cations, the activities of these holoenzymes are drastically reduced. PVP precipitates polyphenols by forming complexes through hydrogen bonding. Increasing the PVP concentration to 2% relative to other conventional CTAB protocols, improved its efficiency in complexing the polyphenols. Acid phenol completes the removal of polyphenols , while the double chloroform wash removes proteins and other secondary metabolites. Whereas LiCl precipitation differentially precipitated large RNA molecules from the mixture of DNA, tRNA and snRNA (Carra et al. 2007), the combination of LiCl and ethanol precipitation greatly increased RNA quality and quantity. This technique has separately been employed in other protocols (Ainsworth 1994, Claros and Canovas 1998, Liu et al. 1998, Scott et al. 1998). Almost the entire extraction process (apart from 2 h of Day 1) was carried out at 4°C to reduce RNA degradation by RNases and to eliminate the chances of secondary metabolite recombining with the RNA, consequently enhancing the quality of the RNA.

The high yield (180 μg total RNA per g of tissue) from this extraction protocol can be attributed to the efficient cleaning and precipitation techniques used. This yield is comparable to that of RNAeasy™ Plant Mini Kit. The quality of the RNA extract was very high and free of proteins contaminants; indicated by the high (2.0–2.2) A_{260}/A_{280} ratio. The A_{260}/A_{230} ratio of the isolate was approximately 2.2, indicating samples free from polysaccharide and phenol contamination. The integrity of the RNA extracted was further assessed by visualization of the ribosomal RNA bands on 1% agarose gel. All the RNA extracts exhibited two distinct bands, corresponding to 28S and 18S rRNA. The intensities of the bands were in the appropriate ratio (28S:18S = 2:1) (Fig. 20.1), indicating very minimum degradation during the extraction process.

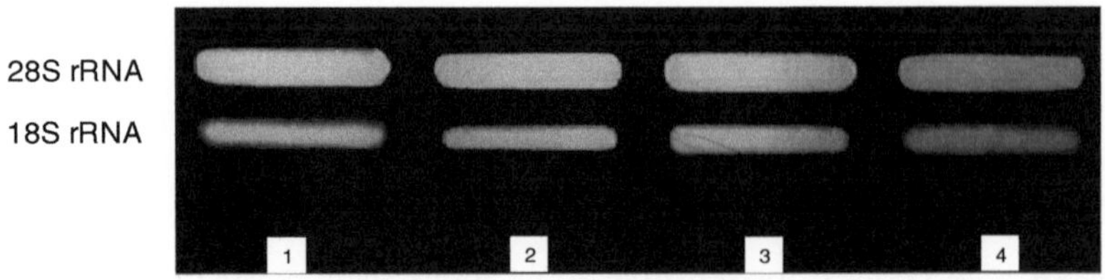

Fig. 20.1 Agarose gel electrophoresis of RNA extracted from 2.3 g of different tissues of *V. vinifera* cv Perllet. Each lane was loaded with 1.0 ug of total RNA. Lane 1: RNA extracted from Young buds; Lane 2: RNA extracted from mature buds; Lane 3: RNA extracted from leaves; Lane 4: RNA extracted from shoot tips

The RNA extracted by this protocol has been successfully used in downstream applications such as Northern blot, dot blot, gene expression studies using Affimetrix Chips, RT-PCR, etc (Or et al. 2000, Ophir et al. 2009). From the RNA extracted, we successfully amplified PCR products 80–3500 bp from cDNA synthesized using standard RT-PCR kit (Promega Corporation, Madison, MI, USA). Though much more time-consuming compared to the CTAB protocol of Gambino et al. (2008), the quality and quantity of RNA extracted by our protocol are significantly higher.

References

Ainsworth C (1994) Isolation of RNA from floral tissue of *Rumex acetosa* (Sorrel). Plant Mol Biol Report 12:198–203.

Asif M, Trivedi P, Solomos T, Tucker M (2006) Isolation of high–quality RNA from apple (*Malus domestica*) fruit. J Agric Food Chem 54:5227–5229

Bilgin DD, DeLucia EA, Clough SJ (2009) A robust plant RNA isolation method suitable for Affymetrix Gene Chip analysis and quantitative real–time RT–PCR. Nature Protocols 4(3):334–340

Boss PK, Davies C, Robinson SP (1996) Analysis of the expression of anthocyanin pathway genes in developing *Vitis vinifera* L. cv Shiraz grape berries and the implications for pathway regulation. Plant Physiol 111:1059–1066

Carra A, Gambino G, Schubert A (2007) A Cetyltrimethyl ammonium bromide–based method to extract low molecular weight RNA from polysaccharide–rich plant tissues. Anal Biochem 360:318–320

Chang S, Puryear J, Cairney J (1993) A simple and efficient method for isolating RNA from pine tree. Plant Mol Biol Report 11:113–116

Chomczynski P, Sacchi N (1987) Single–step method of RNA isolation by acid guanidinium thiocyanate–phenol–chloroform extraction. Anal Biochem 162:156–159

Claros MG, Canovas FM (1998) Rapid high-quality RNA preparation from Pine. Plant Mol Biol Report 16:9–18.

Gambino G, Gribaudo I (2006) Simultaneous detection of nine grapevine viruses by multiplex RT–PCR with coamplification of a plant RNA internal control. Phytopathol 96:1223–1229

Gambino G, Bondaz J, Gribaudo I (2006) Detection and elimination of viruses in callus, somatic embryos and regenerated plantlets of grapevine. Eur J Plant Pathol 114:397–404

Gambino G, Perrone I, Gribaudo I (2008) A rapid and effective method for RNA extraction from different tissues of grapevine and other woody plants. Phytochem Anal 19:520–525

Geuna F, Hartings H, Scienza A (1998) A new method for rapid extraction of high quality RNA from recalcitrant tissues of grapevine. Plant Mol Biol Report 16:61–67.

Iandolino AB, Goes da Silva F, Lim H, Choi H, Williams LE, Cook DR (2004) High–quality RNA, cDNA, and derived EST libraries from grapevine (*Vitis vinifera* L.). Plant Mol Biol Report 22:269–278

Kolosova N, Miller B, Ralph S, Ellis BE, Douglas C, Ritland K, Bohlmann J (2004) Isolation of high–quality RNA from gymnosperm and angiosperm trees. Biotech 36:821–824

Le Provost G, Herrera R, Paiva JAP, Chaumeil P, Salin F, Plomion C (2007) A micromethod for high throughput RNA extraction in forest trees. Biol Res 40:291.297

Li B, Wang B, Tang K, Liang Y, Wang J, Wei J (2006) A simple and convenient approach for isolating RNA from highly viscous plant tissue rich in polysaccharides. Colloids Surf B Biointerfaces 49:101–105

Liu JJ, Goh CJ, Loh CS, Liu P, Pua EC (1998) A method for isolation of total RNA from fruit tissues of banana. Plant Mol Biol Report 16:1–6

Loulakakis KA, Roubelakis–Angelakis KA, Kanellis AK (1996) Isolation of functional RNA from grapevine tissues poor in nucleic acid content. Am J Enol Vitic 47:181–185

Meisel L, Fonseca B, Gonzalez S, Baeza-Yates R, Cambiazo V, Campos R, Gonzalez M, Orellana A, Retamales J, Silva H (2005) A rapid and efficient method for purifying high quality total RNA from peaches (*Prunus persica*) for functional genomics analyses. Biol Res 38:83–88

Moser C, Gatto P, Moser M, Pindo M, Velasco R (2004) Isolation of functional RNA from small amounts of different grape and apple tissues. Mol Biotechnol 26:95–100

Or E, Belausov E, Popilevsky I, Ben Tal Y (2000) Changes in endogenous ABA level in relation to the dormancy cycle in grapevine grown in hot climate. J Hortic Sci Biotechnol 75:190–194

Ophir R, Pang X, Halaly T, Venkateswari J, Lavee S, Galbraith D, Or E (2009) Gene-expression profiling of grape bud response to two alternative dormancy-release stimuli expose possible links between impaired mitochondrial activity, hypoxia, ethylene-ABA interplay and cell enlargement. Plant Mol Biol 71:403–423

Rott ME, Jelkmann W (2001) Characterization and detection of several filamentous viruses of cherry: adaptation of an alternative cloning method (DOP–PCR), and modification of an RNA extraction protocol. Eur J Plant Pathol 107:411–420

Salzman RA, Fujita T, Zhu-Salzman K, Hasegawa PM, Bressan RA (1999) An improved RNA isolation method for plant tissues containing high levels of phenolic compounds or carbohydrates. Plant Mol Biol Report 17:11–17

Scott DL Jr, Clark CW, Deahl KL, Prakash CS (1998) Isolation of functional RNA from periderm tissue of potato tubers and sweet potato storage roots. Plant Mol Biol Report 16:3–8

Tattersall EAR, Ergul A, Alkayal F, DeLuc L, Cushman JC, Cramer GR (2005) Comparison of methods for isolating high–quality RNA from leaves of grapevine. Am J Enol Vitic 56:400–406

Thomas P, Schiefelbein JW (2002) Improved method for purification of RNA from stem tissue of grapevine and its use in mRNA profiling. Am J Enol Vitic 53:231–234

Chapter 21
RNA Extraction from Young, Acidic Berries and Other Organs from *Vitis Vinifera* L

Charles Romieu

Abstract Commercial kits based on RNA extraction in a chaotropic medium followed by purification on disposable silica membrane columns fail in yielding substantial amount of RNA when starting from fleshy fruits tissues. The present protocol was designed in order to overcome the strong acidity and dilution inherent to the flesh tissue, as major difficulties in RNA extraction. Nuclear DNA is eliminated upon centrifugation following extraction in a non chaotropic and higly buffered extraction medium, before RNA concentration by LiCl precipitation. This preliminary concentration and purification step allows to load silica column with sub saturating amounts of RNA.

Contents

Abbreviations

ATA Aurintricarboxylic acid
BET Ethidium bromide

C. Romieu (✉)
Equipe Vigne, UMR DIAPC, INRA-SupAgro, Montpellier 34060, France
e-mail: romieu@supagro.inra.fr

S. Delrot et al. (eds.), *Methodologies and Results in Grapevine Research*,
DOI 10.1007/978-90-481-9283-0_21,

DEPC	Diethylpyrocarbonate
EDTA	Ethylene diamine tetraacetic acid
MOPS	3-(N-morpholino)propanesulfonic acid
PEG	Polyethylene glycol
SDS	Sodium dodecyl sulfate

21.1 Introduction

Recovering high amounts of high quality RNA is a prerequisite for ensuring adequate representation of all transcripts in high throughput expression studies and for the construction of cDNA libraries. Young grape berries are particularly rich in RNA interfering compounds, such as tannins and polysaccharides. Tannins can irreversibly bind proteins and nucleic acids (Newbury and Possingham 1977) while polysaccharides coprecipitate with RNA (Richards et al. 1994) and this is the cause of well-known difficulties concerning RNA extraction and purification from plant tissues. When compared to other plant organs, the true specificity of fleshy fruits relies in the dramatic dilution of cytoplasmic macromolecules like proteins and nucleic acids as a result of pronounced vacuolar expansion along the flesh differentiation process (Ojeda et al. 1999). Moreover, cell expansion during green stage is linked to the accumulation of organic acids and the total acidity reaches considerable values of ca 500 mEq and pH 2.7 in different fleshy fruits like, non exhaustively, grape, citrus, apples, pomegranata, redcurrent and mango. Obviously, such acidity exceeds the buffering capacity of most RNA extraction buffers (see for example Iandolino et al. (2004) and Tattersall et al. (2005) for classical extraction buffers). As the pK of Tris, the prevailing buffer in the literature, is 8.3, its efficiency is dramatically impaired in most extraction media below pH 8.0. This may explain that large dilution of the sample in the extraction buffer is frequently recommended. Commercial kits are designed for the extraction of nucleic acids from 100 mg tissues in the presence of chaotropes, before fixation on silica membrane columns with a nominal capacity of ca 100 μg nucleic acids (i.e. Qiagen RNAeasy® or Macherey Nagel Nucleospin®). Since berries include only ca 25 μg/g FW RNA (Tattersall et al. 2005) a concentration step becomes mandatory in order to load the column within its working range. Moreover, nuclear DNA may be extracted together with RNA in denaturing conditions, and silica membranes do not selectively bind RNA and DNA in the presence of chaotropes. Although on column DNAse treatment seems a convenient way to obtain pure RNA, the final RNA yield may be critically affected by the DNA/RNA ratio in the concentrated extract.

We have designed a protocol for the purification of RNA from acidic fruit tissues, based on RNA extraction in a medium highly buffered at the physiological pH of 7.5. This extraction medium is a modification of the non chaotropic buffer proposed by Tesnière and Vayda (1991), that includes the RNAse inhibitor Aurintricarboxylic acid (Hallick et al. 1977) and a cocktail of anionic and neutral detergents that do not yield to intense DNA extraction. Following the sedimentation of nucleus and polysaccharides, total RNA is precipitated by LiCl, before solubilisation on chaotropic medium and purification on silica membrane.

21.2 Materials and Methods

All glassware, spatulas, mortar and pestle are baked in aluminium foils at 230°C overnight. RNAse free plasticware (tubes, eppendorf, pipette cones) are reserved for RNA extraction, and are used without autoclaving. The pH electrode is preincubated in autoclaved 0.1% SDS; 100 mM NaOH during 1 h, before standardizing and rinsed with the same solution, before use.

21.2.1 Extraction Buffer

Composition: MOPS 200 mM, NaCl 250 mM, LiCl 50 mM, EDTA 25 mM, deoxycholate 1% (w/v), SDS 1.5% (w/v), Tween 20 1% (w/v), aurintricarboxylic acid 1 mM.

Note: This buffer is the extraction buffer Tesnière and Vayda (1991), modified for MOPS, NaCl and LiCl. The detergent cocktail appears as less denaturing than SDS alone that is frequently used for DNA extraction at elevated temperature. The PEG moiety of the Tween 20 molecule probably acts as a scavenger of tanins, preventing them to interact further with nucleic acids. Although we do not recommend it, 0.2% Triton X100 can be substituted for 1% PEG in order to yield a similar molar ratio of neutral detergent in the cocktail.

Preparation: NaCl, LiCl, EDTA, SDS and ca 100 mM NaOH pellets are mixed in ca 80% final volume milliQ water, supplemented with 0.1% DEPC, and incubated overnight at room temperature under intense stirring, before autoclaving. The buffer is then allowed to cool and supplemented with RNAse free MOPS, deoxycholate and Tween, before adjusting the pH to 8.5 with NaOH pellets. The volume is adjusted with autoclaved DEPC treated water, before adding 1 mM aurintricarboxylic acid as a powder.

In case RNA extractions from tissues that are less acidic than green berries are anticipated (e.g. flowers or ripe fruits), an aliquot of the pH 8.5 buffer is adjusted back to pH 7.5 with 14 N HCl.

The extraction buffer is quite stable at room temperature, and can be aliquoted in 50 mL falcon tubes before freezing.

21.2.2 Salt Solutions

5 M NaCl: 14.75 g NaCl are mixed with 50 mL final volume water in a disposable 50 mL tube, supplemented with 50 μL DEPC, extensively shaked and incubated overnight, before autoclaving in vertical position in an aluminium foil.

12 M LiCl: Same preparation as 5 M NaCl, starting from 25.4 g LiCl.

21.2.3 Silica Column Fixation Buffer

Guanidine-HCl 6 M, Tris HCl 50 mM pH 7.0, EDTA 90 mM pH 8.0, 1.5% tween 20, prepared in a disposable 50 mL tube with DEPC treated and autoclaved water. Alternatively, the Qiagen RLC buffer supplemented with 1.5% tween 20 can be used.

21.2.4 Sampling

Young grape berries are dropped in liquid N_2 in the vineyard, immediately once harvested, or following seed removal with a scalpel (10–20 s). Frozen tissues are kept at −80°C and grinded to a fine powder with N_2 liquid frozen mortar and pestle that have been previously baked overnight at 250°C in an aluminium foil to inactivate RNAses, taking great care to obtain an homogeneous and fine powder.

21.2.5 Extraction and Purification Procedures

The amount of tissue necessary to obtain ca 50–100 μg of RNA, close to the saturation of the silica membrane column, is calculated according to Table 21.1 or to a preliminary experiment. The frozen powder is weighed in disposable, RNAse free plastic tubes (2 mL for ca 100 mg FW or 13 mL for ca 1 g FW) with a spatula that was flamed just before use, and frozen in liquid N_2. Tubes are left open(ed) at −20°C within 15 min, in order to prevent freezing of the extraction buffer when it is mixed with the powder. The extraction buffer is supplemented with 1–2% mercaptoethanol immediately before use. 7 mL of RT extraction buffer are added to each g of frozen powder, the tubes are immediately closed and vigorously shaken by hand until all ice clumps have melted. The tubes are then placed on ice and occasionally shaken until all the samples are extracted. Cellular debris including nucleus are then pelletted at 13,000*g*, 10 min in a microfuge or at 3,000 rpm, 10 min in a swinging bucket rotor, for the 13 mL tubes.

Table 21.1 Total RNA expected yield on different grapevine organs

Organ	RNA Yield (μg RNA/g FW)
Floral buds stage 1–2	260
Floral buds, stage 3	170
Floral buds, stage 4	140
Floral buds, stage 5	180
Flower, floraison	150
Berries (0.02 g/berry)	30
Berries (0.06 g/berry)	110
Berries (0.28 g/berry)	42
Berries (0.4 g/berry)	20

Supernatants are dispensed as 1640 μl aliquots in 2 mL eppendorf tubes, before precipitating polysaccharides with 0.18 volume of NaCl 5 M, 10 min incubation on ice, and another centrifugation at 13,000*g*, 10 min. We recently found that 10% v/v (final concentration) ethanol precipitation can be substituted to NaCl precipitation in order to facilitate the solubilisation of RNA pellets, following LiCl precipitation (see below).

The supernatant from previous step is then supplemented with 0.209 volume of LiCl 12 M rapidly mixed, and incubated on ice in 2 mL tubes. Two h incubation at 4°C appeared sufficient for most applications, but overnight incubation may increase RNA yield on viscous samples.

RNA and contaminating proteins are sedimented by centrifugation at 13,000 *g* for 15 min. It is important to mark the size of the tube where the material is expected to sediment, since, depending on the samples, a transparent leaflet covering one third of the internal surface of the tube may form in place of an easily detectable pellet.

The pellets are intensively dispersed in 70% ethanol in DEPC treated water with a P1000, before centrifugation 13,000 rpm 10 min. In such acidic medium, ATA adsorbed on contaminating protein may appear as pink-red. The supernatant is eliminated and the tubes are allowed to stand upsize-down on filter paper in order to eliminate residual 70% ethanol droplets. Pellets are then resuspended in 350 μL column loading buffer supplemented with 1% mercaptoethanol, through extensive pipeting with a P1000 pipet, and occasional resuspension during ca 30 min. incubation at room temperature. Residual material which is difficult to solubilize may be frozen at –20°C, melted, and incubated for 2–3 min in a 50–60°C waterbath, in order achieve a better solubilization.

One hundred μL DEPC-treated water is then added to each 350 μL sample, before adding 225 μL ethanol, loading and centrifugation on silica membrane columns. The eluate is loaded once again on the column, before washing and eluting RNA according to the manufacturer's instructions. We currently elute RNA with 50 μL DEPC treated water prewarmed at 60°C, when using RNAeasy columns. The eluate in loading buffer can be submitted to a second round of extraction on the same column, to check for possible overload, when working for the first time with previously uncharacterized tissues. This is particularly useful when saturating amounts of RNA (100 μg) were obtained from the first column.

21.2.6 RNA Quality

The quality of nucleic acids is assessed by placing 2 μL aliquots in 100 μL DEPC-treated water in 1 cm optical path quartz spectrophotometer cuvettes, and measuring their 300 to 200 nm spectra, assuming that 1 $DO_{260\,nm}$= 2 μg RNA /μL of undiluted sample. Then, 500 ng of RNA are electrophoresed in a 1% RNAse free agar including 0.5 μg/mL BET, to check for RNA degradation and the presence of contaminating DNA (Fig. 21.1). The undegraded 28 S and 18 S ribosomal RNA are clearly visible, with a clear excess of 28 S abundance. A minor 5.8 S rRNA band is also detected provided that migration of BET towards the cathode was limited by

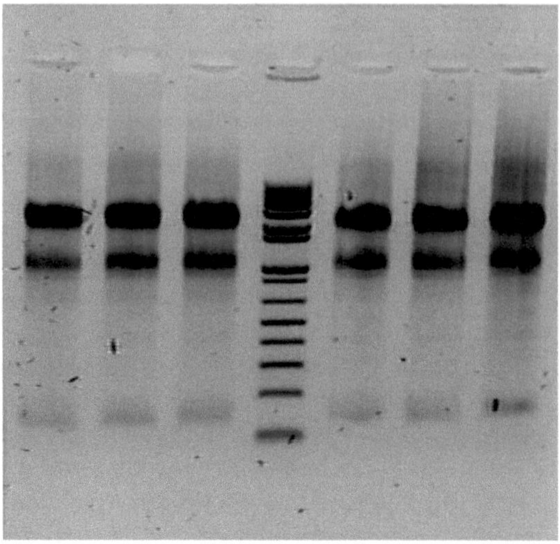

Fig. 21.1 Five hundreds ng of total RNA extracted from young berries were loaded in a 1% gel. Lanes 1, 2, 3: sultanine, 4 MW standard, 5, 6, 7 Gora chirine

stopping the migration of the tracking dye at mid gel. Some diffuse fluorescence of unknown origin can be detected above the 28 S band, that does not disappear upon digestion with DNAse.

21.2.7 Amplification of aRNA

Within 1 μg total RNA were reverse transcribed using a polyT-T7 primer and amplified with T7 polymerase according to the Aminoallyl messageAmp II aRNA amplification kit, using an in vitro transcription incubation time of 14 h. The size of the aRNA ranged from 0.5 to 5 kb in these conditions indicating a good integrity of the mRNA before amplification.

21.2.8 Expected Yield

The expected yield varies considerably depending on the organ, with a 10-fold decrease between flower buds and veraison berries. Organs were sampled from Sultanine cv., berries of increasing size were harvested at different dates before veraison.

References

Hallick RB, Chelm BK, Gray PW, Orozco EM (1977) Use of aurintricarboxylic as an inhibitor of nuclease during nucleic acid isolation Nucleic Acid Res 4:3055–3094

Iandolino AB, Goes da Silva F, Lim H, Choi H, Williams LE, Cook DR (2004) High-quality Rna, cDNA, and derived EST libraries from grapevine (Vitis vinifera L.). Plant Mol Biol Reporter 22:269–278

Newbury HJ, Possingham JV (1977) Factors affecting the extraction of intact ribonucleic acid from plant tissues containing interfering phenolic compounds. Plant Physiol 4:543–547

Ojeda H, Deloire A, Carbonneau A, Ageorges A, Romieu C (1999) Berry development of grapevines: Relations between growth of berries and their DNA content indicate cell multiplication and enlargement. Vitis 38:145–150

Richards E, Reichardt M and Rogers S (1994) Preparation of genomic DNA from plant tissues. In: Ausubel FM, Brent R, Kingston RE, Moore DD, Seidman JG, Smith JA, Stuhl K (eds) Current protocols in molecular biology. Wiley and Sons, New York, pp 231–237

Tattersall EAR, Ergul A, AlKayal F, Deluc L, Cushman J, Cramer GR (2005) Comparison of methods for isolating high-quality RNA from leaves of grapevine. Am J Enol Vitic 56:400–406

Tesnière C, Vayda ME. (1991) Method for the isolation of high quality RNA from grape berry tissues without contaminating tannins or carbohydrates. Plant Mol Biol Rep 3:242–251

Chapter 22
Transcriptomics Analysis Methods: Microarray Data Processing, Analysis and Visualization Using the Affymetrix Genechip® *Vitis Vinifera* Genome Array

Karen A. Schlauch, Jerome Grimplet, John Cushman, and Grant R. Cramer

Abstract The study of transcriptomics is a powerful method of studying the responses of organisms to their environment. The transcriptome consists of the entire set of transcripts that are expressed within a cell or organism at a particular developmental stage or under various environmental conditions. There are various technologies for assaying the transcriptome including hybridization-based microarrays and RNA sequencing. Microarrays have been used extensively to quantify the transcript abundance of grape cells, organs and tissues. Here we provide a practical guide on how to analyze microarray data using a study based on the Affymetrix GeneChip® Vitis vinifera genome array. Microarray studies have proven to be very powerful for the elucidation of molecular response networks and physiological processes. In this Chapter, we have outlined the steps required to process and analyze mRNA expression data. The first step is to check both microarray and data quality. The second step is to remove array and data outliers and reduce the variability of the data with cleansing and normalization techniques. The third step is to perform statistical tests to identify sets of transcripts differentially expressed among conditions under statistical significance. The fourth step is to evaluate these sets of significant transcripts using functional categorization and molecular maps. Such data sets can be compared or integrated with proteomic and metabolomic data sets using a systems biology approach to increase the robustness of the conclusions. From data analysis performed in these ways, hypotheses can be generated for further experimentation and validation, which denotes the fifth, and likely the most important, final step.

Contents

G.R. Cramer (✉)
Department of Biochemistry and Molecular Biology, University of Nevada, Reno, NV 89557, USA
e-mail: cramer@unr.edu

S. Delrot et al. (eds.), *Methodologies and Results in Grapevine Research*,
DOI 10.1007/978-90-481-9283-0_22,

22.1 Introduction

The study of global gene expression (transcriptomics) is part of a repertoire of methods used in functional genomics approaches today and is providing a powerful way to understand and compare the "holistic" responses of organisms to their environment (Trewavas 2006, Wang et al. 2009). The transcriptome consists of the entire set of transcripts that are expressed within a cell or organism at a particular developmental stage or under various environmental conditions. There are various technologies for assaying the transcriptome including hybridization-based microarrays and RNA sequencing (Trewavas 2006, Wang et al. 2009). Microarrays have been used extensively to quantify the transcript abundance of grape cells, organs and tissues. (Terrier et al. 2005, Waters et al. 2005, Espinoza et al. 2006, Waters et al. 2006, Cramer et al. 2007, Deluc et al. 2007, Fernandez et al. 2007, Grimplet et al. 2007, Pilati et al. 2007, Tattersall et al. 2007, Chervin et al. 2008, Figueiredo et al. 2008, Gatto et al. 2008, Lund et al. 2008, Deluc et al. 2009, Mathiason et al. 2009). Here we provide a practical guide on how to analyze microarray data using a study based on the Affymetrix GeneChip® *Vitis vinifera* genome array.

22.2 A Long-Term Growth Experiment as an Example for Microarray Data Analysis

A long-term experiment was designed and conducted to determine the impact of water deficit and salt stress on grapevine physiology and molecular profiles (Cramer et al. 2007). Salt stress, water-deficit stress, and no stress (control) were applied to randomly selected vines for a period of 16 days. Cabernet Sauvignon shoots with and without stress were harvested every 4 days (Day 4, 8, 12 and 16), and shoots without stress were harvested at Day 0. The shoot length and midday stem water potential was recorded every two days as a measure of water-deficit stress. The water potential of plants of both water deficit and salt-stressed populations was kept nearly identical. This was accomplished by monitoring slight differences in the stem water potential between the two stressed populations, using data from a previous experiment, and adjusting on a daily basis the electrical conductivity of the salt application to attain water potentials similar to those measured in water deficit treated vines. Microarray and quantitative RT-PCR transcript profiling were used to define genes and metabolic pathways in *Vitis vinifera* cv. Cabernet Sauvignon with common or divergent responses to long-term (16 days) water-deficit stress and isoosmotic salinity stress.

22.3 Long-Term Growth Experimental Design

The experiment followed a completely randomized factorial design consisting of 3×4 (treatment $\times$ time) conditions, with six to eight individually potted plants per time point and treatment. For each of the experimental conditions, two shoot tips were pooled together to form one of three biological replicates. As an additional quality control measure, six technical microarray replicates were run on one experimental condition (control condition of Day 16) to assess microarray data quality and data reproducibility.

22.4 *Vitis* GeneChip® Design

The Affymetrix GeneChip® *Vitis* genome array was developed by Affymetrix and the UNR *Vitis* research group in 2004. The array included 14,700 Unigenes, for which sequences were selected from GenBank®, dbEST, and RefSeq. Sequence clusters were generated from the UniGene database, Build 7, October 2003. Each sequence was represented by sixteen oligonucleotide probe pairs, each containing a Perfect Match and a Mismatch. Detailed descriptions of probesets can be found elsewhere (Affymetrix 2002a). A number of standard control features were included on the array. A supplemental set of six control features was selected as an additional quality control measure. The extra six controls were placed in a tile that was spotted on the GeneChip® in eight separate instances (Fig. 22.1a). One copy of each control features' probesets was also spotted randomly on the array. This provided an additional mechanism to verify quality control with respect to spatial variation and technical reproducibility. Three of these six controls were positive controls (e.g. actin, GAPDH, and elongation factor 1 alpha); the other three were negative controls (e.g. *aph*A (kanamycin resistance gene); beta-lactamase gene; beta-glucuronidase). Actin, GAPDH, and elongation factor 1 alpha are typically considered as housekeeping genes in that these genes generally do not exhibit large variations in mRNA abundance under many experimental conditions. Additionally, the genes are known to be expressed in many conditions, and thus act also as a positive control. *Aph*A, beta-lactamase gene, and beta-glucuronidase are eubacterial genes that are not expected to be expressed in the genomes of vascular plant species, and thus were chosen as negative controls.

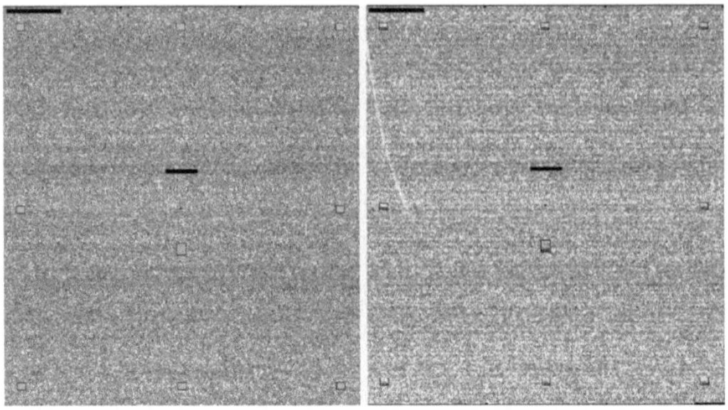

Fig. 22.1 Image examples of the Affymetrix GeneChip® *Vitis vinifera* genome array. **a** (*left*) shows the placement of the eight replicated tilings of control probesets (*green*), and the placement of the tile consisting of the hybridization controls (*orange*) on the array. Remaining control probesets are scattered about the array (single cells in *red* and *green*). **b** (*right*) is the image of a *Vitis* microarray with spatial variation likely caused by a fiber

22.5 Microarray Quality Assessment

To assess the quality and reproducibility of the microarray data prior to performing any analysis procedures, a number of steps should be performed. The following steps are those recommended by Affymetrix and are cited directly from the Affymetrix GeneChip® Data Analysis Fundamentals Manual (Affymetrix 2002a).

(1) 260/280 Absorbance Readings
To ensure that the highest quality RNA is hybridized to the gene expression arrays, users should run the initial total RNA on an agarose gel or an Agilent Technologies 2100 Bioanalyzer to examine the integrity of ribosomal RNA bands. Non-distinct ribosomal RNA bands suggest possible RNA degradation. The 260/280 absorbance readings should be measured for total RNA, and should be within the range between 1.8 and 2.1. Ratios below 1.8 might be a sign of possible protein contamination, whereas ratios above 2.1 might suggest degraded RNA, truncated cRNA transcripts, or an excess of free nucleotides.

(2) Array Background Levels
The average background levels of each array are provided by the Affymetrix GeneChip® Operating Software (GCOS) in the Expression Report files. When arrays are run on 10% PMT scanner settings, background levels should lie between 20 and 100, and should be consistent across all arrays in the study.

(3) Noise Levels
The noise levels of each array are provided in the Expression Report files and are labeled as "RawQ". RawQ values measure the pixel-to-pixel variation on each array. These levels should be consistent across all arrays, and any array deviating grossly from the mean noise level should be examined closely.

(4) Global Normalization and Scaling factors
These factors are provided in the Expression Report files, and should be consistent across all arrays. More than three-fold differences between array scaling factors might indicate notable array variability or sample degradation.

(5) Present Call Rates
The Present Call rates should be uniform across all arrays, with even more stringent uniformity across replicated samples. Arrays with a notably low percentage of Present Calls might have poor sample quality, and should be examined closely with respect to all other quality control metrics.

(6) Microarray Images
All microarray images should be checked for scratches, smeared regions stemming from fibers or air bubbles, exceptionally high or low overall expression levels, and other general spatial variation. Images are provided by the Affymetrix GCOS as jpeg files.

(7) Hybridization Controls
Affymetrix places four hybridization controls on every GeneChip®. The four transcripts bioB, bioC, bioD, cre represent genes in the biotin synthesis pathway of *E. coli,* and are added to the hybridization cocktail at increasing concentrations of 1.5, 5, 25, and 100 pM, respectively. BioB should have Present Calls in at least 50% of the arrays in the experiment; bioC, bioD, cre should have Present Calls in 100% of the arrays. The signal values for bioC, bioD and cre should be increasing, respectively. Absent calls or low signal values of these control probes might indicate a problem with the hybridization reaction or with the washing and staining steps. In this case, it should be verified that the hybridization cocktail was made properly, that the recommended temperature for hybridization and the correct fluidics protocol were used, and that the SAPE staining solution did not deteriorate. See http://www.affymetrix.com/support/help/faqs/ge_assays/faq_19.jsp for more details.

(8) Poly-A Controls
Each Affymetrix GeneChip® includes four poly-A controls used to examine the target labeling process. The controls *lys, phe, thr,* and *dap* are *B. subtilis* genes, which should be present across all arrays, and have increasing signal values, respectively.

(9) Internal Controls
The two internal control genes GAPDH and actin are included on the array as housekeeping genes. The 3'/5' ratios of each measure RNA sample and assay quality. More details can be found at http://www.affymetrix.com/support/help/faqs/ge_assays/faq_17.jsp. Affymetrix guidelines suggest that the 3'/5' ratio be less than three in general. Larger ratios might indicate RNA degradation. More details can also be found in the Affymetrix GeneChip® Data Analysis Fundamentals Manual (Affymetrix 2002a). Any grossly outlying ratios across the group of arrays studied should be inspected more closely.

Step 1 must be verified by your personnel in the microarray facility. Steps 2–5 are verified by a quick inspection of the Expression Report files (.rpt) generated by the Affymetrix processing software (GCOS) in your microarray facility. Some quality control metrics can be generated using the *simpleaffy* package from Bioconductor. Step 6 is confirmed by viewing the image files generated by the GCOS or by generating the images using the Bioconductor's package *affy*. Steps 7–9 are validated by examining the expression levels of the control probes activity upon normalization (see the section below for normalization methods). Additional details regarding these and other quality control protocols can be found in the Affymetrix Expression Analysis Data Analysis Fundamentals Manual (Affymetrix 2002a). Several functions in the *R* programming language are available to perform some of these Affymetrix quality control checks (e.g. *simpleaffy*, *computeRawQ*, *yaqaffy*, *AffyExpress*), and are freely available from the Bioconductor site [http://www.bioconductor.org/]. Please see the website http://bioinformatics.unr.edu/vitis to download some simple *R* scripts that perform these functions.

22.6 Microarray Quality Assessment of the Long-Term Growth Experiment

As this was the first study using the GeneChip® *Vitis* Genome Array, several quality control measures were instated within the experimental design. For each experimental treatment (water-deficit, salt, and control) at each of four time points (Day 4, 8, 12, and 16), a set of biological triplicates was used. Additionally, six technical microarray replicates were run at Day 16 under the control treatment. Expression data were subjected to the quality control steps as described above:

(1) After extraction, RNA quality was assessed by A260/280 absorbance ratios and by the Agilent Bioanalyzer. Samples were found to be of high quality, and identical to results reported by Tattersall et al. (2005).
(2) Average array background levels ranged from 53 to 140, with a mean of 84.4 and a standard deviation of 18.4. A Grubbs' test for outliers (Grubbs 1969) was performed on the 44 background levels, and one array with a background level of 140 proved to be a statistically significant outlier ($p < 0.05$). This array was excluded from further study. No other array was found to have a significant outlying background value. For experiments with less than 30 arrays, a Grubbs' test can be performed using the *R* function *grubbs.test* from the package *outliers*. For larger experiments, formulas and tables are available for the Grubbs' test (Grubbs 1969).
(3) Raw Q noise levels fell between 1.7 and 4.4 with a mean value of 2.6 and standard deviation of 0.5. A Grubbs' test for outliers was performed on the 44 noise levels, and only the microarray exhibiting the highest noise level (4.4) was shown to be a significant outlier, and excluded from further analyses. This array was the same array excluded for high background levels in Step 2 above.
(4) Scaling factors were consistent across all arrays, with a 2-fold difference between the minimum and the maximum value. The mean scaling factor was

0.18, with a standard deviation of 0.032. Scaling factors were computed using the *R* function *qc* in the *outlier* package.

(5) Present Call rates ranged from 74 to 79%. The overall average Present Call rate was 76%, with a standard deviation of 1.12%. Present Call rates across technical and biological replicates exhibited a notably lower standard deviation of 0.74, 0.69%, respectively.

(6) Upon close inspection of all microarray images, one microarray showed spatial variation likely caused by a fiber (Fig. 22.1b), and was excluded from further study.

(7) All four hybridization controls were present on 100% of the arrays, and the normalized signals were increasing with respect to the order *bioC, bioD*, and *cre*.

(8) The behavior of all poly-A controls conformed to the desired recommendations, with the exception of one non-present call on one array for one of the controls, a negligible event.

(9) The internal controls also performed as recommended: 3′–5′ actin ratios were less than 1.01; GAPDH ratios were consistently below 1.2.

In addition to the control steps recommended by Affymetrix as outlined above, the trends in RNA degradation between the 5' end and the 3' end in each probeset were examined. Probes in each probeset are numbered from 1 to 16 from the 5' end to the 3' end for each feature on the array. The trends were computed for each microarray by computing the mean raw Perfect Match expression value for each probe number across all array features. Degradation curves should prove very consistent across all arrays in the experiment, which was the case for the long-term stress study (Fig. 22.2a).

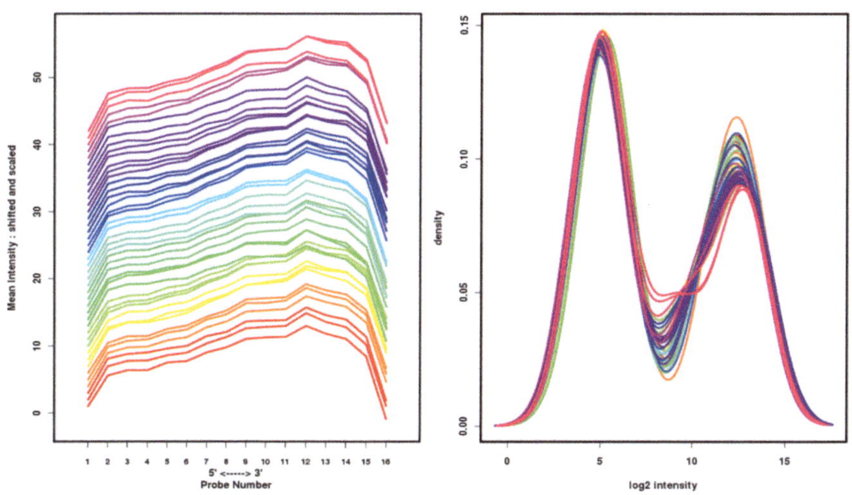

Fig. 22.2 Methods for quality assessment of arrays. **a** (*left*) shows the RNA degradation curves of each array. **b** (*right*) presents the distributions of the normalized expression values (log base 2) of all microarrays

22.7 Normalization of Microarray Data

Data in the long-term experiment were processed using the *R* programming language (version 2.9.0). The *R* language is publicly available at no cost for all operating platforms [http://cran.r-project.org/]. Bioconductor is an Open Source software development project that features many useful, freely available *R* programs for the analysis of microarray (and other) data [http://www.bioconductor.org/].

The long-term microarray data were normalized using the Robust Multi-Array Average (RMA) method (Irizarry et al. 2003a). Other popular normalization methods for Affymetrix data include the Affymetrix MAS 5.0 algorithm (Affymetrix 2002a), and dChip (Li and Hung Wong 2001). The RMA method has been shown to perform better than both MAS 5.0 and dChip in terms of precision, consistency in fold-change estimates, and when detecting differential expression (Irizarry et al. 2003b). More detailed discussions regarding normalization methods can be found in a number of references (Bolstad et al. 2003, Irizarry et al. 2003a). Raw intensity values of the 42 arrays passing quality control standards were processed and normalized by RMA using the Bioconductor *R* package *affy* (Gautier et al. 2004). Upon application of RMA, normalized and log-transformed (base 2) expression values of all 42 arrays exhibited very similar distributions (Fig. 22.2b).

Normalized expression behavior of the six replicated control features was examined first. The coefficient of variation (CV) of expression levels across the nine separate tiles was measured for each of these controls, for each array. The average CV across tiles for the 42 arrays of actin, GAPDH, elongation factor 1 alpha, *aphA* (kanamycin resistance gene), beta-lactamase gene, beta-glucuronidase was 1.0, 1.0, 0.8, 10, 6.1, 2.8%, respectively. This indicated that on average, the variation across the nine separate placements of each control feature was minimal, consistent with the lack of spatial variation portrayed in the array images. Profiles of the control features are graphically presented in Fig. 22.3, which shows the small variation among the nine instances within each probeset, with the *aphA* gene exhibiting the greatest variation. Expression values of the positive controls are of notably greater magnitude than those of the negative controls. Also of interest is the variation of GAPDH across all experimental conditions, which is not consistent with housekeeping behavior, but is consistent with the other *Vitis* microarray studies (VV1, VV2, VV3, VV4, VV10) at www.plexdb.org. In these experiments, GAPDH expression levels range from 6 to 12.5 on the normalized $\log_2$ scale.

Data of the 16,436 non-control probesets on the GeneChip® *Vitis* Genome Array were examined with respect to Present Call rates, as computed by the Affymetrix MAS 5.0 presence/absence detection algorithm (Affymetrix 2002b). The relative intensities between the Perfect Match and Mismatch probes in each of the 16 pairs were combined and used to generate a signal detection Present Call for each probe: P (Present), M (Marginal) and A (Absent), where Present means that the Perfect Match probes show stronger intensity than the Mismatch probes. Another way to think about this is that a Present Call indicates that the signal strength is greater than the background strength. If a probeset has no signal strength, it is difficult to test its relative signal across different conditions. Therefore, the 1981 probesets (12%) with

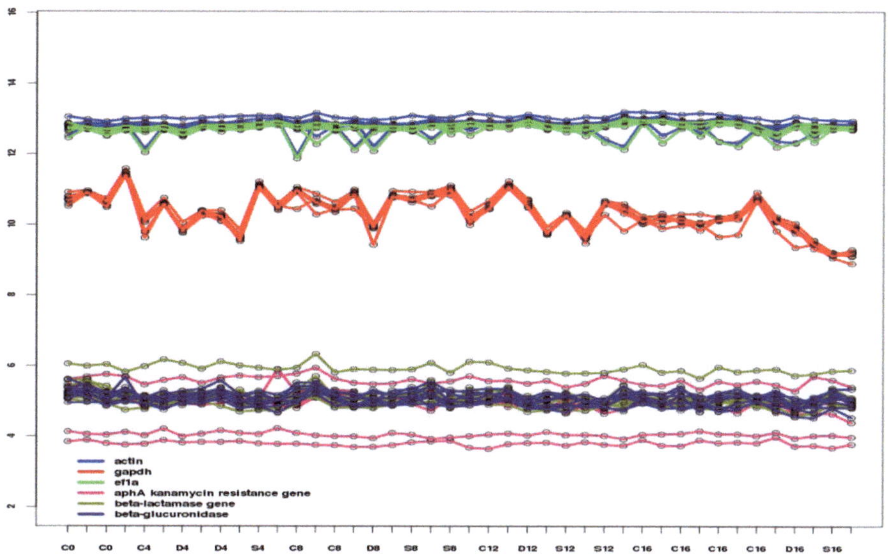

Fig. 22.3 Normalized expression profiles of the six repeated control features across all 42 arrays. Arrays are represented by the horizontal axis, log-transformed (base 2) expression values are represented by the vertical axis

no Present Calls across the 42 arrays were excluded from any further analysis. This left 14,455 non-control probesets.

The normalized data were evaluated for reproducibility. Pair-wise Pearson correlation coefficients were measured across the six technical replicate arrays, all of which were greater than 0.999, and higher than those computed by Baugh et al. (2001), using the Affymetrix *C. elegans* array, and to those of Redman et al. (2004) using the Affymetrix *Arabidopsis* array. Correlation coefficients across biological replicates were also computed and found to be very high: Spearman coefficients ranged from 0.967 to 0.998, and Pearson coefficients ranged between 0.970 and 0.999. For each feature on the array, the coefficient of variation was computed across the six technical replicate arrays: the average technical CV across all probesets was 3.9%, denoting very small amounts of technical variation. Expression values of the six technical replicates were combined into the average for all downstream analyses.

To ensure strict reproducibility standards, all sets of triplicate expression measures were examined for outliers and high levels of variation. The CV of the 187,915 sets of triplicates ($187{,}915 = 14{,}455$ probesets $*$ 13 conditions) was computed on the antilog of normalized expression values; the average triplicate CV was 8.6%. The distribution of the triplicate CV measures can be seen in Fig. 22.4a. Figure 22.4b shows that larger coefficients of variation are generally associated with lower expression measures. These CV measures are exceptionally low as compared to within-group CV computations of 25% using the HGU95A GeneChip® and the MAS 5.0 normalization method (Welle et al. 2002, Affymetrix 2002b); 22 and 27%

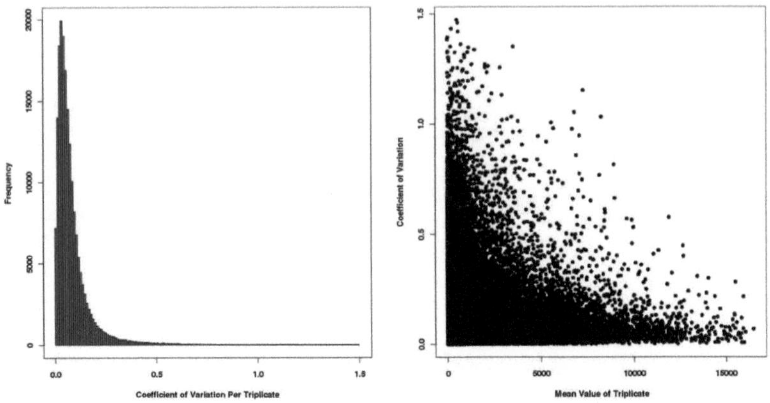

Fig. 22.4 Estimation of the variation of expression measures on the arrays. **a** (*left*) shows the distribution of the CV values for all 187,915 triplicate sets of expression measures. **b** (*right*) shows the relationship between mean antilog triplicate expression values on the horizontal axis, and the associated CV on the vertical axis

using the arrays U95Av2 and U133A, respectively, and the MAS 5.1 normalization method (Affymetrix 2002c, Daly et al. 2005); 5–15% for technical replicates and greater than 20% for biological and technical replicates of the Murine Genome U74Av2 array using RMA (Han et al. 2004). Any set of triplicates in which one of the measures exhibited a standard deviation of more than 1.14 (the maximum possible standard deviation for three measures is 1.1547; 10% of all expression measures exhibited one outlying element with this great a standard deviation), and a CV greater than 0.25 (3% of all triplicates exhibited this large of a CV) for the triplicate set was scrutinized. These thresholds were chosen to represent the 1% most variant triplicate sets having one deviating outlier. If one single measure in a triplicated set was near 1.1547, this indicated that the remaining two measures were nearly identical, and that the third triplicate was at its maximum outlying capacity, and thus this one triplicate was removed. This procedure left two replicates within the set of which the mean was used for subsequent analyses. Only 1% of all measurements were excluded using this rule (2,066 single measures). Additionally, any remaining triplicate set having a CV greater than 0.5 was removed. This included only 414 triplicate sets (0.26% of all triplicates), and reduced the mean coefficient of variation of all triplicates to a very low level of 7.3%. We found that these thresholds allowed us to identify gross outlying individual measurements within triplicates.

22.8 Data Organization

Principal component analysis (PCA) was used to simplify and define associations between stress conditions and temporal stages. The first two principal components represented 97.9 and 1.6% of the overall variation in the data, respectively, and

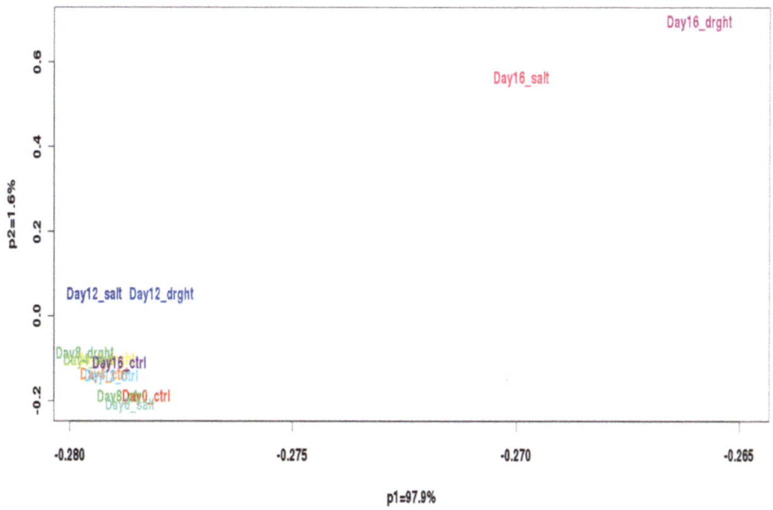

Fig. 22.5 Principal component analysis of the normalized and averaged microarray data. The horizontal axis represents the first principal component, which explains 98% of total variation of the data; the second component is represented by the vertical axis, and represents 1.6% of the total data variation

presented a distinct separation between Day 16 stress conditions and all other conditions, as shown in Fig. 22.5. To examine genes differentially expressed among stress conditions at any of the four time points an ANOVA was performed on the RMA expression values. We first fit a model with 12 coefficients corresponding to all conditions excluding the Day 0 measurements and extracted comparisons of interest by using contrasts. Day 0 measurements consisted of only the non-stress condition, and created an unbalanced design, making an ANOVA application difficult to assess properly. The Bioconductor package *limma* was used for ANOVA methods (Smyth 2005). A multiple testing adjustment (Benjamini and Hochberg 1995) was performed on the *t*-statistics of each contrast to adjust the false discovery rate. Differentially expressed genes with adjusted p-value < 0.05 were extracted for further inspection. Most of the differences in expression were exhibited between Day 16 stress and control conditions, consistent with the PCA results. Lists of significantly differentially expressed probesets among stress and control conditions at all days can be found at http://bioinformatics.unr.edu/vitis.

22.9 Functional Annotation and Categorization

Data were then organized by assigning functions in order to understand the biological significance of the observed changes in mRNA abundance. Functional annotation and categorization can be complicated as transcripts might have multiple functions at many different levels of organization. This problem is analogous to the question of which airport (hub) has the most traffic. What kind of traffic? Is it people, helicopters, planes or jets? Is it in New York, Paris or Beijing, and which geographic category level should be used? Is it at the city, state, country or continent level? To extend this analysis to plants, is RUBISCO categorized as an enzyme or in photosynthesis or energy or carbon or sugar metabolism and in what organelle, cell, tissue or organ (i.e. stroma, chloroplast, mesophyll, leaf or shoot)? Such gene ontology (GO) assignments (Ashburner et al. 2000) have been made and can be adapted to grapevine. However, with multiple assignments how does one quantify categories of transcript abundance? The assignment of one probeset to multiple categories is problematic, as this results in the associated transcript being counted more than once in different functional categories. For this study, annotation from the Munich Information Center for Protein Sequences (MIPS, ver. 2.1) catalog of top *Arabidopsis* BLAST hits (Ruepp et al. 2004, Schoof et al. 2004) was used, as we thought it gave the best and most complete description of metabolic pathways for plants at the time we used it. Each probeset on the Affymetrix *Vitis* array was manually assigned to exactly one MIPS (Deluc et al. 2009) category. Associations were arbitrarily assigned based on the perspective of the analyst, and can be changed upon a different perspective. Our functional category assignments are available at the grape annotation page of PLEXdb, which is a public resource for gene expression for plants and plant pathogens. The database can be found at http://www.plexdb.org and the *Vitis* annotation at http://www.plexdb.org/modules/PD_probeset/annotation.php?genechip=Grape. Please note that these assignments were not meant to be permanent and should be updated periodically as the annotation status of the *Vitis* genome improves. Others have also recognized the multiple-category annotation problem and have attempted to simplify GO assignments. An example is the BiNGO plugin (Maere et al. 2005) with GOSlim ontologies [www.geneontology.org/GO.slims.shtml] for the Cytoscape visualization software platform (Maere et al. 2005), which can be used to discover over-representation of functional categories in a data set. It might be wise to use both approaches in order to get different perspectives of your data set.

22.10 Data Visualization and Integration

An important challenge of microarray studies is to provide biological meaning to the plethora of data produced by annotating genes and integrating them within their biological context. Several visualization tools for molecular pathways are available including AraCyc (Maere et al. 2005), MapMan (Thimm et al. 2004) and VitisNet (Grimplet et al. 2009). In contrast to these general models for Arabidopsis, VAt this time only VitisNet [http://vitis-dormancy.sdstate.org] has been manually annotated extensively and specifically for visualizing microarray expression data on grapevine molecular pathways. Recently, a MapMan ontology has been developed for Vitis including Vitis specific pathways for carotenoids, terpenoids and phenylpropanoids (Rotter et al. 2009). The sequences from the *Vitis* genome sequencing projects (Jaillon et al. 2007, Velasco et al. 2007) and ESTs [www.ncbi.nlm.nih.gov] from the *Vitis* genus have been combined and the resulting 39,424 unique sequences have been manually annotated and mapped to molecular networks. To date, 13,145 genes have been assigned to 219 networks, including networks for metabolic, hormone, transport, and transcriptional pathways. More will be added in the future. Only 4,755 unique Affymetrix probesets (and an additional 2,110 redundant probesets) can be matched to the grapevine genome. This will be greatly improved when a whole genome array for grapevine is produced in the near future. Proteins and metabolites can also be visualized in VitisNet. The VitisNet tool incorporates Cytoscape, a versatile and customizable visualization software platform for molecular and interaction networks (Maere et al. 2005).

We have used the VV2 data at www.plexdb.org here to demonstrate network visualization using VitisNet. Networks can be downloaded from http://vitis-dormancy.sdstate.org and uploaded to Cytoscape. To begin to understand and identify ABA regulation in grapevine (Cramer 2010), a selection of probesets was created by taking *Arabidopsis* genes identified to be regulated by ABA (Huang et al. 2007, Matsui et al. 2008) and matching them with grapevine genes that were significantly induced by water deficit (Cramer et al. 2007, Tattersall et al. 2007, Cramer 2010). The genes have been assigned functional categories and were initially visualized and quantitatively analyzed with bar graphs (Fig. 22.6). This type of analysis allowed a broad visualization of the data. One change in functional categories was notable: water deficit substantially decreased the number of transcripts involved with protein synthesis. The large decrease in protein synthesis transcripts was a rather striking result and consistent with an earlier proteomic study (Vincent et al. 2007), in which a decrease in proteins involved in protein synthesis was highly correlated with the inhibition of growth.

To extend the analysis with more detail and meaning, probesets were mapped in VitisNet to the transcript box items corresponding to the grapevine genes using the Cytoscape software. A mapping text file was created. It contained in the first column a grapevine gene obtained from the genome sequencing project (Jaillon et al. 2007) and corresponding Affymetrix probesets in the following columns. The probeset with the highest average expression across all grapevine microarray experiments in PLEXdb [www.plexdb.org] was listed as the first probeset in the text file. In the case of multiple probesets per gene, additional probesets were presented in the following columns. The mapping file was loaded through the "Import Annotation File" window in Cytoscape, which created a probeset attribute for every transcript associated to a probeset. The expression data were then assigned to the transcript box item by linking them to the probeset attribute through the "Import Annotation File" window in Cytoscape. The probesets differentially expressed at Day 16 under water stress were loaded on the networks. Among them, 4,018 can be mapped on transcripts present on the networks (3,717 unique and 1,301 redundant transcripts). Once the expression data were loaded, nodes corresponding to transcripts with differential

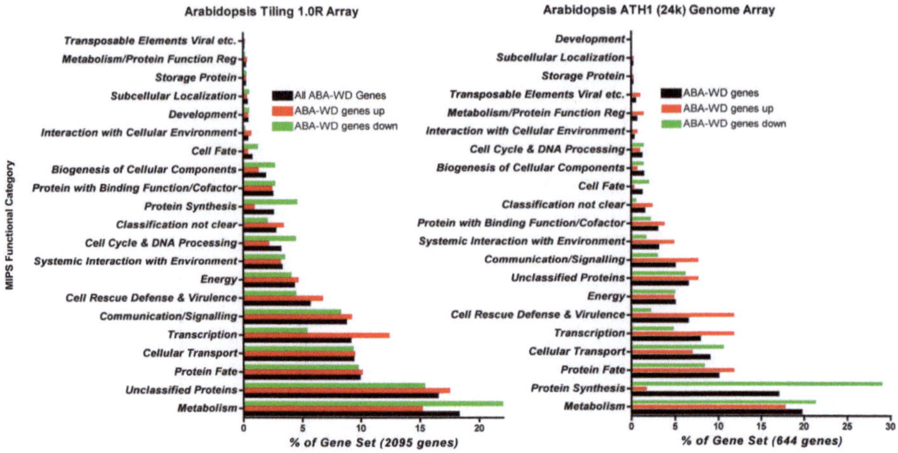

Fig. 22.6 Functional categorization of a subset of grape transcripts responsive to water deficit and also linked with Arabidopsis genes that are affected by ABA. This subset of grape transcripts is referred to as WD-ABA transcripts. ABA responsive genes in *Arabidopsis* were identified using the *Arabidopsis* tiling array data set from (Matsui et al. 2008) and the *Arabidopsis* genome array data set from (Huang et al. 2007). The initial subset of genes in grapes responsive to water deficit and which were also identified as ABA-responsive genes in *Arabidopsis* were functionally categorized (*black bars*) and then separated into those genes with increased (*grey bars*) or decreased (*white bars*) transcript abundance during water deficit. Each subset is display separately with the subset of 2095 grape transcripts based on the tiling array data on the left and the subset of 644 grape transcripts based upon the genome array data on the right. Functional categories are organized along the horizontal axis from the highest to the lowest number of transcripts in a functional category (*bottom* to *top*). The percent of transcript set refers to the percent of total transcripts in that particular set of transcripts that were placed into that functional category. The tiling array data are from (Matsui et al. 2008) and the genome array data are from (Huang et al. 2007)

abundance were presented with colors representing their respective abundance with the VizMapper tool in Cytoscape. A subset of the genes induced by water deficit and ABA in the ABA signaling pathway can be visualized in Fig. 22.7.

A simple analysis of the ABA signaling pathway indicates that many, but not all water deficit induced genes are also induced by ABA. Other factors besides ABA might have a role (e.g. water potential or the level of stress-induced reactive oxygen species). This mapping effort serves to generate new hypotheses that need to be tested in subsequent experiments. One would have to test these hypotheses with carefully designed experiments. For example, one could separate water-deficit effects from ABA effects by conducting water deficit experiments with ABA mutants or applications of ABA to non-stressed vines. Although mapping of the molecular pathways is incomplete mapping efforts will continue with more research and annotation progress.

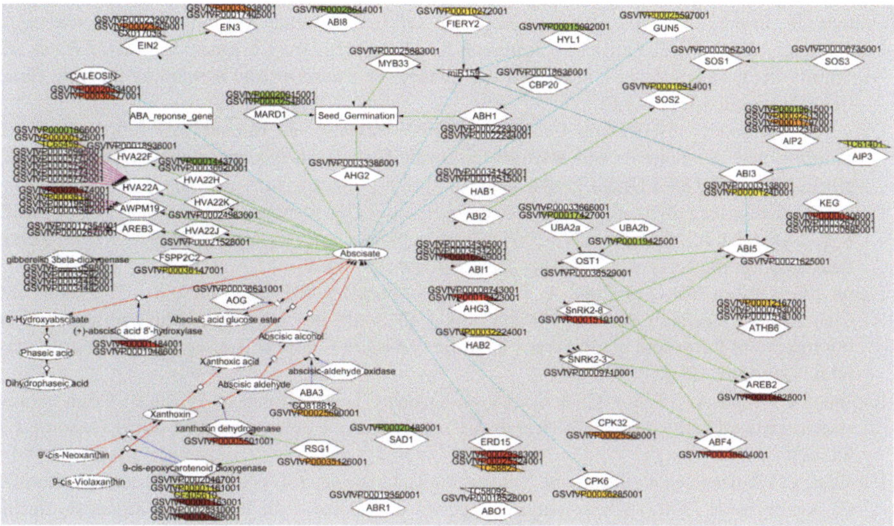

Fig. 22.7 The effect of water deficit on Cabernet Sauvignon shoot transcripts as visualized on the ABA signaling pathway in VitisNet. Changes in transcript abundance (decreased abundance to increased transcript abundance; a gradient of dark green to dark red, respectively) are displayed by colored boxes. White boxes indicate no changes or the transcripts were not detected. Circles denote ABA-induced transcripts in *Arabidopsis*. *Arrows* are colored to indicate the kind of reaction: (*red*) metabolic; (*blue*) enzymatic catalysis; (*green*) trigger; (*yellow*) inhibition; (*dark green*) unknown interaction

References

Affymetrix (2002a) GeneChip® Data Analysis Fundamentals Manual. Affymetrix, Santa Clara

Affymetrix (2002b) Affymetrix Microarray Suite 5.0 User's Guide. Affymetrix, Santa Clara

Affymetrix (2002c) Affymetrix Microarray Suite 5.1 User's Guide. Affymetrix, Santa Clara

Ashburner M, Ball CA, Blake JA, Botstein D, Butler H, Cherry JM, Davis AP, Dolinski K, Dwight SS, Eppig JT, Harris MA, Hill DP, Issel-Tarver L, Kasarskis A, Lewis S, Matese JC, Richardson JE, Ringwald M, Rubin GM, Sherlock G (2000) Gene ontology: tool for the unification of biology. The gene ontology consortium. Nat Genet 25:25–29

Baugh LR, Hill AA, Brown EL, Hunter CP (2001) Quantitative analysis of mRNA amplification by in vitro transcription. Nucleic Acids Res 29:E29

Benjamini Y, Hochberg, Y (1995) Controlling the false discovery rate: a practical and powerful approach to multiple testing. J R Stat Soc Ser B 57:289–300

Bolstad BM, Irizarry RA, Astrand M, Speed TP (2003) A comparison of normalization methods for high density oligonucleotide array data based on variance and bias. Bioinformatics 19: 185–193

Chervin C, Tira-Umphon A, Terrier N, Zouine M, Severac D, Roustan JP (2008) Stimulation of the grape berry expansion by ethylene and effects on related gene transcripts, over the ripening phase. Physiol Plant 134:534–546

Cramer GR (2010) Abiotic stress & plant responses from the whole vine to the genes. Aus J Grape Wine Res 16:86–93

Cramer GR, Ergul A, Grimplet J, Tillett RL, Tattersall EA, Bohlman MC, Vincent D, Sonderegger J, Evans J, Osborne C, Quilici D, Schlauch KA, Schooley DA, Cushman JC (2007) Water and salinity stress in grapevines: early and late changes in transcript and metabolite profiles. Funct Integr Genomics 7:111–134

Daly TM, Dumaual CM, Dotson CA, Farmen MW, Kadam SK, Hockett RD (2005) Precision profiling and components of variability analysis for Affymetrix microarray assays run in a clinical context. J Mol Diagn 7:404–412

Deluc LG, Grimplet J, Wheatley MD, Tillett RL, Quilici DR, Osborne C, Schooley DA, Schlauch KA, Cushman JC, Cramer GR (2007) Transcriptomic and metabolite analyses of Cabernet Sauvignon grape berry development. BMC Genomics 8:429

Deluc LG, Quilici DR, Decendit A, Grimplet J, Wheatley MD, Schlauch KA, Mérillon JM, Cushman JC, Cramer GR (2009) Water deficit alters differentially metabolic pathways affecting important flavor and quality traits in grape berries of Cabernet Sauvignon and Chardonnay. BMC Genomics 10:212

Espinoza C, Vega A, Medina C, Schlauch K, Cramer G, Arce-Johnson P (2006) Gene expression associated with compatible viral diseases in grapevine cultivars. Funct Integr Genomics 7: 95–110

Fernandez L, Torregrosa L, Terrier N, Sreekantan L, Grimplet J, Davies C, Thomas MR, Romieu C, Ageorges A (2007) Identification of genes associated with flesh morphogenesis during grapevine fruit development. Plant Mol Biol 63:307–323

Figueiredo A, Fortes AM, Ferreira S, Sebastiana M, Choi YH, Sousa L, Acioli-Santos B, Pessoa F, Verpoorte R, Pais MS (2008) Transcriptional and metabolic profiling of grape (Vitis vinifera L.) leaves unravel possible innate resistance against pathogenic fungi. J Exp Bot 59: 3371–3381

Gatto P, Vrhovsek U, Muth J, Segala C, Romualdi C, Fontana P, Pruefer D, Stefanini M, Moser C, Mattivi F, Velasco R (2008) Ripening and genotype control stilbene accumulation in healthy grapes. J Agric Food Chem 56:11773–11785

Gautier L, Cope L, Bolstad BM, Irizarry RA (2004) affy – analysis of Affymetrix GeneChip data at the probe level. Bioinformatics 20:307–315

Grimplet J, Cramer GR, Dickerson JA, Mathiason K, Fennell AY (2009) VitisNet: "Omics" integration through grapevine molecular networks. PLoS ONE (submitted)

Grimplet J, Deluc LG, Tillett RL, Wheatley MD, Schlauch KA, Cramer GR, Cushman JC (2007) Tissue-specific mRNA expression profiling in grape berry tissues. BMC Genomics 8:187

Grubbs F (1969) Procedures for detecting outlying observations in samples. Technometrics 11: 1–21

Han ES, Wu Y, McCarter R, Nelson JF, Richardson A, Hilsenbeck SG (2004) Reproducibility, sources of variability, pooling, and sample size: important considerations for the design of high-density oligonucleotide array experiments. J Gerontol A Biol Sci Med Sci 59:306–315

Huang D, Jaradat MR, Wu W, Ambrose SJ, Ross AR, Abrams SR, Cutler AJ (2007) Structural analogs of ABA reveal novel features of ABA perception and signaling in Arabidopsis. Plant J 50:414–428

Irizarry RA, Bolstad BM, Collin F, Cope LM, Hobbs B, Speed TP (2003a) Summaries of Affymetrix GeneChip probe level data. Nucleic Acids Res 31:e15

Irizarry RA, Hobbs B, Collin F, Beazer-Barclay YD, Antonellis KJ, Scherf U, Speed TP (2003b) Exploration, normalization, and summaries of high density oligonucleotide array probe level data. Biostatistics 4:249–264

Jaillon O, Aury JM, Noel B, Policriti A, Clepet C, Casagrande A, Choisne N, Aubourg S, Vitulo N, Jubin C, Vezzi A, Legeai F, Hugueney P, Dasilva C, Horner D, Mica E, Jublot D, Poulain J, Bruyere C, Billault A, Segurens B, Gouyvenoux M, Ugarte E, Cattonaro F, Anthouard V, Vico V, Del Fabbro C, Alaux M, Di Gaspero G, Dumas V, Felice N, Paillard S, Juman I, Moroldo M, Scalabrin S, Canaguier A, Le Clainche I, Malacrida G, Durand E, Pesole G, Laucou V, Chatelet P, Merdinoglu D, Delledonne M, Pezzotti M, Lecharny A, Scarpelli C, Artiguenave F, Pe ME, Valle G, Morgante M, Caboche M, Adam-Blondon AF, Weissenbach J, Quetier F,

Wincker P (2007) The grapevine genome sequence suggests ancestral hexaploidization in major angiosperm phyla. Nature 449:463–467

Li C, Hung Wong W (2001) Model-based analysis of oligonucleotide arrays: model validation, design issues and standard error application. Genome Biol 2: RESEARCH0032

Lund ST, Peng FY, Nayar T, Reid KE, Schlosser J (2008) Gene expression analyses in individual grape (*Vitis vinifera* L.) berries during ripening initiation reveal that pigmentation intensity is a valid indicator of developmental staging within the cluster. Plant Mol Biol 68:301–315

Maere S, Heymans K, Kuiper M (2005) BiNGO: a Cytoscape plugin to assess overrepresentation of gene ontology categories in biological networks. Bioinformatics 21:3448–3449

Mathiason K, He D, Grimplet J, Venkateswari J, Galbraith DW, Or E, Fennell A (2009) Transcript profiling in Vitis riparia during chilling requirement fulfillment reveals coordination of gene expression patterns with optimized bud break. Funct Integr Genomics 9:81–96

Matsui A, Ishida J, Morosawa T, Mochizuki Y, Kaminuma E, Endo TA, Okamoto M, Nambara E, Nakajima M, Kawashima M, Satou M, Kim JM, Kobayashi N, Toyoda T, Shinozaki K, Seki M (2008) Arabidopsis transcriptome analysis under drought, cold, high-salinity and ABA treatment conditions using a tiling array. Plant Cell Physiol 49:1135–1149

Pilati S, Perazzolli M, Malossini A, Cestaro A, Dematte L, Fontana P, Dal Ri A, Viola R, Velasco R, Moser C (2007) Genome-wide transcriptional analysis of grapevine berry ripening reveals a set of genes similarly modulated during three seasons and the occurrence of an oxidative burst at veraison. BMC Genomics 8:428

Redman JC, Haas BJ, Tanimoto G, Town CD (2004) Development and evaluation of an Arabidopsis whole genome Affymetrix probe array. Plant J 38:545–561

Rotter A, Camps C, Lohse, Kappel C, Pilati S, Hren M, Stitt M, Coutos-Thévenot P, Moser C, Usadel B, Delrot S, Gruden K (2009) Gene expression profiling in susceptible interaction of grapevine with its fungal pathogen Eutypa lata: extending MapMan ontology for grapevine. BMC Plant Biology 9:104

Ruepp A, Zollner A, Maier D, Albermann K, Hani J, Mokrejs M, Tetko I, Guldener U, Mannhaupt G, Munsterkotter M, Mewes HW (2004) The FunCat, a functional annotation scheme for systematic classification of proteins from whole genomes. Nucleic Acids Res 32: 5539–5545

Schoof H, Ernst R, Nazarov V, Pfeifer L, Mewes HW, Mayer KF (2004) MIPS *Arabidopsis thaliana* Database (MAtDB): an integrated biological knowledge resource for plant genomics. Nucleic Acids Res 32:D373–376

Smyth G (2005) Limma: linear models for microarray data. In: Gentlemen R, Carey V, Dudoit S, Irizarry R, Huber W (eds) Bioinformatics and computational biology solutions using R and bioconductor. Springer, New York, pp 397–420

Tattersall E, Ergul A, Al-Kayal F, Cushman JC, Cramer GR (2005) A comparison of methods for isolating RNA from leaves of grapevine (Vitis vinifera). . Am J Enol Vitic 56:400–406

Tattersall EA, Grimplet J, Deluc L, Wheatley MD, Vincent D, Osborne C, Ergul A, Lomen E, Blank RR, Schlauch KA, Cushman JC, Cramer GR (2007) Transcript abundance profiles reveal larger and more complex responses of grapevine to chilling compared to osmotic and salinity stress. Funct Integr Genomics 7:317–333

Terrier N, Glissant D, Grimplet J, Barrieu F, Abbal P, Couture C, Ageorges A, Atanassova R, Leon C, Renaudin JP, Dedaldechamp F, Romieu C, Delrot S, Hamdi S (2005) Isogene specific oligo arrays reveal multifaceted changes in gene expression during grape berry (*Vitis vinifera* L.) development. Planta 222:832–847

Thimm O, Blasing O, Gibon Y, Nagel A, Meyer S, Kruger P, Selbig J, Muller LA, Rhee SY, Stitt M (2004) MAPMAN: a user-driven tool to display genomics data sets onto diagrams of metabolic pathways and other biological processes. Plant J 37:914–939

Trewavas A (2006) A Brief History of Systems Biology: "Every object that biology studies is a system of systems." Francois Jacob (1974). Plant Cell 18:2420–2430

Velasco R, Zharkikh A, Troggio M, Cartwright DA, Cestaro A, Pruss D, Pindo M, Fitzgerald LM, Vezzulli S, Reid J, Malacarne G, Iliev D, Coppola G, Wardell B, Micheletti D, Macalma T,

Facci M, Mitchell JT, Perazzolli M, Eldredge G, Gatto P, Oyzerski R, Moretto M, Gutin N, Stefanini M, Chen Y, Segala C, Davenport C, Dematte L, Mraz A, Battilana J, Stormo K, Costa F, Tao Q, Si-Ammour A, Harkins T, Lackey A, Perbost C, Taillon B, Stella A, Solovyev V, Fawcett JA, Sterck L, Vandepoele K, Grando SM, Toppo S, Moser C, Lanchbury J, Bogden R, Skolnick M, Sgaramella V, Bhatnagar SK, Fontana P, Gutin A, Van de Peer Y, Salamini F, Viola R (2007) A high quality draft consensus sequence of the genome of a heterozygous grapevine variety. PLoS One 2: e1326

Vincent D, Ergul A, Bohlman MC, Tattersall EA, Tillett RL, Wheatley MD, Woolsey R, Quilici DR, Joets J, Schlauch K, Schooley DA, Cushman JC, Cramer GR (2007) Proteomic analysis reveals differences between *Vitis vinifera* L. cv. Chardonnay and cv. Cabernet Sauvignon and their responses to water deficit and salinity. J Exp Bot 58:1873–1892

Wang Z, Gerstein M, Snyder M (2009) RNA-Seq: a revolutionary tool for transcriptomics. Nat Rev Genet 10:57–63

Waters DL, Holton TA, Ablett EM, Lee LS, Henry RJ (2005) cDNA microarray analysis of developing grape (*Vitis vinifera* cv. Shiraz) berry skin. Funct Integr Genomics 5:40–58

Waters DLE, Holton TA, Ablett EM, Slade Lee L, Henry RJ (2006) The ripening wine grape berry skin transcriptome. Plant Sci 171:132–138

Welle S, Brooks AI, Thornton CA (2002) Computational method for reducing variance with Affymetrix microarrays. BMC Bioinformatics 3:23

Chapter 23
Visualisation of Transcriptomics Data in Metabolic Pathways

Ana Rotter, Matjaž Hren, Björn Usadel, and Kristina Gruden

Abstract When performing genome wide transcriptomics analysis life scientists quite often face problems with interpretation of huge amounts of data obtained. MapMan is a tool developed to help plant scientists in this aspect. Microarray data (or other omics data) are plotted onto diagrams representing metabolic and signaling pathways or alternatively large protein families. Application is flexible thus specificities of grapevine metabolism could have been implemented when adopting the system for this species. Overview of different analytical approaches using MapMan and few examples of analytical results are presented within this chapter.

Keywords Data analysis · Transcriptomics · DNA microarrays · Visualization · Pathway analysis · Secondary metabolism

Contents

Abbreviations

ATP Adenosine-5′-triphosphate
CDD Conserved Domain Database
CHO Carbohydrate
DNA Deoxyribonucleic acid
GO Gene Ontology

A. Rotter (✉)
Department of Biotechnology and Systems Biology, National Institute of Biology, Ljubljana 1000, Slovenia
e-mail: ana.rotter@nib.si

S. Delrot et al. (eds.), *Methodologies and Results in Grapevine Research*,
DOI 10.1007/978-90-481-9283-0_23,

PPAP	Plant Proteome Annotation Program
RNA	Ribonucleic acid
TC	Tentative contig
TCA	Tricarboxylic acid
TIGR	The Institute for Genomic Research
VvGI	*Vitis vinifera* Gene Index

23.1 Introduction

The last step in analysis of microarray data is biological interpretation of the results. The output of microarray data statistical analysis is normally a list of genes with certain description (annotation), expression values and statistical significance of the result (most often a p-value). Depending on the null hypothesis that is being tested and the experimental treatments that have been carried out, these lists can vary in length, but these are often too long for rigorous manual inspection. This poses a problem of complexity of interpretation. Lists can be condensed by organising them according to gene function, which requires systematically organised gene ontologies.

Gene Ontology (GO) is a system of controlled descriptions of gene functions according to the biological process, molecular function, and cellular component of individual genes (Ashburner et al. 2000). This system enables a more rapid, accurate, and consistent assignment of predicted functions to genes and permits the development of more accurate relationships between genes in different organisms. It is, however, primarily designed for mammals and therefore several plant specific pathways are not well represented or are completely absent.

One step further in data interpretation is the visualisation of results in biological context, e.g. in metabolic or signaling pathways. Various visualization tools that help the data analysts and biologists are available today from GenMapp, Pathway Processor and GeneXpress to KaPPA-View and VANTED (reviewed in Rotter et al. 2007). By using them, it is possible to find trends that would be less apparent directly from lists of genes. While they are suitable for some organisms, their usefulness for plant organisms is again limited because they have often been developed for microbial or animal systems and thus have categories that are irrelevant for plant systems and lack plant-specific pathways and processes. On the other hand the AraCyc database (Mueller et al. 2003) displays computationally predicted *Arabidopsis thaliana* metabolic pathways that are largely manually curated. It is, however, available only as an on-line tool thus the user is dependent on the availability of a fast internet connection and host server workload at the time of the analysis. Somehow different tools also exist like the VirtualPlant (Gutierrez et al. 2007) and ONDEX (Kohler et al. 2006), which have created graph-based integrations of knowledge and gene functional inferences that may be queried, filtered and further analyzed with different appended tools.

23.2 The Principles of MapMan Organisation

MapMan introduced a hierarchical ontology different from GO terms, that can be used for visualizing large data sets onto metabolic pathways and other biological processes (Thimm et al. 2004). The ontology was originally built for the model species *A. thaliana*, and furthermore extended to cover also maize (Doehlemann et al. 2008), Medicago (Tellström et al. 2007), tomato (Urbanczyk-Wochniak et al. 2006), potato (Rotter et al. 2007), rice (Howell et al. 2009), soybean (Leakey et al. 2009) and barley (Sreenivasulu et al. 2009). Mappings for wheat and poplar are also available at MapMan website (mapman.gabipd.org).

MapMan consists of two modules, (i) Scavenger module and (ii) Image-Annotator. The Scavenger module is a gene ontology scavenger, in which genes are assigned based on their annotation into largely nonredundant and hierarchically organised BINs. Each BIN consists of items of similar biological function and can be further split into subBINs corresponding to subnodes of the biological function (Thimm et al. 2004). The original BIN assignments for *A. thaliana* were based on publicly available gene annotations from TIGR (The Institute for Genomic Research) using a process which involved alternation between automatic recruitment and manual correction (Thimm et al. 2004). The resulting BINs are shown in Table 23.1; these are broken down by current versions into > 1,200 subBINs.

Table 23.1 Hierarchical organisation of MapMan ontologies into BINs. Each BIN is further split into several subBINs. Additionally, number of manual corrections made when annotating grapevine Gene Index (VvGI 5.0) for each BIN is reported. Taken from Rotter et al. (2009)

BIN	BIN name	No. of clones in the BIN	No. of corrected clones	% of corrected clones
1	Photosynthesis	494	23	4.6
2	Major CHO metabolism	165	8	4.8
3	Minor CHO metabolism	162	13	8
4	Glycolysis	123	9	7.3
5	Fermentation	52	0	0
6	Gluconeogenese/glyoxylate cycle	22	2	9
7	Oxidative pentose phosphate pathway	42	1	2.4
8	TCA/org. transformation	123	8	6.5
9	Mitochondrial electron transport/ATP synthesis	156	4	2.6
10	Cell wall	595	4	0.7
11	Lipid metabolism	459	27	5.9
12	N-metabolism	59	4	6.8
13	Amino acid metabolism	459	17	3.7
14	S-assimilation	15	0	0
15	Metal handling	142	14	9.9
16	Secondary metabolism	543	92	16.9
17	Hormone metabolism	502	29	5.8
18	Co-factor and vitamine metabolism	45	3	6.7
19	Tetrapyrrole synthesis	56	14	25

Table 23.1 (continued)

BIN	BIN name	No. of clones in the BIN	No. of corrected clones	% of corrected clones
20	Stress	948	456	48.1
21	Redox	282	15	5.3
22	Polyamine metabolism	18	0	0
23	Nucleotide metabolism	147	6	4.1
24	Biodegradation of xenobiotics	24	1	4.2
25	C1-metabolism	33	0	0
26	Miscellaneous enzyme groups	1219	69	5.7
27	RNA	2296	85	3.7
28	DNA	422	43	10.2
29	Protein	3628	157	4.3
30	Signalling	1157	81	7
31	Cell	655	12	1.8
33	Development	405	31	7.6
34	Transport	951	32	3.4
35	35.1. not assigned. no ontology	3276	437	13.3
	35.2. not assigned. unknown	15571	31	0.2
	$\sum$	35246	1728	4.9

The ImageAnnotator module uses the classifications from the Scavenger module in the form of mapping files in order to display data on various diagrams of the user choice (Thimm et al. 2004). A mapping file for an organism includes, but is not limited to, these categories: BIN code, BIN name, identifier and description. The latter two includes gene IDs or clone IDs, i.e. the names which link the mapping file with the experiment file with their descriptions (microarray results file). The ImageAnnotator also uses diagrams for data display. They can be obtained with the MapMan software or can be added independently from MapMan website.

Later on statistical tools (Usadel et al. 2005) were added to this software package to get an unbiased overview of changed pathways or processes (Fig. 23.1).

23.3 Specificities of Grapevine MapMan

Recently, annotation of *Vitis vinifera* Gene Index (VvGI version 5) to MapMan ontology was set up (Rotter et al. 2009). Information on Arabidopsis proteome and plant protein domains (found in SwissProt/Uniprot PPAP), the Conserved Domain Database CDD, Eukaryotic Orthologous Groups KOG and InterPro was included. Additionally, manual annotation was performed by several expert groups. A manual correction usually consisted of blasting the appropriate tentative contig (TC) sequence, followed by classification using expert knowledge and literature search. Altogether, 1728 manual corrections of automated annotation were made using this approach (Table 23.1). Due to specificities of grapevine physiology, we have created new pictorial representations focusing on three selected pathways: carotenoid pathway, terpenoid pathway and phenylpropanoid pathway, the products

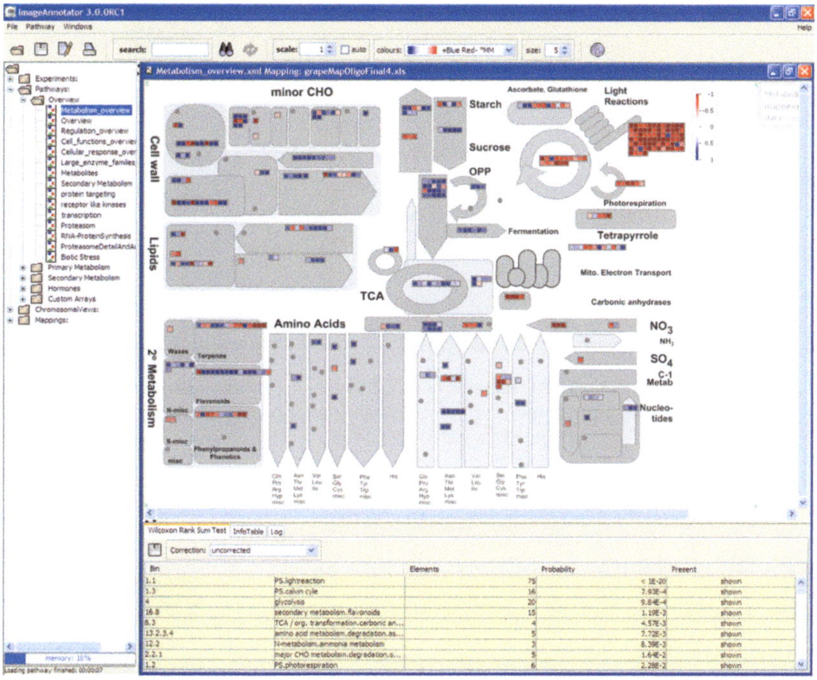

Fig. 23.1 A snapshot of ImageAnnotator in an analysis of grapevine phytoplasma dataset (Hren et al. 2009). An overview of changes in metabolism can first be checked. In the main part of the screen, a pictorial representation of different metabolism related BINs is shown with expression values for genes presented as coloured squares (see the colour scale on the image). At the bottom of the screen, the results of statistical analysis for significantly altered processes (BINs) are presented, which is also helpful for further steps of analysis

of these pathways being important for wine aroma, flavour and colour, as well as plant defence against pathogens. The carotenoid biosynthesis pathway was adopted from potato carotenoid pathway (Diretto et al. 2006) with the help of already published MapMan representation for potato (Rotter et al. 2007). The flavonoid biosynthesis pathway was constructed based on Grimplet et al. (2007) and Ageorges et al. (2006) where grapevine gene names, playing roles in flavonoid biosynthesis, were directly used for creating our improved, grapevine-based flavonoid pathway. The existing pathway mapping file for terpenoid biosynthesis from the MapMan website which included only the mevalonate pathway was modified in order to include the non-mevalonate pathway, too.

Currently, two mappings are available for visualisation of results of transcriptomics experiments on grape, one to be used with Operon oligoset-based microarrays and the other to be used with Affymetrix microarrays. The mappings for the genome-data based microarrays are currently being prepared and will be available on the MapMan website. An example of analysis of grapevine phytoplasma interaction (Hren et al. 2009) is taken here to show different features of application (Figs. 23.1, 23.2, 23.3).

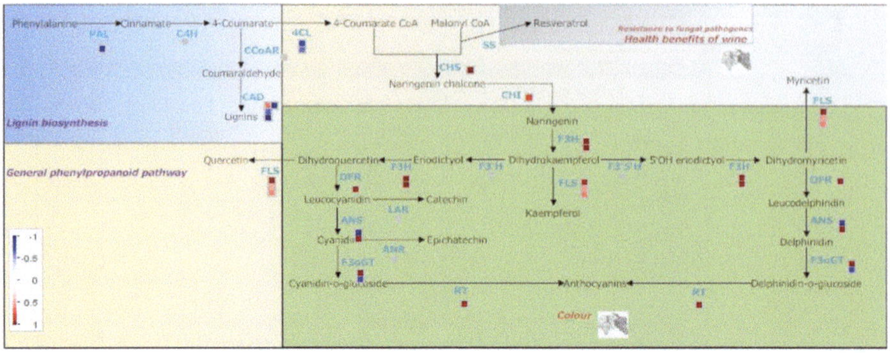

Fig. 23.2 Analysis of secondary metabolites. Changes in expression of genes involved in phenylpropanoid pathway for grapevine-phytoplasma interaction (dataset from Hren et al. 2009)

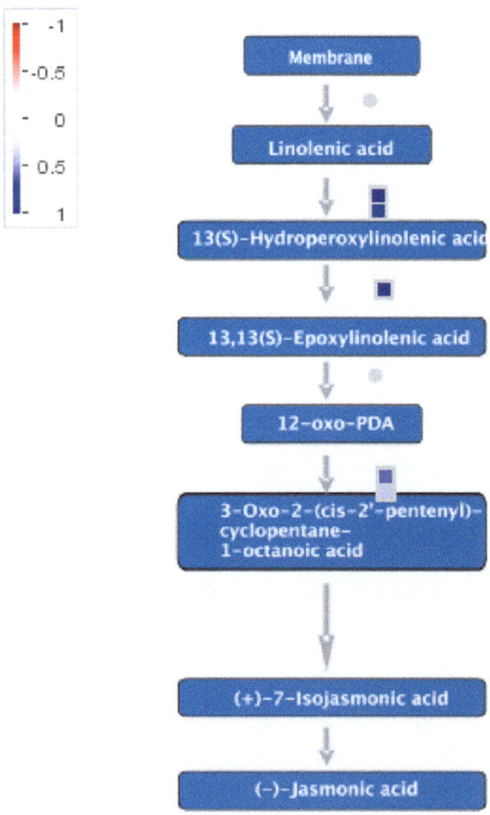

Fig. 23.3 Changes in metabolism of hormones. Example of jasmonic acid synthesis pathway induction in phytoplasma infected grapevine leaves (dataset from Hren et al. 2009)

23.4 Conclusions

Tools that integrate results of microarray data analysis with biological processes are extremely important for faster, more comprehensive and relevant interpretation of obtained gene expression results. Several tools that are specific for plant metabolism have already been developed but many of them are Arabidopsis-specific. As plants are remarkably diverse, additional pathways may need to be implemented for specific species as well as corrections of some existing pathways. The extension of MapMan ontology to grapevine together with the newly constructed pictorial representations of carotenoid, terpenoid and phenylpropanoid metabolism thus provide a step forward to the analysis of grapevine gene expression experiments performed with Affymetrix or Operon oligoset-based microarrays in a user friendly format. What is also important for the grapevine research community is that MapMan can be used to visualise other types of omics data in addition to transcriptomic, e.g. metabolomic and proteomic, and can thus provide a knowledge base for the integrative approach in data analysis.

References

Ageorges A, Fernandez L, Vialet S, Merdinoglu D, Terrier N, Romieu C (2006) Four specific isogenes of the anthocyanin metabolic pathway are systematically co-expressed with the red colour of grape berries. Plant Sci 170:372–383

Ashburner M et al (2000) Gene ontology: Tool for the unification of biology. The Gene Ontology Consortium. Nat Genet 25:25–29

Diretto G, Tavazza R, Welsch R, Pizzichini D, Mourgues F, Papacchioli V, Beyer O, Giuliano G (2006) Metabolic engineering of potato tuber carotenoids through tuber-specific silencing of lycopene epsilon cyclase. BMC Plant Biol 6:13

Doehlemann G, Wahl R, Horst RJ, Voll LM, Usadel B, Poree F, Stitt M, Pons-Kühnemann J, Sonnewald U, Kahmann R, Kämper J (2008) Reprogramming a maize plant: transcriptional and metabolic changes induced by the fungal biotroph Ustilago maydis. Plant J 56: 181–195

Grimplet J, Deluc LG, Tillett RL, Wheatley MD, Schlauch KA, Cramer GR, Cushman JC (2007) Tissue-specific mRNA expression profiling in grape berry tissues. BMC Genomics 8:187

Gutierrez RA, Lejay LV, Dean A, Chiaromonte F, Shasha DE, Coruzzi GM (2007) Qualitative network models and genome-wide expression data define carbon/nitrogen-responsive molecular machines in Arabidopsis. Genome Biol 8:R7

Howell K, Narsai R, Carroll A, Ivanova A, Lohse M, Usadel B, Millar AH, Whelan J (2009) Mapping metabolic and transcript temporal switches during germination in Oryza sativa highlights specific transcription factors and the role of RNA instability in the germination process. Plant Physiol 149:961–980

Hren M, Nikolić P, Rotter A, Blejec A, Terrier N, Ravnikar M, Dermastia M, Gruden K (2009) 'Bois noir' phytoplasma induces significant reprogramming of the leaf transcriptome in the field grown grapevine. BMC Genomics 10:460

Kohler J, Baumbach J, Taubert J, Specht M, Skusa A, Ruegg A, Rawlings C, Verrier P and Philippi S (2006) Graph-based analysis and visualization of experimental results with ONDEX. Bioinformatics 22:1383–1390

Leakey AD, Xu F, Gillespie KM, McGrath JM, Ainsworth EA, Ort DR (2009) Genomic basis for stimulated respiration by plants growing under elevated carbon dioxide. Proc Natl Acad Sci USA. 2009 106:3597–3602

Mueller LA, Zhang P, Rhee SY (2003) AraCyc: A biochemical pathway database for Arabidopsis. Plant Physiol 132:453–460

Rotter A, Usadel B, Baebler Š, Stitt M, Gruden K (2007) Adaptation of the MapMan ontology to biotic stress responses: application in solanaceous species. Plant Methods 3:10

Rotter A, Camps C, Lohse M, Kappel C, Pilati S, Hren M, Stitt M, Coutos-Thévenot P, Moser C, Usadel B, Delrot S, Gruden K (2009) Gene expression profiling in susceptible interaction of grapevine with its fungal pathogen Eutypa lata: Extending MapMan ontology for grapevine. BMC Plant Biol 9:104–117

Sreenivasulu N, Usadel B, Winter A, Radchuk V, Scholz U, Stein N, Weschke W, Strickert M, Close TJ, Stitt M, Graner A, Wobus U (2008) Barley grain maturation and germination: Metabolic pathway and regulatory network commonalities and differences highlighted by new MapMan/PageMan profiling tools. Plant Physiol 146:1734–1761

Tellström V, Usadel B, Thimm O, Stitt M, Küster H, Niehaus K (2007) The Lipopolysaccharide of Sinorhizobium meliloti Suppresses Defense-associated gene expression in cell cultures of the host plant medicago truncatula. Plant Physiol 143:825–837

Thimm O, Bläsing O, Gibon Y, Nagel A, Meyer S, Krüger P, Selbig J, Müller LA, Rhee SY and Stitt M (2004) MAPMAN: A user-driven tool to display genomics data sets onto diagrams of metabolic pathways and other biological processes. Plant J 37:914–939

Urbanczyk-Wochniak E, Usadel B, Thimm O, Nunes-Nesi A, Carrari F, Davy M, Bläsing O, Kowalczyk M, Weicht D, Polinceusz A, Meyer S, Stitt M, Fernie AR (2006) Conversion of MapMan to allow the analysis of transcript data from Solanaceous species: effects of genetic and environmental alterations in energy metabolism in the leaf. Plant Mol Biol 60:773–792

Usadel B, Nagel A, Thimm O, Redestig H, Blaesing OE, Palacios-Rojas N, Selbig J, Hannemann J, Conceição Piques M, Steinhauser D, Scheible WR, Gibon Y, Morcuende R, Weicht D, Meyer S, Stitt M (2005) Extension of the visualization tool MapMan to allow statistical analysis of arrays, display of coresponding genes, and comparison with known responses. Plant Physiol 138:1195–1204

Chapter 24
Small RNA Extraction and Expression Analysis by Northern Blot

Giorgia Batelli, Giorgio Gambino, Erica Mica, Andrea Schubert, and Andrea Carra

Abstract Small non-coding RNAs are important regulatory elements able to cause endogenous gene silencing at both the transcriptional and post-transcriptional levels. Here we describe in detail the procedure of small RNAs extraction from grape tissues, which was developed as a modification of an RNA extraction method initially used for pine tree tissues. Polyacrylamide gel electrophoresis in denaturing conditions (obtained with high concentrations of urea) is used to efficiently separate small RNAs. In hybridization of small RNA northern blot, a single stranded DNA or LNA oligonucleotide in antisense orientation to the target small RNA is used as probe. The technique reported here was used for the isolation and expression analysis of small RNAs from grape berries at different stages of maturation and allowed the identification of several non-conserved miRNAs and siRNAs. Some major constraints of the procedure are discussed, including the comparably high amounts of tissues needed for the isolation of small RNAs, especially when berries are used as samples. Some recently developed methods to address these problems are briefly discussed, which are based on Real-Time (or quantitative) PCR.

Contents

A. Carra (✉)
Department of Arboriculture and Pomology, University of Torino, Grugliasco 10095, Italy
e-mail: andrea.carra@unito.it; andrea.m.carra@gmail.com

S. Delrot et al. (eds.), *Methodologies and Results in Grapevine Research*,
DOI 10.1007/978-90-481-9283-0_24,

24.1 Introduction

The regulation of gene expression in plants is pivotal for the tight control of biological processes, including developmental phase transition, tissue differentiation as well as response to environmental cues, either biotic or abiotic. Even though the fine-tuning of gene expression is largely entrusted to transcription factors able to induce or repress transcription of loci, additional modes of regulation are being elucidated over the years.

Since their discovery in plants 10 years ago (Hamilton and Baulcombe 1999), small non-coding RNAs (sRNAs) have emerged as important regulatory elements able to cause endogenous gene silencing at both the transcriptional and post-transcriptional levels. sRNAs are single stranded RNAs ranging in length from 18 to 24 nucleotides (nt) and can be divided into two classes, based on the nature of the precursor and the enzymes involved in their biogenesis.

microRNAs (miRNAs) are produced by cleavage, operated in the nucleus by members of the Dicer-like family of ribonucleases (DCLs), of hairpin precursors which are derived from longer primary transcripts encoded in the genomes. The cleavage by DCL enzymes leads to the formation of a~21 nt double stranded RNA (dsRNA) duplex which is exported to the cytoplasm, where one of the two strands, referred to as the mature miRNA, is loaded into the RNA-induced silencing complex (RISC), guiding it to complementary sites on target mRNAs. In higher plants the degree of complementarity between the miRNA and target transcripts is nearly 100% and the pairing can cause repression of gene expression via translational inhibition or, most frequently, mRNA cleavage (Chen 2005, Jones-Rhoades et al. 2006).

Short interfering RNAs (siRNAs) are produced from longer dsRNAs which can be generated through transcription of inverted repeats or conversion from ssRNAs operated by RNA-dependent RNA polymerases (RDRs). The dsRNAs are then cleaved by DCL proteins in 21–24-nt increments to generate multiple siRNAs from a single precursor. The siRNAs can then be incorporated into the RISC to drive post-transcriptional gene silencing (PTGS) or guide DNA methylation to target loci through RNA-dependent DNA methylation (Vaucheret 2006). Subclasses of siRNAs have been characterised, including *trans*-acting siRNAs (ta-siRNAs), able to regulate the expression of target genes other than their originating loci, and natural *cis*-antisense RNAs (nat-siRNAs) derived from convergent transcription of overlapping, antisense loci (Borsani et al. 2005, Katiyar-Agarwal et al. 2006).

sRNAs have been implicated in regulation of development and patterning (Jones-Rhoades et al. 2006). sRNAs are also able to repress gene expression in response to abiotic stresses such as oxidative stress (Sunkar et al. 2006, Yamasaki et al. 2007), nutrient deficiency (Jones-Rhoades and Bartel 2004, Fujii et al. 2005), drought stress (Zhao et al. 2007), high salinity (Borsani et al. 2005) or biotic stresses (reviewed by Ding and Voinnet 2007).

The use of high-throughput sequencing techniques has revealed a vast world of small RNAs in the model plant *Arabidopsiss* (Rajagopalan et al. 2006), in crop species (Sunkar et al. 2008, Johnson et al. 2007), and in fruit plants (Moxon et al.

2008), suggesting that the expression of a high number of genes might be regulated, at least in part, by small RNA-related processes. Thus, the characterization of small RNAs appears to be a key step for the understanding of the biological processes that influence the performances and the quality parameters of agronomically important species, including grapevine.

The method described in this chapter deals with the isolation of small RNA followed by northern blot analysis.

24.2 Description of the Procedure

24.2.1 Extraction of Small RNAs from Grapevine Tissues

Here we describe in details the procedure of small RNAs extraction from grape tissues, which was developed (Carra et al. 2007) as a modification of an RNA extraction method initially used for pine tree tissues (Chang et al. 1993). The method described by Chang and colleagues has proven successful for the isolation of total RNA from a variety of tissues with high concentrations of polysaccharides and polyphenols, such as sweet potato, and, with some modifications, grape berries (Iandolino et al. 2004). This method includes a LiCl precipitation step, which efficiently removes the polysaccharide fraction but causes loss of the low molecular weight (LMW) RNA. The procedure was adapted to the isolation of small RNAs by replacing the LiCl precipitation step with an isopropanol precipitation. Additional chloroform extractions steps were introduced to remove polysaccharides. A PEG precipitation step was added to enrich the LMW RNA fraction and to further purify the extracts.

24.2.1.1 Materials and Reagents

Pestles, mortars and all glassware are kept overnight at 180°C; plasticware is autoclaved before use, whereas solutions are prepared with RNase-free reagents in diethyl pyrocarbonate (DEPC)-treated water.

Extraction buffer: 100 mM Tris–HCl (pH 8.0), 25 mM EDTA (pH 8.0), 2% w/v CTAB (Sigma), 2% w/v PVP (Sigma, MW 40,000), 2.5 M NaCl, 0.05% w/v spermidine (Fluka), autoclaved at 120°C for 20 min.

- β-mercaptoethanol (Sigma)
- Sand (Fluka)
- Chloroform–isoamyl alcohol (24:1 v/v) (Sigma)
- Isopropanol
- 80% ethanol and 70% ethanol prepared with DEPC-treated water
- Phenol–chloroform (5:1 v/v) acid-equilibrated pH 4.7 (Sigma)
- 5 M NaCl
- 15% w/v PEG (Fluka, MW 8000)-2 M NaCl

- 50-mL polypropylene centrifuge tubes
- 14-mL polypropylene centrifuge tubes
- 2-mL microcentrifuge tubes
- 2-mL siliconized microcentrifuge tubes (Sigma)

24.2.1.2 Procedure

- Prewarm 12.5 mL of extraction buffer at 65°C in a 50-mL polypropylene tube. Add β-mercaptoethanol (250 μL) to each buffer-containing tube right before use.
- Grind grapevine tissue (0.6–2 g) in liquid nitrogen using a mortar and pestle. For better sample homogenisation, sand (0.5 g/ sample) is added to each sample. The ratio of buffer volume to fresh weight is optimised depending on the tissue. To achieve high RNA purity and avoid RNA degradation, we use 2 g of tissue for grape berries, 1 g for roots, while 0.6 g are used for grapevine leaves. Transfer ground tissue to the prewarmed, β-mercaptoethanol containing extraction buffer.
- Incubate the samples at 65°C for 10 min, vortexing every 2–3 min.
- Add an equal volume of chloroform:isoamyl alcohol (24:1) and immediately invert and vortex vigorously the tube. Centrifuge at 3,900*g* for 20 min at 4°C.
- Recover the aqueous phase and divide it into two 14-mL polypropylene tubes. Re-extract adding to each tube one volume of chloroform:isoamyl alcohol and centrifuge at 8,000*g* for 15 min at 4°C.
- Transfer the supernatant to a new 14-mL polypropylene tube, re-extract twice with an equal volume of chloroform:isoamyl alcohol and centrifugation at 8,000*g* for 15 min at 4°C.
- Transfer the aqueous phase to a new 14-mL polypropylene tube. Add 0.8 volume of isopropanol. Mix by inverting the tube and precipitate at 4°C for 1 h.
- Centrifuge at 8,000*g* for 30 min at 4°C.
- Remove the supernatant, wash the pellets with 80% ethanol and centrifuge at 8,000*g* for 5 min at 4°C.
- Completely remove the ethanol using a pipette and briefly air-dry the pellet.
- Dissolve the pellet in 350 μL of DEPC-treated water and pool the RNA from each biological sample into a 2-mL microcentrifuge tube.
- Add an equal volume of acidic phenol:chloroform (5:1) and immediately vortex vigorously. Centrifuge at 22,000*g* for 10 min at 4°C.
- Transfer the supernatant to a new microcentrifuge tube and re-extract with an equal volume of acidic phenol:chloroform at 22,000*g* for 10 min at 4°C.
- Transfer the upper phase (500 μL) to new microcentrifuge tubes. Add 50 μL of 5 M NaCl and 550 μL of chloroform:isoamyl alcohol. Extract by inverting the tube and vortexing. Centrifuge at 22,000*g* for 10 min at 4°C.
- Transfer the upper phase to new microcentrifuge tubes and incubate at 65°C for 5 min.
- Add an equal volume of 15% PEG 8000-2 M NaCl solution to the warm sample, mix by inversion, and incubate on ice for 1 h.

- Centrifuge at 19,000*g* for 10 min at 4°C. Transfer the LMW RNA-containing supernatant to 2-mL siliconized microcentrifuge tubes. Save the pellet if you wish to extract high-molecular-weight (HMW) RNA as well.
- Add an equal volume of isopropanol and precipitate overnight at 4°C.
- Centrifuge at 22,000*g* for 30 min at 4°C.
- Discard supernatant, wash the pellets with 80% ethanol and centrifuge at 22,000*g* for 5 min at 4°C.
- Remove the ethanol and briefly air-dry the pellet.
- Dissolve the pellet in 20 μL of DEPC-treated water.

High Molecular Weight RNA Recovery

- HMW RNA is obtained resuspending the pellets from the PEG precipitation in 600 μL of DEPC-treated water and performing overnight precipitation with 300 μL of 8 M LiCl at 4°C; pellet by centrifugation at 22,000*g* for 30 min at 4°C, wash with 70% ethanol, and resuspend in 30 μL of DEPC-treated water.

The extraction protocol described allowed efficient extraction of LMW RNA from different tissues of grapevine "Nebbiolo" plants: leaves, roots, deseeded berries collected 4 weeks post-flowering (green, immature) and 15 weeks post-flowering (black, mature).

Spectrophotometric A_{260}/A_{230} and A_{260}/A_{280} ratios indicate low polysaccharide and protein contamination (Table 24.1). The intact 5.8S and 5S ribosomal bands and intense tRNA smears on denaturing polyacrylamide gel electrophoresis (PAGE) (Fig. 24.1) indicate that little or no RNA degradation occurred during extraction. A good recovery of LMW RNA molecules is essential for accurate small RNA expression analysis by northern blotting and for small RNA library construction.

Table 24.1 Typical yields and quality of LMW RNA from grapevine tissues (Adapted from Carra et al. 2007)

	LMW RNA		
Tissue	Yield μg/gFW	A_{260}/A_{280}	A_{260}/A_{230}
Leaves	62.08 ± 14.82	1.79 ± 0.05	2.32 ± 0.05
Roots	26.6 ± 7.35	1.81 ± 0.19	1.97 ± 0.06
Immature berries	4.1 ± 0.19	1.71 ± 0.00	2.04 ± 0.04
Mature berries	1.52 ± 0.01	1.49 ± 0.01	1.98 ± 0.04

24.2.2 Small RNA Electrophoresis and Blotting

Polyacrylamide gel electrophoresis in denaturing conditions (obtained with high concentrations of urea) is used to efficiently separate small RNAs, which would not be efficiently resolved in agarose gels.

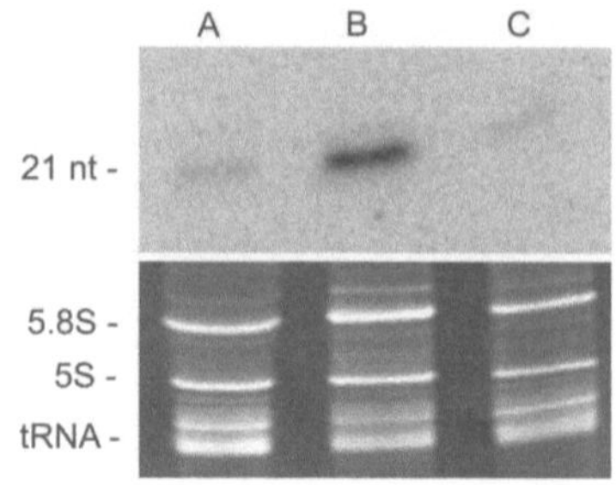

Fig. 24.1 Expression of miR160 in mature berries (**a**), roots (**b**) and immature berries (**c**) of *Vitis vinifera* cv Nebbiolo. Ethidium bromide staining of 5.8S and 5S rRNA and tRNA is shown as loading controls

The amount of RNA loaded differs depending on the expected expression of the target sRNA, but we do not recommend to load more than 15 μg/lane when using LMW RNA extracted from berries, to ensure correct running behaviour.

For the blotting to nylon membrane, electrophoretical transfer is preferred to capillarity transfer due to the small size of the polyacrylamide gel pores.

24.2.2.1 Materials and Reagents

- Standard format vertical gel unit (Amersham) with glass plates (18 x 16 cm) and 1–2 mm thick spacers and comb
- High voltage power supply (up to 600 V)
- Trans-blot SD semi-dry transfer cell (Bio-Rad)
- PowerPac HC high-current power supply (Bio-Rad)
- Extra thick blotting paper (Bio-Rad)
- Positively charged nylon membrane (Roche)
- Urea (Fluka)
- Acrylamide:N,N′-methylenebisacrylamide 19:1 solution 40% in water (Fluka)
- TBE 5X
- TEMED (Fluka)
- Ammonium persulfate (Fluka): prepare a fresh 10% solution in DEPC-treated water
- 21 nt and 24 nt 5′-end labeled RNA oligonucleotides can be used as markers

24.2.2.2 Procedure

Electrophoresis

- Clean glasses, spacers and comb with ethanol, then with isopropanol and assemble on the gel casting.
- Prepare denaturing gel: dissolve Urea (8 M), Acrylamide:N,N′-methylenebisacrylamide (15%) in TBE (0.5X) (for 1 mm thick gels, final volume is 25 mL with DEPC-treated water per gel), then add APS (0.07%), TEMED (0.035%) and pour the gel in the casting unit. Allow to polymerize for at least 2 h.
- Wash thoroughly the wells with 0.5X TBE and pre-run for 1 h at 350 V.

- Bring RNA samples (3–15 μg) to equal volume with DEPC-treated water (we use a final volume of 40 μL), then add an equal volume of loading buffer (98% formamide, 10 mM EDTA (pH 8.0), 0.01% bromophenol blue, 0.01% xylene cyanol).
- Denature the samples and markers at 65°C for 10 min and cool down on ice.
- Stop the pre-run and wash again the wells with 0.5X TBE.
- Load the samples and run 30 min at 350 V, then increase the voltage to 450 V and run for additional 2 h or until the bromophenol dye runs out of the gel.
- Disassemble the sandwich, stain the gel in a 0.025% ethidium bromide solution in TBE 0.5X for 10 min and check for RNA integrity and equal loading on a transilluminator.

Blotting

- Cut nylon membrane and two pieces of blotting paper according to gel size.
- Equilibrate the membrane with 0.5 X TBE and soak thoroughly the blotting paper in 0.5 X TBE.
- Assemble the sandwich in the semi-dry transfer unit in the following order: (from the bottom, i.e. from the positive electrode) one layer of blotting paper, membrane, gel, blotting paper. Make sure no air bubbles are left between the gel and the membrane.
- Run at 400 mA for 40–50 min. Voltage will increase during run. If voltage rises above 20 V, stop the run, lower the amperage to 300 mA and run longer (about 1 h).
- Disassemble the sandwich, cross-link the membrane with UV and bake at 120°C for 30 min.
- Store the membrane dry at 4°C until use.

24.2.3 Small RNA Hybridization

Hybridization for small RNA northern blot differs from classical northern in many aspects. Given the nature of small RNAs to be analyzed, a single stranded oligonucleotide is used as probe, that is easily labelled at its 5'-end adding a [γ^{32}P]ATP. The probe, designed in antisense orientation to the target small RNA, may be of variable length (19–22 nt) and may be ordered as a standard DNA oligo (desalted oligo) or as an LNA (Locked Nucleic Acid) oligo (produced by Exiqon).

The choice between DNA oligos and LNA oligos is driven by qualitative and economical considerations. LNA (Koshkin et al. 1998, Valoczi et al. 2004) is a nucleic acid analogue in which the ribose ring is locked by a methylene bridge. The presence of the methylene bridge confers to LNA oligos very high affinity and excellent specificity toward complementary RNA. The company synthesizes the probes incorporating a variable number of LNA nucleotides at the most stable positions. The LNA::RNA hybrids are more stable than DNA::RNA ones, conferring higher sensitivity, allowing higher hybridization temperatures and lowering background and aspecific signals. As a drawback, it has to be taken into account that LNA oligos are far more expensive than traditional DNA oligos.

The reduced length of both probes and target RNAs forces us to use lower hybridization temperature and less stringent conditions for washes compared to traditional northerns.

24.2.3.1 Materials and Reagents

- PerfectHyb Plus buffer 1x (Sigma)
- OptiKinase™ (Usb-Affymetrix)
- 5'-[γ^{32}P]ATP solution 10 μCi/μL (Perkin-Elmer)
- Micro Bio-Spin® Columns P30 Tris RNase free (BioRad)
- Wash buffer 2X: 2X SSC, 0.2% SDS in deionized water
- Wash buffer 1X: 1X SSC, 0.1% SDS in deionized water
- Wash buffer 0.5X: 0.5X SSC, 0.1% SDS in deionized water
- 20% SDS in deionized water

24.2.3.2 Procedure

5' End Probe Labelling Reaction

Add in a 1.5 mL tube the following:
- 10 pmol probe (0.5 μL from 10 μM stock solution)
- 5 μL Buffer 10X
- 2 μL Optikinase, 20U
- 2.5 μL [γ^{32}P]ATP
- Deionized H_2O to 50 μL

Incubate the reaction for at least 1 h at 37°C or over-night at room temperature.

If LNA oligos are used, it is recommended to scale down the labelling reaction mix, as 25 μL with only 5 pmol of probe are sufficient.

To eliminate unincorporated 5'-[γ^{32}P]ATP, purify the probe using Micro Bio-Spin® Columns (Biorad) following manufacturer's instructions. Add 200 μL of hybridization buffer to the column-purified probe and heat at 95°C for 5 min, to denature it. Chill on ice and add the purified, labelled probe to the hybridization buffer.

Hybridization Procedure

Prehybridize the membranes with 20–25 mL of hybridization buffer in a hybridization oven for at least 2 h at 42°C before adding the labelled and purified probe.

Hybridize over night at a temperature calculated on the basis of the probe melting temperature; we usually use (probe Tm-15)°C, where the probe Tm is, for both DNA and LNA probes, the melting temperature indicated by the manufacturer.

Washes

Prewarm wash buffers at hybridization temperature. Washes are performed at the same temperature as hybridization. Add to the hybridization tube 50 mL of 2X wash buffer, and wash briefly at room temperature. Discard the liquid and add 50 mL fresh 2X wash buffer. Put the tube back in the hybridization oven for 20 min, at hybridization temperature. Discard the liquid again, and repeat the washing step once.

Check the counts of the membranes and, if needed, repeat the washing steps with 1X and 0.5X wash buffers. At the end, dry the membranes briefly on Whatman paper and prepare it for detection.

Membrane Stripping

Prepare 1 L of boiling water, and add 5 mL of 20% SDS (0.1%. final concentration). Pour carefully the solution in a glass box and add the membrane, keeping in oscillation until cooled down.

Wrap the membrane in Saran wrap and expose overnight to check if the previous signal was deleted.

Each membrane can be reused 3 to 4 times, depending on the background signals that develop and accumulate each time. When LNA oligos are used, it may be more difficult to strip the membrane, however, we obtained good results with the procedure described.

24.3 Conclusions and Perspectives

In this chapter, we have described the procedure for the extraction of small RNAs from grapevine tissues followed by northern blot hybridization.

The technique reported here was used for the isolation and expression analysis of small RNAs from grape berries at different stages of maturation and allowed the identification of several non-conserved miRNAs and siRNAs which appear, so far, to be grapevine-specific (Carra et al. 2009). In particular, a previously unreported mechanism which may play a significant role in grape berry maturation has been described. Among the small RNAs identified, siRNA id65 accumulated specifically in mature berries and targeted a gene which encodes a cytokinin synthase. Accordingly, the transcript of the target gene was shown to be dramatically reduced at post-véraison stage (Carra et al. 2009), suggesting that PTGS is involved in the regulation of cytokinin content in berries. Deeper sequencing of small RNAs expressed in berries may bring us to better understand the complex events taking place during ripening.

However, the procedure described here has some major constraints which are basically common to all methods involving northern blot analysis.

Large amounts of tissue are needed for the isolation of small RNAs, especially when berries are used, because several rounds of chloroform extraction have to be performed in order to obtain high quality LMW RNA. Furthermore, it is advisable that at least three micrograms of LMW RNA are loaded in a small RNA gel and, even so, the expression of the target small RNA, especially if a siRNA, may be below detection limit in a northern blot analysis. The sensitivity problems can be partially overcome by the use of LNA probes and by RNA cross-linking with l-ethyl-3-(3-dimethylaminopropyl) carbodiimide (EDC) (Pall et al. 2007, Pall and Hamilton 2008). Another important issue can be the specificity: most of the miRNAs are part of families whose members may differ by as little as one or two bases and are therefore difficult to discriminate.

To address these problems, recently developed methods based on Real-Time (or quantitative) PCR (qPCR) have adapted the powerful qPCR technology to the small RNA world. In one approach (Shi and Chiang 2005), the first step is the in vitro polyadenylation of the small RNA, operated by the Poly(A) Polymerase enzyme, to produce poly(A)sRNAs. After the polyadenylation step, poly(A)sRNAs are reverse-transcribed using a poly(T) adapter primer and the target sRNA is then amplified using SYBR Green with a sequence specific forward primer and a reverse primer complementary to the adapter sequence. This method, effective even on total, non fractionated RNA, was used for the detection of several *Arabidopsis* miRNAs and was proven suitable for the discrimination of different miRNA species within a single miRNA family (Shi and Chiang 2005). In an alternative strategy, the reverse transcription step is carried out using a sequence specific primer containing a stem-loop structure that extends the 3' end of the small RNA, making the resulting cDNA suitable for a standard Real-Time PCR using TaqMan Assay (Chen et al. 2005) or SYBR Green (Pant et al. 2008).

The expected future wide use of PCR based technologies, together with "classical" expression analysis methods for sRNA detection and quantification, promises to produce significant improvement of our knowledge of this important area of cell biology.

References

Borsani O, Zhu J, Verslues PE, Sunkar R, Zhu JK (2005) Endogenous siRNAs derived from a pair of natural cis-antisense transcripts regulate salt tolerance in Arabidopsis. Cell 123: 1279–1291

Carra A, Gambino G, Schubert A (2007) A cetyltrimethylammonium bromide-based method to extract low-molecular-weight RNA from polysaccharide-rich plant tissues. Anal Biochem 360:318–320

Carra A, Mica E, Gambino G, Pindo M, Moser C, Pe ME, Schubert A (2009) Cloning and characterization of small noncoding RNAs from grape. Plant J 59:750–763

Chang S, Puryear J, Cairney J (1993) A simple and efficient method for isolating RNA from pine trees. Plant Mol Biol Rep 11:113–116

Chen C, Ridzon DA, Broomer AJ, Zhou Z, Lee DH, Nguyen JT, Barbisin M, Xu NL, Mahuvakar VR, Andersen MR, Lao KQ, Livak KJ, Guegler KJ (2005) Real-time quantification of microRNAs by stem-loop RT-PCR. Nucleic Acids Res 33:e179

Chen X (2005) MicroRNA biogenesis and function in plants. FEBS Lett 579:5923–5931

Ding SW, Voinnet O (2007) Antiviral immunity directed by small RNAs. Cell 130:413–426

Fujii H, Chiou TJ, Lin SI, Aung K, Zhu JK (2005) A miRNA involved in phosphate-starvation response in Arabidopsis. Curr Biol 15:2038–2043

Hamilton AJ, Baulcombe DC (1999) A species of small antisense RNA in posttranscriptional gene silencing in plants. Science 286:950–952

Iandolino AB, Goes da Silva G, Lim H, Choi H, Williams LE, Cook DR (2004) High-quality RNA, cDNA, and derived EST libraries from Grapevine (Vitis vinifera L.). Plant Mol Biol Rep 22:269–278

Johnson C, Bowman L, Adai AT, Vance V, Sundaresan V (2007) CSRDB: a small RNA integrated database and browser resource for cereals. Nucleic Acids Res 35:D829–833

Jones-Rhoades MW, Bartel DP (2004) Computational identification of plant microRNAs and their targets, including a stress-induced miRNA. Mol Cell 14:787–799

Jones-Rhoades MW, Bartel DP, Bartel B (2006) MicroRNAS and their regulatory roles in plants. Annu Rev Plant Biol 57:19–53

Katiyar-Agarwal S, Morgan R, Dahlbeck D, Borsani O, Villegas A, Jr., Zhu JK, Staskawicz BJ, Jin H (2006) A pathogen-inducible endogenous siRNA in plant immunity. Proc Natl Acad Sci USA 103:18002–18007

Koshkin AA, Nielsen P, Meldgaard M, Rajwanshi VK, Singh SK, Wengel J (1998) LNA (Locked Nucleic Acid): An RNA Mimic Forming Exceedingly Stable LNA:LNA Duplexes. J Am Chem Soc 120:13252–13253

Moxon S, Jing R, Szittya G, Schwach F, Rusholme Pilcher RL, Moulton V, Dalmay T (2008) Deep sequencing of tomato short RNAs identifies microRNAs targeting genes involved in fruit ripening. Genome Res 18:1602–1609

Pall GS, Codony-Servat C, Byrne J, Ritchie L, Hamilton A (2007) Carbodiimide-mediated cross-linking of RNA to nylon membranes improves the detection of siRNA, miRNA and piRNA by northern blot. Nucleic Acids Res 35:e60

Pall GS, Hamilton AJ (2008) Improved northern blot method for enhanced detection of small RNA. Nat Protoc 3:1077–1084

Pant BD, Buhtz A, Kehr J, Scheible WR (2008) MicroRNA399 is a long-distance signal for the regulation of plant phosphate homeostasis. Plant J 53:731–738

Rajagopalan R, Vaucheret H, Trejo J, Bartel DP (2006) A diverse and evolutionarily fluid set of microRNAs in Arabidopsis thaliana. Genes Dev 20:3407–3425

Shi R, Chiang VL (2005) Facile means for quantifying microRNA expression by real-time PCR. Biotechniques 39:519–525

Sunkar R, Kapoor A, Zhu JK (2006) Posttranscriptional induction of two Cu/Zn superoxide dismutase genes in Arabidopsis is mediated by downregulation of miR398 and important for oxidative stress tolerance. Plant Cell 18:2051–2065

Sunkar R, Zhou X, Zheng Y, Zhang W, Zhu JK (2008) Identification of novel and candidate miRNAs in rice by high throughput sequencing. BMC Plant Biol 8:25

Valoczi A, Hornyik C, Varga N, Burgyan J, Kauppinen S, Havelda Z (2004) Sensitive and specific detection of microRNAs by northern blot analysis using LNA-modified oligonucleotide probes. Nucleic Acids Res 32:e175

Vaucheret H (2006) Post-transcriptional small RNA pathways in plants: mechanisms and regulations. Genes Dev 20:759–771

Yamasaki H, Abdel-Ghany SE, Cohu CM, Kobayashi Y, Shikanai T, Pilon M (2007) Regulation of copper homeostasis by micro-RNA in Arabidopsis. J Biol Chem 282:16369–16378

Zhao B, Liang R, Ge L, Li W, Xiao H, Lin H, Ruan K, Jin Y (2007) Identification of drought-induced microRNAs in rice. Biochem Biophys Res Commun 354:585–590

Chapter 25
A Rapid and Efficient Method for Isolating High-Quality Total Proteins from Mature Buds and Other Woody Tissues of *Vitis Vinifera*

Atiako Kwame Acheampong, Ariel Rotman, Chuanlin Zheng, Alexandra Keren, Tamar Halaly, Omer Crane, Aliza Ogrodovitch, and Etti Or

Abstract Protocols from most published plant proteomics studies use Trichloroacetic acid (TCA), acetone and phenols to extract and precipitate the proteins. However, these protocols are generally less effective for protein extraction from woody grapes tissues. The major challenges to protein extraction from these tissues include, among others, the accumulation of high levels of proteases and secondary metabolites, which interfere with the efficient extraction and proper separation of proteins. This section presents a fast and easy protocol for the extraction of total proteins from woody tissues of grapes.

Contents

Abbreviations

TCA Trichloroacetic acid
SDS Sodium dodecyl sulfate
DTT Dithiothreitol
PAGE Polyacrylamide gel electrophoresis

A.K. Acheampong (✉)
Institute of Plant Sciences, A.R.O. Volcani Center, Bet Dagan 50250, Israel
e-mail: vhettior@agri.gov.il

S. Delrot et al. (eds.), *Methodologies and Results in Grapevine Research*,
DOI 10.1007/978-90-481-9283-0_25,

25.1 Introduction

Studying plant protein profiles and functions depends, to a large extent, on efficient extraction, separation and identification of proteins. Most common interfering substances are proteolytic and oxidative enzymes, terpenes, pigments, organic acids, inhibitory ions, carbohydrates and phenolic compounds such as phenols, tannins, lignins and flavonoids (Gorg and Weiss 2000, Carpentier et al. 2005). To reduce co-extraction and precipitation of these non-protein component, TCA, acetone, sodium dodecyl sulfate (SDS) and phenols have been widely employed for protein extraction in a range of plant species (Saravanan and Rose 2004, Carpentier et al. 2005, Song et al. 2006, Zheng et al. 2007, Maldonado et al. 2008). While different protocols have been designed for isolating proteins from grape berries (Tesnière and Robin 1992, Sarry et al. 2004, Vincent et al. 2006), none of these is efficient for extracting total proteins from mature buds and woody tissues. A simple, fast and efficient protocol has therefore been developed for the isolation of total protein from mature buds and other hard tissues of grapes. The protocol produces good quality proteins that can be used for one- and two-dimensional sodium dodecyl sulfate-polyacrylamide gel electrophoresis (SDS-PAGE), protein phosphorylation activity studies, as well as Western analyses (Pang et al. 2007).

25.2 Materials and Methods

25.2.1 Plant Materials

Mature buds and dry tissues were collected from field-grown *V. vinifera* cv. Perlette, snap-frozen in liquid nitrogen and stored at –80°C prior to protein extraction.

25.2.2 Reagents

- Phosphate Buffer, which component and preparation are detailed below:
  - Solution A: 0.2 M KH_2PO_4
  - Solution B: 0.2 M K_2HPO_4
  - Solution C: 6.6 mL of solution A added to 33.6 mL of solution B. The pH of the mix is adjusted to 7.5
  - To 25 mL of Solution C was added the following:
    - 10 mg Ascorbate
    - 2.3 g NaCl
    - 18.66 mg EDTA

  The volume of the buffer was adjusted to 50 mL using sterile double distilled water.

  This buffer can be stored at 4°C for up to 2 weeks.

- Triton X-100
- Complete^{TM} Protease inhibitor (Roche Diagnostics GmbH, Mannheim, Germany)
- Dithiothreitol (DTT) (Cleland's reagent)
- Bradford reagent
- Sample Buffer (5X) which component are listed below:

  0.35 M Tris–HCl (pH 6.8)
  50% (v/v) Glycerol
  10% (w/v) SDS
  7.7% (w/v) DTT
  0.5% (w/v) Bromophenol Blue

Just before use, 1.0 mL of the Phosphate Buffer was mixed with 3.0 μL Triton X–100, 1.0 μL of 10 M DTT and 10 μL of Complete^{TM} Protease inhibitor solution (1/4 pill dissolved in water) to produce the Extraction Buffer.

25.2.3 Glassware and Plasticware

All glassware and plasticware were sterilized by autoclaving. Mortar and pestle were washed and cleaned with acetone and ethanol (100%).

25.2.4 Procedure

25.2.4.1 Protein Extraction

- Plant material (2.0 g) was ground into fine powder with pre-chilled mortar and pestle in liquid nitrogen.
- The frozen powder was transferred into a 40 mL Sorvall centrifuge tube (Thermofisher Scientific, Waltham, MA), containing 3 mL ice-cold extraction buffer, and mixed well by inverting tube several times.
- Mixture was centrifuged at 10,000 rpm (Rotor SS34, Sorvall RC6 centrifuge) for 5 min at 4°C.
- Supernatant was decanted into sterile 2.0 mL Eppendorf tubes.
- The supernatant, containing the soluble proteins, was again centrifuged at 10,000 rpm for 5 min at 4°C to remove residual particles, decanted and filtered through two layers of microcloth (Calbiochem).
- The filtrate was divided into 100 μL aliquots, snap frozen in liquid nitrogen, and stored at –80°C.

If protein is to be used for activity studies, further purification is required. The protein extract is precipitated by 25% saturation of ammonium sulphate, centrifuged at 13,000 g for 15 min at 4°C, and the supernatant dialysed against 10 mM Tris (pH 7.2) at 4 °C overnight.

- Using Bradford reagent (BioRad), the concentration of total proteins in the extract was determined from calibration curve of standards according to the manufacturer's instructions.

25.2.4.2 Gel Electrophoresis

The protein extract was resolved by SDS-PAGE. The stacking gel (4.75%) and the resolving gel (12.5%) used for the electrophoresis contained 0.1% SDS. Samples were mixed with 5X sample buffer, heat-denatured at 95°C for 5 min and resolved at constant voltage (120 V) for 1 h in a Bio-Rad Mini-Protean II apparatus. After electrophoresis, the gel was stained with standard staining buffer and washed several times in destaining buffer.

25.3 Results and Discussion

Plant materials were mechanically disrupted in liquid nitrogen to increase extraction yield (Wang et al. 2003) and also to reduce proteolysis during sample pulverization. Contrary to reports by Zheng et al. (2007), but similar to that of Song et al. (2006), we found this SDS protocol to be most suitable for extracting proteins from the hardy tissues of grapes. Good quality proteins of appreciable yields (2.5 mg protein per gram of plant tissue) have been isolated from bud tissues by this protocol. The protein extracts have been used for a wide range of applications including the determination of endogenous protein phosphorylation activities (Fig. 25.1; Pang et al. 2007), and for one- and two-dimensional gel electrophoresis.

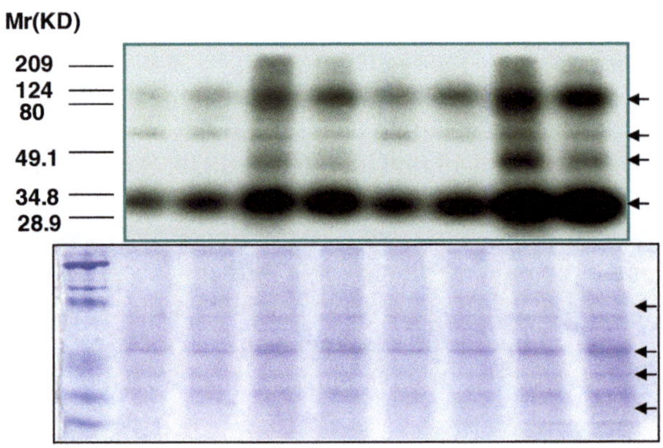

Fig. 25.1 SDS-PAGE showing endogenous protein phosphorylation activities in total protein extracts (20 μg) from mature buds of *Vitis vinifera* (*grapes*) subjected to different treatments at different time points. The phosphorylated proteins are identified by arrows, and their relative positions highlighted on the Comassie stained gels in the lower panels. The pre-stained molecular mass markers used were: myosin (206 kDa), β-galactosidase (124 kDa), bovine serum albumin (80 kDa), carbonic anhydrase (49.1 kDa), soybean trypsin inhibitor (30.6 kDa), lysozyme (17.8 kDa), and aprotinin (6.9 kDa) (Source: Pang et al. 2007)

References

Carpentier SC, Witters E, Laukens K, Deckers P, Swennen R, Panis (2005) Preparation of protein extracts from recalcitrant plant tissues: an evaluation of different methods for two-dimensional gel electrophoresis analysis. Proteomics 5:2497–2507

Gorg A, Weiss W (2000) Proteome Research: two–dimensional electrophoresis and identification methods. Springer, Berlin

Maldonado AM, Echevarría-Zomeño S, Jean-Baptiste S, Hernández M, Jorrín-Novo JV (2008) Evaluation of three different protocols of protein extraction for *Arabidopsis thaliana* proteome analysis by two–dimensinal electrophoresis. J Proteomics 71:461–472

Pang X, Halaly T, Crane O, Keilin T, Keren-Keiserman A, Ogrodovitch A, Galbraith D, Or E (2007) Involvement of calcium signaling in dormancy release of grape buds. J Exp Bot 58:3249–3262

Saravanan RS, Rose JK (2004) A critical evaluation of sample extraction techniques for enhance proteomic analysis of recalcitrant tissues. Proteomics 4:2522–2532

Sarry JE, Sommerer N, Sauvage FX, Bergoin A, Rossignol M, Albagnac G Romieu C (2004) Grape berry biochemistry revisited upon proteomic analysis of mesocarp. Proteomics 4:201–215

Song J, Braun G, Bevis E, Doncaster K (2006) A simple protocol for protein extraction of recalcitrant fruit tissues suitable for 2–DE and MS analysis. Electrophor 27:3144–3151

Tesnière C, Robin JP (1992) Two–dimensional electrophoresis of the total polypeptides in ripe red grape berries. Electrophor 13:93–96

Vincent D, Wheatley MD, Cramer GR (2006) Optimization of protein extraction and solubilization for matured grape berry clusters. Electrophor 27:1853–1865

Wang W, Scali M, Vignani R, Spadafora A, Sensi E, Mazzuca S, Cresti M (2003) Protein extraction for two-dimensional electrophoresis from olive lead, a plant tissue containing high levels of interfering compounds. Electrophor 24:2369–2375

Zheng Q, Song J, Doncaster K, Rowland E, Byerss DM (2007) Qualitative and quantitative evaluation of protein extraction protocols for apple and strawberry fruit suitable for two-dimensional electrophoresis and mass spectrometry analysis. J Agric Food Chem 55:1663–1673

Chapter 26
Expression Analysis in Grapevine by In Situ Hybridization and Immunohistochemistry

Steven Colas, Lucile Jacquens, Sébastien Manteau, Jérome Devy, Geneviève Conéjéro, Christophe Clément, Fabienne Baillieul, Florence Mazeyrat-Gourbeyre, and Laurence Monti-Dedieu

Abstract In situ hybridization (ISH) associated to immunohistochemistry has become a powerful tool for the examination of gene expression. Application of these techniques in grapevine tissues is limited mainly because of technical difficulties with this plant material. Here we present detailed protocols for ISH and immunohistochemistry, recommended controls and troubleshooting, along with examples of several applications to grapevine tissues.

Keywords Grapevine · In situ hybridization · Immunohistochemistry

Contents

F. Mazeyrat-Gourbeyre (✉)
Laboratoire de Stress, Défenses et Reproduction des Plantes, URVVC-SE EA 2069, Université de Reims, Champagne-Ardenne, UFR Sciences Exactes et Naturelles, Reims, 51687, France
e-mail: florence.gourbeyre@univ-reims.fr

S. Delrot et al. (eds.), *Methodologies and Results in Grapevine Research*,
DOI 10.1007/978-90-481-9283-0_26,

Abbreviations

AP	Alkaline phosphatase
DIG	Digoxigenine
GS primer	Gene specific primer
ISH	In situ hybridization
NTE	NaCl-Tris-EDTA buffer
PAL	Phenylalanine ammonia-lyase
PBS	Phosphate buffer saline
Rib	26S ribosomic RNA
SSC	Saline sodium citrate
STS	Stilbene synthase

26.1 Introduction

RNA in situ hybridization (ISH) is a widely used technique that enables the detection and localization of RNAs within tissue sections and therefore answers questions about gene expression. It combines cytology, molecular biology and microscopy and is based on the base pairing between two complementary nucleic acid sequences. Briefly, a labelled single-stranded RNA (probe) is applied on tissue sections and allowed to hybridize to complementary mRNAs. Hybrids formed are detected using a colorimetric or a fluorescence method and localized within the tissue after microscopic observations.

In plant biology, the technique has been adapted from the technique developed for animal systems considering the specificity of plant tissues (Cox and Goldberg 1988). The successful application of ISH in plant tissues is limited by both the presence of the cell wall and the vacuole (Brugeon 1996). The cell wall constitutes a rigid and semi-permeable barrier that impedes diffusion of fixative and probe penetration. The vacuole which is bound to the membrane and filled with water has a high solute concentration and must not be altered by treatments involved in tissue preparation.

Immunohistochemistry offers a complementary approach to ISH to expression analysis. Proteins are localized in tissues using an antigen-antibody reaction. Antibodies are often conjugated to fluorochromes.

ISH and immunohistochemistry in plants then require compromise between the appropriate preservation of tissue morphology and the accessibility of target RNAs/proteins to the probe/specific antibodies.

This chapter focuses on applications of ISH and immunohistochemistry to grapevine (*Vitis vinifera* L.) in order to study gene expression and protein localization in various organs of this plant.

26.2 Historical Background

To date only few works using ISH have been reported for grapevine. Among them, we can mention the studies on bud development and the localization of transcripts of VLF, the grapevine Floricaula/leafy ortholog gene (Carmona et al.

2002). Expression was detected in reproductive meristems and developing floral organs but also in other meristematic tissues involved in the differentiation of leaves and tendrils. ISH has been also successfully applied to young grape berry tissues. In the Merlot cultivar, the expression of a gene coding for a proline-rich protein associated with ripening was evidenced in 8 to 14-weeks post-flowering berries (Burger et al. 2004). Accumulation of transcripts was detected in the mesocarp and exocarp cells. Fernandez et al. (2007) observed in young berries (7 to 10 days after anthesis) the expression of genes known to be associated to developmental processes in other species. Fouquet et al. (2008) detected in green berry tissues (20 days after flowering) the expression of both the plasma membrane aquaporin *VvPIP2*; 1 and the tonoplast aquaporin *VvTIP2*; 1 genes. The former was strongly expressed in the outer integument of developing seeds while the latter was expressed in well-defined cell types of the exocarp. Immunohistochemical studies were also performed on young grape berries to localize enzymes involved in sugar, organic acid and amino acid metabolism (Walker et al. 1999, Famiani et al. 2000). Several enzymes (phosphoenolpyruvate carboxylase and enzymes involved in amino acid metabolism) were shown to be present in vascular tissues whereas other enzymes (NADP-malic enzyme and soluble acid invertase involved in the metabolism of sugars and organic acids) were abundant in flesh cells. Vignault et al. (2005) studied in berries too, the expression of a gene coding for a monosaccharide transporter (*VvHT1*). ISH and immunolocalization experiments showed that transcripts were present in the phloem region of the vascular bundles and that transporters were localized in the plasma membrane of the sieve elements and of the flesh cells. Application of ISH to root system has also been reported concerning aquaporin gene expression (Vandeleur et al. 2009). Transcripts of two genes encoding for proteins of the plasma membrane intrinsic protein (PIP) family were observed in the exodermis, the cortex and the stele mainly in the root apex of both Grenache and Chardonnay cultivars. A similar distribution for the corresponding proteins was observed using immunohistochemical studies.

Only few references to ISH and immunohistochemistry in grapevine are then available, mainly because of technical difficulties with this plant material as mentioned by all authors. The aim of this chapter is so to present protocols for both techniques that could be applied to grapevine tissues.

26.3 Methods and Materials

26.3.1 PCR Generation of T7 Promoter-Tailed DNA

A PCR strategy for generation of template DNA for synthesis of RNA probes was described by Ambion (TB 154, Ambion). This step synthesized 2 specific DNA sequences (antisense and sense) template with T7 RNA polymerase binding site at the 5' end of the antisense or sense strand.

The 26S rRNA distribution was studied as a marker of general transcriptional activities and a positive control of labelling reaction. The 26S rRNA (Rib) primers (reverse: 5'-ccaagtcagacgaacgatttgcacg-3', forward: 5'-ccgacctcgatcttatgagaaggg-3') were designed based on sequence information on *Vitis vinifera* 26S rRNA (accession number AY847628.1) and allowed to amplify a 150 bp PCR product subcloned into pGEM T Easy vector (Promega) and sequenced.

For target gene, a probe of 200–400 bp including 3' UTR of target DNA is recommended. The fragment was amplified by PCR with specific primers (20 to 24 nt length and 60°C minimum Tm) and subcloned into pGEM T Easy vector (Promega), and then sequenced.

The target sequence (a 362 bp fragment of an acidic class IV chitinase *VvChi4D* accession number AF532966 in e.g.) was amplified from plasmid using specific primers for 30 cycles (one cycle was 30 s at 94°C, 30 s at 65°C, 30 s at 72°C). The resulting fragment was gel-purified using NucleoSpin Extract II (Macherey-Nagel), eluted with 50 μL and diluted 1/200 to 1/400 (final concentration around 0.1 ng/μL).

In the second reaction, 2 separate amplifications were performed using the diluted amplification products from the first reaction. One PCR product had T7 promoter at the 5' end of the antisense strand (makes antisense RNA probe) and one product had T7 promoter at the 5' end of the sense strand (makes sense probe). Since each PCR product was individually transcribed to produce an RNA probe, it was not necessary to have different promoters for the sense and antisense strands. For each primer pair, one primer had no additional nucleotides (target sequence only) and the other gene specific (GS) primer was tailed with 30 bases, 23 bases representing the T7 RNA promoter site, 7 bases were added 5' to the T7 promoter to stabilize the protein-DNA interaction of the polymerase with the promoter site: 5'-gcgaaat-taatacgactcactatagggaga-GS-3'. The following PCR conditions were used: 94, 69 and 72°C, each for 30 s, for 25 cycles in a final volume of 25 μL.

In the third reaction (PCR3), the 1/200 diluted amplification products (1 ng/μL) of the second reaction were reamplified to extend and complete the full-length T7 promoter. Each of the amplifications contains one specific primer as used in the first PCR and a second universal T7 primer containing 13 nucleotides of the T7 promoter and the additional 7 nucleotides at the 5' end (5'-gcgaaattaatacgactcac-3'). The following PCR conditions were used: 94°C, 69°C and 72°C, each for 30 s, for 25 cycles in a final volume of 50 μL. Five amplifications were carried out for each sample (antisense and sense), and the 5 amplifications product were pooled, precipitated in 1/10 V sodium acetate 3 M pH 5.2 with 2 V of cold ethanol 100% and resuspended in 30 μL of H_2O_2 (final concentration around 110 ng/μL).

To test the specificity of each matrix (antisense and sense) a PCR reaction was realized: each product of the PCR3 was amplified with 2 primer pairs (a positive and a negative control): the forward specific primer + the reverse T7 primer and the reverse specific primer + the forward T7 primer for 30 cycles at 94°C 30 s, 65°C 30 s and 72°C 30 s.

26.3.2 In Vitro Transcription Labelling of RNA

DNA amplicons of PCR3 containing the T7 polymerase promoter were used as templates for in vitro transcription. Sense and antisense probes were labelled during transcription using UTP-digoxigenin (UTP-DIG) according to Ambion protocol with minor modifications. Briefly, the transcription reaction was performed on 0.5–1 μg of DNA template. Afterwards, 1–2 μL of DNase, depending on the probe, was added to remove the template DNA. The presence of a single band corresponding to RNAs was checked by gel electrophoresis analysis. Transcripts were precipitated in 1/10 V sodium acetate 3 M pH 5.2 with 2 V of cold ethanol 100%. Labelled RNAs were resuspended in 20 μL of nuclease-free water. Addition of a RNase inhibitor (SUPERase·In^{TM}) at a final concentration of 1 U/μL was performed to prevent RNA degradation. After labelling, the incorporation of labelled nucleotides into the probe was checked by dot blotting on a nylon membrane with an alkaline phosphatase conjugate antibody raised against digoxigenin (Roche Diagnostics). The concentration of labelled RNAs was estimated by spectrophotometry.

26.3.3 Preparation of Sectioned Plant Material

Various grapevine organs were prepared: leaves, berries, wood and inflorescences. Grape berries of *V. vinifera* L. cv. Pinot Noir were harvested during the growing seasons 2007 and 2008 in vineyard. Three stages of development according to the BBCH scale (Meier 2001) were collected: pea-sized (BBCH 75), closed cluster (BBCH 77) and "post-véraison" (BBCH 83). Wood of *V. vinifera* L. cv. Chardonnay was collected in vineyard in February 2009. It corresponded to 1 year old lignified stems. Leaves and inflorescences were collected from cuttings of *V. vinifera* L. cv. Chardonnay. Cuttings were obtained from canes of vineyard according to the protocol improved by Lebon et al. (2005). Inflorescences were separated in flowers and bunchstems. Bunchstem includes the peduncle corresponding to the stem of the panicle and the pedicels that wear the flowers (Jackson and Coombe 1995).

At all times of plant sample preparation, precautions are needed to prevent the RNA degradation. It is necessary to wear gloves constantly and use sterile glassware or plasticware.

Fixation was performed using fresh paraformaldehyde 4% in potassium phosphate buffer (PBS 10 mM, pH 7.4). Plant tissues (leaves, wood and berries) were cut in small pieces of 5 mm × 5 mm or in small sections of 5 mm (bunchstems) whereas flowers were left intact. Samples were placed into fixative with a drop of detergent (Tween 20) in a vacuum desiccator to favour fixative penetration. One h later, fixative was replaced and samples were left at 4°C overnight or up to 18 h for wood samples to ascertain the fixative penetration in all tissues. Flowers were immediately transferred into fixative to 4°C without vacuum infiltration that could cause stamen burst. After fixation, samples were washed twice in PBS 10 mM with glycine 0.1 M for 15 min, twice in PBS 10 mM for 15 min and then dehydrated with graded ethanol series in water (50, 70, 95 and 100%) for 3 h. Ethanol was then replaced by butanol. Samples were left in butanol between 1 and several days, up to 12 days for wood samples. Butanol was gradually replaced by the solvent of the paraffin wax (Paraplast X-TRA, Labonord) and then by molten paraffin wax. Samples were left in paraffin wax at 58°C for 2 days with replacing paraffin every day. Using warmed forceps, samples were deposited and arranged in plastic moulds containing molten paraffin wax. After solidification, blocks were stored at –20°C until use. Wax sections at 8–15 μm thickness were obtained using a microtome and deposited on silanized slides to favour tissue adherence. Slides were left to dry on a hotplate at 45°C for 1 h and then overnight at 40°C in an oven.

26.3.4 Hybridization, Washes and Detection of Hybridization Sites

Slides were dewaxed, rehydrated and then incubated at 37°C for 30 min in proteinase K (3 μg/mL) to remove nucleic acid linked proteins that may impede probe access. Sections were washed for 10 min in $CaCl_2$ 2 mM, $MgCl_2$ 50 mM, Tris–HCl pH 7.5, for 2 min in PBS 10 mM glycine 0.02 M, and twice for 2 min in PBS 10 mM. After dehydration in ethanol series (50, 70 and 100%), sections were hybridized with an hybridization buffer containing 5 μg/mL of probe, 10% dextran sulfate, 1X Denhardt's solution, 50% formamide, 1 μg/μL tRNA and 2X SSC buffer (NaCl 0.3 M, sodium citrate 30 mM pH 7.0). Hundred μL of the hybridization buffer were applied on each slide and covered with a coverslip. Slides were then incubated overnight at 45°C in a humid chamber. After hybridization, washes were carried out to remove unbound or weakly bound probes and the hybridization mixture. Coverslips were removed carefully and slides incubated in 2X SSC for 5 min at room temperature, and for 45 min at 50°C. To remove unhybridized RNA, slides were transferred in NTE buffer (0.01 M Tris–HCl pH 7.5, NaCl 0.5 M, EDTA 1 mM) for 5 min at room temperature and twice for 5 min at 37°C. Each section was then incubated for 30 min at 37°C with 20 μg/mL of an RNase solution. Slides were washed in NTE buffer twice for 5 min at room temperature and then in 1X SSC twice for 30 min at 55°C. Incubation in PBS 10 mM (10 min at room temperature) was followed by a colorimetric detection using alkaline phosphatase (AP). First, slides were pretreated with a blocking solution (Roche Diagnostics) to prevent non specific binding of antibodies for 20–60 min at room temperature. Slides were then incubated with 500 μL of a detection solution for 45 min at 37°C. The detection solution was prepared in the blocking solution at a final concentration of 2 μL/mL of anti-digoxigenin (Fab fragment) conjugated to AP. Slides were washed in PBS 10 mM 3-fold for 10 min and twice in Tris–HCl 100 mM pH 8.2 for 5 min at room temperature. Detection was performed with the AP-substrate solution from the Vector Blue Substrate Kit (Vector Laboratories, Burlingame, CA, USA) for 3–9 h in dark replacing the substrate solution every 3 h. Sections were then mounted in Mowiol.

26.3.5 Immunohistochemistry

Immunohistochemical studies were performed on grape berries at the 75 and 83 stages according to the BBCH scale (Meier 2001) using antibodies against the class IV chitinase CHV5 (Derckel et al. 1998). The antiserum specific for CHV5 enzyme was raised in rabbit against the enzyme purified from *V. vinifera* L. cv. Pinot Noir berries. Berry samples were fixed, embedded and mounted as described above for ISH. Sections were deparaffinised, rehydrated and incubated in PBS 10 mM for 5 min. Slides were treated with sodium citrate buffer ($C_6H_8O_7$, H_2O 1.8 M, $C_6H_5O_7Na_3$, $2H_2O$ 8.2 M, pH 6) at 94–97°C for 40 min to release the antigens. After washing in distilled water 4-fold for 5 min and PBS 10 mM for 5 min, slides were incubated in blocking solution (PBS containing 3% bovine serum albumin, 0.1% triton, 0.05% tween 20) at room temperature for 1 h in a humidified chamber. Slides were then incubated overnight at 4°C with the primary antibody at a dilution of 1/200. Primary antibodies were then drained off sections in PBS 10 mM (3 x 10 min) and immunolabelling was revealed after 1 h-incubation at room temperature with the secondary antibodies conjugated to AlexaFluor 488 at a dilution of 1/250. After washing in PBS 10 mM (3 × 10 min), sections were mounted in Mowiol. A Leica SP2 confocal system coupled with a DRM2 optical microscope (Leica Microsystems, Heidelberg, Germany) was used for acquisitions. All acquisitions were made using UPlan FI × 20, 0.4 numerical aperture objective. We used the 488-nm of an air-cooled 100 mW argon laser for excitations of AlexaFluor 488. Emitted fluorescence was detected through the combination of the appropriate filter set. Images were treated with IMAGE J.

26.4 Results

26.4.1 PCR Generation of T7 Promoter-Tailed DNA

The T7 promoter tailed DNA template to be used for RNA probe synthesis was prepared by 3 rounds of amplification.

The first round was performed with specific primers flanking the region of interest. A 362 bp fragment of *VvChi4* and 150 bp fragment of *Rib* were successfully amplified by PCR1 without non-specific amplification (Fig. 26.1a, b, lanes 1). The second round was carried out with primer containing the T7 promoter and specific primer to incorporate the T7 promoter into the PCR2 product. Agarose gel electrophoresis revealed that the addition of the T7 promoter sequence to the primer did not impair the amplification of DNA. Amplicons of the expected size (362 + 30 bp) for *VvChi4* and (150 + 30 bp) for *Rib* were obtained (lanes 2–3, Fig. 26.1a, b). The third round allowed to extent and complete the full-length T7 promoter with specific primer in combination with T7 primer. Amplicons of 392 bp for *VvChi4* and 180 bp for *Rib* were obtained (Fig. 26.1a, b, Lanes 4–5).

Verification of the specificity of each DNA template, tested by reamplification with 2 primer pairs, was assessed by migration on agarose gel showing

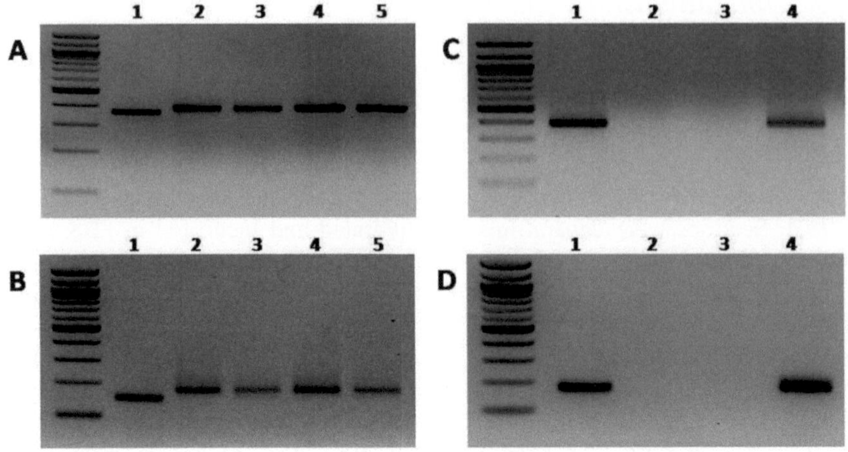

Fig. 26.1 Generation of PCR templates. Aliquots of each step of template generation were analyzed by electrophoresis in 1.8% agarose gel. **a**: Generation of *VvChi4* PCR template. *Lane 1*: PCR1 with *VvChi4* specific primers. *Lane 2*: PCR2 with reverse *VvChi4* specific primer and forward *VvChi4* primer tailed with T7 promoter at the 5' end. *Lane 3*: PCR2 with forward *VvChi4* specific primer and reverse *VvChi4* primer tailed with T7 promoter at the 5' end. *Lane 4*: PCR3 with reverse *VvChi4* specific primer and T7 specific forward primer. *Lane 5*: PCR3 with forward *VvChi4* specific primer and T7 specific reverse primer. **b**: Generation of *Rib* PCR template. *Lane 1*: PCR1 with *Rib* specific primers. *Lane 2*: PCR2 with reverse *Rib* specific primer and forward *Rib* primer tailed with T7 promoter at the 5' end. *Lane 3*: PCR2 with forward *Rib* specific primer and reverse *Rib* primer tailed with T7 promoter at the 5' end. *Lane 4*: PCR3 with reverse *Rib* specific primer and T7 specific forward primer. *Lane 5*: PCR3 with forward *Rib* specific primer and T7 specific reverse primer. **c**: Verification of *VvChi4* PCR template specificity. *Lane 1*: Reamplification of sense matrix (a *Lane 4*) with reverse *VvChi4* specific primer and T7 specific forward primer (positive control). *Lane 2*: Reamplification of sense matrix (a *Lane 4*) with forward *VvChi4* specific primer and T7 specific reverse primer (negative control). *Lane 3*: Reamplification of antisense matrix (a *Lane 5*) with reverse *VvChi4* specific primer and T7 specific forward primer (negative control). *Lane 4*: Reamplification of antisense matrix (a *Lane 5*) with forward *VvChi4* specific primer and T7 specific reverse primer (positive control). **d**: Verification of *Rib* PCR template specificity. *Lane 1*: Reamplification of sense matrice (b *Lane 4*) with reverse *Rib* specific primer and T7 specific forward primer (positive control). *Lane 2*: Reamplification of sense matrice (b *Lane 4*) with forward *Rib* specific primer and T7 specific reverse primer (negative control). *Lane 3*: Reamplification of antisense matrix (b *Lane 5*) with reverse *Rib* specific primer and T7 specific forward primer (negative control). *Lane 4*: Reamplification of antisense matrix (b *Lane 5*) with forward *Rib* specific primer and T7 specific reverse primer (positive control)

a single band in both positive controls and no band in both negative controls (Fig. 26.1c, d).

26.4.2 RNA Probe Synthesis

After in vitro transcription, a single band with the expected size was observed after agarose gel electrophoresis for *VvChi4* and *Rib* (Fig. 26.2a, b). DIG incorporation

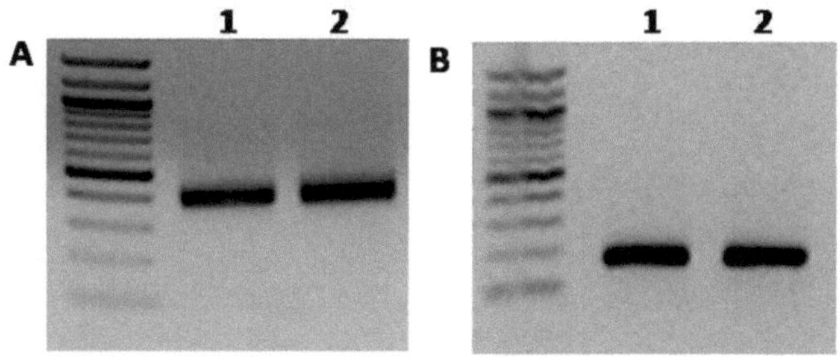

Fig. 26.2 Probe synthesis by in vitro transcription. Aliquots from in vitro transcription reaction were analyzed by 1% agarose gel electrophoresis. **a**: Probe synthesis of *VvChi4*. *Lane 1*: sense probe. *Lane 2*: antisense probe. **b**: Probe synthesis of *Rib*. *Lane 1*: sense probe. *Lane 2*: antisense probe

was confirmed by NBT/BCIP staining of nylon membrane spotted with RNA probe. The positive staining was obtained for all RNA solution (data not shown).

26.4.3 In Situ Hybridization on Various Grapevine Organs

For hybridization to RNA targets, controls are essential to assess the hybridization signal and to distinguish it from background. For applications to grapevine tissues, we used the 26S ribosomic RNA sequences (Rib) as controls. Antisense and sense probes were generated and were expected to give signals in all cellular types in the former case and no signal in the latter one.

In situ hybridization on leaf sections of *V. vinifera* L. (cv. Chardonnay) with the Rib antisense probe showed hybridization signals on the whole tissues including the epidermis, the parenchyma and the vascular bundles. Stronger hybridization was observed in the parenchyma limb compared to the vein (Fig. 26.3a). No signal was observed on sections hybridized with the Rib sense probe (Fig. 26.3b).

In situ hybridization on grape berries of V. *vinifera* L. (cv. Pinot Noir) was performed on the following three stages of development: pea-sized (BBCH 75), closed cluster (BBCH 77) and "post-véraison" (BBCH 83). At the BBCH 75, signals with Rib antisense probe were observed on the whole berries but were more intense in the exocarp and in the outer and medium integuments of the seed according to the histological study of Cadot et al. (2006) (Fig. 26.3c). No signal was observed on berry sections with the *Rib* sense probe. At the BBCH 77 (Fig. 26.3e, f) and BBCH 83 (Fig. 26.3 g, h) stages, hybridization studies gave similar results than at the BBCH 75 stage. Signal was however more diffuse in the mesocarp of "post-véraison" berries in relation to the structural changes affecting this particular tissue during berry development. Moreover, in this case, unspecific signals were detected. A purple staining was observed with both antisense and sense probes at the transition between the exocarp and the mesocarp (Fig. 26.3 g, h).

In situ hybridization technique was also applied to 1 year old lignified stems of *V. vinifera* L. (cv. Chardonnay). Using Rib antisense probe, strong hybridization was localized in the secondary xylem (Fig. 26.3i), whereas no signal was detected in the

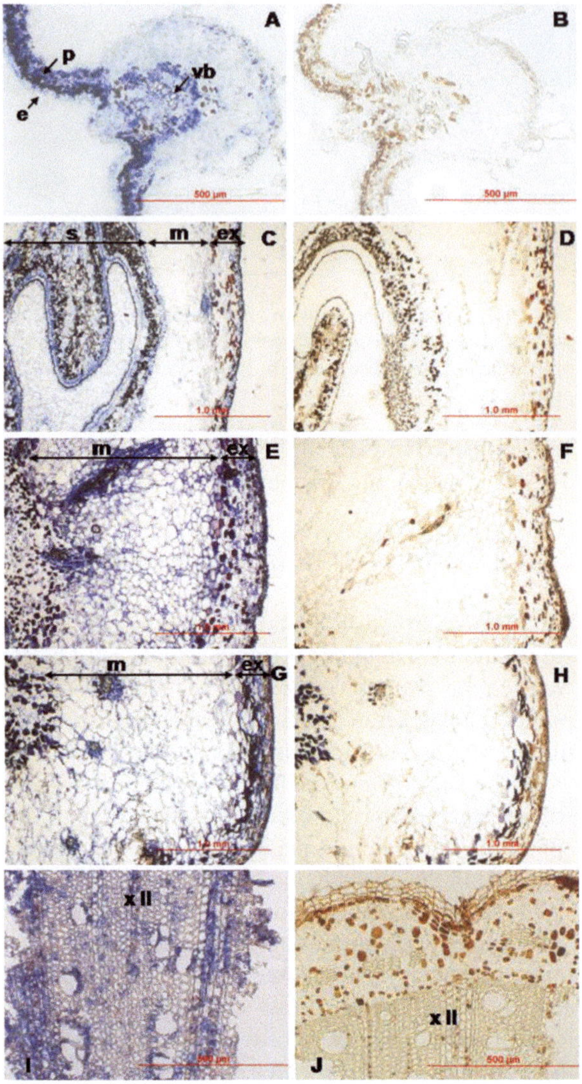

Fig. 26.3 In situ hybridization with *Rib* RNA antisense (**a, c, e, g, i**) and sense (**b, d, f, h, j**) probes on Chardonnay leaves from cuttings (**a, b**); Pinot noir berries of BBCH 75 (**c, d**), BBCH 77 (**e, f**), BBCH 83 (**g, h**); Chardonnay 1 year old lignified stems (**i, j**). *e* epidermis; *p* parenchyma; *vb* vascular bundles; *s* seed; *ex* exocarp; *m* mesocarp; *x II* secondary xylem

medullar parenchyma (data not shown). As expected, no signal was observed with the *Rib* sense probe.

In situ hybridization experiments were also performed on inflorescences (bunch-stems and flowers) of *V. vinifera* L. (cv. Chardonnay) using the control *Rib* probes. Results (not shown) corresponded to the expected ones: staining covering all tissues with the antisense probe and lack of hybridization with the sense probe.

Once in situ hybridization conditions were validated with the *Rib* control probes on the various grapevine organs mentioned above, the localization of transcripts corresponding to several other genes was investigated.

26.4.4 Expression of a Class IV Chitinase Gene in the Developing Grape Berry

In situ hybridization experiments using *VvChi4* labelled RNA probes were carried out on grape berries of *V. vinifera* L. (cv. Pinot Noir) from BBCH 75 and 83 stages. No transcript was detected at the BBCH 75 stage (Fig. 26.4a) while accumulation of transcripts was observed at the BBCH 83 stage mainly around the vessels in the endocarp (Fig. 26.4c) and in the exocarp (data not shown). No transcript was detected in the exocarp epidermis (data not shown). Hybridization with the sense probe showed no signal in both stages (Fig. 26.4b, d).

Localization of CHV5 proteins was also investigated using immunohistochemistry with an antiserum specific for CHV5. No protein was detected in berry sections of the BBCH 75 stage (Fig. 26.5a) while later during berry development, CHV5 was dectected in the grape berry tissues (Fig. 26.5b). It was mainly associated to the epidermal cell layer of the exocarp and to cells belonging to the xylem parenchyma.

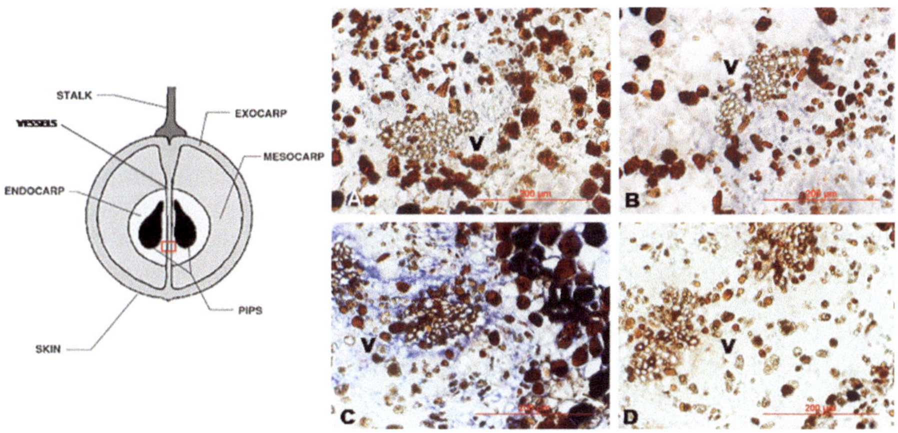

Fig. 26.4 In situ hybridization of the *VvChi4* class IV chitinase gene in developing grape berries of BBCH 75 (**a, b**) and BBCH 83 (**c, d**) stages. **a, c**: sections hybridized with the *VvChi4* antisense probe. **b, d**: sections hybridized with the *VvChi4* sense probe. Pictures **a, b, c** and **d** correspond to sections from the red boxed insert. *V* vessels

26.5 Discussion

In this study, the preparation of RNA probes was performed using PCR based method instead of the conventional vector-based probe preparation.

As mentioned by Roche documentation, some polylinker regions in DNA templates are homologous to portions of the ribosomal RNA sequences. Therefore, labelled probes may generate specific, but untargeted signal in samples that contain the prominent RNAs.

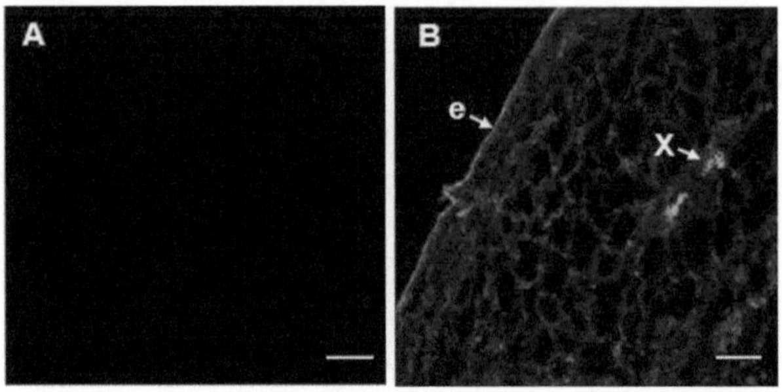

Fig. 26.5 Immunolocalization of the CHV5 protein (**a, b**) in developing grape berries of BBCH 75 (**a**) and BBCH 83 (**b**) stages. *e* epidermis, *X* xylem. The green fluorescence indicates the presence of the enzyme. Bars = 100 μm

The selection of the primer carrying T7 promoter determines the direction of the transcription reaction. If the forward primer carries the promoter, the synthesized probe is complementary to the sense strand. If the promoter sequence is attached to the reverse primer, then a probe complementary to the antisense strand is produced.

As mentioned by Ambion documentation, unexplained extra bands sometimes arise in PCR. To overcome mismatches of primers, high annealing temperatures were applied and allowed us to obtain a single band for each PCR. With the PCR method it is possible that primers used for a prior PCR remain in the subsequent one. This might cause the antisense DNA template to be mixed with sense DNA templates. This in turn might result in signal detection in the section hybridized with the sense probe. So, as previously described (Suzuki et al. 2005), after running of the first PCR, product was cut out of the gel and purified. To ensure that antisense DNA template was not mixed with sense DNA templates, each matrix (antisense and sense) was reamplified with the 2 combinations primers pair to obtain a positive and a negative control. The DNA was used as template in an in vitro transcription for digoxygenin-labelled antisense RNA probe only if these ultimate controls were validated. In a previous study (Vandeleur et al. 2009), all sense probe controls generated with PCR DNA template, showed minimal background hybridization. Generation of RNA probes validated by this last control allowed us to obtain specific antisense probe with a corresponding negative control sense probe which never gave background hybridization for all probes tested.

To confirm our approach, RNA probes were generated using primers for different genes (glucanase, chitinase, PAL, STS…).

Several control procedures must be carried out using in situ hybridization experiments to be sure that the observed labelling is specific to the target sequence. One such control procedure consists in using the corresponding sense probe as a negative control. Other negative controls are often performed such as omitting probe slides or pretreating sections with RNase. Hybridization with Rib antisense RNA sequences may be used as a positive control as described. Controls are also important for immunohistochemistry studies and usually include slides without antibody and slides with the secondary antibody alone. When using a fluorescent detection, slides without probe (for ISH) or without antibody (for immunolocalization) are useful to estimate the autoflorescence (often important) in plant tissues.

One of the major difficulty in applying ISH and immunohistochemistry to grapevine is due to tissue preparation. All tissues are not equally easy to prepare and some of them set more problems. Berries collected after "véraison" during the ripening period contain high amounts of water and sugars that impede fixation procedure (Diakou and Carde 2001, Vignault et al. 2005, Fouquet et al. 2008). Diakou and Carde (2001) showed that the injection of fixative into the exocarp instead of the usual immersion protocols improved fixative diffusion and preservation of cell ultrastructure. Presence of tannins and phenolic compounds is another obstacle to ISH and immunohistochemistry giving sometimes strong background (Fernandez et al. 2007). In the grape berry of some cultivars, tannins are abundant in the exocarp and the endocarp and tend to darken all along treatments which may impede signal identification. In our experiments, we observed a black-purple staining in the exocarp cells containing vacuolar tannins.

Using the usual fixative procedure on berries from both BBCH 75 and BBCH 83, we showed that *VvChi4* gene was expressed during ripening. None of the related transcript and protein were detected at pea-sized stage, whereas mRNAs and CHV5 protein accumulations were observed after "véraison" mainly around vascular bundles. These results are consistent with previous studies that have shown that *VvChi4* genes and CHV5 protein were constitutively expressed in grapevine berries during ripening (Derckel et al. 1998, Jacobs et al. 1999).

In situ hybridization in grapevine is an interesting approach to investigate and localize gene expression. This technique could also be useful for temporal and spatial expression studies which have been poorly explored until now in plants (Tavares et al. 2008). Studies dealing with the expression of defence genes in response to pathogen infections, can however be mentioned. Transcript accumulation of some defence-related genes could then be correlated with fungus infection progress in tissues of parsley leaves and maize embryos (Reinold and Halbrock 1996, Bravo et al. 2003).

To conclude, in situ hybridization combined to immunohistochemistry provides a powerful tool to explore expression of genes and investigate their spatial and temporal patterns.

Acknowledgments The authors thank the EMOA laboratory UMR INRA FARE 614 Reims for giving us access to their confocal equipment.

References

Bravo JM, Campo S, Murillo I, Coca M, San Segundo B (2003) Fungus- and wound-induced accumulation of mRNA containing a class II chitinase of the pathogenesis-related protein 4 (PR-4) family of maize. Plant Mol Biol 52:745–759

Brugeon J (1996) *In situ* hybridization to RNA in plant biology. In: Nicole M, Gianinazzi-Pearson V (eds) Histology, ultrastructure and molecular cytology of plant – microorganism interactions. Kluwer Academic Publishers, The Netherlands

Burger AL, Zwiegelaar JP, Botha FC (2004) Characterisation of the gene encoding the Merlot ripening-induced protein 1 (mrip 1): evidence that this putative protein is a distinct member of the plant proline-rich protein family. Plant Sci 167:1075–1089

Cadot Y, Miñana-Castelló MT, Chevalier M (2006) Anatomical, histological, and histochemical changes in grape seeds from *Vitis vinifera* L. cv Cabernet franc during fruit development. J Agric Food Chem 54:9206–9215

Carmona MJ, Cubas P, Martinez-Zapater JM (2002) VFL, the grapevine Floricaula/Leafy ortholog, is expressed in meristematic regions independently of their fate. Plant Physiol 130:68–77

Cox KH, Goldberg RB (1988) Analysis of plant gene expression. In: Shaw CH (ed) Plant molecular biology: a practical approach. Oxford IRL Press, Oxford

Diakou P, Carde JP (2001) *In situ* fixation of grape berries. Protoplasma 218:225–235

Derckel JP, Audran JC, Haye B, Lambert B, Legendre L (1998) Characterization, induction by wounding and salicylic acid, and activity against *Botrytis cinerea* of chitinases and β-1,3-glucanases of ripening grape berries. Physiol Plant 104:56–64

Famiani F, Walker RP, Técsi L, Chen ZH, Proietti P, Leegood RC (2000) An immunohistochemical study of the compartmentation of metabolism during the development of grape (*Vitis vinifera* L.) berries. J Exp Bot 345:675–683

Fernandez L, Torregrosa L, Terrier N, Sreekantan L, Grimplet J, Davies C, Thomas MR, Romieu C, Ageorges A (2007) Identification of genes associated with flesh morphogenesis during grapevine fruit development. Plant Mol Biol 63:307–323

Fouquet R, Léon C, Ollat N, Barrieu F (2008) Identification of grapevine aquaporins and expression analysis in developing berries. Plant Cell Rep 27:1541–1550

Jacobs AK, Dry IB, Robinson SP (1999) Induction of different pathogenesis-related cDNAs in grapevine infected with powdery mildew and treated with etephon. Plant Pathology 48: 325–336.

Jackson DI, Coombe BG (1995) Early bunchstem necrosis-a matter of nomenclature. Am. J Enol Vit 46:579–580

Lebon G, Duchêne E, Brun O, Clément C (2005) Phenology of flowering and starch accumulation in grape (*Vitis vinifera* L.) cuttings and vines. Ann Bot 95:943–948

Meier U (2001) Grapevine. In: Meier U (ed) Growth stages of mono- and dicotyledonous plants. BBCH monograph, federal biological research centre for agriculture and forestry. Blackwell Wissenschafts-verlag, Berlin

Reinold S, Halbrock K (1996) Biphasic temporal and spatial induction patterns of defense-related mRNAs and proteins in fungus-infected parsley leaves. Plant Physiol 112:131–140

Suzuki T, Akimoto M, Mandai M, Takahashi M, Yoshimura N (2005) A new PCR-based approach for the preparation of RNA probe. Biochem Biophysic Methods 62:251–258

Tavares LS, de O Santos M, Viccini LF, Moreira JS, Miller RNG, Franco OL (2008) Biotechnological potential of antimicrobial peptides from flowers. Peptides 29:1842–1851

Vandeleur RK, Mayo G, Shelden MC, Gilliham M, Kaiser BN, Tyerman SD (2009) The role of plasma membrane intrinsic protein aquaporins in water transport through roots: diurnal and drought stress responses reveal different strategies between isohydric and anisohydric cultivars of grapevine. Plant Physiol 149:445–460

Vignault C, Vachaud M, Cakir B, Glissant D, Dedaldechamp F, Buttner M, Atanassova R, Fleurat-Lessard P, Lemoine R, Delrot S (2005) VvHT1 encodes a monosaccharide transporter expressed in the conducting complex of the grape berry phloem. J Exp Bot 56:1409–1418

Walker RP, Chen ZH, Técsi L, Famiani F, Lea PJ, Leegood RC (1999) Phosphoenolpyruvate carboxykinase plays a role in interactions of carbon and nitrogen metabolism during grape seed development. Planta 210:9–18

Chapter 27
Marker Development for Important Grapevine Traits by Genetic Diversity Studies and Investigation of Differential Gene Expression

Achim Schmitt, Martina Rex, Stefan Ebert, Wolfgang Friedt, Reinhard Töpfer, and Eva Zyprian

Abstract Grapevine (*Vitis vinifera* L.) needs to be improved for disease resistance to allow the development of sustainable viticulture. Resistance genes from American or Asian wild relatives acting against downy and powdery mildew pathogens have to be introduced by cross breeding. Physical attributes like the density of fruit clusters also contribute to the variability of susceptibility and need to be considered in breeding.

The long-term breeding process is facilitated and accelerated through marker-assisted selection. Genetic markers in linkage to resistance traits therefore need to be identified. The strategies to develop such markers through different approaches are described. They include genetic mapping, quantitative trait locus identification, differential gene expression analysis yielding candidate genes for genetic diversity studies and association mapping.

Contents

E. Zyprian (✉)
Julius Kühn-Institut, Federal Research Centre for Cultivated Plants, Institute for Grapevine Breeding Geilweilerhof, Siebeldingen 76833, Germany
e-mail: eva.zyprian@jki.bund.de

S. Delrot et al. (eds.), *Methodologies and Results in Grapevine Research*,
DOI 10.1007/978-90-481-9283-0_27, © Springer Science+Business Media B.V. 2010

Abbreviations

FRET	Fluorescence resonance energy transfer
MAS	Marker assisted selection
qRT-PCR	Quantitative real time polymerase chain reaction
QTL	Quantitative trait locus
RT	Reverse transcription
SNP	Single nucleotide polymorphism
SSR	Simple sequence repeat

27.1 Introduction

Grapevine (*Vitis vinifera* L.) is a plant with long cultural history. Viticulture forms part of the European tradition and has spread to other areas in the world. However, commercial viticulture is based on cultivars with severe drawbacks such as high susceptibility to some infectious diseases. Especially the protection from pathogens like powdery mildew (*Erysiphe necator*, an ascomycete fungus) and downy mildew (*Plasmopara viticola*, a fungus-like oomycete), both introduced accidentally in the nineteenth century from America to Europe, requires substantial chemical treatments to ensure good quality production in every growing season. Another significant pathogen in viticulture is *Botrytis cinerea* (*Botryotinia fuckeliana*, an ascomycete). It infects grapes thereby severely damaging young, developing fruit clusters and also almost mature grape bunches. In contrast to the two mildew pathogens, that both are obligate parasites specific to grape, *Botrytis* is ubiquitous and infects many other plants in addition to grapevine.

Pathogens like these require the introgression of resistance traits into high quality grapevine cultivars by breeding to reduce the heavy fungicide applications currently unavoidable.

27.2 The Need to Develop Molecular Markers for Breeding

European grapevines (*Vitis vinifera* L.) do not carry any resistance traits in their gene pool that could defend them against the two mildew pathogens. The only way to improve these cultivars is to use wild *Vitis* spec. relatives from America and Asia as resistance donors in crosses for resistance breeding. In this approach the effects of linkage drag, bringing in not only pathogen resistance genes but also negative quality characteristics, requires the backcrossing to high quality *V. vinifera* cultivars followed by the careful evaluation of the progeny over several generations. This is a successful strategy, but a very time consuming process. It can be accelerated by employing molecular markers genetically linked to resistance traits to speed up the identification of resistance carriers within large numbers of seedlings. Furthermore, resistance-linked markers (e.g. Welter et al. 2007) are very useful to characterize and select the best parental genotypes for breeding and to score the accumulation of multiple independent sources for durable resistance (Eibach et al. 2007), provided that different sources are identified and labelled with molecular markers.

In other cases of pathogenesis, such as in *Botrytis* infections, physical attributes of the fruit cluster evidently influence the level of susceptibility. One example is high compactness of bunches; it leads to extended maintenance of small water deposits within the corners formed between the fruits of a dense cluster after rain falls. These reservoirs of humidity facilitate *Botrytis* infection. In contrast, cultivars with loose fruit clusters will dry much faster after rain falls and therefore will get less infected.

Morphological traits like fruit cluster architecture can show diverse expression within the cultivars of traditional *V. vinifera* grapevines or even within different clones of one single cultivar. In these cases as well it is advantageous to identify markers in genetic linkage to the trait of interest for cultivar or clone description in the frame of the characterization of genetic resources, clone selection and cross breeding. For the "classical" evaluation of any kind of fruit or cluster characteristic after a cross, the breeder needs to wait at least 3 years to be able to score such traits, as grapevine has a long juvenile period and the progeny will not carry fruits until the 3rd or 4th year of cultivation.

27.3 Strategies to Develop Molecular Markers

27.3.1 Genetic Mapping and QTL Analysis

Development of molecular markers in genetic linkage to traits is thus an important task. It can be achieved in different, complementary ways. One option is to go through the elaboration of genetic maps from a trait- and marker-segregating cross population. In this approach several hundreds of molecular markers are analysed for their linkage and recombination frequencies in comprehensive F1 progeny of around 200 or more individual plants. Scoring the recombination rates of markers segregating from the maternal and the paternal genotype separately, following the double pseudo testcross strategy for highly heterozygous plants (Weeden 1994), will result in two parental maps. These can further be integrated by double heterozygous markers that segregate from both parents (Fig. 27.1).

If the segregation of the agronomic trait of interest can be scored in a quantitative way in the same progeny (over at least two or three growing seasons) and shows a normal distribution (eventually after some mathematical transformation of the original data), statistic analysis for quantitative trait loci (QTL) can identify regions in the genetic map that carry factors determining the traits expression. This is done by "interval mapping" based on the genetic map (Lander and Botstein 1989) and its advanced methods. Molecular markers genetically close to a QTL will show different phenotypic trait distributions in the corresponding genotypic classes present in the mapping population. Marker alleles positively linked to the trait of interest will become evident and can further be applied for breeding. Several QTL of varying strength may be present scattered around the genetic map. In these cases, marker

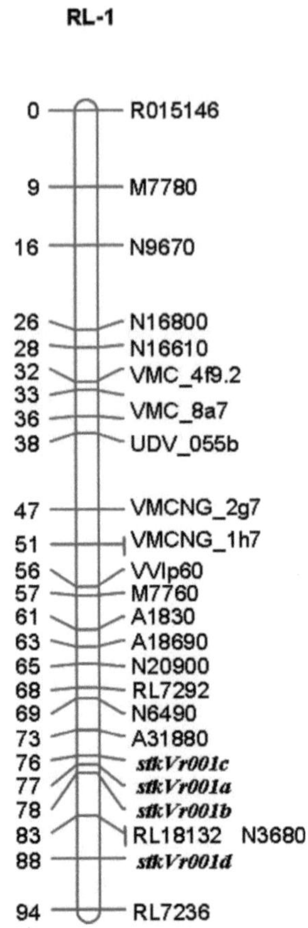

Fig. 27.1 Part of an integrated genetic map exemplifying one linkage group (LG 1) from the cross of "Regent" x "Lemberger". For details see Welter el al. 2007

combinations from the different QTL regions can be applied for marker-assisted selection or, alternatively, only the most important "major" QTL may be chosen and used for selection purposes.

27.3.2 Genetic Diversity Studies

A second option of marker identification is currently entering plant genetics. This approach uses non-family based genetic diversity studies to look for associations between specific genetic variants and the trait of interest. Molecular variation around the genome or in selected genomic regions delimited by a major QTL or candidate genes related to a trait by their genetic position, functional annotation or specific expression patterns (see below) is analysed over large plant sample sets. Correlations can be established between a trait and a specific genotypic constitution, e.g. with the help of the software TASSEL (Bradbury et al. 2007). In order to minimize the number of false positive associations, the genetic structure (the amounts of relatedness) of the individuals in the sample set has to be known and taken into account. This can be achieved by investigating a representative number of co-dominant SSR (simple sequence repeat) markers. Computation with programs such like STRUCTURE (Pritchard et al. 2000) may then deliver the genetic groupings within the sample set. This information should be considered in the following step of searching for significant associations. SSR and SNP (single nucleotide polymorphisms) markers can be employed to capture the genetic diversity.

27.3.2.1 SSR Markers

The core of an SSR (simple sequence repeat) or microsatellite motif is a very short repetitive DNA sequence, with repeat units generally one to six base pairs long, but longer and more complex repeats are also possible. Flanking such a repetitive sequence with primers complementary to neighboring unique sequences yields efficient PCR (polymerase chain reaction)-based markers. SSR markers are highly length polymorphic, ubiquitous elements of the genome and genetically most informative due to their co-dominance (i.e. Arnold et al. 2002). Amplified SSR fragments usually have a size of 100–400 base pairs. Due to the small differences in length between the two alleles of the diploid and largely heterozygous grapes, their length determination is done with sequencing level accuracy, e.g. with an automatic sequence/fragment analyzer. Commonly one of the SSR primers in a pair is labeled with a fluorescent dye. Depending on how many different dyes the analyzer instrument can detect, this opens up possibilities to combine several PCR reactions in multiplexes. For this efficient way of analysis, different fluorescent dyes are used in combination with different PCR product size ranges.

Advanced SSR marker technology has high automation feasibility, high reliability and low cost per analysis (Beyermann et al. 1992, Powell et al. 1996, Goldstein et al. 1999). SSR markers have wide ranging applications in genetic investigations including mapping studies (i.e. Doligez et al. 2006, Welter et al. 2007) genotyping (i.e. Macaulay et al. 2001, Hayden et al. 2008), population studies (i.e. Selkoe and Toonen 2006) and studies of genetic relationships (i.e. Queller et al. 1993, Blouin et al. 1996).

27.3.2.2 SNP Markers

With sequencing technology becoming cheaper and more reliable, allowing even complete genome sequencing of a number of organisms, another marker technology has gained major importance. These are SNP (single nucleotide polymorphism) markers, single nucleotide exchanges or small insertions and deletions (“indels”) between various alleles or DNA sequences (Edwards et al. 2007). They share the high reproducibility, automation potential and low running cost per data point with SSR markers and are abundant across the genome. SNPs can be observed both in coding and non-coding regions. In the human genome a genetic diversity of one SNP per 100–300 base pairs was counted. SNP frequencies in plants show a wide range, with less than 0.05% in cotton loci (Small et al. 1999) and up to 3% in maize (Ching et al. 2002). In grapevine they are quite frequent (Salmaso et al. 2004, Lijavetzky et al. 2007, Velasco et al. 2007, Vezzulli et al. 2008).

The rapid advancements of new methods and technical platforms like Pyrosequencing (Ronaghi et al. 1998), the Illumina GoldenGate SNP Genotyping Assay (http://www.illumina.com/technology/sequencing_technology.ilmn) or the Fluidigm Dynamic Array for SNP Genotyping (http://www.fluidigm.com/applications/genotype-profiling.html) result in high-throughput SNP analysis. Today, different databases are already established that contain SNP information for a large number of organisms (e.g. NCBI [http://www.ncbi.nlm.nih.gov/projects/SNP/] or SNPSTR [http://www.sbg.bio.ic.ac.uk/~ino/cgi-bin/SNPSTRdatabase.html]).

27.3.3 Differential Gene Expression Analysis

A third option is to perform differential gene expression studies to identify candidate genes for the traits of interest. At first, a well defined sample set of several genotypes with variable phenotypic trait expression is investigated to find differentially expressed genes specific to the investigated physiological situation. In follow-up studies, the sequence diversity of the resulting differentially expressed candidates is investigated in association to the phenotypic trait of interest over large sample numbers.

For differential gene expression studies in the context of disease resistance, susceptible and resistant plants have to be challenged with the pathogen under controlled conditions. Untreated (uninfected) or mock-inoculated plants are used for control. The expression patterns are then compared from total RNA extractions for each of the experimental settings and can reveal genes highly expressed (up-regulated) or down-regulated predominantly in the resistant sample under attack of the pathogen. For other traits like those related to morphological features and physical resistance factors the follow-up of developmental stages and RNA extractions at different growth time points may be necessary to reveal the most promising stages for analysis.

27.3.3.1 Global Screening with Micro-Arrays

Global gene expression patterns can be investigated by hybridisation to micro-arrays carrying probes that represent the majority of genes known for the species. Arrays are used because they allow a researcher to look in an unbiased fashion at how experimental manipulation might affect any of thousands of genes represented on the array. Such micro-array chips are available for grapevine since several years (e.g. Affymetrix GeneChip® *Vitis vinifera* Genome Array). Usually these hybridisations reveal a long list of potentially regulated genes, up- or down-regulated in their RNA abundance in certain developmental or physiological situations. The problem with micro-arrays is that there can be some artefacts, and it is often difficult to obtain reliable quantitative data or adequate statistical power with current array technology.

In any case, the results have to be validated. A set of promising candidates like e.g. genes for transcription factors, is usually selected and analysed in more detail for differential expression. The method of choice generally applied is quantitative Real Time PCR (qRT-PCR).

27.3.3.2 Quantitative Real Time PCR

The quantitative real time polymerase chain reaction (qRT-PCR), which is based on PCR, uses fluorescent reporter molecules to monitor the accumulation of amplification products during each cycle of a PCR reaction. It combines the DNA amplification and DNA detection steps into one homogeneous assay. In contrast to regular PCR, analysis of amplification products by gel electrophoresis is dispensable (unless the specificity of the amplification reaction needs to be controlled by checking amplification of a single product). The main areas of application of qRT-PCR are mRNA expression level measurements, DNA copy number determinations and the validation of micro-array hybridisation results. Currently, qRT-PCR is primarily used for gene expression studies like absolute or relative quantification of nucleic acids such as DNA, mRNA or miRNA. The technique of qRT-PCR is the most sensitive method for the detection and quantification of gene expression levels. Successful qRT-PCR assays are depending on many factors, especially the quality of RNA extraction, the reverse transcriptase (RT) reaction, the choice of the chemistry and the experimental conditions.

RNA Extraction

RNA is the starting material for all gene expression analyses. qRT-PCR needs high-quality, non-degraded RNA, free of genomic DNA. RNA is less stable than DNA, and can be easily degraded by cleavage of contaminating RNases during tissue sampling, RNA purification, and RNA storage (Swift et al. 2000). The amount and the quality of RNA depend also on the source of RNA, the sampling technique as well as the isolation method. Purified RNA should have an $A_{260/280}$ ratio greater than 1.9 and should appear intact when visualized by gel electrophoresis. Most RNA preparations are contaminated with genomic DNA and proteins at very low levels (Mannhalter et al. 2000). To get rid of genomic DNA, it may be necessary to treat the RNA samples with RNase–free DNase. To confirm the absence of genomic DNA either a "minus-reverse transcription" or "water control" should be included in the experimental design.

Reverse Transcription

The production of single-stranded complementary DNA copies of the RNA through reverse transcription (RT) is the next step in qRT-PCR experiments. This step is mostly responsible for the variability encountered in qRT-PCR experiments (Wong et al. 1998). The RT efficiency can be negatively affected by salt contamination, phenol, alcohol and other inhibitors carried over from the RNA isolation procedure

(Schwabe et al. 2000). Another source of RT variability is the choice of the priming method used to initiate cDNA synthesis. It is possible to use target gene-specific primers, or target gene-unspecific primers like poly (dT) primers, random hexamer, octamer or decamer primers (Freeman et al. 1996, Raja et al. 2000). If target gene-specific primers are used, a separate RT reaction has to be performed for each gene of interest. The use of poly (dT) primers or target gene-unspecific primers allows to synthesize a pool of cDNA of many different genes, derived from one mixed RNA sample. This cDNA pool can be split afterwards into a number of different target-specific qRT-PCR assays.

The yield and the quality of cDNA can be highly variable. Thus, reliable internal quality control of cDNA synthesis is essential. These controls are normally performed by PCR amplification of reference genes, mostly common "housekeeping genes", like the genes encoding GAPDH (glyceraldehyde-3-phosphate dehydrogenase), actins, tubulins or 18S ribosomal RNA. It is important to perform the controls for each cDNA synthesis prior to the qRT-PCR experiment.

Detection Chemistry

Today, two general methods for quantitative detection of amplicons are established, based on fluorescence resonance energy transfer (FRET) (Wittwer et al. 1997). The first method uses 5′ nuclease assays like in the application of TaqMan probes (Gibson et al. 1996). The 5′ nuclease assay generates a fluorescent signal by cleavage of a fluorescent molecule at the 5′-end of a target-specific oligo nucleotide. In native (unprocessed) form the 5′-end attached fluorescence is quenched by a second molecule at the 3′-end of the probe. The second common method is the use of non-specific fluorescent DNA binding dyes intercalating in double stranded DNA, like SYBR Green I or Ethidium bromide (Morrison et al. 1998), (Fig. 27.2).

SYBR Green I can only be used for single reactions, while TaqMan probes allow to perform multiplex reactions. There is a lot of other chemistry available for qRT-PCR, like Molecular Beacons or Scorpions. Among the qRT-PCR detection chemistry, SYBR Green I and the TaqMan assays produce a comparable dynamic range and sensitivity, while SYBR Green I detection is more precise and produces a more linear decay plot than the TaqMan probe detection (Schmittgen et al. 2000).

Quantification Methods in qRT–PCR

Two different quantification methods can be performed in qRT-PCR. The levels of gene expression can be measured by absolute or relative qRT-PCR. Absolute quantification measures the nucleic acid copy number in a given sample. This requires a reference sample of known quantity (copy number) of the gene of interest that can be diluted to generate a standard curve. Unknown samples are compared with this standard curve for absolute quantification. The primary limitation of this approach is the necessity to obtain an independent reliable standard for each gene to be analysed. Concurrent standard curves have to be run during each assay (Kuhne et al.2002).

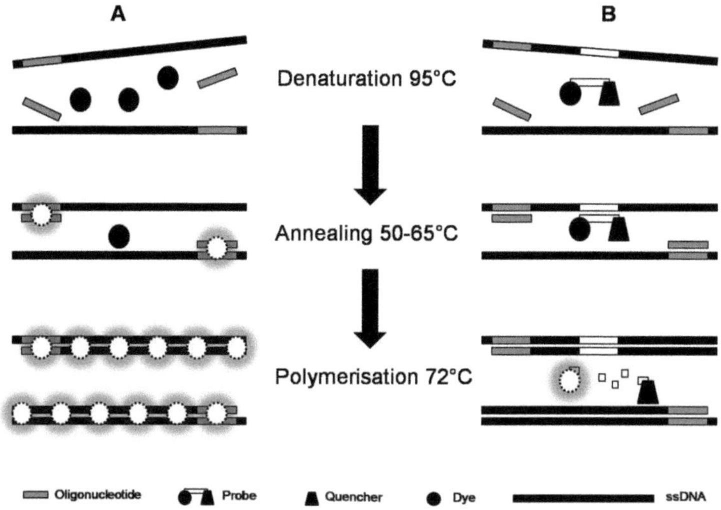

Fig. 27.2 Real Time PCR products detection chemistries. (**a**) Nonspecific detection (e.g. SYBR Green I). In solution, the unbound dye exhibits little fluorescence; during the PCR assay, increasing amounts of dye bind to the emerging ds DNA. (**b**) Specific detection (e.g. TaqMan; Bustin, 2002). Although the fluorophore (*circle*) and a quencher (*trapezoid*) are carried by the same probe molecule, any light emitted by the fluorophore upon excitation is quenched. When the polymerase displaces and cleaves the probe, the fluorophore and the quencher become physically separated, and light emission from the fluorophore can be detected

Relative quantification measures changes in the steady-state levels of a gene of interest relative to an invariant gene (Dheda et al. 2004). "Housekeeping genes" (as described before) that are not expected to change under the experimental conditions serve as convenient internal standard. As the absolute quantity of the internal standard is not known, only relative changes can be determined by this method. This is not a problem for most research applications because the fold change is informative irrespective of the absolute value. However, this approach is limited by the lack of absolute quantification and the necessity for stably expressed "housekeeping genes" to be used as internal standards.

Advantages and Limitations of Quantitative Real Time PCR

There are many methods, like Northern hybridisation, RNase protection assays or in situ hybridisation in molecular biology for measuring quantities of target nucleic acid sequences. Most of these methods exhibit one or more of the following shortcomings: they are time consuming, labour intense, insufficiently sensitive, non-quantitative, require the use of radioactivity, or have substantial probability of cross contamination (Reischel et al. 2002). Real time PCR has distinct advantages over these earlier methods. Perhaps the most important one is its ability to quantify nucleic acids over an extraordinarily wide dynamic range (at least five log units). This is coupled with extreme sensitivity, allowing detection of less than five copies (perhaps only one copy in some cases) of a target sequence. With appropriate internal standards and calculations, mean variation coefficients are 1–2%, allowing reproducible analysis of subtle gene expression changes even at low levels of transcription (Klein 2002, Luu-The et al. 2005). In addition, all real time platforms are relatively quick, with some offering high throughput automation. Finally, real time PCR is performed in a closed reaction vessel that requires no post PCR manipulations, thereby minimizing the chances for cross contamination in the laboratory.

But there are also some limitations to real time PCR methods. Most of them reside in the PCR or qRT-PCR-based techniques. Real time PCR is susceptible to PCR inhibition by compounds present in certain biological samples that may interfere with the activity of the thermo-stable DNA polymerase. Because of the necessary conversion of RNA into DNA with an extra enzymatic step of reverse transcription (RT), more problems have the opportunity to occur (see above). The conversion of RNA to cDNA is subject to variability because multiple reverse transcriptase enzymes with different characteristics exist, and different classes of oligo nucleotides can be used to prime RT (see above). Probably the largest limitation of real time PCR at present, however, is not inherent to the technology but rather resides in human errors: improper assay design, incorrect data analysis, or unwarranted conclusions.

Real time PCR gene expression analysis measures mRNA levels. This, only suggests possible changes on protein levels or function rather than demonstrating them. Further gene regulation at post-transcriptional level may occur. Although there is generally a tight correlation between gene expression and gene product function (Brown and Botstein 1999), a direct correlation is certainly not always given. Conclusions based on data derived from q RT-PCR are best utilized when the biological context is well understood (Bustin 2002).

Conversion of qRT-PCR Results into Informative Markers

Once a candidate gene has been found up- or down-regulated under specific physiological conditions linking it to a trait of interest, its sequence including the promoter region can be analysed in comparative diversity studies. This approach can reveal mutations potentially causative for the differential expression observed. Such sequence variants may include SNPs or small “indels” that can be scored as markers over large sample sets, again associating these sequence polymorphisms with the level of a specific quantitative trait. Differentially expressed genes therefore provide a good source for informative markers.

In addition, the expression of a candidate gene can be studied in trait-segregating mapping populations, localizing factors that determine the genes expression to a genetic region (expression QTL analysis, eQTL). In this case, transcriptional activity of the candidate can be labelled by eQTL-flanking genetic markers such as SSR markers mapping nearby.

Model whole genome sequences are available for grapevine (Jaillon et al. 2007, Velasco et al. 2007). If the genomic sequence of the eQTL-delimited region shows the candidate gene coding frame itself, this has good chances to be further useful as a completely linked marker, the perfect situation for application in marker-assisted selection. If the genomic region of the eQTL shows candidate genes putatively involved in gene regulation, this may give hints revealing the regulatory molecular cascade behind a trait. Understanding the regulatory networks governing the important traits in grapevine on molecular level is the final aim of all molecular analysis.

27.4 Conclusions and Summary

Molecular markers in linkage to traits are important tools in grapevine breeding. They will be helpful to characterize genetic resources. Marker assisted selection (MAS) of parentals and individuals of crosses will allow fast selection and identify combinations of desired traits in progeny in the near future. This genetic analysis is much more efficient than traditional evaluation and independent from environmental and developmental factors. Trait-linked markers for MAS can be developed by genetic mapping and QTL analysis, by association mapping in large sample sets of known structure (levels of relatedness) and through differential gene expression studies addressing specific physiological situations. SSR markers have been the best choice for genetic analysis over the last years. The next generation markers will be SNP markers that are amenable to high throughput, minimizing the costs of analysis per data point.

Acknowledgments We wish to express our thanks to the German Federal Ministry for Nutrition, Agriculture and Consumers Protection, the German Federal Ministry of Education and Research and the German Federal Agency for Agriculture and Nutrition that provided funding for the research applying the strategies outlined in this chapter.

References

Arnold C, Rossetto M, McNally J, Henry RTJ (2002) The application of SSRs characterized for grape (*Vitis vinifera*) to conservation studies in *Vitaceae* 1. Am J Bot 89:22–28

Beyermann B, Nürnberg P, Weihe A, Meixner M, Epplen JT, Börner T (1992) Fingerprinting plant genomes with oligonucleotide probes specific for single repetitive DNA sequences. Theor Appl Genet 83:691–694

Blouin MS, Parsons M., Lacaille V, Lotz, S (1996) Use of microsatellite loci to classify individuals by relatedness. Mol Ecol 5:393–401

Bradbury PJ, Zhang Z, Kroon DE, Casstevens TM, Ramdoss Y, Buckler ES (2007) TASSEL, software for association mapping of complex traits in diverse samples. Bioinformatics. 23:2633–2635

Brown PO, Botstein D (1999) Exploring the new world of the genome with DNA microarrays. Nat Genet 21:33–37

Bustin BA (2002) Quantification of mRNA using real-time reverse transcriptase PCR (RT-PCR): trends and problems. J. Mol. Endocrinol 29:169–193

Ching A, Caldwell KC, Jung M, Dolan M, Smith OS, Tingey S, Morgante M, Rafalski AJ (2002) SNP frequency, haplotype structure and linkage disequilibrium in elite maize inbred lines. BMC Genetics 3:19.

Dheda K, Huggett JF, Bustin SA, Johnson MA, Rook G, Zumla A (2004) Validation of housekeeping genes for normalizing RNA expression in real time PCR. Biotechniques 37:112–119

Doligez A, Adam-Blondon AF, Cipriani G, Laucou V, Merdinoglu D, Meredith CP, Riaz S, Roux C, This P, Di Gaspero G (2006) An integrated SSR map of grapevine based on five mapping populations. Theor Appl Genet 113:369–382.

Edwards D, Forster JW, Chagné D, Batley J (2007) What are SNPs? In: Oraguzie NC, Rikkerink EHA, Gardiner SE, De Silva NH (eds): Association mapping in plants. Springer Science + Buissness Media, LCC. New York, ISBN-10:0-387-35844-7

Eibach R, Zyprian E, Welter L, Töpfer R (2007) The use of molecular markers for pyramiding resistance genes in grapevine breeding. Vitis 45(2):120–124

Freeman WM, Vrana SL, Vrana KE (1996) Use of elevated reverse transcription reaction temperatures in RT – PCR. Biotechniques 20:782–783

Gibson U, Heid C, Williams P (1996) A novel method for real time quantitative RT-PCR. Genome Res 6:995

Goldstein DB, Schlötterer C (eds) (1999) Microsatellites: evolution and applications. Oxford University Press, Oxford.

Hayden MJ, Nguyen TM, Warterman A, Chalmers KJ (2008) Multiplex ready PCR: a new method for multiplexed SSR and SNP genotyping. BMC Genomics 9:80

Jaillon O, Aury JM, Noel B, Policriti A, Clepet C, Casagrande A, Choisne N, Aubourg S, Vitulo N, Jubin C, Vezzi A, Legeai F, Hugueney P, Dasilva C, Horner D, Mica E, Jublot D, Poulain J, Bruyère C, Billault A, Segurens B, Gouyvenoux M, Ugarte E, Cattonaro F, Anthouard V, Vico V, Del Fabbro C, Alaux M, Di Gaspero G, Dumas V, Felice N, Paillard S, Juman I, Moroldo M, Scalabrin S, Canaguier A, Le Clainche I, Malacrida G, Durand E, Pesole G, Laucou V, Chatelet P, Merdinoglu D, Delledonne M, Pezzotti M, Lecharny A, Scarpelli C, Artiguenave F, Pè ME, Valle G, Morgante M, Caboche M, Adam-Blondon AF, Weissenbach J, Quétier F, Wincker P; French-Italian Public Consortium for Grapevine Genome Characterization (2007) The grapevine genome sequence suggests ancestral hexaploidization in major angiosperm phyla. Nature 449, 463–467

Klein D (2002) Quantification using real-time PCR technology; applications and limitations. Trends Mol Med 8:257–260

Kuhne BS, Oschmann P (2002) Quantitative real time PCR using hybridization probes and imported standard curves for cytokine gene expression analysis. Biotechniques 33:1078–1084

Lander ES, Botstein D (1989) Mapping Mendelian factors underlying quantitative traits using RFLP linkage maps. Genetics 121:185–199

Lijavetzky D, Cabezas JA, Ibáñez A, Rodríguez V, Martínez-Zapater JM (2007) High throughput SNP discovery and genotyping in grapevine (*Vitis vinifera* L.) by combining a re-sequencing approach and SNPlex technology. BMC Genomics 8:424

Luu-The V, Paquet N, Calvo E, Cumps J (2005) Improved real-time PCR method for high-throughput measurements using second derivative calculation and double correction. Biotechniques 38:287–293

Macaulay M, Ramsay L, Powell W, Waugh R (2001) A representative, highly informative 'genotyping set' of barley SSRs. Theor Appl Genet 102:801–809

Mannhalter C, Koizar D, Mitterbauer G (2000) Evaluation of RNA isolation methods and reference genes for RT-PCR analyses of rare target RNA. Clin Chem Lab Med 38:171–177

Morrison TB, Weis JJ, Wittwer CT (1998) Quantification of low-copy transcripts by continuous SYBR Green I monitoring during amplification. Biotechniques 24:954–962

Powell W, Machray GC, Provan J (1996a) Polymorphism revealed by simple sequence repeats. Trends Plant Sci 1:215–222

Pritchard JK, Stephens M, Donnelly P (2000) Inference of population structure using multilocus genotype data. Genetics 155:945–959

Queller DC, Strassman JE, Hughes CR (1993) Microsatellites and Kinship. Trends Ecol Evolu 8:285–288

Raja S, Luketich JD, Kelly LA, Ruff DW, Godfrey TE (2000) Increased sensitivity of one-tube, quantitative RT-PCR. Biotechniques 20:702–708

Reischel U, Wittwer CT, Cockerill F (2002) Rapid cycle real time PCR:methods and applications. In: Microbiology and food analysis. Springer, New York
Ronaghi M, Uhlén M, Nyrén P (1998) DNA SEQUENCING: a sequencing method based on real-time pyrophosphate. Science 281:363–365
Salmaso M, Faes G, Segala C, Stefanini M., Salakhutdinov I., Zyprian E, Toepfer R, Grando MS, Velasco R. (2004): Genome diversity and gene haplotypes in the grapevine (*Vitis vinifera* L.) as revealed by single nucleotide polymorphisms. Mol Breed 14:385–395
Schmittgen TD, Zakrajsek BA, Mills AG, Gorn V, Singer MJ, Reed MW (2000) Quantitative reverse transcription-polymerase chain reaction to study mRNA decay: comparison of endpoint and real time methods. Anal Biochem 285:194–204
Schwabe H, Stein U, Walther W (2000) High – copy cDNA amplification of minimal total RNA quantities for gene expression analyses. Mol Biotech 14:165–172
Selkoe KA, Toonen RJ (2006) Microsatellites for ecologists: a practical guide to using and evaluating microsatellite markers. Ecol Lett 9:615–629
Small RL, Ryburn JA, Wendel JF (1999) Low levels of nucleotide diversity at homoeologous *Adh* loci in allotetraploid cotton (*Gossypium* L.). Mol Biol Evol 16:491–501
Swift GM, Peyton MJ, MacDonald RJ (2000) Assessment of RNA quality by semi-quantitative RT PCR of multiple regions in long ubiquitous mRNA. Biotechniques 28:524–531
Velasco R, Zharkikh A, Troggio M, Cartwright DA, Cestaro A, Pruss D, Pindo M, Fitzgerald LM, Vezzulli S, Reid J, Malacarne G, Iliev D, Coppola G, Wardell B, Micheletti D, Macalma T, Facci M, Mitchell JT, Perazzolli M, Eldredge G, Gatto P, Oyzerski R, Moretto M, Gutin N, Stefanini M, Chen Y, Segala C, Davenport C, Demattè L, Mraz A, Battilana J, Stormo K, Costa F, Tao Q, Si-Ammour A, Harkins T, Lackey A, Perbost C, Taillon B, Stella A, Solovyev V, Fawcett JA, Sterck L, Vandepoele K, Grando SM, Toppo S, Moser C, Lanchbury J, Bogden R, Skolnick M, Sgaramella V, Bhatnagar SK, Fontana P, Gutin A, Van de Peer Y, Salamini F, Viola R (2007) A high quality draft consensus sequence of the genome of a heterozygous grapevine variety. PloS One 2(12):e1326
Vezzulli S, Micheletti D, Riaz S, Pindo M, Viola R, This P, Walker A, Troggio M, Velasco R (2008) A SNP transferability survey within the genus *Vitis*. BMC Plant Biol 8:128
Weeden, NF (1994) Approaches to mapping in horticultural crops. In: Gresshoff PM (ed) Plant Genome analysis. CRC Press, Boca Raton, Ann Arbor, London, Tokyo, pp 57–68
Welter LJ, Göktürk-Baydar N, Akkurt M, Maul E, Eibach R, Töpfer R, Zyprian EM (2007) Genetic mapping and localization of quantitative trait loci affecting fungal disease resistance and leaf morphology in grapevine (*Vitis vinifera* L.). Mol Breed 20:359–374
Wittwer CT, Herrmann MG, Moss AA, Rasmussen RP (1997) Continuous fluorescence monitoring of rapid cycle DNA amplification. Biotechniques 22:130–138
Wong L, Pearson H, Fletcher A, Marquis CP, Mahler S (1998) Comparison of the efficiency of Moloney Murine Leukaemia Virus (M-MuLV) reverse transcriptase, RNase H-M-MuLV reverse transcriptase and Avian Myeloblastoma Leukaemia Virus (AMV) reverse transcriptase for amplification of human immunoglobulin genes. Biotechniques 12:485–489

Chapter 28
Phenolic Maturity in Red Grapes

Montserrat Nadal

Abstract The criteria for estimating optimal maturity in red grapes are complex. The ratio sugars/acids do not give enough information to determine precisely the date of harvesting. The accumulation of phenolic compounds depends on climate, soil, genetics and cultural practices. Grape phenolics, which are structurally diverse, therefore have a variable extraction potential (called extractability). This chapter discusses the methods for determining phenolic maturity based on anthocyanin extraction from the whole berries. First, the description and modifications done by different authors are explained. Second, the interpretation of values and indexes is exposed. Finally, the effects of environmental and genetic factors on phenolic content of grapes and the colour extractability are commented.

Contents

M. Nadal (✉)
Facultat d'Enologia de Tarragona, Departmento Bioquímica i Biotecnologia, Universitat Rovira i Virgili. Campus Sescelades, Tarragona 43007, Spain
e-mail: montserrat.nadal@urv.cat

S. Delrot et al. (eds.), *Methodologies and Results in Grapevine Research*,
DOI 10.1007/978-90-481-9283-0_28,

Abbreviations

AOC	Wine county region, (*Appellation d'Origine Controlée*)
DAD	Diode Array Detection
DMACH	*p*-dimethylaminocinnamaldehyde
HPLC	High Performance Liquid Chromatography
HPLC-MS	Liquid Chromathography coupled to Mass Spectrometry
IPT	Absorbance at 280 nm, Total Phenol Index
NIR	Near Infrared
MS	Mass Spectrometry
PAL	Phenylalanine Ammonia Lyase
PAR	Photosynthetic Active Radiation
UV-Vis	Ultraviolet Visible

28.1 Introduction

The criteria for estimating optimal maturity in red grapes are multi-faceted and complex. Several important classes of compounds change during ripening and maturation of the fruit on the vine. The importance of harvesting at a given ripeness stage in order to obtain a quality product has been widely recognized (González-San José et al. 1991, Calò et al. 1996, Ribéreau-Gayon et al. 2000, Bisson 2001, Hunter et al. 2004, Nadal and Hunter 2007). Each year, berry composition is analyzed during development and ripening in numerous vineyards, in order to determine the harvest date with respect to changing climatic conditions. The most commonly analyzed solutes are: sugars, total titratable acidity, organic acids, potassium, anthocyanins and sometimes tannins and amino acids. Anthocyanin pigments and tannins are particularly important for red wine quality. Therefore, many studies have investigated pulp and skin composition during ripeness in order to determine the optimal harvest date.

Winemakers have been mostly using the ratio between sugars and acids as an indication of the degree of ripeness for establishing when grapes are ready to be harvested. Nowadays, a great interest arises in understanding the evolution of the phenolic compounds in the skins of red varieties. Frequently, optimal physiological ripeness of the pulp may differ from optimal ripeness in the skins.

The total berry phenolic concentration slowly increases during maturation until a maximum is reached one or two weeks before harvest, depending on variety, climatic conditions and cultural practices (González San José et al. 1990, Cacho et al. 1992, Jordao et al. 1998, Vivas et al. 2001, Harbertson and Adams 2002, Downey et al. 2006). Before veraison, there is no significant increase of phenolic compounds in the berries. The concentration of tannins, phenols synthesised during development of the green berries, decreases during the ripening period (Kennedy et al. 2001, Ojeda et al. 2002, Valls 2004). In the skin, the catechins and esters of hydroxycinnamic acids progressively decrease due to the increase in berry size. In the seeds, this decrease may also be attributed to oxidation processes (Kennedy et al. 2000). In contrast, the rate of anthocyanin synthesis increases significantly during veraison, contributing to the total increase in phenolics during this time.

Climate, soil, cultivation and biology are some of the most relevant factors affecting synthesis and concentration of phenols in berries (Donwey et al. 2006). Accumulation of phenolic compounds depends on soil type and fertility (Van Leeuwen et al. 2004), inherent soil water-holding capacity, and the annual amount and distribution of the rainfall in a specific terroir (Choné et al. 2001). These factors significantly affect the vigour and yield of the plant (Jackson and Lombard 1993). Cultivation techniques, such as trellising system, pruning method and canopy management are critical for optimal performance of the canopy and grapes, and the establishment of a balance between vegetative and reproductive growth (Hunter et al. 1995). Furthermore, both photosynthetic active radiation (PAR) and phenylalanine ammonia lyase (PAL) activities are essential for the synthesis of phenols in berries (Roubelakis-Angelakis and Kliewer 1986). Moreover, the ratio of skins and seeds to berry size, has an important role for the extraction of phenolics on wine (Roson and Moutounet 1992).

The phenolic composition of grapes is highly complex. Methods for analysing polyphenols include extraction with organic solvents, fractionation, HPLC chromatography, and DAD or MS detection (Roggero et al. 1986, Nagel and Glories 1991, Roson and Moutonnet 1992, de Freitas et al. 1998 ...etc). However, even sophisticated techniques like these are unable to completely elucidate the phenolic composition. Many compounds have high molecular weights; they can consist of very similar isomeric structures, and thus are difficult to separate using known methods. For example, in the case of flavan-3-ols, monomers (catechin, epicatechin, catechin-gallate, epicatechin-gallate), and even their dimeric forms can be relatively easy identified, but for trimeric, oligomeric and polymeric forms it becomes more complicated. Concerning anthocyanins, the monoglucosides of 5 anthocyanidins (malvidin, delphinidin, peonidin, cyanidin and petunidin) and their acylated forms have been described in *Vitis vinifera*, each variety having a different profile that is also affected by other parameters (maturation, soil, irrigation...). Although acylated forms can co-elute, HPLC separation of the main 5 anthocyanidin-monoglucosides is well established. Nevertheless, chromatographic techniques are expensive and difficult to use in a winery to follow the phenolic maturity during ripening.

28.2 Determination of Phenolic Maturity

Grape phenolics are structurally diverse, from simple molecules to oligomers and polymers that are usually designated as *tannins*. It is possible to define the phenolic composition of grapes according to the stage of maturity, taking into account the high concentration of phenolic compounds, anthocyanins and tannins, and especially their reactivity with proteins which indicate the progress of ripening. Extraction of anthocyanins and tannins depends on their location in the berry and their solubility. The analysis of phenolic maturity during the ripening period can be used to evaluate the development of these molecules and classify vineyards, terroir, grape varieties or even individual plots (Ribéreau-Gayon et al. 2000).

Phenolic maturity should cover not only the overall concentration of substances of this family, but also their structure and capacity to be extracted from grapes during vinification (wine-making). Theoretically, under similar winemaking techniques, grapes with higher anthocyanin content should produce wines with more colour but this is not always the case (La Notte et al. 1992, Boulton et al. 1996, Mattivi et al. 2003, Nadal et al. 2004, Cagnasso et al. 2008). Grapes, therefore, have a variable extraction potential or extractability.

Various methods for assessing the total phenols content in grapes have been suggested. Most methods are based in quantifying phenols after extraction. A traditional method for assessing phenolic maturity relies on the determination of total phenols (and in some cases, also the colour) in an extract obtained just by crushing berries. Samples taken during maturity permit to follow the evolution until reaching its maximum value. Another fair method, that could be easier to interpret, has been suggested by Glories and Augustin (1993), Saint-Cricq et al. (1998). Modifications and improvement of this method have been proposed by different authors as Lamadon (1995), Venencie et al. (1997), Peyron (1998), Iland et al. (1996, 2000), Di Stefano et al. (2000), Celotti and Carcereri (2000), Mateos (2003) and, Jensen et al. (2007). All of them follow analogous procedures to determine phenolic maturity by improving different steps or modifying the length of the method.

Lately, NIR (near infrared) technology has been used to determine berry colour reading directly on a homogenized sample. The NIR database is obtained by calibrating the instrument against a standard method. This approach offers a rapid means for measuring berry colour (Dambergs et al. 2003). Celotti et al. (2007) have been using the spectroscopy on whipped grapes to determine an index of phenolic maturity (Phenolic Maturity Trend, PMT) derived from the elaboration of the reflectance signals from six light emitting diodes in the spectral range 525–880 nm. The PMT index allows to determine the progress of the overall maturity of berry polyphenols in the laboratory. Celotti et al. (2007) conclude that this method is equivalent to the traditional methods used for the identification of polyphenols. On the other hand, Sethuramasamyraja et al. (2007) used the dataset obtained with a portable near-infrared spectrometer to establish a zone map in order to harvest the vineyard by quality. Two different zones were defined by the high and low anthocyanin berry content measured in the field (precision viticulture).

28.3 Phenolic Maturity Methods Based on the Phenol Extraction

The quantitative method consists of extracting the anthocyanins from the whole berries in two different acidic solutions in order to facilitate first a mild extraction and then, a second extraction under more extreme conditions. The acid medium disrupts the proteophospholipid membrane, breaking the protein bonds and releasing the contents of the vacuoles. Glories and Augustin (1993) proposed two different solutions for extracting phenols. All the anthocyanins diffuse in the more acidic solution at pH 1 (extracting Solution A), while part of the easily extractable anthocyanins is solubilized in solution at pH 3.2 (extracting Solution B). The pH 3.2 is close to the pH value of wine, and therefore comparable to that occurring in the wine-making processes.

Finally, phenols are analyzed in the resulted extract: total anthocyanins measured according to Ribéreau-Gayon and Stonestreet (1965), and the total phenols index by absorbance at 280 nm (Ribéreau-Gayon 1970). Total tannins may be also analyzed on the extracts (determination according to Ribéreau-Gayon and Stonestreet 1966). Recently, another method called MCP (Methyl Cellulose Precipitable) tannin assay allows tannin quantification in wine and grape samples. MCP method (developed by Sarneckis et al. 2006) is based on tannin precipitation with methyl cellulose and processed extracts are analyzed by UV-Vis at 280 nm wavelength for total phenolics. Tannin concentration is expressed as g/L epicatechin equivalents.

Reference methods for determining phenols have been described by different authors: Kramling and Singleton (1969), Somers and Evans (1977), Glories (1984), Roggero and Archier (1989), Di Stefano et al. (1991), Singleton and Trousdale (1992) and Ribéreau-Gayon et al. (2000).

28.3.1 Glories Method – Protocol for Phenol Extraction

28.3.1.1 Sampling

Two samples of 200 whole berries each are processed. Grapes are picked a random from several vines representing the total berry population. The must of one sample is obtained by pressing and used for determination of the sugar content, acidity and pH, while the second one is used for the determination of phenolic maturity (Glories and Augustin 1993, Saint-Cricq et al. 1998).

28.3.1.2 Procedure of Phenolic Maturity Determination

1. Crush the sample of 200 berries in a blender to obtain a homogeneous mixture, for 2 min at slow speed (number 1), in order to get a visually uniform extract.
2. Maceration: The homogenate is weighed and placed in two tubes, X g each, calculated by $X =$ (50 mL $\times$ density)/1000. Add the different pH solutions. Solutions are aqueous: 50 mL of solution (A) at pH 1.0 with HCl N/10; and solution (B) at pH 3.2, prepared with 5 $g{\cdot}L^{-1}$ tartaric acid neutralized by 1/3. As a consequence, the homogenate after crushing is diluted by half. Both samples, A and B are stirred manually and left for 4 h.
3. Filtering through glass wool.
4. Analyses of both the extract A and B supernatants, determination of the anthocyanin content in A and B and; measurement of absorbance at 280 on (B) supernatant. Extractability (%EA) and Seed maturity (%SM) indexes, and the pomace/juice ratio, are calculated.

28.3.1.3 Calculation

The total anthocyanin content is determined in A and B extracts in order to calculate the potential or capacity of phenolic extraction after maceration. Subsequently, the difference between the total potential anthocyanins A (pH = 1.0) and the extractable anthocyanins B (pH = 3.2) related to the total potential anthocyanins, give us a percentage of the *extractability (EA).*

$$\%EA = ((AntpH1.0 - AntpH3.2) \times 100)/AntpH1.0$$

Total potential anthocyanins (A) may vary from 500 to over 2,000 mg L^{-1} according to grape variety; lower extractability meaning easier extractable anthocyanins. Extractability decreases during ripening. Another term, *seed maturity* (*SM*) estimates the contribution of tannins of seeds to the ripeness evaluation. Seed tannin content diminishes progressively along the ripening because of the process of polymerization, and consequently, green tannin astringency decrease. Phenol content (measured at A280 in pH 3.2 extract), correspond to the sum of skins anthocyanins, skins tannins; and seed tannins. Assuming a similar extractability for anthocyanins and tannins, it is possible to establish that:

$$\begin{aligned} \text{Abs280pH3.2} &= \text{skinAnthocyanins} - \text{skinTannins} \\ &= \text{AnthocyaninspH3.2} \times \alpha \end{aligned}$$

Hence, the ratio (α) of A280/anthocyanins in the extract at pH 3.2 is used to calculate the SM index. A value of α = 40, considered as an average, was obtained by analysing several cultivars. This index decreases along ripening. Values lower than 60% indicate that seed maturity has been reached.

$$\%\text{SM} = (\text{Abs280pH3.2} - (\text{AntpH3.2}\times 40))\times 100)/\text{Abs280pH3.2}$$

The CASV method (*Chambre d'Agriculture de la Gironde, Service Vigne 1998*) proposes to determine the optimal harvest date by analysing berry phenolic maturity (Glories method) together with the determination of pH, probable alcoholic degree and titratable acidity.

28.3.2 Other Methods Based in Glories Protocol

28.3.2.1 Lamadon Method

The evaluation of phenol content on grapes is carried out in only one acid solution (15%vol of ethanol).

Sample size: pick at random in the field between 300 and 500 berries.

Two hundreds berries are ground in a blender for 2 min at slow speed.

Weigh 50 g of homogenate in a vessel of 250 mL and add 15 mL of ethanol 96%. At this point, the homogenate should be kept frozen (–18°C) before the analysis.

After addition of 85 mL of acid solution to the homogenate (HCl 0.1%), maceration is done for 2 h in a shaker at room temperature.

Filtering: Determination of anthocyanins (blenching with SO_2) and total phenols (A280), both expressed by g/kg berry (Lamadon 1995).

This method was proposed by ITV (Institut Technique de la Vigne, France).

28.3.2.2 Peyron Method

Different solutions for phenols (total and extractable anthocyanins) and tannins extractions are used:

Solution A (pH = 1.0). Two different solutions are prepared: A1 is an aqueous solution of acetone/water/0.1 HCl (70/30/0.1) and A2 is a solution of methanol + 0.1% HCl.

Solution B (pH = 3.2) is a synthetic wine solution: 12% vol of ethanol + 5 $g{\cdot}L^{-1}$ of tartaric acid.

Analysis: determination of total anthocyanins at pH = 1.0 and pH = 3.2; and determination of total tannins.

The extraction of total anthocyanins and tannins is much better with acetone solution (A1) than with methanol (A2). Nevertheless, the extraction with acetone is more laborious because any trace of acetone remaining in the homogenate should be removed before analysis. When the relation between the values of phenol extraction on Pinot noir grapes and the values of phenols in the wine was studied, the best correlation was found when the sample was analyzed at pH = 3.2 (Peyron 1998).

28.3.2.3 Nadal and Mateos Method

Differences with Glories method: the solutions A and B are buffers; pH 3.2 is changed to pH 3.6, in order to be closer to the pH of wines produced in warmest regions (and basic soils). Shorter time of maceration is used: 1 h instead of 4 h. Statistical differences on the length of maceration were not found analyzing phenolic maturity in different cultivars. Results were statistical equals using pH =3.6 instead to pH = 3.2. The maceration of the homogenate with solutions A and B is carried out with 20 mL (Mateos 2003).

Sampling: Samples are picked at random from several vines. Bunches or collected berries are put into plastic bags and transported to the laboratory on ice, taking care to prevent juice loss during transport. It is recommended to analyze samples by duplicate or triplicate. For duplicates, the smallest sample recommended is 500–600 berries, depending on berry size, and; from 15 to 20 bunches, depending on the genotype of the variety. To determine pulp maturity, a sample of 100 berries is taken to determine sugar content, acidity, pH and berry weight. These samples (bunches or berries) should be analyzed after sampling or kept in the refrigerator for not more than 24 h.

For the determination of phenolic maturity (Fig. 28.1), berries should be analyzed directly after sampling, but bunch samples can be carefully frozen for a period of 3 months before the analysis.

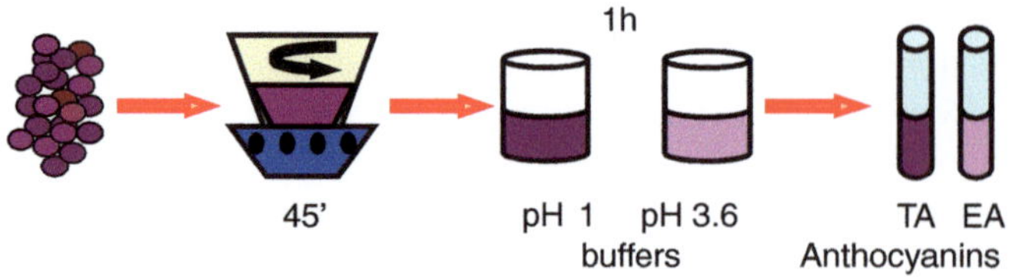

Fig. 28.1 Scheme for phenol extraction (Nadal and Mateos). The homogenate is shaked in plastic pots, with buffers at pH 1.0 and pH 3.6 for 1 h, at room temperature. TA = Total Anthocyanins; EA = Extractable Anthocyanins

1. Crush the sample of 100 berries in a blender to obtain a homogeneous mixture at moderate speed for 1 min. If berries are ripe, 45 s is enough. When working with small-berry varieties such as Cabernet, the sample size should be increased to 120 berries in order to obtain a uniform homogenate.
2. Add 20 mL of buffered A and B solutions according to:
   Solution A: KCl buffer (3.72 g KCl + 4.10 mL HCl), aqueous hydrochloric acid solution (0.1 N), 10% of ethanol, pH adjusted at 1.0. Weight X (g) = (20 × density/1000) of homogenate.
   Solution B: Citrate buffer (14.4 g citric acid + 9.28 g sodium citrate), aqueous acid solution, 10% of ethanol, pH adjusted at 3.6. Weight X (g) = (20 × density/1000) of homogenate.
3. Maceration for 1 h shaking at room temperature (20–22ºC) and keep the samples in the dark.
4. Centrifugation at 8,000 rpm; 8 min. The supernatant is kept in tubes for analyzing.
5. Determinations: Total and extractable anthocyanins, and total phenols determination by UV/VIS spectrophotometry; Extractability (EA (%)) and Seed Maturity (SM (%)) indexes calculation. For a better understanding, total potential anthocyanins extracted at pH = 1 are called *Total anthocyanins* and the anthocyanins extracted at pH =3.6 are called *Extractable anthocyanins*

Determination of anthocyanins is done according to the method of decolouration by bisulphite ($NaHSO_3$), Ribéreau-Gayon and Stonestreet (1965); Total phenols index (IPT, absorbance at 280 nm).

Anthocyanins and tannins are extracted under similar conditions from the skins (Glories and Augustin 1993). Determination of procyanidins by the DMACH (*p*-dimethylaminocinnamaldehyde) method in solution (B) is complementary to the previous methods mentioned above. The DMACH can give a better understanding of the stage of berry ripeness and also provide the knowledge for improving phenol extraction in wine-making. DMACH decreases during maturity. Total tannins are analysed as described by Ribéreau-Gayon and Stonestreet (1966) method and procyanidins according to the DMACH method (Vivas et al. 1994).

The phenolic potential is calculated according to Glories method:

$$\%EA = (\text{TotalAnt} - \text{ExtractableAnt}) \times 100/\text{TotalAnt}$$

$$\%SM = (\text{AbspH3.6} - (\text{ExtractAnt}) \times 0.40)) \times 100)/\text{AbsPH3.6}$$

28.3.2.4 Iland Method

This method describes the measurement of total anthocyanins in red grape berries based on the method described by Iland et al. (1996) in the Australian Wine Research Institute. It involves the extraction of these compounds from a homogenized grape sample, expression of their colour at low pH and quantification based on their absorbance in the visible region of the light spectrum. The results are expressed in malvidin equivalents only for comparative purposes.

Procedure

1. Homogenization of whole berries sample. Berry samples are homogenised with an ultra-Turrax: a sample size of 200 berries is processed at maximum speed (24,000 rpm setting) for approximately 30 s; scrape the homogenate from the homogenizer shaft into the vessel, and then homogenise for a further 30 s. Ensure that seeds are thoroughly macerated and all homogenate is scraped from the homogeniser shaft and collected in the homogenising vessel. Other homogenates may be obtained with different blenders; Retsch (processing the sample at a speed of 8,000 rpm for 20 s) or Waring (processing the sample for 60 s in high vessel). Nevertheless, the latter does not homogenize as effectively as the other types.
   A minimum sample size of 50 g (that approximately represents 50 berries) is required for the determination. The samples can be analysed fresh or can be frozen (−20ºC) prior to analysis.
2. Extraction:
   Samples must be extracted within 4 h. Mix the homogenate and transfer 1 g to a preweighed plastic 10 mL centrifuge tube A and 1 g to tube B.
   Solution A: Add 10 mL of 50% v/v aqueous ethanol to the portion of homogenate and adjust at pH = 2.0 with HCl, cap the tube and agitate on a mixing device for one h.
   Solution B: Add 10 mL of a solution of tartaric (5 g of tartaric acid in 1 L of water) and adjust at pH = 3.2 with sodium hydroxide. Alternatively mix by inverting the tube regularly (every 10 min) over a period of one h. Ensure that the pellet does not lodge in the bottom of the tube.
3. Centrifugation
   Centrifuge the macerated mixture at 1,800*g* for 10 min. The supernatant is named extract. If 1 g of homogenate was used, verify the final extract volume at 10.5 mL. The extract can be stored in the freezer (at −20ºC) for up to 3 months without significant loss of colour.

4. Analyses: Total anthocyanins
   Transfer 200 μL to an acrylic cuvette (10 mm pathlength). Add 3.8 mL of 1.0 M HCl, cover with *parafilm* and mix. Incubate at room temperature for a minimum of 3 h (not longer than 24 h). This time can be reduced if the HCl is used for adjustment.
   Measure absorbance of the acidified diluted extract at 520 nm using a 1.0 M HCl blank.
5. Calculations of colour: Anthocyanins (Ant) and total phenols (Tphenols).

$$\text{AntBerry(mg/Berry)} = \text{Abs520} \times \text{DF} \times \text{finalExtract(ml)} \times \text{50berriesW(g)} \times 1000/ \\ (500 \times 100 \times \text{homogenateW(g)} \times 50$$

$$\text{Ant(mg/g)} = \text{AntBerry/berryweightaverage(g)}$$

$$\text{Tphenols(mg/Berry)} = \text{Abs280} \times \text{DF} \times \text{finalextract(ml)} \times \text{50berriesW(g)} \times 1000/ \\ (100 \times \text{homogenateW(g)} \times 50)$$

$$\text{Tphenols(mg/g)} = \text{(Totalphenols/berry)/berryW(g)}$$

Notes: Factor 1/500 corresponds to the extinction coefficient of malvidin. DF: the dilution factor in this case is 20. T = Total; W = weight; Ant = Anthocyanins; Abs = Absorbance.

28.4 Helpful Hints, Results and Comments About Phenolic Maturity

28.4.1 Caution at Bunch and/or Berry Sampling

Sampling berries or bunches in the vineyard should be very accurate because the major source of variation within a vineyard is between vines. Berry or bunch sampling should be performed by randomly choosing vines in the vineyard; within the vine (take sunny and shaded bunches) on both sides of the row; in the bunch, from the top, middle and bottom.

Fresh samples must be stored cool (approximately 4°C) and analysed within 24 h of collection. For field samples, this can be achieved, for example, by storing freshly collected grapes in cooler boxes with ice packs until delivered to the laboratory. There is not significant colour loss during storage of frozen whole grapes for at least 3 months. If frozen, thaw berries (usually overnight in a refrigerator at 4°C) and process them cold (less than 10°C) to minimize oxidation of colour components.

Processing duplicates of samples (or triplicates for statistical analyses) is recommended in order to avoid the heterogeneity of berries and improve the results of the analysis. This fact becomes even more important in unhealthy, unripe and over-ripe berries.

Whole berries without pedicel are crushed. Crushing the whole berry allows obtaining a representative sample of the combination of pulp, skin and seed components. Crushing seeds also results in a partial extraction of their tannins, which is necessary to assess the characteristics of grapes.

In precision viticulture, it is necessary to compare the results obtained by the near-infrared method with a quantified method based on phenol extraction. Changes of terroir and varieties influence significantly the prediction of phenol content.

28.4.2 Interpretation of Values and Indexes

The interpretation of EA and SM indexes can facilitate the decision-making process regarding the adaptation of winemaking techniques to the specific characteristics of grapes. Unhealthy or overripe berries have more porous membranes and in consequence, the extractability increases. On the other hand, the extraction process becomes more difficult in unripe berries compared to berries in optimal ripeness. Moreover, in shrivelled berries the extraction process shows high variability and overall, decreases at the end of maturity.

EA index varies between 10% and 80% and decreases along maturity (extractability increases). Smaller differences between total (potential) anthocyanins and extractable anthocyanins correspond to a higher extractability, generally reached at harvest. Values of %EA ≤ 30 and total anthocyanins > 1,200 mg·L^{-1} are the optimal range for the colour extraction into the wine. When %EA is between 50 and 60%, the extraction becomes slow. In this case, for a higher extraction it is recommended to work with low temperatures at the beginning of the fermentation. Healthy berries with a smaller amount of colour (total anthocyanins < 1,000 mg·L^{-1}) have less capacity to release anthocyanins into the wine; hence, it is advisable to improve the winery techniques in order to increase colour extraction (Ribéreau-Gayon et al. 2000).

SM index presents wider ranges from 0 to 60%, according to grape variety, number of seeds and ripeness stage. It decreases along the ripening process. Higher values of SM could have a negative influence on the structure when sensory evaluation is carried out on the wine tasting. Higher values of %SM characterize seeds with tannins not polymerized that give high green astringency. The variation of %EA and %SM indexes are linked to the stage of ripeness, terroir and genetics (Peyron 1998, Saint-Cricq et al. 1998, Mateos et al. 2001).

28.4.3 Effect of the Terroir, Variety and Water Availability on Grape and Wine Phenol Content

Anthocyanin content is heavily influenced by the vintage, soil type, mesoclimate and cultural practices.

Figure 28.2 reports differences in maturity of Syrah grapes due to the terroir. Grapes grown in a stony soil formed by shale (Fig. 28.3) and in a dry

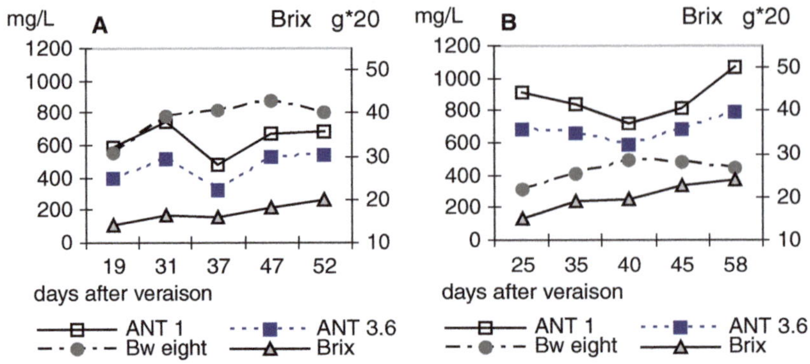

Fig. 28.2 Ripeness of Syrah in Tarragona AOC (**graph a**) and Priorat AOC (**graph b**), Spain. Vintage 2004. Weight of 20 berries (Bweight), sugar concentration (Brix°) and anthocyanin concentration (ANT) at pH 1.0 and pH 3.6 in mg·L^{-1}

summer region (Priorat) accumulate higher concentrations of sugars and anthocyanins (total and extractables) than in Tarragona, a region characterized by a deep loamy-clay soil and Mediterranean mesoclimate, with high humidity and mild temperatures in summer. Grenache and Syrah grapes in deep soils (quaternary soils in Tarragona) reached low values of anthocyanins. Merlot and Tempranillo grown in these soils (deep with enough water resources) display the higher SM index (Mateos et al. 2001).

In stony and dry soils (Priorat), Cabernet Sauvignon reached the highest phenolic potential, compared to Syrah and Grenache. Despite the fact that Grenache is poor in colour, high values were shown just in old grapevines. Small berry size, high

Fig. 28.3 Terrace in Priorat. Shallow soil of schists. Carignan

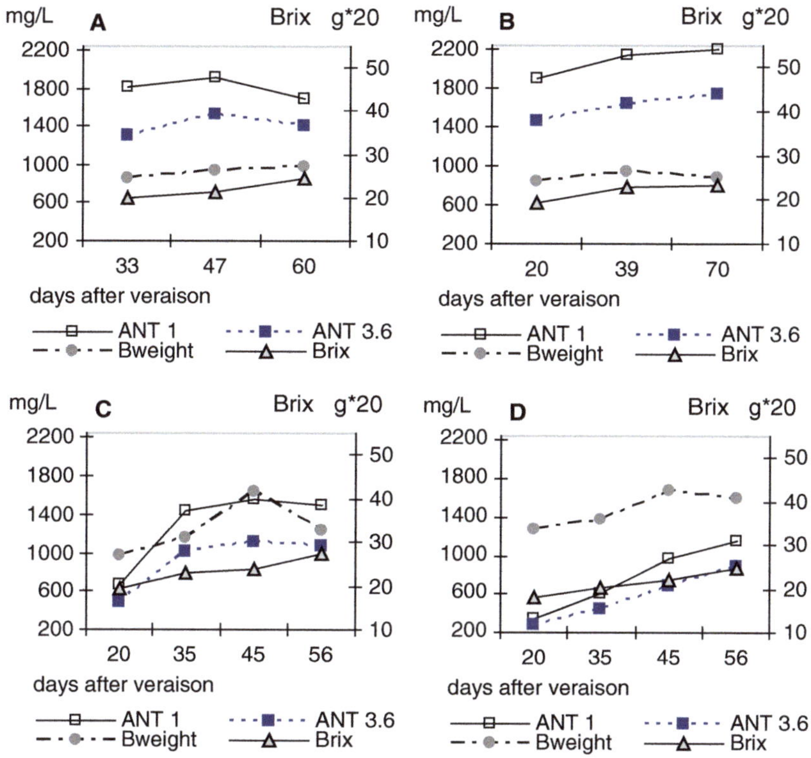

Fig. 28.4 Ripeness of Cabernet Sauvignon, parcel 1 (**graph a**) and parcel 2 (**graph b**). Grenache 60 years old (**graph c**) and Grenache 15 years old (**graph d**). Vintage 2004. Weight of 20 berries (Bweight), sugar concentration (Brix°) and anthocyanin content (ANT) at pH 1.0 and pH 3.6 in mg·L^{-1}. Priorat AOC, Spain

sugars content, and high levels of anthocyanins were associated to healthy berries in balanced vines (Nadal 2002). Only Cabernet Sauvignon reached almost 2,000 mg of anthocyanins, but, the difference between total and extractable anthocynins is generally higher than in the other varieties (Fig. 28.4).

Topography and mesoclimate effects were evaluated on Grenache cultivar in Priorat region (Fig. 28.5) comparing early and late mesoclimate; two villages in two different mesoclimates, early and late ripening zones (temperate because of the sea wind influence), and, two different terraces in each, topographically located up and down the slope. Data were obtained in two vintages; 2002 (temperate) and 2003 (warm).

The results revealed higher concentration of anthocyanins, lower yield and lower berry weight associated to the terraces located in the upper side of the hill in temperate vintage. No differences were found in warm vintage for anthocyanin content. The warm vintage in 2003 affected grape composition by decreasing anthocyanins and yields. Extractable anthocyanins in berries increased during ripening in earlier

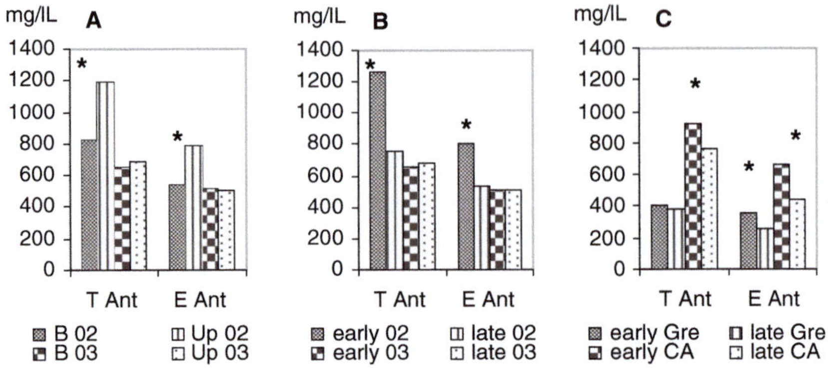

Fig. 28.5 Results are the averages at harvest time: anthocyanin content (mg·L^{-1}) at pH 1 (T Ant) and pH 3.6 (E Ant). **a** and **b** graphs display, respectively, the effect of topography and mesoclimate in Priorat AOC on Grenache (vintages 2002 and 2003). The terraces are situated up and bottom (b), in early and late zones. Mesoclimate effect on Terra Alta AOC is graphed in c. Data are the average of Grenache (Gre) and Carignan (CA) in 2007 and 2008. Both AOC located in Tarragona, Spain: Priorat, inside land and shallow soil; Terra Alta (TA) inside land and deep soil. ANOVA and test de Fisher: (∗) significant differences at $p \leq 0.05$

regions and cooler vintages; nevertheless, the accumulation stopped or decreased a week before harvest in those early regions (Nadal et al. 2008).

Another study was performed during 2007 and 2008 vintages in Terra Alta (Tarragona), a Mediterranean warm region characterized by loam and deep soils (Fig. 28.6).

The trial establishes two parcels (early and late mesoclimates) within two varieties, Grenache and Carignan. Deep soils have higher berry yield and less

Fig. 28.6 Deep loamy soil in Terra Alta. Grenache

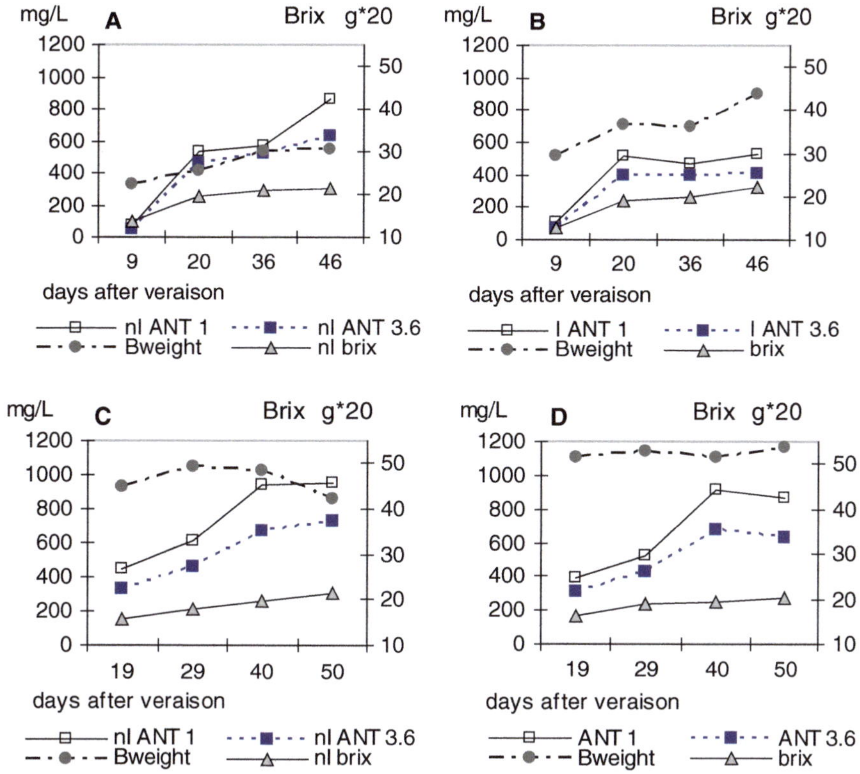

Fig. 28.7 Ripeness of Tempranillo. **a** = non irrigated (2003); **b** = irrigated (2003); **c** = non irrigated (2004); **d** = irrigated (2004). Weight of 20 berries (Bweight), sugar concentration (Brix°) and anthocyanin content (mg·L^{-1}) at pH 1.0 (Ant T) and pH 3.6 (Ant E). Irrigation trial; irrigated (I) and non irrigated (nI) plots in sandy soil. Location: Montsant AOC, in Tarragona, Spain

extractable anthocyanins than vines growing in dry conditions (Edo et al. 2009). Both, Grenache and Carignan accumulate higher extractable anthocyanin content in early regions.

Cultural practices such as irrigation generally decrease the anthocyanin content. Tempranillo vines growing in granitic soil and dry conditions (non irrigated: nI) showed higher values of berry anthocyanins than irrigated (I) trial grapes. Comparing vintages (2003 and 2004), the severity of dry conditions in 2003 translated into a depletion of the plant and a consistent decrease of anthocyanins (Fig. 28.7). The results of wine composition presented significant differences between vines (I) and (nI) except for total acidity (Nadal and Lampreave 2007). Irrigation decreases the anthocyanin content in wine in both vintages. The wines produced in the warmest vintage showed very low anthocyanin content when comparing with the grape anthocyanin concentration.

Comparing grapes and wines: Anthocyanin content in Cabernet Sauvignon is quite high in berries, but the wines obtained are characterized by lower anthocyanin

Table 28.1 Anthocyanins and total phenols on berry and wine Data from Priorat AOC (Spain), stony and dry soil. T ANT = total anthocyanins (pH1), E ANT = extractable anthocyanins (pH3.6), Tan = total tannins, b = berry, w = wine, A280 = total phenol content index (IPT). Cabernet S = Cabernet sauvignon. Irrigation trial in sandy soil: Temp = Tempranillo, I = irrigated, nI = non irrigated. Numbers 1, 2, 3 corresponds to different vineyard plots. α′= *anthocyanins wine/extractable anthocyanins berry*, β = *Absorbance280 wine/Absorbance280 berry* (Unpublished data by Nadal)

Vintage 2004	T ANT berry	E ANT berry	A280 b	A280 w	ANT w	Tan w	α′	β
Cabernet S 1	1,696	1,417	85	64	357	3.5	25	76
Cabernet S 2	2,251	1,777	85	74	425	3.1	24	87
Cabernet S 3	2,214	1,754	90	69	417	3.7	24	77
Carignan 1	1,938	1,208	78	58	589	2.4	49	74
Carignan 2	1,228	747	60	58	442	2.9	59	96
Grenache 1	1,495	1,071	67	58	282	3.2	26	87
Grenache 2	866	723	46	46	266	2.3	37	100
Syrah	1,066	798	73	61	457	2.4	57	84
Temp I	873	630	79	42	594	2.3	94	53
Temp nI	955	833	87	47	792	2.1	95	54
2003	–	–	–	–	–	–	–	–
Temp I	522	412	72	41	261	2.3	63	57
Temp nI	858	632	84	47	356	2.4	56	56

concentrations, and higher IPT (Absorbance 280) and tannin content (the alcoholic degree of the wine is between 13.7 and 15.5%vol). The anthocyanin tendency to be extracted is low, however the tannin extraction is high (Table 28.1).

Tempranillo shows similar anthocyanin concentration in both, grape and wine (except for dry and warm vintages as 2003); therefore, it is associated to high extractability (Mateos et al. 2001). However, the values of total phenols in wine are much lower than the concentration found on grapes (Table 28.1).

Taking into account the anthocyanin content extracted from grapes and released into the wine and calculating the anthocyanin extraction coefficient in wine: $\alpha=$ *anthocyanins wine/skin anthocyanins* (expressed in mg. L^{-1}), Glories found percentages between 57 and 85% in Merlot and Cabernet Sauvignon (Ribéreau-Gayon et al. 2000).

In our studies (Table 28.1), if another ratio α' = *anthocyanins in wine/extractable anthocyanins (pH3.6) in berries* (expressed as mg. L^{-1}) is considered, percentages of extraction varies widely, between 24 and 95%. Anthocyanin coefficients are different depending on the berry maturity and environment conditions. Less variability was found taking into consideration the total phenol content (Absorbance at 280 nm) and calculating the coefficient β as a ratio of (*Absorbance280 wine/Absorbance280 berry*), representing the phenolic extraction into the wine.

Most values of coefficient β vary between 75 and 85% for Cabernet sauvignon, syrah and Grenache, but the percentage of extraction decrease at 50–60% for Tempranillo. Values of Grenache and Carignan from Terra Alta (inland and deep soils) reached between 80 and 90%, being higher in Grenache.

In order to predict wine colour attributes by a multivariate approach, Jensen et al. (2008) found low ratios wine/grape for anthocyanins but, high ratio for gallic acid.

28.4.4 Stage of Ripeness Influence on Phenol Composition

Along the maturity the tannins become softer because of the polymerization process. Unripe grapes have skins with low concentrations of anthocyanins and relatively simple tannins that have not lost their reactivity, and seeds with a high content of little polymerized and therefore highly reactive tannins. Unhealthy berries have high extractability, but also release few colour into the wine. If the skins are damaged, enzymes increase the extractibility.

Phenolic extraction on wine depends on genetics (variety), climate, soil, the process of synthesis and accumulation in berries during maturity and the final concentration are achieved at harvest. During fermentation, extraction occurs fastest with flavonols and slower with flavan-3-ols catechins and procyanidins. At the end of fermentation, the content falls as phenols bind and precipitate with proteins and cell remnants: about 25% of the anthocyanins may have polymerized with tannins. This level may rise to 40% or more within 1 year (Singleton and Trousdale 1992, Boulton et al. 1996).

The anthocyanins in ripe or just slightly overripe berries are easily extracted and as a consequence, the anthocyanin content is higher in the wine than it would have been prior to maturity. At phenolic maturity, grapes have both a high pigmentation potential and a good capacity for releasing these substances into wine. In overripe berries, Total phenol index and colour decrease in both, berry and wine values (Fig. 28.8).

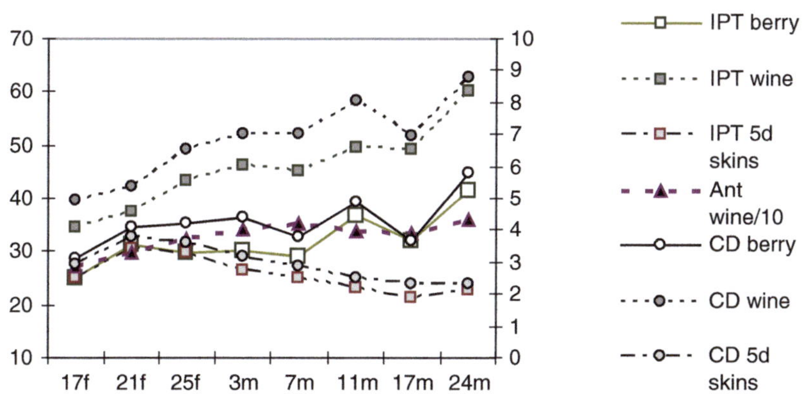

Fig. 28.8 Evolution of total phenol content index IPT (absorbance at 280), CD= Colour density (absorbance at 420 +520) in berries, wine and skins after fermentation. Ant wine/10= Anthocyanin content in wine. Grapes and wines analyzed at different stages of ripeness from the 17th of February (f) to 24th of March (m). Length of alcoholic fermentation is five days (5d). From Syrah vineyard in Stellenbosch, (Institute Nietvoorbij)

The dehydration continues at late harvest allowing a concentration of solutes, thus increasing the phenol content at harvest. Last two harvests of this experiment presented certain amount of shrivelled berries and high alcoholic degree in the wine. After tasting wines, best products came from the 7th and 11th of March, classified as more balanced and complex. Overripe berries with symptoms of shrinkage did not accumulate more phenols and, in addition, a reduction of phenols released into the wine is observed. In overripe grapes, phenol extraction from the skins is limited, despite an increase of values in the wine. However, more extraction of flavan-3-ols from the seeds may occur in such grapes, explaining the values obtained in wine (Nadal et al. 2004).

28.4.5 Concluding Remarks

The variability of the phenol concentration of grapes depends on the variety, terroir and cultural practices; also we should consider the effect of vintage. All these factors may affect the phenol extraction into the wines.

Many studies are carried out in order to know the phenolic content in berries; skins, pulp and seeds as an indicator of the optimal ripeness. Nowadays, the research is mainly based on a better understanding of the phenol extraction into the wine.

Analysing the phenolic maturity on the whole berry, the total phenol content on grapes – measured as the absorbance at 280 nm, – gives a better approach than anthocyanins for predicting the phenol extraction and content in wines.

Acknowledgments The author acknowledges the partial financial support of Spanish Ministry of Science and Education through the projects AGL2005-06927-CO2-02 and AGL2008-04525-CO2-02 and to the CDTI project with UNIÓ Cooperatives Company in Tarragona, Spain. We thank the scientific collaboration of Felicidad de Herralde and Josep Valls. We thank the collaboration of Miriam Lampreave and Sumpta Mateos for laboratory work.

References

Bisson L (2001) In search of optimal grape maturity, Dept of Enology and Viticulture, UC Davis, July/Ag, 32–43

Boulton R, Singleton V, Bisson L, Kunkee R (1996) Principles and practices in wine making. The Chapman and Hall Enology Library

Cacho J, Fernández P, Ferreira V, Castells JE (1992) Evolution of five Anthocyanidin-3-glucosides in the skin of the Tempranillo, Moristel, and Garnacha grape varieties and influence of climatological variables. Am J Enol Vitic 43:244–248

Cagnasso E, Rolle L, Caudana A, Gervi V. (2008) Relationship between grape phenolic maturity and red wine phenolic composition. Ital J Food Sci 20:365–380

Calò A, Tomasi D, Crespan M, Costacurta A. (1996) Relationship between environmental factors and the dynamics of growth and composition of the grapevine. Proc. workshop strategies to optimize wine grape quality. Acta Hort 427:217–231

Celotti E, and Carcereri G. (2000) La qualità fenolica delle uve rosse: valutazzione oggettiva mediante misura del colore. Ind Bev 29:168–138

Celotti E, Vedova T. della, Ferrarini R, Martinand S. (2007) The use of reflectance for monitoring phenolic maturity curves in red grapes. Ital J Food Sci 19:91–100

Choné X, Van Leeuwen C, Chéry P, and Ribereau-Gayon P. (2001) Terroir influence on water status and nitrogen status of non-irrigated Cabernet sauvignon (*Vitis vinifera*): Vegetative development, must and wine composition (example of Medoc top Estate vineyard, Saint Julien Area, Bordeaux, 1997. S Afr J Enol Vitic 22:8–15

Dambergs RG, Cozzolino D, Esler MB, Cynkar WU, Kambouris A, Francis IL, Hoi PB, Ginshen M. (2003) The use of infrared spectroscopy for grape quality measurement. Aust NZ Grapegrowers Winemaker 473:76–69

De Freitas VP, Glories Y, Bourgeois G, Vitry C. (1998) Characterization of oligomeric and polymeric procyanidins from grape seeds by liquid secondary ion mass spectometry. Phytochemistry 49:1435–1441

Di Stefano R, Cravero MG, Gentilini N (1991) Metodi per lo studio dei polifenoli dell'uva, Riv Vitic Enol 44:37

Di Stefano R, Borsa D, Bosso A. (2000) Documento tecnico sul significato e sui metodi di determinazione dello stato di maturità dei polifenoli. L'enologo 12:73–76

Downey MO, Dokoozlian NK, Krstic MP (2006) Cultural practices and environmental impacts on the flavonoid composition of grapes and wine: a review of recent research. Am J Enol Vitic 57:257–268

Edo M, Lampreave M, Nadal M (2009) Effects of mesoclimate and vintage on grape composition in a warm summer region, GESCO 12–16 July, Davis CA, 2009

Glories Y (1984) La couleur des vins rouges II. Mesure, origine et interpretation. Conn Vigne Vin 18:253–271

Glories Y, Augustin M (1993) Maturité phénolique du raisin, consequences technologiques: application aux millesimes 1991 et 1992. Actes du colloque: Journée technique du CIVB. Bordeaux, 21 janvier, 56–61

González-San Jose ML, Santa-Maria G, Diez C (1990) Anthocyanins as parameters for differentiating wines by grape variety, wine-growing region, and wine-making methods. J Food Composit Anal 3:54–66

González-San José ML, Barrón LJ, Junquera B, Robredo LM (1991) Application of principal component analysis to ripening indices for wine grapes. J. Food Composit Anal 4: 245–255

Harbertson F, Adams JA (2002) Tannins in skins and seeds of Cabernet sauvignon, Syrah and Pinot noir berries during ripening. Am J Enol Vitic 53:54–59

Hunter JJ, Ruffner HP, Volsench C (1995) The effect of partial defoliation on growth characteristics of Vitis vinifera cv Cabernet Sauvignon II. Reproductive growth in grapevine. S Afr J Enol Vitic 121:869–874

Hunter JJ, Pisciotta, A, Volschenk, CG, Archer E, Novello V, Kraeva E, Deloire A, Nadal M. (2004) Role of harvesting time/optimal ripeness in zone/terroir expression. *Proc. Joint OIV, GESCO, SASEV Intl Conference on Viticultural Zoning*, 15–19 November 2004, Cape Town, South Africa, 466–478

Iland PG, Cynar W, Francis LI, Williams PJ, Coombe BG (1996) Optimisation of methods for determination of total and red free glycosyl- glucose in black grape berries of *Vitis vinifera*. Aust J Grape Wine R 2:171–178

Iland PG, Ewart A, Sitters J, Markides A, Bruer N (2000) Techniques for chemical analysis and quality monitoring during winemaking, Patrick Iland Wine Promotions, Campbelltown, SA

Jackson JI, Lombard PB (1993) Environmental and management practices affecting grape composition and wine quality – A review. Am J Enol Vitic 44:409–429

Jensen JS, Blachez B, Egebo M, Meyer AS (2007) Rapid extraction of polyphenols from red grapes. Am J Enol Vitic 58:451–461

Jensen JS, Damiray S, Egebo M, Meyer AS (2008) Prediction of wine colour attributes from the phenolic profiles of red grapes (*Vitis vinifera*). J Agric Food Chem 56:1105–1115

Jordao A. M, Ricardo da Silva JM, and Laureano O. Evolution of anthocyanins during grape maturation of two varieties (*Vitis vinifera* L.). (1998) Castelao Frances and Touriga Francesa. Vitis 37:93–94

Kennedy JA, Hayasaka Y, Vidal S, Waters EJ, Jones GP (2001) Composition of grape skin tannins at different stages of berry development. J Agric Food Chem 49:5348–5355

Kennedy JA, Matthews MA, Waterhouse AL (2000) Changes in grape seeds polyphenols during fruit ripening. Phytochemistry 55:77–85.

Kramling TE, Singleton VL (1969) An estimate of the nonflavonoid phenols in wines. Am J Enol Vitic 20:86–92

La Notte E, Liuzzi VA, Esti M. (1992) The polyphenolic wine components 2. Note: Anthocyanins content and composition in relation to different winemaking techniques. Vignevini 19:49–55

Lamadon F (1995) Protocole pour l'évaluation de la richesse phénolique des raisins. Revue des Enologues 76:37–38

Mateos S (2003) Determinació de la maduresa fenòlica en varietats de raïm negres, DEA, Universitat Rovira i Virgili

Mateos S, Valls J, Nadal M, Arola L (2001) Estudio de la madurez fenólica sobre diferentes variedades tintas y su relación con el color de los vinos. Tecnología del vino 2:45–50

Mattivi F, Prast A, Nicolini G, Valentine L (2003) Ilpotenziale polyfenolico delle uve rosse e la sua aplicazione in enologia 39:105

Nadal, M. (2002) Els vins del Priorat. Ed. Cossetània, Tarragona, Spain

Nadal M, Lampreave M (2007) Fundamentos, aplicación y consecuencias del riego en la vid. Capítulo 5.2. – Influencia del riego en la maduración polifenólica de las bayas. In: Montsant DO (ed) Experiencia del riego en la comarca del Priorat, Agrícola Española, 231–256

Nadal M, Volschenk N, Hunter JJ (2004) Phenolic extraction during fermentation as affected by ripeness level of Syrah/R99 grapes. *Proc. Joint OIV, GESCO, SASEV Intl Conference on Viticultural Zoning*, 15–19 November 2004, Cape Town, South Africa

Nadal M, Hunter JJ (2007) Different wine styles as related to ripeness level of Syrah/R99 grapes, *Intervitis-Interfructa, 59th German Viticulture Congress*, 22–26 April 2007, Stuttgart, Germany

Nadal M, Mateos S, Lampreave M. (2008) Influence de la topographie et du mésoclimat sur la composition des raisins et rendement dans le terroir de l'AOC Priorat, VII Congres Int. terroirs viticoles 19–23 mai 2008 Nyon, Suïssa, 2:590–595

Nagel CW, Glories Y (1991) Use of a modified Dimethylaminocinnamaldehyde reagent for analysis of flavanols. Am J Enol Vitic 42:364–366

Ojeda H, Andary C, Kraeva E, Carbonneau A, Deloire A (2002) Influence of pre- and post-veraison on water deficit, synthesis and concentration of skin phenolic compounds during berry growth of *Vitis vinifera* cv. Shiraz. Am J Enol Vitic 53:261–267

Peyron D. (1998) Le potentiel phénolique du pinot noir. Rev Fr œnol 170 :42–45

Ribéreau-Gayon P (1970) Le dosage des composes phénoliques totaux dans les vins rouges. Chemist-Analyst 52:627–631

Ribéreau-Gayon P, Stonestreet E (1965) Le dosage des anthocyanes dans les vins rouges. Bull Soc Chim 9 :2649–2651

Ribéreau-Gayon P, Stonestreet E (1966) Dosage des tannins dans le vin rouge et détermination de leur structure. Chim Anal 48:188–196

Ribéreau-Gayon P, Glories Y, Maujean, A, Dubourdieu D. (2000) Handbook of enology, vol 2: The chemistry of wine and stabilization and treatments. Wiley, New York

Roggero JP, Archier P (1989) Mise au point d'une méthode de dosage des phénols simples des vins. Application à des vins d'origines et d'âges différents. Conn Vigne Vin 23:25–27

Roggero JP, Coen S, Ragonnet B (1986) High peformance liquid chromatography survey on changes in pigment content in ripening grapes of syrah. An approach to anthocyanin metabolism. Am J Enol Vitic 36:77–83

Roson JP, Moutounet M (1992) Quantités d'anthocyanes et de tanins des raisins de quelques cepages du sud-ouest en 1988 et 1989. Rev Fr Oenol 32:17–27

Roubelakis-Angelakis KA, Kliewer WM (1986) Effect of exogenous factors on phenylalanine ammonia lyase activity and accumulation of anthocyanins and total phenolics in grape berries. Am J Enol Vitic 53:261–267

Saint-Cricq N, Vivas N, Glories Y (1998) Maturité phénolique: définition et côntrole. Rev Fr Oenol 173:22–25

Sarneckis CJ, Dambergs P, Jones M, Mercurio M, Herderich MJ, Smith PA (2006) Quantification of condensed tannis by precipitation with methyl cellulose; development and optimised tool for grape and wine analysis, Aust J Grape Wine Res 12:39–49.

Sethuramasamyraja B, Sachidhananthan S, Yen M, Wample R (2007) Interpolation of wine grape quality indicators (anthocyanin and Brix) and development of differential harvest attachment. Am Soc Agric Bio Eng Michigan, ASAE Annual Meeting

Singleton VL, Trousdale EK. (1992) Anthocyanin-Tannin interactions explaining differences in polymeric phenols between white and red wines. Am J Enol Vitic 43:63–70

Somers TC, Evans ME. (1977) Spectral avaluation of young wines anthocyanin equilibria, toal phenolics, free and molecualr SO_2, chemical age. J Sci F Agric 28:68–74

Valls J (2004) Composició fenòlica en varietats negres de *Vitis vinifera*. Estudi de la influencia de diferents factors. PhD. University Rovira i Virgili, Tarragona, Spain

van Leeuwen C, Friant P, Choné X, Tregoat O, Koundouras S, Dubourdieu D (2004) Influence of climate, soil, and cultivar on terroir. Am J Enol Vitic 55:207–217

Venencie C, Uveira MN, Guiet S (1997) Maturité polyphénolique du raisin mise en place d'une méthode d'analyse de routine. Rev Fr. Oenol 167:36–41

Vivas N, Glories Y, Lagune L, Saucier C, Agustin M (1994) Estimation of the polymerisation level of procyanidins from grapes and wines by use of p-dimethylaminocinnamaldehyde. J Int Sci Vigne Vin 28:319–336

Vivas de Gaulejac N, Nonier M.F, Guerra C, Vivas N (2001) Anthocyanin in grape skins during maturation of *Vitis vinifera* L. cv. Cabernet Sauvignon and Merlot noir from Bordeaux terroirs. J Int Sci Vigne Vin 35:149–156

Chapter 29
Aromas in Grape and Wine

Denis Rusjan

Abstract Aromatic compounds are an important quality parameter of grape and wine, which has gained further importance with consumer demand. The extraction methods and analytical techniques used should be quick, inexpensive, with high reproducibility and sensibility, and require a low volume of sample and automated for the major part. A procedure of solid-phase microextraction (DVB/CAR/PDMS fibre) for few aroma compounds from grape, musts and wines has been developed and the analyses were performed by Gas chromatography-Mass spectrometry (GC-MS). Different hydrolytic approaches were also tested; the most successful were enzymatic hydrolysis (with two different enzymes) and acidic hydrolysis at pH 3.00. Non-hydrolysed terpene glycosides were extracted from the musts using solid-phase extraction. The extract was analysed with Liquid chromatography-Mass spectrometry (LC-MS) and some compounds were tentatively identified as terpene glycosides. The selection of the methods and techniques of aroma determination in grape or wine should be made according to laboratory capacity, equipment, time, number of samples, budget and the group of aroma compounds to evaluate.

Contents

D. Rusjan (✉)
Biotechnical Faculty, University of Ljubljana, Ljubljana 1001, Slovenia
e-mail: denis.rusjan@bf.uni-lj.si

S. Delrot et al. (eds.), *Methodologies and Results in Grapevine Research*,
DOI 10.1007/978-90-481-9283-0_29,

Abbreviations

GC	Gas chromatography
GC-MS	Gas chromatography-Mass spectrometry
HS-SPME-GC-MS	Head space-Solid Phase Microextraction-Gas chromatography-Mass Spectrometry
LC-MS	Liquid chromatography-Mass spectrometry
LLE	Liquid-liquid extraction
LLME	Liquid-liquid microextraction
MWE	Microwave extraction
SBSE	Stir bar sorptive extraction
SDSE	Simultaneous distillation-Solvent extraction
SPE	Solid-phase extraction
SPME	Solid-phase microextraction
SCFE	Supercritical fluid extraction
USE	Ultrasonic extraction

29.1 Introduction

Grape and wine have followed the main civilisations in the last millenniums and thus have to be considered as one of the oldest products of mankind. Broad interest in grape cultivation and wine production may be linked to the adaptation of grapevine to different climates and soils, but also to the diverse pleasures offered by wine to consumers belonging to many cultures worldwide. The popularity of wine may be explained by appropriate marketing campaigns, the abundance of wine magazines and journals, but also with the positive influence on human health. Grapes but especially wines are the most complex fruits and beverages, where respectively thousands of different substances (phenols, aromatic compounds, acids, sugars etc.) result in a diversity of sensorial variations. A wide number of chemical substances is synthesised in grape, as many different grapevine varieties are grown in different *terroirs*. Furthermore during processing, musts can be affected by several species of microorganisms (yeast, bacteria, fungus etc.), different vinification techniques and storage methods (type of vessels, corking, blending, aging period etc.).

The styles of wine may change according to the demands of the consumers, and this may induce plantation of new varieties, recent viticulture managements and new accessions to oenological practices. Additionally climate changes – global warming, unexpected and frequent storms, flooding, appearance of longer periods of drought or snow – are pushing winegrowers to search new solutions to overcome these factors; as for instance, replacement of varieties and rootstocks with recent combinations that are more adapted to the new conditions in *terroirs*, new viticulture practices to either decrease unsuitable weather conditions or better exploit suitable conditions to limit abiotic and biotic stress. Knowledge on the response of chemical compounds to various levels of stresses has improved in last decades. Stress may be accompanied by sensory deviations in grape and wine quality (taste, odour etc.).

The aroma compounds have become more and more important in winemaking, as they contribute to a cocktail of chemical compounds which give an aroma, taste, odour or bouquet to the grape, must and wine. The contribution depends on the grape variety which probably stands for the greater part of the odoriferous compounds specific for a certain wine (Sánchez-Palomo et al. 2005, Prosen et al. 2007a), but this depends also on the environment (climate, soil, location) (Smart and Robinson 1991, Miklósy and Kerény 2004), winemaking (fermentation conditions and aging) (Spillman et al. 2004, Francis and Newton 2005, Swiegers et al. 2005, Comuzzo et al. 2006, Piñeiro et al. 2006, Chalier et al. 2007, Košmerl et al. 2008) as well as vine growing practices (Bureau et al. 2000, Darriet et al. 2001, Sala et al. 2004, Peyrot des Gachons et al. 2005, Sánchez-Palomo et al. 2007). It is well known that apart from free aroma compounds, a significant part of flavour compounds is accumulated in the grape berries as odourless non-volatile glycosides (Günata et al. 1985, Sánchez-Palomo et al. 2005). The precursors of important primary volatile aroma compounds of monoterpenes (monoterpenoids; Luan et al. 2005; Fig. 29.1) and C_{13}-norisoprenoids (as carotenoids; Baumes et al. 2002) are synthesized in the earlier phase of berry development. The complexity of the biochemical and chemical pathway of aroma precursors and their final products becomes clearer when we take into account that more than 800 different aromas have already been determined in wine (Ribéreau-Gayon et al. 2000) (Table 29.1), but only 30–40 proven to be odour active (Baumes et al. 1986, Rapp and Mandery 1986). These compounds are present as free volatiles, which directly affect odour and non-volatile sugar bound glycosidic conjugates. The main groups of aroma compounds are aglycone moieties of glycosides including terpenes, straight-chain alcohols, benzene derivates, C13 norisoprenoids, esters and volatile phenols, where the sugar part is represented by glucose or disaccharides (Günata et al. 1985, Sánchez-Palomo et al. 2005, Prosen et al. 2007a). Some components are present in high concentrations (hundreds of mg L^{-1}), whereas most others at a rather low $\mu g\ L^{-1}$ level. The enhancement of volatile aroma compounds in wine can be released with generally enzymatic hydrolysis (Schneider et al. 2001, Castro-Vázquez et al. 2002, Sánchez-Palomo et al. 2005,

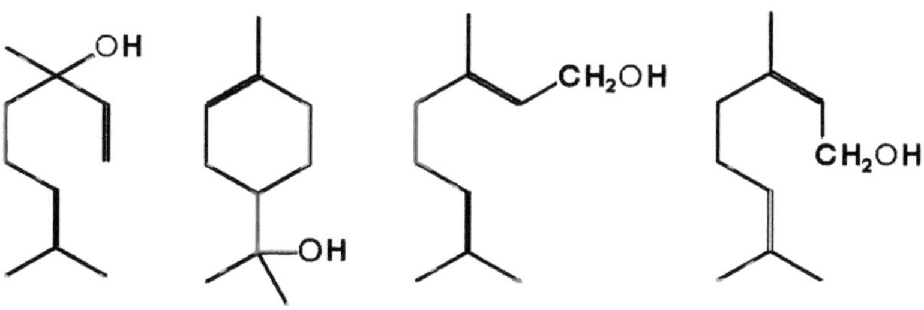

Fig. 29.1 The most frequent primary aroma compounds in grape and wine

Prosen et al. 2007a). Because of the different concentrations and sources in plant tissues, some compounds need to be extracted and concentrated before analysis, whereas others can be analysed with direct injection in GC. Grape, must or wine, contain a cocktail of major compounds, which behave as a complex matrix causing frequent interferences in the analysis of trace compounds. The quantification of compounds present in micro concentrations, as for example terpenoids, is rather difficult in grape as well as wine (Ortega-Heras et al. 2002, Sousa-Câmara et al. 2007).

29.2 Analytical Methods

The first analyses of aroma compounds have started in the middle of the twentieth century, since chemical compounds of high concentrations were detected and quantified not only in wines but also in other foods (milk, meat, fruits etc.). The extraction methods and analytical techniques have made considerable progress since then, especially in the last decades, as different analytical methods have been studied and described, as for instance the liquid-liquid extraction ((LLE); Hardy 1969, Ferreira et al. 1993, Wada and Shibamoto 1997, Prosen et al. 2007b), liquid-liquid microextraction ((LLME); Ferreira et al. 1993, Ortega et al. 2001), solid-phase extraction ((SPE); Günata et al. 1985, López et al. 2002), microwave extraction ((MWE); Bureau et al. 2006), stir bar sorptive extraction ((SBSE); Zalacain et al. 2007, Caven-Quantrill and Buglass 2006), simultaneous distillation-solvent extraction ((SDSE); Nunez et al. 1984); supercritical fluid extraction ((SCFE); Blanch et al. 1995) and ultrasonic extraction ((USE); Cocito et al. 1995). The latter four analytical methods listed are less known and seldom used; reasons for this may be linked to a relatively low reproducibility, high risks of solvent contamination, inadequate selectivity and length of analysis. On the other hand, some positive characteristics can be ascribed to SPE and LLME methods, especially low costs and rapidity, but loss of some volatile compounds during the evaporations must be expected because of further steps of sample preparations. LLME is an extraction method based on the contact between the liquid sample and an organic solvent. The principle of the extraction is the solubility of the volatile compounds in a solvent whose density differs from that of water and its immiscibility with water. The method is quite difficult, labour-intensive and time-consuming. Losses of volatile compounds can be observed during sample preparation procedure and multistage operations. The SPE technique is based on an adsorption of volatile compounds on a stationary phase and their subsequent separation by elution with organic solvents. Also this method requires a multi-step process, especially where the use of packed silica C18 and Amberlite XAD-2 is frequent (Engel and Tressl 1983). SDSE is a distillation system with a continuous extraction, which is known as the most appropriate isolation and recovery method for a wide range of compounds (Chaintreau 2001).

Table 29.1 The list, identification technique, odour descriptor, threshold and media of odour active compounds from recent studies

Volatile compounds	Chemical compounds	Identification (method)	Technique	Odour descriptor	Odour threshold (mg L^{-1})	Material	References
Monoterpenes	Citronellol	D, G	SDSE	Citrusy, citron	0.1[a]	RW	Selli et al. (2004)
	α-Terpineol	D	SPME	Lily		WW	Prosen et al., (2007)
	Linalool	D	LLE	Flower, lavender	0.025[d]	RW, WW	Ferreira et al. (2000)
	Geraniol	D	LLME	Rose, geranium	0.030[e]	RW, WW	Escudero et al. (2004)
	cis-Rose oxide	D	LLME	Green, flower, lychee, rose	0.002[a]	WW	Guth (1997)
Alcohols	2-Heptanol	A, B, C	LLE	Earthy, oily		RW	Rocha et al. (2004)
	1-Hexanol	A, B, C, D, G	LLE, SDSE	Herbaceous, resin, flower green (cut grass)	4.80[c] 8.00[a]	RW	Rocha et al. (2004) Selli et al. (2004)
	2-Nonanol	A, B, C	LLE			RW	Rocha et al. (2004)
	3-methylthio-1-propanol	A, B, C, D, G	LLE	Sweet, potato, fruity, alcohol	1.0[d]	WW, RW	Cullere et al. (2004) López et al. (2003) Fretz et al. (2005)
	Methionol	A, B, C	LLE	Sweet, potato	1.0[d]	WW, RW	Rocha et al. (2004) Escudero et al. (2004)
	Benzyl alcohol	A, B, C	LLE	Flowery-sweet		RW	Rocha et al. (2004)
	2-phenylethanol	A, B, C, D, G	LLE, SDSE	Flowery, rose, honey, lavender, spice, lilac	10.0[a,b,c,] 14.0[d]	WW, RW	López et al. (2003) Rocha et al. (2004) Selli et al. (2004) Escudero et al. (2004) Cullere et al. (2004)
	1-butanol	D, G	LLE	Grass		RW	Selli et al. (2004)

Table 29.1 (continued)

Volatile compounds	Chemical compounds	Identification (method)	Technique	Odour descriptor	Odour threshold (mg L^{-1})	Material	References
	3-mercaptohexanol	A, B, C, D, E	LLE	Passion fruit, cat urine, rhubarb, fruity	0.0006[n]	WW, RW	López et al. (2003) Murat et al. (2001) Fretz et al. (2005)
	Isoamyl alcohol	D, G	SDSE, LLME	Red fruit, framboesia, whisky, malt, burn	40.0[b] 30.0[a]	WW, RW	Selli et al. (2004) Guth (1997)
	(Z)-3-hexenol	D, G	LLE, SPME	Green (cut grass)	0.4[f]	WW, RW	Cullere et al. (2004)
	Isobutanol	D, G	LLME, SPME, SDSE	Wine, solvent, bitter	40.0[f]	WW, RW	Guth (1997) López et al. (2003)
	3-methyl butanol	A, B, C	LLE	Rancid almonds, pungent		WW	Fretz et al. (2005)
Acids	Acetic acid	A, B, C, D, G	LLE	Vinegar, sour, pungent	200.0[f]	WW, RW	Rocha et al. (2004) Cullere et al. (2004)
	Propanoic acid	A, B, C, D	LLE	Fruity, acid, rancid, soapy-sweet	8.1[i]	WW, RW	Rocha et al. (2004) Etiévant (1991)
	Iso – butyric acis	A, B, C, D	LLE	Fruity, cheesy, sweaty, rancid, butter	0.40[c] 2.30[d]	WW, RW	Rocha et al. (2004) López et al. (2003)
	Butyric acid	A, B, C, D, G	LLE	Fatty-rancid, cheesy, sweaty, rancid	0.40[c] 0.17[b]	WW, RW	Rocha et al. (2004) Escudero et al. (2004) Selli et al. (2004)
	3-methylbutanoic acid	B, C	LLE	Fatty-rancid, cheesy	0.25	RW	Rocha et al. (2004)

Table 29.1 (continued)

Volatile compounds	Chemical compounds	Identification (method)	Technique	Odour descriptor	Odour threshold (mg L^{-1})	Material	References
	Hexanoic acid	A, B, C, D, G	LLE, LLME, SPME	Rancid, grass, fruity, cheese, sweet	6.70[c] 0.42[d]	RW	Rocha et al. (2004) Curelle et al. (2004) Guth (1997) López et al. (2003)
	Octanoic acid	A, B, C, D, G	LLE	Fatty acid, dry, dairy, cheese, sweet	2.20[c] 0.50b	WW, RW	Rocha et al. (2004) Selli et al. (2004) Escudero et al. (2004) López et al. (2003)
	Decanoic acid	A, B, C, D, G	LLE	Fatty acid, dry, woody, fat	1.40[c] 0.10[d]	WW, RW	Rocha et al. (2004) Selli et al. (2004) Escudero et al. (2004) López et al. (2003)
	Phenylacetic acid	A, B, C	LLE	Honey, pungent, floral		RW	Rocha et al. (2004)
	Isovaleric acid	D, G	SPME, LLE	Sweat, acid, rancid	0.33[d]	WW, RW	Escudero et al. (2004) Cullere et al. (2004)
Esters	Ethyl 2-hydroxypropanoate	A, B, C	LLE			RW	Rocha et al. (2004)
Lactones							
	Ethyl octanoate	A, B, C, D, E, G	LLE, SPME	Ripe fruits, pear, sweety, floral, fruity, exotic and passion fruits	0.24[c] 0.002[a] 0.005[e]	WW, RW	Rocha et al. (2004) Selli et al. (2004) Escudero et al. (2004)

Table 29.1 (continued)

Volatile compounds	Chemical compounds	Identification (method)	Technique	Odour descriptor	Odour threshold (mg L^{-1})	Material	References
	Ethyl 3-hydroxybutyrate	B, C	LLE	Fruity		RW	Rocha et al. (2004)
	Ethyl hexanoate	A, B, C, D, E, G	LLE, LLME	Floral, fruity, apple peel, pear, ester	0.005[a] 0.014[d]	WW, RW	Selli et al. (2004) Guth (1997) Cullere et al. (2004) López et al. (2003) Fretz et al. (2005)
	Ethyl decanoate	A, B, C, D, E	LLE	Sweety, fruity, dry fruits, grape	1.10[c] 0.2[d]	WW, RW	Rocha et al. (2004) Selli et al. (2004)
	Diethyl succinate	A, B, C	LLE	Cheese, earthy, spicy		RW	Rocha et al. (2004)
	2-phenylethyl acetate	A, B, C	LLE	Fruity, floral, rose, honey	0.25[a]	RW	Rocha et al. (2004)
	Ethyl 4-hydroxybutanoate	B, C	LLE			RW	Rocha et al. (2004)
	Diethyl malate	A, B, C	LLE	Fruity		RW	Rocha et al. (2004)
	Diethyl 2-hydroxypentanodioate	B, C	LLE			RW	Rocha et al. (2004)
	Ethyl 2-hydroxy-3-phenylpropanoate	B, C	LLE			RW	Rocha et al. (2004)
	Ethyl vanillate	B, C	LLE	Vanilla, chocolate		RW	Rocha et al. (2004)
	Ethyl 2-methylbutyrate	A, B, D, E	LLE	Apple	0.018[d]	WW, RW	Selli et al. (2004) Escudero et al. (2004) López et al. (2003)
	Isoamyl acetate	A, B, C, D, G	LLE, SPME	Fruity, banana	0.030[a, b]	WW, RW	Selli et al. (2004) Cullere et al. (2004) López et al. (2003) Murat et al. (2001) Guth ans Sies (2002) Fretz et al. (2005)

Table 29.1 (continued)

Volatile compounds	Chemical compounds	Identification (method)	Technique	Odour descriptor	Odour threshold (mg L^{-1})	Material	References
	Ethyl phenyl acetate	D	LLE	Honey	0.25[a]	RW	Selli et al. (2004)
	Ethyl isobutyrate	A, D, E	LLME, LLE	Sweet, rubber	0.015[a]	WW, RW	Guth (1997) Escudero et al. (2004)
	Ethyl butyrate	A, B, C, D, E	LLME, LLE	Apple, sweet caramel	0.020[a]	WW, RW	Guth (1997) López et al. (2003) Fretz et al. (2005)
	3-mercaptohexyl acetate	A, B, D, E	LLE	Grapefruit, passionfruit, cat urin	0.00004[n]	WW, RW	López et al. (2003) Murat et al. (2001)
	γ-butyrolactone	A, B, C	LLE	Sweety, buttery		RW	Rocha et al. (2004)
	3-methyl-2(*5H*)-furanone	A, B, C	LLE	Caramel		RW	Rocha et al. (2004)
	Tetrahydro-2*H*-pyran-2-one	B, C	LLE	Caramel		RW	Rocha et al. (2004)
	Dihydro-4-methyl-2(3*H*)-furanone	B, C	LLE	Caramel		RW	Rocha et al. (2004)
	γ-nonalactone	A, B, C,	LLE, SPME	Coconut, fruity, almond-like	0.065[c] 0.025[j]	WW, RW	Rocha et al. (2004) Escudero et al. (2004)
	γ-decalactone	D, E	SPME	Peach, fat	0.088[i]	RW	Ferreira et al. (2004)
	γ-dodecalactone	A, D, E	SPME	Sweet, fruit, flower	0.007[j]	RW	Ferreira et al. (2004)
	pentolactone	B, C	LLE		1.60	RW	Rocha et al. (2004)
	5-oxo-γ-hexalactone	B, C	LLE	Alcoholic	1.60	RW	Rocha et al. (2004)
	4-ethoxycarbonyl-γ-butyrolactone	B, C	LLE	Red fruits, cherry	0.40	RW	Rocha et al. (2004)
	4-(1-hydroxyethyl)-γ-butyrolactone	B, C	LLE	Red fruits	1.60	RW	Rocha et al. (2004)
	4-mercaptomethyl pentan-2-one	A, D, E	LLME, LLE	Passion fruit, cat urine	0.0006[n]	WW	Guth (1997) López et al. (2003)

Table 29.1 (continued)

Volatile compounds	Chemical compounds	Identification (method)	Technique	Odour descriptor	Odour threshold (mg L^{-1})	Material	References
Amides	*N*-ethylacetamide	B, C	LLE			RW	Rocha et al. (2004)
	N-(3-methylbuthyl) acetamide	B, C	LLE			RW	Rocha et al. (2004)
	N-(2-phenylethyl) acetamide	B, C	LLE			RW	Rocha et al. (2004)
Phenols	2-methoxyhenol (guaiacol)	A, B, C, E, G	LLE	Smoky, burning, sweet, phenolic	0.075^{c} 0.095^{d}	RW	Rocha et al. (2004) Cullere et al. (2004)
	Phenol	A, B; C	LLE	Phenolic	25.0^{c}	RW	Rocha et al. (2004)
	2-methoxy-4-vinylphenol	A, B, C	LLE	Black pepper, species, clove-like	0.38^{c}	RW	Rocha et al. (2004)
	4-ethylguaiacol	A, B, D, E	LLE	Spice, clove	0.033^{d}	RW	Guth and Sies (2002)
	eugenol	A, B, E, G	LLE	Clove, honey	0.006^{d}	WW, RW	Cullere et al. (2004) Guth and Sies (2002)
	4-ethyl phenol	A, E, G	LLE	Phenol, spice	0.440^{l}	RW	Cullere et al. (2004)
Norisoprenoids	β-damascenone	A, B, E, G	LLE	Apple, rose, honey	0.00005^{a}	WW, RW	López et al. (2003) Guth and Sies (2002)
	β-ionone	A, B, C, E, G	LLE	Seaweed, violet, flower, raspberry	0.00009^{d}	WW; RW	Escudero et al. (2004) Murat et al. (2001) Fretz et al. (2005)
Others	3-hydroxy-2-butanone	A, B, C	LLE	Buttery		RW	Rocha et al. (2004)
	3-hydroxy-2-methyl-4H-pyran-4-one	B, C	LLE	Caramel		RW	Rocha et al. (2004)
	5-hydroxymethylfurfural	A, B, C	LLE	Sweet, caramel	1.60	RW	Rocha et al. (2004)
	Vanillin	D, E, G	LLE	Vanilla, chocolate	0.2^{1a}	RW	Selli et al. (2004)
	3-methyl-3-furan-thiol	A, B, E	LLE	meat	0.00005^{o}	WW	López et al. (2003)
	acetoin	A, B, E, G	LLE	Butter, cream	150^{d}	RW	Cullere et al. (2004)

Table 29.1 (continued)

Volatile compounds	Chemical compounds	Identification (method)	Technique	Odour descriptor	Odour threshold (mg L^{-1})	Material	References
	Phenyl-acetaldehyde	A, B, E, G	LLE	Hawthorne, honey, sweet	0.005[d]	RW	Cullere et al. (2004)
	3-isobutyl-2-methoxypyrazine	A, B, E, G	LLE	Earth, spice, green pepper	0.00002[i]	RW	
	acetaldehyde	A, B, E, G	LLE	Pungent, ether	0.5[d]	RW	Cullere et al. (2004)

A – mass spectrum and retention time consistent with those of an authentic standard; B – structural proposals are given on the basis of mass spectral data; C – mass spectrum consistent with spectra found in the literature; D – GC retention and MS data agreement with that of pure compound; E – GC retention and MS data in agreement with spectra found in the library; F – tentatively identified by MS matching with library spectra only; G – GC/O – gas chromatography/olfactometry; WW – white wine; RW – red wine.

[a]In water/ethanol (90/10, w/w) – white wine according to Guth (1997).
[b]Lopez et al. (1999).
[c]Rocha et al. (2004).
[d]Ferreira et al. (2000).
[e]Escudero et al. (2004).
[f]Cullere et al. (2004).
[g]Lopez et al. (2003).
[h]Selli et al. (2004)..
[i]Etiévan (1991).
[j]Ferreira et al. (2004).
[k]Guth and Sies (2002).
[l]Boidron et al. (1988).
[m]Murat et al. (2001).
[n]Tominaga et al. (1998).
[o]Ferreira et al. (2002).

29.2.1 Analyses of Aromatic Compounds in the Past

In the last decades, many studies of aroma compounds in wines, but also in fruits, milk, vinegars, juices etc. were conducted with a solid-phase microextraction (SPME) (Arthur and Pawliszyn 1990, Goncalves and Alpendurada 2002). This technique has many advantages, like high sensibility, reproducibility, low cost, short time of extraction, simplicity, low volume of samples required and possible automation. Other techniques frequently require solvent and also pre-treatment of samples. The SPME offers a wide range detection of wine aroma compounds grouped and known as monoterpenes, C_{13} norisoprenoids, esters, volatile (di-)sulphides, methoxypyrazines, lactones, but also organochlorine insecticides. The analysis can be performed on wines with different characteristics as colour (red, white), age (young, aging), sweetness (sweet, semi-sweet, dry etc.), style (sparkling, still), barrique, etc. The headspace solid-phase microextraction (HS-SPME) provides a new technique, where a fibre is placed into the headspace above an equilibrated sample (Figs. 29.2 and 29.3). The efficiency of this technique for the concentration of volatile aroma compounds from media depends on fibre composition, temperature of media and on the time of extraction (Prosen et al. 2007a, Solís-Solís et al. 2007).

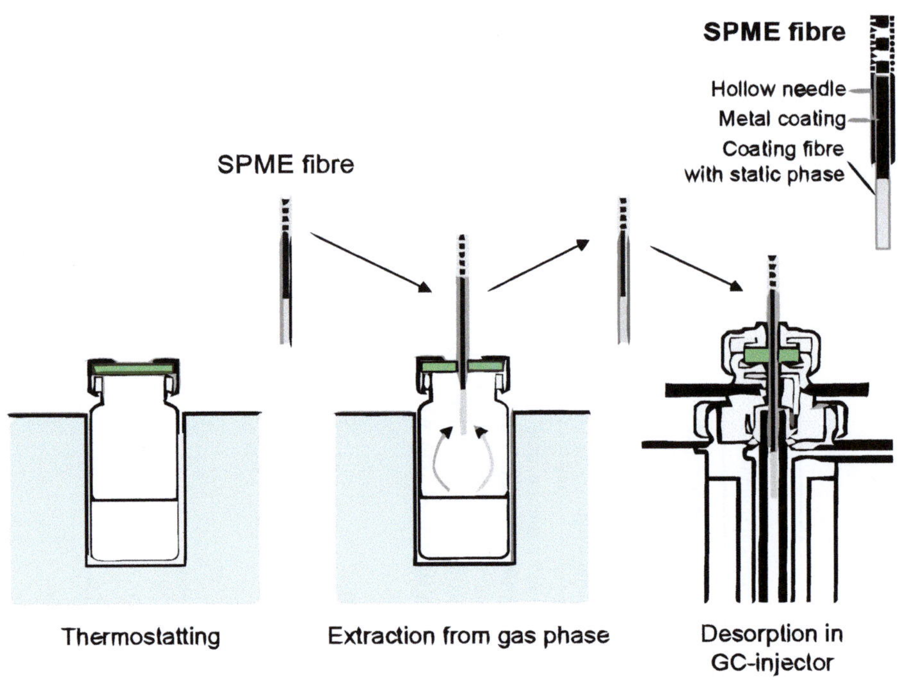

Fig. 29.2 Scheme of HS-SPME technique for volatile compounds adsorption (Auwärter 2006; http://edoc.hu-berlin.de)

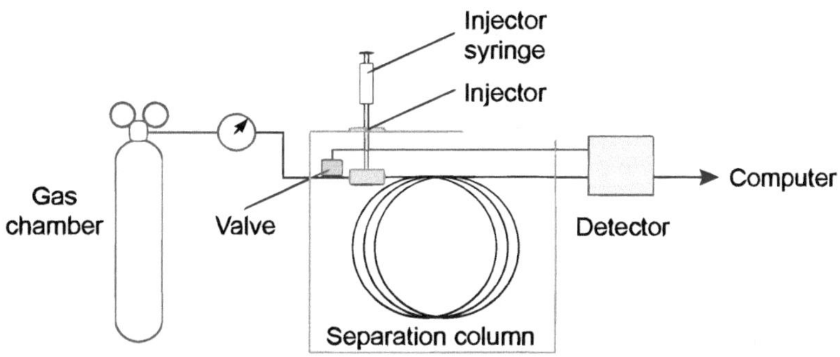

Fig. 29.3 Scheme of fibre application on GC-MS

29.3 Extraction Technique for Volatile Compounds in Grape and Wine

Extraction and analysis of volatile aroma compounds in grape and wines were made as described by Prosen et al. (2007a).

Grape is abundant with primary aromatic compounds, especially with monoterpenes which are mostly synthesised in berry skins (Bayonove 1993, Prosen et al. 2007a) (Fig. 29.1, Tables 29.2 and 29.3). According to the content of monoterpenes grape varieties can also be grouped in aromatic, less aromatic and non-aromatic varieties and according to the total of free monoterpenes as group of intensively muscats, non-muscats but aromatic and neutral varieties (Table 29.4).

Table 29.2 Concentration of some volatile aroma compounds (μg L^{-1}) in grape of some grapevine varieties *Vitis vinifera* L. (Prosen et al. 2007a)

Variety Compound	Muscat blanc	Pinot noir	Danijela	Aurora
Hexanal	34.40	43.90	32.90	25.90
(E)-2-hexen-1-al	7.80	9.00	11.50	6.90
Benzaldehyde	0.04	0.02	0.03	n.d.
Linalool	0.15	n.d.	n.d.	0.45
α-terpineol	n.d.	n.d.	n.d.	0.007
Nerol	n.d.	n.d.	n.d.	0.006
Geraniol	0.11	n.d.	n.d.	0.01

n.d. – not detected.

29.3.1 Extraction and Analysis of Volatile Compounds from Grape

One hundred g of grape berries of different varieties of *Vitis vinifera* L. (aromatic varieties Aurora, Beograjska rana, Muscat blanc and less aromatic varieties

Table 29.3 Differences in concentrations (μg L^{-1}) of two aliphatic aldehydes in berries skins and pulps of some grapevine varieties *Vitis vinifera* L. (Prosen et al. 2007a)

	Hexanal		(E)-2-hexen-1-al	
Grapevine variety	Skin	Pulp	Skin	Pulp
Muscat blanc	92.0	28.6	11.9	3.3
Pinot noir	96.2	14.8	19.3	2.0
Perlette	41.7	13.8	17.4	4.0
Aurora	23.7	14.0	8.1	1.3
Beograjska rana	44.8	18.9	18.5	3.2

Table 29.4 The classification of some grapevine varieties (*Vitis vinifera* L.) according to total free monoterpene (TFM) contents (mg L^{-1}) (Mateo and Jiménez 2000)

Muscat varieties (TFM > 6 mg L^{-1})	Non-muscat varieties (1 < TFM < 4 mg L^{-1})	Neutral varieties (TFM no important for the aroma)
Gewürztraminer	Kerner	Cabernet sauvignon
Muscat d'Alexandria	Müller-Thurgau	Carignan
Muscat Hamburg	Riesling	Chardonnay
Muscat Ottonel	Schurebe	Chasselas
Muscat blanc	Sylvaner	Chenin blanc
	Traminer	Clairette
		Malvasia d'Istria
		Merlot
		Pinot gris
		Rebula (Ribolla)
		Sauvignon
		Syrah
		Trebbiano
		Verdelho
		Viognier

Pinot noir, Perlette and Danijela) were randomly sampled and stored in plastic bags in freezer at −30°C until analysis. The samples for aroma compounds analyses were sorted out separately from skins (berries were peeled), pulps (berries were pressed and then filtered) and grape (berries without seeds were mixed). In all samples, the seeds were removed.

Sample preparations:

Skin; 2 g of skins were diluted in 10 mL of MilliQ purified water then mixed for 10 s at 6.000 min^{-1} with Ultra-torrax (IKA T25 digital, Germany) and homogenized for 15 min in 20 mL headspace vials, crimped and afterwards thermostated for 15 min at 50°C in water-bath.

Pulp; 10 mL of grape juice was pipetted into 20 mL headspace vials, crimped and thereupon thermostated for 15 min at 50°C in water-bath.

Grape; grape berries without seeds were homogenised (20 g diluted to 100 mL with MilliQ purified water; 10 s at 6.000 min^{-1} with Ultra-torrax (IKA T25 digital, Germany), additionally 10 mL of homogenate was put into 20 mL headspace vials, crimped and later thermostated for 15 min at 50°C in water-bath (Figs. 29.4 and 29.5).

After 15 min, a SPME fibre was inserted in the headspace vials (Fig. 29.2), where the volatile compounds were sampled for 35 min at 50°C still. The fibre was subsequently inserted into the injector port of a gas chromatograph (Fig. 29.3) and desorbed for 10 min. The process for aliphatic aldehydes was the same except that they were sampled at room temperature (25°C) for 15 min. Different SPME fibres were also tasted to prove their efficiency for volatile aroma compounds analyses (Fig. 29.6).

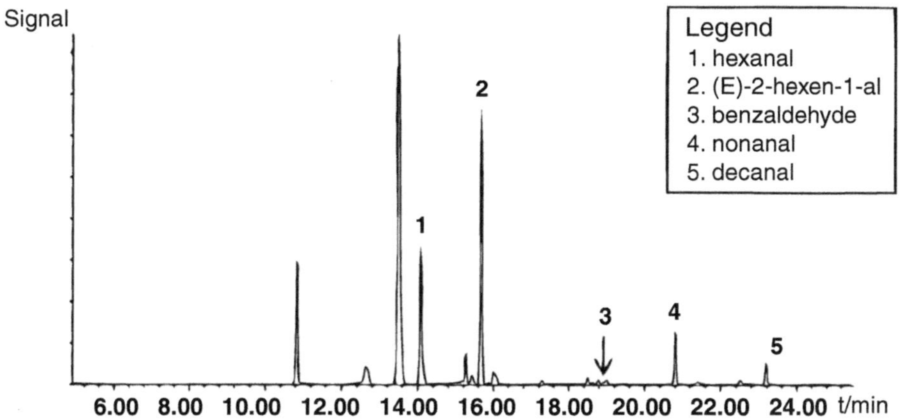

Fig. 29.4 Chromatogram after HS-SPME extraction of aroma compounds from grape of Pinot noir (Prosen et al. 2007a)

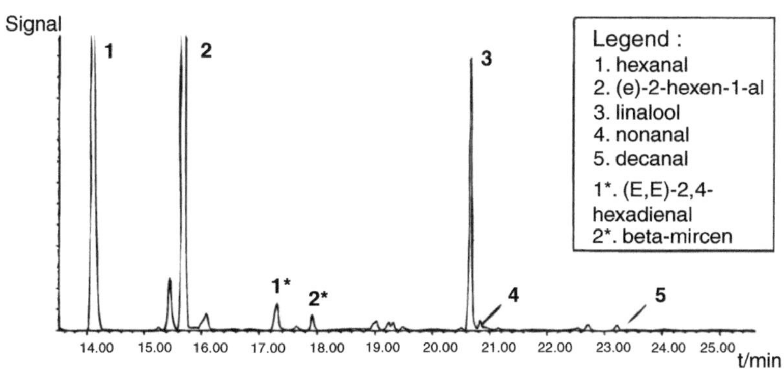

Fig. 29.5 Chromatogram after HS-SPME extraction of aroma compounds from grape of Aurora (Prosen et al. 2007a)

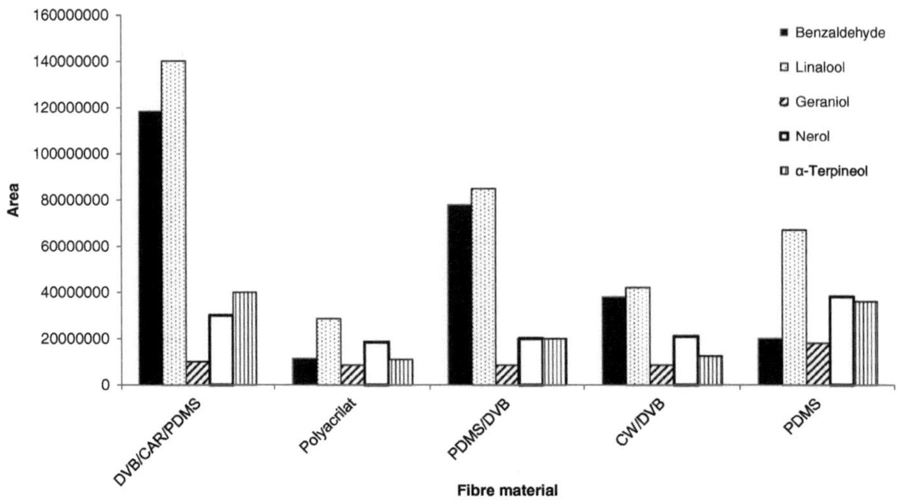

Fig. 29.6 The differences in efficiency of stationary phases (different fibers) on SPME for some aroma compounds (Prosen et al. 2007a)

29.3.2 Extraction and Analysis of Volatile Compounds from Wine

The 10 mL musts and wines were placed into 20 mL headspace vials, and the HS-SPME-GC-MS analysis of volatile compounds was made according to Prosen et al. (2007a).

In the mass spectrometer, electron impact (EI) ionization was used, chromatograms were recorded in the total ion current (TIC) mode. Compounds were identified based on their retention time (comparison with standards) and spectra, using the searchable EI-MS spectra library database (NIST02). The peak area for quantisation was measured either in a TIC chromatogram or in an extracted ion chromatogram in case of a co-elution with other compounds.

29.3.2.1 Experimental

Chemicals

Standard compounds used in this study were monoterpenes α-terpineol (99% purity), nerol (90% purity), geraniol (96% purity), linalool (97% purity), all from Fluka (Buchs, Switzerland); as well as benzaldehyde (puris., Riedel-de Haën, Seelze, Germany), hexanal (98% purity, PolyScience, Niles, IL, USA), and (E)-2-hexen-1-al (97% purity, Fluka, Buchs, Switzerland). Solvents *n*-hexane, acetone, methanol and acetonitrile were of HPLC grade purity, obtained from Sigma-Aldrich (Steinheim, Germany). Other chemicals used were of p.a. grade quality from different producers. For solid-phase extraction Supelclean C18 (1 g) extraction cartridges from Supelco (Bellefonte, PA, USA) were used. For stir bar sorptive extraction (SBSE) a stir bar coated with polydimethylsiloxane (PDMS) coating, with the dimensions 20 × 1.0 mm, from Gerstel (Mülheim an der Ruhr, Germany) was used. For solid-phase microextraction a manual holder was used and fibres with different coatings: polydimethylsiloxane (PDMS), 100 μm, polyacrylate (PA), 85 μm, polydimethylsiloxane-divinylbenzene (PDMS/DVB), 65 μm, Carbowax-divinylbenzene (CW/ DVB), 65 μm, divinylbenzene-Carboxen-polydimethylsiloxane (DVB/CAR/PDMS), 50/30 μm, all from Supelco (Bellefonte, PA, USA). For gas chromatography, helium (>99.999%) from Messer (Gumpoldskirchen, Austria) was used. The analytical capillary column used was VOCOL, dimensions 60 m × 0.25 mm (i.d.), film thickness 1.5 μm, from Supelco (Bellefonte, PA, USA).

Instrumentation

Gas chromatograph used was a HP 5890 Series with mass spectrometric detector (MSD) 6890 from Hewlett-Packard (Palo Alto, CA, USA). For LC-MS/MS experiments a liquid chromatograph Perkin Elmer Series 200 from Perkin Elmer (Shelton, CT, USA) and 3200 QTRAP LC/MS/MS System equipped with ESI and APCI ion sources from Applied Biosystems/MDS Sciex (Foster City, CA, USA) were used. The analytical balance was Mettler Toledo MX5 (Mettler Toledo, Kuesnacht, Switzerland). We used a Visiprep SPE Vacuum Manifold from Supelco (Bellefonte, PA, USA).

Preparation of Standard Solutions

Stock standard solutions of compounds were prepared by dissolving the weighed solid standard in methanol to obtain the concentration of 0.8–1.2 g L^{-1}. These solutions were kept in the refrigerator and were stable for several months. They were further diluted in *n*-hexane or methanol to obtain working solutions for injection in gas chromatograph or in water for extraction optimisation.

Solid-Phase Extraction (SPE)

Extraction cartridge was conditioned with 5 mL of methanol and 5 mL of deionised water. A filtered homogenisate of grape berries was passed under vacuum through the cartridge using the vacuum manifold. The cartridge was rinsed with 20 mL of deionised water. Free aroma compounds were eluted with 8 mL of *n*-hexane and the remaining water was sorbed on solid sodium sulphate. The solvent was transferred into a conical test tube and evaporated to 2 mL under the stream of nitrogen on a water bath (approx. 50°C). The glycoside fraction was eluted with 10 mL of methanol. The solvent was transferred into a conical test tube and evaporated to dryness under the stream of nitrogen on a water bath (approx. 50°C). The dry residue was reconstituted in 1 mL of methanol for further LC analysis.

Stir-Bar Sorptive Extraction (SBSE)

A tentative procedure for SBSE without the thermal desorption unit was as follows: a sample solution (5 mL) was extracted for 30 min on a magnetic stirrer; stir bar was transferred to 2 mL of *n*-hexane and placed in an ultrasonic bath for 10 min. The hexane extract (1 μL) was injected into the gas chromatograph.

Optimization of SBSE Conditions: Due to the lack of suitable thermal desorption unit (TDU) for the sorptive stir-bar thermal desorption, we conducted a series of experiments in SB sorptive extraction, the results of which gave us a hint at the method's real potential. The sorption time was 1 h after which the stir-bar was transferred into 2 mL of *n*-hexane and placed in an ultrasonic bath to desorb the analytes into the organic solvent. Although different variations of this procedure had already been attempted, the best extraction recoveries obtained were 18–40% for the analysed terpenes, 17% for hexanal and an inexplicably high value of 208% for benzaldehyde (possible carry-over effect). The extraction recoveries for the same terpenes using TDU are reported to be 72–78%, but only 6% for benzaldehyde and 7% for hexanal. In another study, no extraction recoveries for wine aroma compounds using SBSE had been reported, although authors conclude it is more efficient than SPME (Prosen et al. 2007a).

Solid-Phase Microextraction (SPME)

For the extraction of aroma compounds from grapes according to the method by Prosen et al. (2007a), a DVB/CAR/PDMS coated fibre was used. The procedure for aliphatic aldehydes was similar, except that the sampling was performed for 15 min at room temperature (25°C).

Optimisation of SPME Conditions: Several different fibres were initially tested to assess their efficiency for the extraction of the selected terpenes and benzaldehyde from the headspace of homogenised grapes (Table 29.5). We assessed the upcoming (mixed) phase to extract the highest amount of all analytes except geraniol (Fig. 29.6), for which the PDMS fibre was slightly more efficient. These findings are in agreement with the results obtained by Tat et al. (2005) and Solís-Solís et al. (2007). These authors found the DVB/CAR /PDMS phase to be most suitable for slightly less volatile and more polar compounds of wine aroma (*t*R > 15 min),

Table 29.5 Main characteristics of different tasted SPME fibres for extraction of volatile compounds from gas media (Prosen et al. 2007a)

Fibre (colour code)	Mobile phase	Thickness 'deposit' per μm
DVB/CAR/PDMS (grey)	Divinylbenzene Carboxan Polydimethylsiloxan	50/30
PA (white)	Polyacrilat	85
PDMS/DVB (blue)	Polydimethylsiloxan Divinylbenzene	65
CW/DVB (orange)	Carbovax Divinylbenzene	65
PDMS (red)	Polydimethylsiloxan	100

while for those with *t*R < 15 min, the CAR/PDMS fibre (not tested by us) happened to be more suitable. The sample temperature and duration of headspace sampling with SPME were optimised, when temperatures ranged between 25 and 50°C. The concentration of the volatile compounds still increased above 50°C, but for safety reasons the temperature was not further increased (Fig. 29.7). The amount of primary aroma compounds on the fibre (at 50°C) was screened between 25 and 35 min of extraction, whereby the 35 min revealed the best results (Fig. 29.8). However, for the SPME of the two aliphatic aldehydes, the selected temperature (50°C) turned out to be too high, which can be linked to the higher volatility of these analytes and their relatively high concentration in the grape aroma. The optimal conditions for SPME of hexanal and (*E*)-2-hexenal proved to be the extraction at room temperature (25°C) for 15 min. At optimal sampling conditions, a part of the method evaluation was performed using the synthetic aqueous solutions of the analytes. It is also evident that the method shows good repeatability and

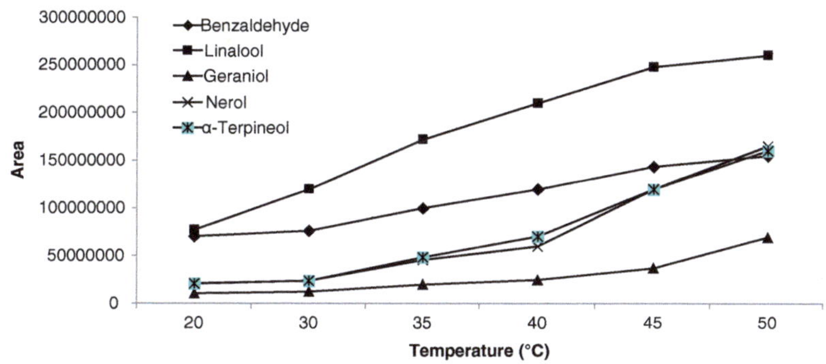

Fig. 29.7 Influence of temperature (°C) on SPME sorption of some aroma compounds from wine (Prosen et al. 2007a)

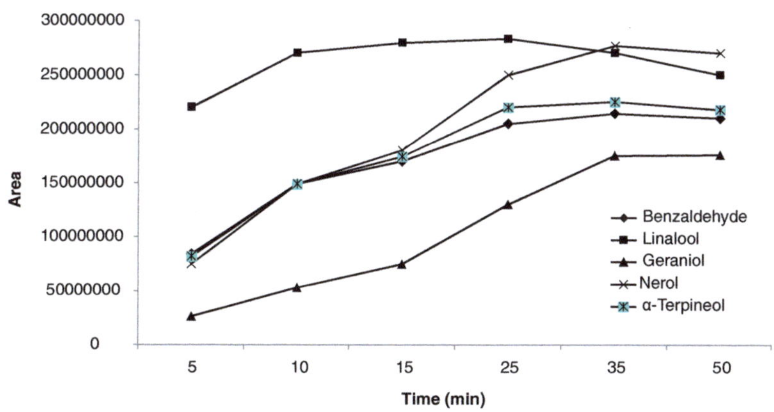

Fig. 29.8 Influence of sorption time (min) with HS-SPME on extraction efficiency of some aroma compounds from wine (Prosen et al. 2007a)

excellent linearity over a wide linear range, making it suitable for determination of concentrations of aroma compounds both in aromatic and non-aromatic grape varieties.

GC-MS Conditions: Volatile grape aroma compounds were analysed using a gas chromatograph with mass spectrometric detector. The temperature programme was 50°C (2 min) – 10°C min^{-1} – 210°C (40 min). Temperature of the injector was 250°C, that of the detector 280°C. Injection volume of *n*-hexane/methanol solutions or extracts was 1 μL (splitless). SPME fibre was left in the injector for 10 min. In the mass spectrometer, electron impact (EI) ionisation was used; chromatograms were recorded in the total ion current (TIC) mode. Compounds were identified according to their retention times (comparison with standards) and spectra, using the searchable EI-MS spectra library database (NIST02). The peak area for quantitation was measured either in a TIC chromatogram or in an extracted ion chromatogram in the case of conformance with other compounds.

LC-MS Conditions: Extracts of glycosides in methanol were analysed using a liquid chromatograph coupled to a mass spectrometer through an electrospray ion source. The HPLC column was Hypersil ODS (Agilent Technologies) with the dimensions 250 mm x 4 mm, 5 μm particles. The mobile phase was composed of acetonitrile (phase A) and 0.5% acetic acid in deionised water (phase B). Gradient elution was applied: 0–5 min 0% A, 5–60 min from 0 to 90% A and hold for 10 min. Mobile phase flow was 0.8 mL min–1, injection volume was 10 μL. Electrospray ion source (ESI) voltage was 5500 V and the temperature was 400°C (Prosen et al. 2007b).

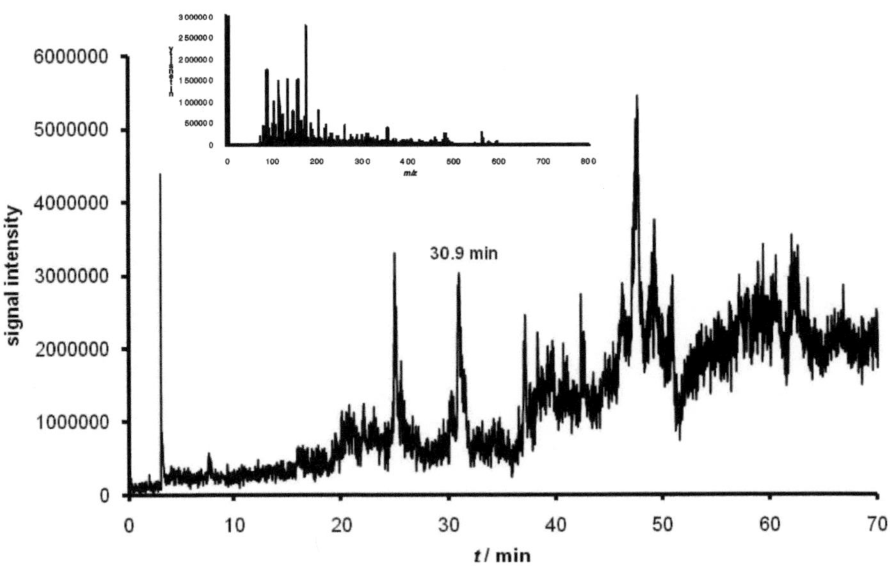

Fig. 29.9 LC-MS chromatogram (extracted *m/z*: 310–314) of a glycoside extract from the grape variety Muscat blanc. Insertion: mass spectrum of peak at 30.9 min (Prosen et al. 2007b)

Some LC-MS analyses (electrospray interface) of non-hydrolysed terpene glycosides extracted from the grape berries were conducted. An extracted LC-MS chromatogram (*m/z* 310–314) of the grape extract is shown in Fig. 29.9. The ions at *m/z* 310–314 interval were chosen as possible fragments of the disaccharide part of terpene glycosides. At least two compounds eluting at *t*R 25.0 min and *t*R 30.9 min could be identified as terpene glycosides based on mass spectra (example for the compound eluting at *t*R 30.9 min shown as an insertion in Fig. 29.9). Typical ions featuring in these spectra were at *m/z* 156, 174, which could be associated to the aglycone (terpene) part of the terpene glycosides. Ions at *m/z* 115 and 133 could be fragments of the monosaccharide moieties. However, despite of these interesting results, the content of terpene glycosides in the extract obviously resulted to be quite low, also the peaks were not clearly visible in the chromatogram. Therefore, the extraction procedure revealed to be less than satisfactory and has to undergo further optimisation.

Hydrolysis of Terpene Disaccharides

Two protocols for the hydrolysis of terpene disaccharides were adopted: with pectinolytic enzymes for wine clarification having side glycosidase activity and acid hydrolysis. Enzyme hydrolysis: an enzyme preparation (50 mg) was added to the homogenisate of grape berries (20 g), diluted to 100 mL with deionised water, pH adjusted to 4.5. The flask was sealed and placed in a water bath (40°C) for 24 h. Acid hydrolysis: an homogenate of grape berries (20 g) diluted to 100 mL with deionised water was prepared, and the pH was adjusted to either 3.00 or 1.00 by addition of H_2SO_4 solution. The flask was then sealed and placed in a thermostated oven (100°C) for 30 min. Pectinolytic enzymes for wine clarification with side glycosidase activity were Rohavin MX from AB Enzymes (Darmstadt, Germany) and Lallzyme BETA from Lallemand (St. Simon, France) (Table 29.6, Figs. 29.10, 29.11, 29.12, 29.13).

Glycosidically bound terpenes can be released by enzymatic treatment (glycosidase) (Bureau et al. 2000, Maicas and Mateo 2005) or by acidic hydrolysis (Maicas and Mateo 2005). The enzymatic hydrolysis of terpene diglycosides is usually a two-step process, involving two different enzymes, although it has also been described as a one-step hydrolysis. The glycosidases can be of plant or microbial origin, but are not immediately available. A frequently used practical approach is to apply pectinase enzymatic preparations with substantial glycosidase side-activity (Bureau et al. 2000, Maicas and Mateo 2005). The efficiency of two enzyme preparations was tested: Rohavin MX from AB Enzymes and Lallzyme BETA from Lallemand that are recommended to enhance the aroma potential of wine. First, an hydrolysis was conducted with the amount of enzymes recommended by the producer (5 mg/100 mL), but no change in the aroma profile was observed. Glycosidases in general are inhibited by a pH value below 4.00 (Sarry and Günata 2004), by the presence of a high level of phenolic compounds (Bureau et al. 2000), and are strongly inhibited by glucose (Sarry and Günata 2004). Most researches on the enzymatic hydrolysis of the terpene disaccharides focused on more or less

Table 29.6 Concentrations (μg g^{-1}; RSD 2–7%) of some aroma compounds in different grapevine varieties *Vitis vinifera* L. before and after different hydrolytic treatments (Prosen et al. 2007a)

Variety	Compound	Before hydrolysis	Rohavin MX hidrolysis	Lallzyme-β hidrolysis	Acid hydrolysis at pH 3	Acid hydrolysis at pH 1
Muscat blanc	Linalool	0.15	4.40	5.27	2.79	0.02
	Geraniol	0.11	0.33	1.00	0.31	n.d.
	Nerol	n.d.	0.06	0.23	0.09	n.d.
	α-terpineol	n.d.	0.09	0.12	0.41	0.63
	Benzaldehyde	0.04	n.d.	n.d.	n.d.	n.d.
	Hexanal	34.4	15.9	20.9	23.2	4.31
Danijela	Linalool	n.d.	n.d.	n.d.	n.d.	n.d.
	Geraniol	n.d.	n.d.	n.d.	n.d.	n.d.
	Nerol	n.d.	n.d.	n.d.	n.d.	n.d.
	α-terpineol	n.d.	n.d.	n.d.	n.d.	n.d.
	Benzaldehyde	0.03	0.06	0.03	0.03	0.04
	Hexanal	32.9	1.08	11.6	4.45	1.68
Pinot noir	Linalool	n.d.	n.d.	n.d.	n.d.	n.d.
	Geraniol	n.d.	0.01	0.03	n.d.	n.d.
	Nerol	n.d.	n.d.	n.d.	n.d.	n.d.
	α-terpineol	n.d.	n.d.	n.d.	n.d.	n.d.
	Benzaldehyde	0.02	0.05	0.02	0.02	0.01
	Hexanal	43.9	8.39	19.0	13.2	1.72

n.d. – not detected.

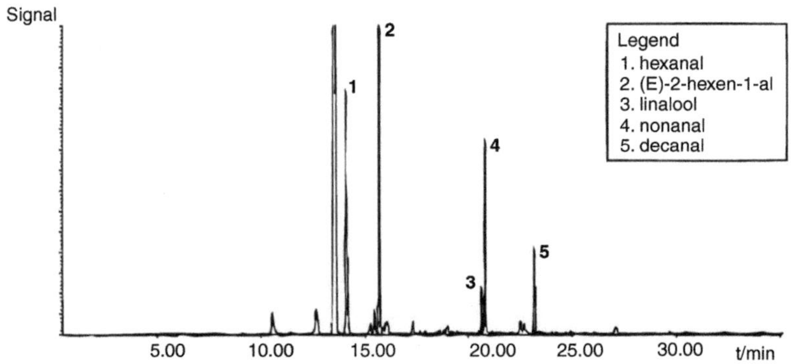

Fig. 29.10 Chromatogram of HS-SPME extract from grape of grapevine variety Muscat blanc before enzymatic hydrolysis (Lallzyme-β) (Prosen et al. 2007a)

glucose-free wine samples, therefore there are no clear guidelines available about the amount of enzymes that should be used in the treatment of non-fermented musts. Bearing in mind the possibility of problems with different enzymes, we used a ten-fold recommended amount of enzyme, incubating the macerate for 24 h at 40°C. This approach was successful in releasing a substantial part of terpenes,

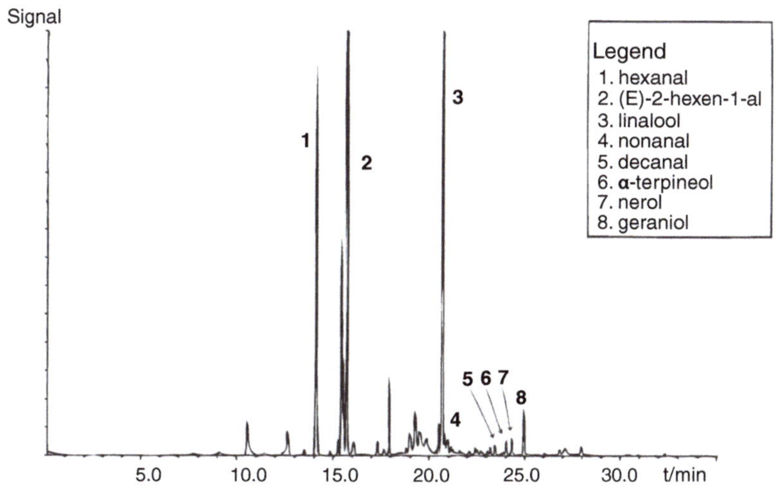

Fig. 29.11 Chromatogram of HS-SPME extract from grape of grapevine variety Muscat blanc after enzymatic hydrolysis (Lallzyme-β) (Prosen et al. 2007a)

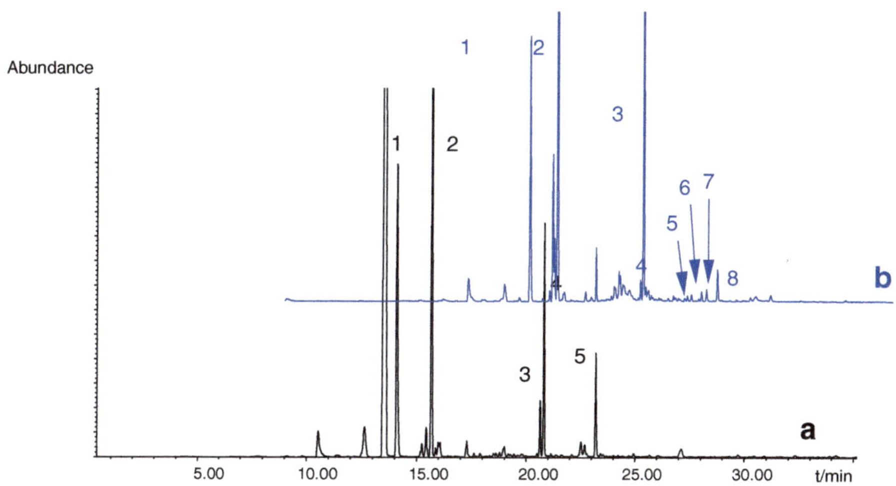

Fig. 29.12 Comparison of HS-SPME-GCMS chromatograms of the variety Muscat blanc before (**a**) and after (**b**) the enzymatic treatment (Lallzyme BETA). Peak assignments: 1-hexanal, 2-(*E*)-2-hexen-1-al, 3-linalool, 4-nonanal, 5-decanal, 6-α-terpineol, 7-nerol, 8-geraniol (Prosen et al. 2007b)

as can be seen from Fig. 29.12 for the Muscat blanc grape variety. The linalool content was substantially increased, while at least three new terpenes appeared in the GC-MS chromatogram: nerol, geraniol and α-terpineol. A part of the quantitative results is shown in Table 29.6. The results obtained with Lallzyme BETA were slightly better than those for Rohavin MX, which is not marketed for the purpose of aroma enhancement. As is also obvious from Table 29.6 and Fig. 29.12, the enzymatic treatment had practically no effect on the benzaldehyde content, while

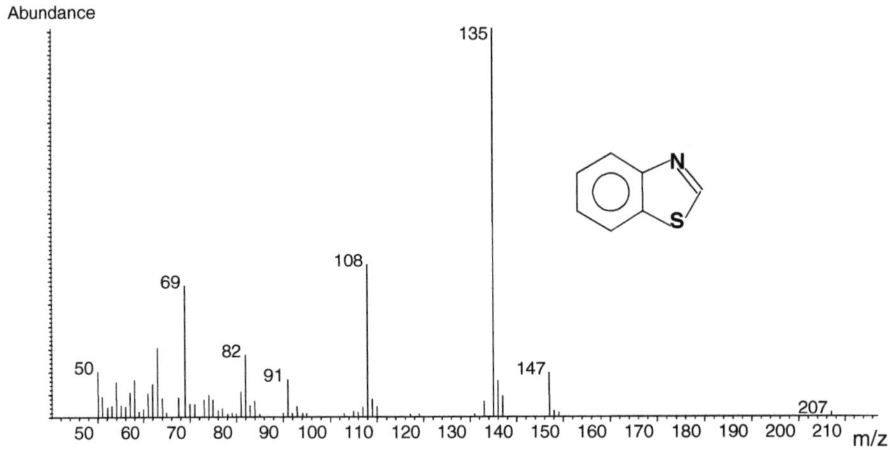

Fig. 29.13 Mass spectrum of the compound present in the aroma of several grape varieties after enzymatic hydrolytic treatment with either Rohavin MX or Lallzyme-β (identified from its mass spectrum as benzothiazol) (Prosen et al. 2007a)

the content of the aliphatic aldehydes was significantly decreased. The latter fact is also favourable, as these compounds contribute to the unpleasant, "herbaceous" smell in wines. Another observation was that several new peaks appeared in the GC-MS chromatogram after hydrolysis. The compounds yielding these peaks were preliminary identified by means of the mass spectra library database. An example for rarely identified benzothiazole is shown in Fig. 29.13. Table 29.7 is listing some compounds that were most often observed solely after the enzymatic treatment.

Table 29.7 Some compounds identified in the grape aroma only after hydrolytic treatment (identification from mass spectra) (Prosen et al. 2007a)

Compound	Hydrolytic treatment	Grapevine variety (*Vitis vinifera* L.)
Benzothiazole	EH-L, EH-R	Beograjska rana, Muscat blanc, Pinot noir, Perlette, Danijela
Furfural	EH-L AH3	Perlette
Thymine	EH-L	Perlette
2-phenylethyl acetate	EH-R	Perlette
Linalool oxide (pyranic form)	EH-L, EH-R, AH1 AH1	Muscat blanc Aurora
2,2-dimethyl-5-(1-methyl-1-propenyl)-tetrahydrofuran	AH1	Muscat blanc, Aurora, Danijela
3,6-dihydro-4-methyl-2-(2-methyl-1-propenyl)-pyran	AH1, AH3	Muscat blanc
Eucalyptol / 1,8-cineol	AH1	Aurora, Muscat blanc

EH-R = enzymatic hydrolysis with Rohavin MX; EH-L = enzymatic hydrolysis with Lallzyme BETA, AH1 = acid hydrolysis at pH 1; AH3 = acetic hydrolysis at pH 3.

Some of them have been described in the aroma profile of musts and wines, e.g. 2-phenylethyl acetate, linalool oxide and furfural. These compounds contribute to a sweet and fruity aroma of wines. Acidic hydrolysis was also performed to release bound terpenes. This procedure is actually supposed to simulate the processes taking place during the storage and ageing of wines. The hydrolysis was conducted at two different pH values: pH 3.00, resembling pH conditions of wine and pH 1.00, chosen as these extreme conditions probably ensure the rearrangement reactions of terpenes. A part of the results is shown in Table 29.6. Acidic hydrolysis at pH 3.00 obviously resulted in a release of terpenes through glycosides, although with a lower yield than enzymatic hydrolysis. Benzaldehyde concentration was not significantly affected, while the concentration of aliphatic aldehydes decreased. After hydrolysis at pH 1.00, the content of monitored terpenes was even lower than before hydrolysis, which is consistent with general data from literature. Aliphatic aldehydes were also significantly decomposed under these conditions. However, it was interesting to observe the release (or formation) of several compounds, which had been unnoticed after enzymatic hydrolysis (examples in Table 29.6), e.g. eucalyptol (or 1.8-cineol), a typical product of extremely harsh hydrolytic conditions (Prosen et al. 2007b).

29.3.3 Extraction Method for Carotenoids in Grape

The sample of green berries was manually crushed in liquid nitrogen with pestles. Around 2 g of grape powder (with the exact weight specified) were homogenised in ice cold aceton for 3 min (dark ambient), mixed for 15 s with Ultra-torrax (IKA T25 digital, Germany), then centrifuged at 4.200 rpm for 4 min ($T < 4°C$), filtered through 0.45 μm syringe filter (Minisaet RC15, Sartorius, Germany), decanted into 2 mL dark vials and analysed quickly as possible (Fig. 29.14).

29.3.3.1 HPLC Conditions for Carotenoids Analysis

Acetone extracts have undergone an HPLC gradient analysis (column Spherisorb S5 ODS-2 250 × 4.6 mm with precolumn S5 ODS-2 50 × 4.6 mm). The pigments were separated using the following solvents: solvent A: acetonitrile/methanol/water (100/10/5, v/v/v); solvent B: acetone/ethylacetate (2/1, v/v), at a flow rate of 1 $mL{\cdot}min^{-1}$, linear gradient from 10% solvent B to 70% solvent B in 18 min was applied, run time 30 min, photometric detection at 440 nm.

29.3.3.2 Enzymes Enhancing of Primary Aroma Compounds

White aromatic variety Traminer (Gewürztraminer, Dišeči traminec) and less aromatic variety Furmint (Šipon, Moslavac) (*Vitis vinifera* L.) were chosen to study the impacts of several enzyme preparations in winemaking to enhance volatile aroma compounds in wine.

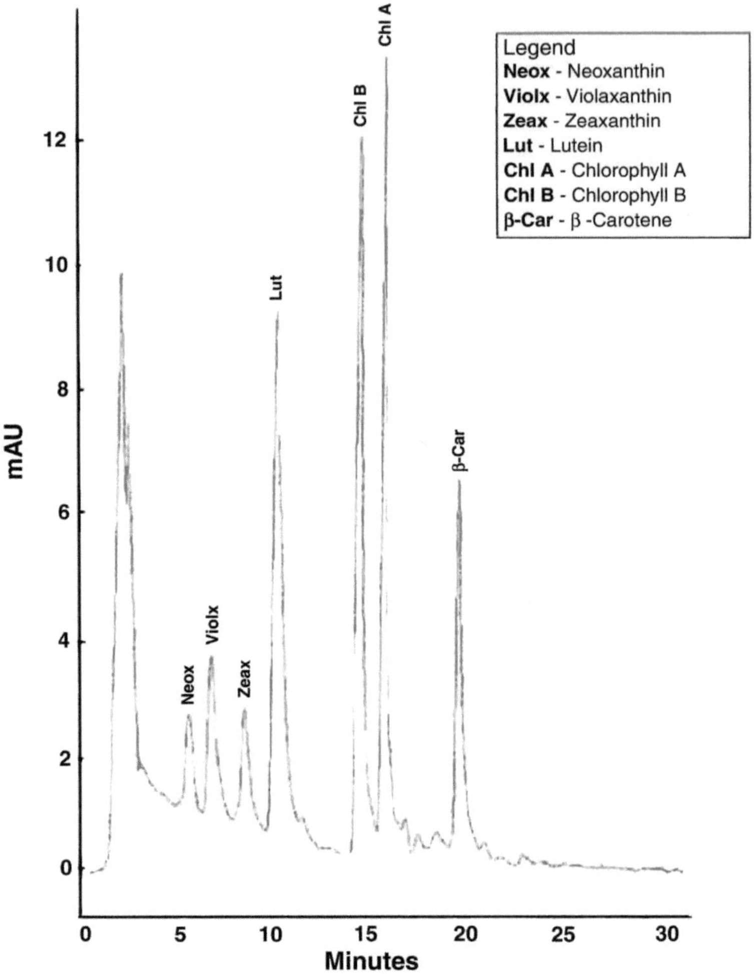

Fig. 29.14 The pigments (some of precursors of C_{13}-norisoprenoids) in grape (*Vitis vinifera* L.) extracted from green berries

Material

Around 300 kg of grapes per variety were harvested manually at appropriate ripe stage. Around 1 kg of grape of each sample was immediately frozen and stored at −20°C until analysis. The main parts of the grapes were crushed, pressed and decanted; 10 L of marc into each of the 24 equal vessels (3 repetitions for each enzyme preparation) for each variety and furthermore left on berry skins during the night at 17°C. Immediately after decantation, the enzyme preparations were added in vessels as prescribed by producers. The same yeast (*Saccharomyces cerevisiae*) and quantity (0.25 g per L of must) was added to prevent the influence of yeasts on aroma compounds. The musts obtained were also treated with sulphur (50 mg of $K_2S_2O_5$ per L of must with skins) to prevent oxidation. The musts and wine were sampled during and after fermentation. Exactly 10 mL of samples were pipetted into 20-mL headspace glass vials, crimped and frozen (−20°C) until analysis.

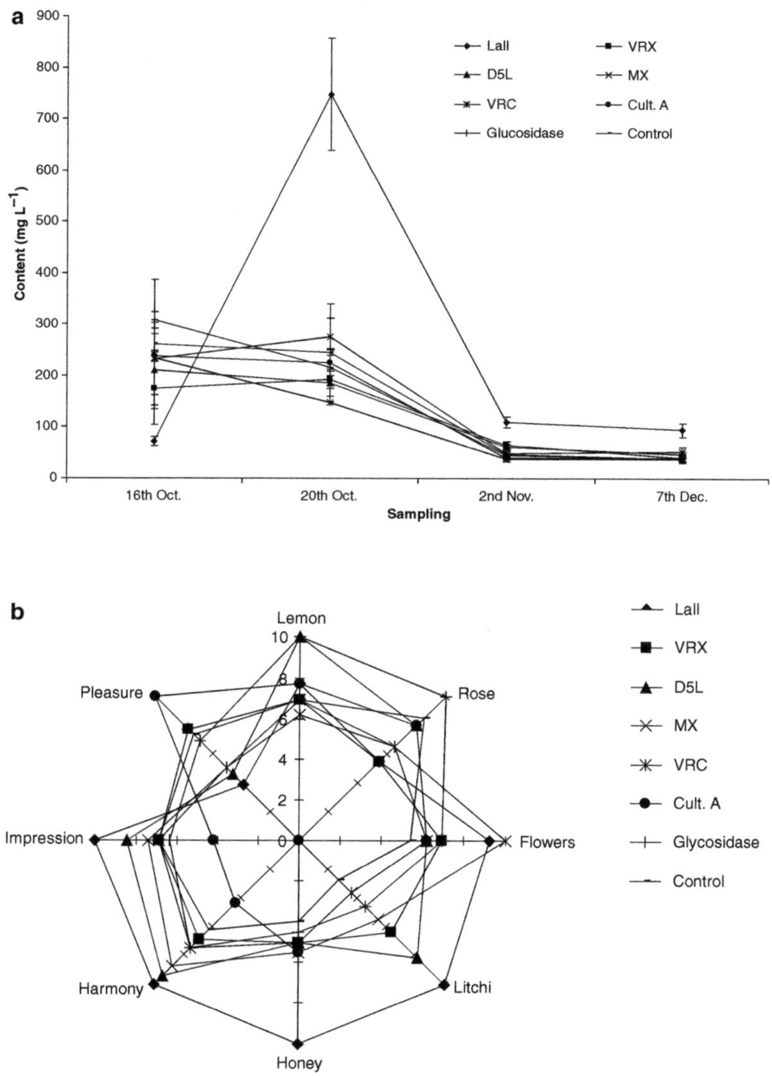

Fig. 29.15 The average contents of total terpene compounds (TT; $\mu g\ L^{-1}$) in musts and wines (left) and wine aroma profile (right) of Gewürztraminer during vinification according to enzyme treatments. The standard error bars are given at each data point. Legend: ∗, ∗∗, ∗∗∗: significant at $p \leq 0.05, 0.01, 0.001$, respectively (Rusjan et al. 2009)

Chemicals and Instrumentations

Standard compounds used in this study were monoterpenes α-terpineol (99% purity), nerol (90% purity), geraniol (96% purity), linalool (97% purity), all from Fluka (Buchs, Switzerland), as well as benzaldehyde (puris., Riedel-de Haën, Seelze, Germany). Stock standard solutions of compounds were prepared by dissolving the weighed solid standard in methanol to obtain the concentration of 0.8–1.2 g L^{-1}. These solutions were kept in the refrigerator and were stable for several months. Solvents n-hexane and methanol, obtained from Sigma-Aldrich (Steinheim, Germany), were of HPLC grade purity. Other chemicals used were of p.a. grade quality from different producers. For solid-phase microextraction a manual holder and divinylbenzene-Carboxen-polydimethylsiloxane (DVB/CAR/PDMS), 50/30 μm fibre from Supelco (Bellefonte, PA, USA) were used. For gas chromatography, helium (>99.999%) from Messer (Gumpoldskirchen, Austria) was used. The analytical capillary column was VOCOL, dimensions 60 m × 0.25 mm (i.d.), film thickness 1.5 μm, from Supelco (Bellefonte, PA, USA). Gas chromatograph was HP 5890 Series with mass spectrometric detector (MSD) 6890 from Hewlett-Packard (Palo Alto, CA, USA).

	Benzaldehyde	Linalool	α-terpineol	Nerol	Geraniol
Lall	156.61	6.39	10.51	45.9	31.8
VRX	138.15	4.35	3.76	16.26	14.53
D5L	22.95	4.96	4.43	21.1	15.46
MX	17.26	4.86	5.01	22.06	20.81
VRC	36.08	4.33	4.01	16.53	12.61
Cult. A	38.28	4.06	4.75	17.61	12.93
β-glycosidase	31.47	4.17	2.8	15.33	15.18
Control	131.35	4.21	4.17	15.32	13.28

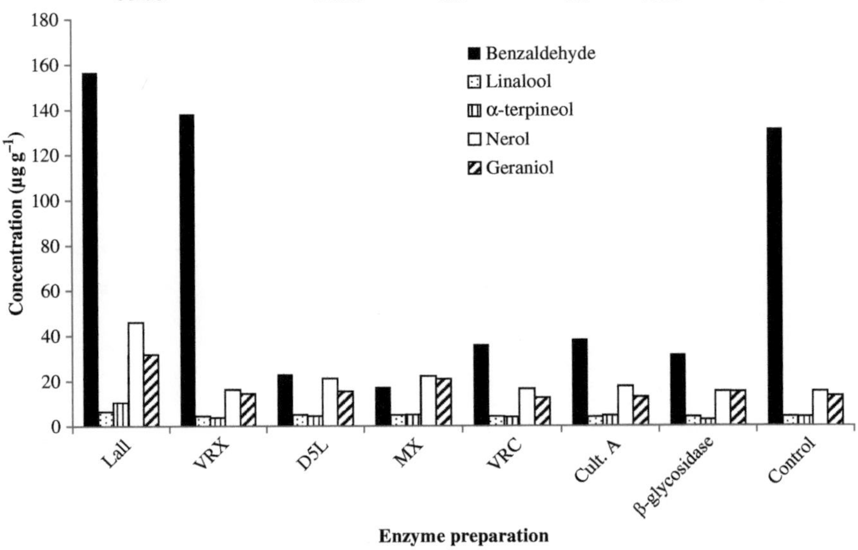

Fig. 29.16 The average concentrations (μg L^{-1}) of primary aroma compounds in wine according to used enzyme preparation

Extraction and Analysis of Volatile Compounds

The must in the vial was defrosted at room temperature and then subjected to HS-SPME-GC-MS analysis. The sample vial was thermostated for 15 min at 50°C. An SPME fibre was inserted in the headspace and the compounds were sampled for 35 min. The fibre was subsequently inserted into the injector port of a gas chromatograph and desorbed for 10 min. The temperature program in GC was: 50°C (2 min) – 10°C min^{-1} – 210°C (40 min). Temperature of the injector was 250°C, while temperature of the detector 280°C. In the mass spectrometer, electron impact (EI) ionization was used, chromatograms were recorded in total ion current (TIC) mode. Compounds were identified on the basis of their retention times (comparison with standards) and spectra using the searchable EI-MS spectra library database (NIST02). The peak area for quantitation was measured either in TIC chromatogram or in an extracted ion chromatogram in case of coelution with other compounds (Prosen et al. 2007a, Figs. 29.15 and 29.16).

References

Arthur CL, Pawliszyn J (1990) Solid phase microextraction with thermal desorption using fused silica optical fibres. Anal Chem 62:2145–2148

Baumes R, Cordonnier R, Nitz S, Drawert F (1986) Identification and determination of volatile constituents in wine from different vine cultivars. J Sci Food Agric 37:927–943

Baumes R, Wirth J, Bureau S, Gunata Y, Razungles A (2002) Biogeneration of C_{13}-norisoprenoid compounds: experiments supportive for an apo-carotenoid pathway in grapevine. Anal Chim Acta 45:3–14

Bayonove C (1993) Les composés terpéniques. In: Donèche B (ed) Les acquisitions récentes en chromatographie du vin. Applications á l'analyse sensorielle des vins, 1st edn. Lavoisier, Paris

Blanch GP, Reglero G, Herraiz M (1995) Analysis of wine aroma by off-line and on-line supercritical fluid extraction-gas chromatography. J Agric Food Chem 43:1251–1258

Boidron JN, Chatonnet P, Pons M (1988) Influence du bois sur certaines substances odorantes des vins. Conn Vigne Vin 22:275–294

Bureau SM, Baumes RL, Razungles AJ (2000) Effects of vine or bunch shading on the glycosylated flavour precursors in grapes of *Vitis vinifera* L. cv. Syrah. J Agric Food Chem 48: 1290–1297

Bureau SM, Razungles A, Baumes R, Bayonove C (2006) Glycosylated flavour precursor extraction by microwaves from grape juice and grapes. J Food Sci 31:557–560

Castro-Vázquez L, Pérez Coello MS, Cabezudo MD (2002) Effects of enzyme treatments and skin extraction on varietal volatile in Spanish wines made from Chardonnay, Muscat, Airen and Macabeo grapes. Anal Chim Acta 1:1–6

Caven-Quantrill D, Buglass A (2006) Comparison of micro-scale simultaneous distillation-extraction and stir bar sorptive extraction for the determination of volatile organic constituents of grape juice. J Chromatogr A 1117:121–131

Chaintreau A (2001) Simultaneous distillation-extraction: from birth to maturity-review. Flavour Fragr J 16:136–148

Chalier P, Angot B, Delteil D, Doco T, Gunata Z (2007) Interactions between aroma compounds and whole mannoprotein isolation from *Saccharomyces cereviseae* strains. Food Chem 100: 22–30

Cocito C, Gaetano G, Delfini C (1995) Rapid extraction of aroma compounds in must and wine by means of ultrasound. Food Chem 52:311–320

Comuzzo P, Tat L, Tonizzo A, Battistutta F (2006) Yeast derivates (extracts and autolysates) in winemaking: release of volatile compounds and effects on wine aroma volatility. Food Chem 99:217–230
Cullere L, Escudero A, Cacho J, Ferreira V (2004) Gas chromatography-olfactometry and chemical quantitative study of the aroma of six premium quality Spanish aged red wines. J Agric Food Sci 52:1653–1660
Darriet P, Bouchilloux P, Poupot C, Bugatet Y, Clerjeau M, Sauris P, Medina B, Dubourdieu D (2001) Effects of copper fungicide spraying on volatile thiols of the varietal aroma of Sauvignon blanc, Cabernet Sauvignon and Merlot wines. Vitis 40:93–99
Engel KH, Tressl R (1983) Formation of aroma components from non volatile precursors in passion fruit. J Agric Food Chem 31:998–1002
Escudero A, Gogorza B, Melus MA, Ortin N, Cacho J, Ferreira V (2004) Characterization of the aroma of a wine from Maccabeo. Key role played by compounds with low odour activity values. J Agric Food Chem 52:3516–3524
Etiévant PX (1991) Wine. In: Maarse H (ed) Volatile compounds in food. Marcel Dekker Inc, New York
Ferreira V, Rapp A, Cacho JF, Hastrich H, Yavas I (1993) Fast and quantitative determination of wine flavour compounds using microextraction with freon 113. J Agric Food Chem 41: 1413–1420
Ferreira V, Lopez R, Cacho JF (2000) Quantitative determination of the odorants of young red wines from different grape varieties. J Sci Food Agric 80:1659–1667
Ferreira V, Jarauta I, Ortega L, Cacho J (2004) Simple strategy for the optimization of solid-phase extraction procedures through the use of solid-liquid distribution coefficients. Application to the determination of aliphatic lactones in wine. J Chromatogr A 1025:147–156
Francis IL, Newton JL (2005) Determining wine aroma from compositional data. Aust J Grape Wine Res 11:114–126
Fretz C, Känel S, Luisier JL, Amadò R (2005) Analysis of volatile components of Petite Arvine wine. Eur Food Res Technol 221:504–510
Gonçalves C, Alpendurada MF (2002) Solid-phase microextraction-gas chromatography-(tandem) mass spectrometry as a tool for pesticide residue analysis in water sample at high sensitivity and selectivity with confirmation capabilities. J Chromatogr A 1026:239–250
Günata YZ, Bayonove CL, Baumes RL, Cordonnier RE (1985) The aroma of grape. I. Extraction and determination of free and glycosidically bound fraction of some grape aroma components. J Chromatogr 331:83–90
Guth H (1997) Quantitation and sensory studies of character impact odorants of different white wines varieties. J Agric Food Chem 45:3027–3032
Guth H, Sies A (2002) Flavour of wines: Towards an understanding by reconstitution experiments and an analysis of ethanol's effect on odour activity of key compounds. In: Proceedings of Eleventh Australian Wine Industry Technical Conference, Adelaide
Hardy P (1969) Extraction and concentration of volatiles from dilute aqueous and aqueous-alcoholic solution using trichloromethane. J Agric Food Chem 17:656–658
Košmerl T, Jakončič M, Kralj Cigić I, Strlič M, Prosen H (2008) Aroma compounds in "sur lies" produced and aged Chardonnay wines. In: Organisation Internationale de la Vigne et du Vin (OIV) (ed) 31° Congresso mondiale della vigna e del vino, 6a Assemblea Generale dellO.I.V. Ministero delle Politiche Agricole Alimentari e Forestali, Verona
López R, Ortín N, Pérez-Trujillo JP, Cacho J, Ferreira V (2003) Impact odorants of different young white wines from the Canary Islands. J Agric Food Chem 51:3419–3425
Lopez R, Aznar M, Cacho J, Ferreira V (2002) Quantitative determination of minor and trace volatile compounds in wine by Solid-phase Extraction and Gas Chromatography with Mass Spectrometric detection. J Chromatogr A 966:166–177
Fang L, Mosandl A, Münch A, Wüst M (2005) Metabolism of geraniol in grape berry mesocarp of *Vitis vinifera* L. cv. Scheurebe: demonstration of stereoselective reduction, *E/Z*-isomerization, oxidation and glycosylation. Phytochemistry 66:295–303

Maicas S, Mateo JJ (2005) Hydrolysis of terpenyl glycosides in grape juice and other fruit juices: a review. Appl Microbiol Biotechnol 67:322–335

Mateo JJ, Jimenez M (2000) Monoterpenes in grape juice and wines. J Chromatogr A 881:557–567

Miklósy E, Kerény Z (2004) Comparison of the volatile aroma components in noble rotted grape berries from two different locations of the Tokaj wine district in Hungary. Anal Chim Acta 513:177–181

Murat ML, Tominaga T, Dubourdieu D (2001) Impact of some components on Bordeaux rosés and clairets aroma. J Int Sci Vigne Vin 35:99–105

Núñez AJ, Bemelmans JMH, Maarse H (1984) Isolation methods for the volatile components of grapefruit juice. Distillation and solvent extraction methods. Chromatographia 18:153–158

Ortega C, Lopez R, Cacho J, Ferreira V (2001) Fast analysis of important wine volatile compounds. Development and validation of a new method based on gas chromatographic-flame ionisation detection analysis of dichloromethane microextracts. J Chromatogr A 923:205–214

Ortega-Heras M, González-SanJosé ML, Beltrán S (2002) Aroma composition of wine studied by different extraction methods. Anal Chim Acta 458:85–93

Gachons CP, Leeuwen C, Tominaga T, Soyer JP, Gaudillère JP, Dubourdieu D (2005) Influence of water and nitrogen deficit on fruit ripening and aroma potential of *Vitis vinifera* L. cv Sauvignon blanc in field conditions. J Sci Food Agric 85:73–85

Piñeiro Z, Natera R, Castro R, Palma M, Puertas B, Barroso CG (2006) Characterisation of volatile fraction of monovarietal wines: influence of winemaking practices. Anal Chim Acta 563: 165–172

Prosen H, Janeš L, Strlič M, Rusjan D, Kočar D (2007a) Analysis of free and bound aroma compounds in grape berries using headspace solid-phase microextraction with GC-MS and a preliminary study of solid-phase extraction with LC-MS. Acta Chim Slov 54:25–32

Prosen H, Kočar D, Strlič M, Rusjan D (2007b) In vino veritas : LC-MS in wine analysis. LC GC Eur 12:617–621

Rapp A, Mandery H (1986) Wine aroma. Cell Mol Life Sci 43:873–884

Ribéreau-Gayon P, Glories Y, Maujean A, Dubourdieu D (2000) Varietal aroma. In: Ribéreau-Gayon P, Glories Y, Maujean A et al (eds) Handbook of enology. The chemistry of Wine Stabilization and Treatments, Lavoisier, Paris

Rocha SM, Rodrigues F, Coutinho P, Delgadillo I, Coimbra MA (2004) Volatile composition of Baga red wine assessment of the identification of the would-be impact odorants. Anal Chim Acta 513:257–262

Rusjan D, Strlič M, Košmerl T, Prosen H (2009) The response of Monoterpenes to different enzyme preparations in Gewürztraminer (*Vitis vinifera L.*) wines. S Afr J Enol Vitic 30:56–64

Sala C, Busto O, Guasch J, Zamora F (2004) Influence of vine training and sunlight exposure on the 3-alkyl-2-methoxypyrazines content in musts and wines from the *Vitis vinifera* variety Cabernet Sauvignon. J Agri Food Chem 52:3492–3497

Sánchez-Palomo E, Díaz-Maroto HMC, González-Viñas MA, Pérez-Coello MS (2005) Aroma enhancement in wines from different grape varieties using exogenous glucosidases. Food Chem 92:627–635

Sánchez-Palomo E, Díaz-Maroto MC, González-Vinas MA, Soriano-Perez A, Perez-Coello MS (2007) Aroma profile of wines from Albillo and Muscat grape varieties at different stages of ripening. Food Control 18:398–403

Sarry JE, Günata Z (2004) Plant and microbial glycoside hydrolases: Volatile release from glycosidic aroma precursors. Food Chem 87:509–521

Schneider R, Razungles A, Augier C, Baumes R (2001) Monoterpenic and norisoprenoidic glycoconjugates of Vitis vinifera L. cv. Melon B as precursors of odorant in Muscated wines. J Chromatogr A 936:145–157

Selli S, Cabaroglu T, Canbas A, Erten H, Nurgel C, Lepoutre JP, Gunata Z (2004) Volatile composition of red wine from cv. Kalecik Karasi grown in central Anatolia. Food Chem 85:207–213

Smart R, Robinson M (1991) Sunlight into wine. Winetitles, Adelaide

Solis-Solis HM, Calderón-Santoyoa M, Schorr-Galindob S, Luna-Solanoc G, Ragazzo-Sáncheza JA (2007) Characterization of aroma potential of apricot varieties using different extraction techniques. Food Chem 105:829–837

Sousa-Câmara J, Arminda-Alves M, Carlos-Marques J (2007) Classification of Boal, Malvazia, Sercial and Verdelho wines based on terpenoid patterns. Food Chem 101:475–484

Spillman PJ, Sefton MA, Gawel R (2004) The contribution of volatile compounds derived during oak barrel maturation to the aroma of a Chardonnay and Cabernet Sauvignon wine. Aust J Grape Wine Res 10:227–235

Swiegers JH, Bartowsky EJ, Henschke PA, Pretorius IS (2005) Yeast and bacterial modulation of wine aroma and flavour. Aust J Grape Wine Res 11:139–173

Tat P, Comuzzo Stolfo I, Battistutta F (2005) Optimization of wine headspace analysis by solid-phase microextraction capillary gas chromatography with mass spectrometric and flam ionization detection. Food Chem 93:361–369

Wada K, Shibamoto T (1997) Isolation and identification of volatile compounds from a wine using solid phase extraction, gas chromatography, and gas chromatography/mass spectrometry. J Agric Food Chem 45:4362–4366

Zalacain A, Marin J, Alonso GL, Salinas MR (2007) Analysis of wine primary aroma compounds by stir bar sorptive extraction. Talanta 71:1610–1615

Index

DOI 10.1007/978-90-481-9283-0,

MIX
Papier aus verantwortungsvollen Quellen
Paper from responsible sources
FSC® C105338

If you have any concerns about our products,
you can contact us on
ProductSafety@springernature.com

In case Publisher is established outside the EU,
the EU authorized representative is:
Springer Nature Customer Service Center GmbH
Europaplatz 3, 69115 Heidelberg, Germany

Printed by Libri Plureos GmbH
in Hamburg, Germany